AF780598

Implementing the 3Rs in Laboratory Animal Research—From Theory to Practice

Implementing the 3Rs in Laboratory Animal Research—From Theory to Practice

Editor

Garikoitz Azkona

MDPI

Basel • Beijing • Wuhan • Barcelona • Belgrade • Novi Sad • Cluj • Manchester

Editor
Garikoitz Azkona
University of the Basque
Country (UPV/EHU)
San Sebastian, Spain

Editorial Office
MDPI
St. Alban-Anlage 66
4052 Basel, Switzerland

This is a reprint of articles from the Special Issue published online in the open access journal *Animals* (ISSN 2076-2615) (available at: https://www.mdpi.com/journal/animals/special_issues/3Rs_Laboratory_Animal).

For citation purposes, cite each article independently as indicated on the article page online and as indicated below:

Lastname, A.A.; Lastname, B.B. Article Title. *Journal Name* **Year**, *Volume Number*, Page Range.

ISBN 978-3-0365-9837-6 (Hbk)
ISBN 978-3-0365-9838-3 (PDF)
doi.org/10.3390/books978-3-0365-9838-3

© 2023 by the authors. Articles in this book are Open Access and distributed under the Creative Commons Attribution (CC BY) license. The book as a whole is distributed by MDPI under the terms and conditions of the Creative Commons Attribution-NonCommercial-NoDerivs (CC BY-NC-ND) license.

Contents

About the Editor

Garikoitz Azkona

Garikoitz Azkona, DVM, PhD, has a Bachelor's degree in Veterinary Medicine (University of Zaragoza (UZ), 2002), a Master's degree in Science and Laboratory Animal Science and Welfare (Autonomous University of Barcelona (UAB), 2011) and a PhD in Neuroscience (University of the Basque Country (UPV/EHU), 2009). His training has always been closely linked to animal behaviour and welfare, combining neuroscience research with laboratory animal science. At the neuroscience level, he has worked in different research centres in Barcelona, the Centre for Genomic Regulation (CRG) and the Institut d'Investigacions Biomèdiques August Pi i Sunyer (IDIBAPS) in Barcelona. In the field of laboratory animal science, he has worked as a Designated Veterinarian and Animal Welfare Officer in different animal facilities; the Inbiomed Foundation in San Sebastian, the University of Barcelona (UB) and the Barcelona Biomedical Research Park (PRBB) in Barcelona. Since 2019 he has been a lecturer in psychobiology at the School of Psychology of the UPV/EHU. His current research focuses on the neurobiological basis of social stress and possible sex differences, as well as on human–animal interaction.

Preface

The 3Rs principle (replacement, reduction and refinement) is not only the cornerstone of current legislation on the use of laboratory animals, but also the framework that allows for us to think about and evaluate the benefits and harms of using animals in biomedical research. It is, therefore, of great interest to be able to share both the theoretical and practical advances that have been made in this area. With this in mind, he has written this reprint for all those who work with laboratory animals. This Reprint presents a range of perspectives on current research into the implementation of the 3Rs, from practical applications to theoretical frameworks, all with the common aim of improving the welfare of laboratory animals.

Garikoitz Azkona
Editor

animals

MDPI

Editorial

Implementing the 3Rs in Laboratory Animal Research—From Theory to Practice

Garikoitz Azkona

Department of Basic Psychological Processes and Their Development, Euskal Herriko Unibertsitatea (UPV/EHU), Tolosa Hiribidea 70, 20018 Donostia, Spain; garikoitz.azkona@ehu.eus

The regulatory framework for the use of animals in research in many countries is based on the 3Rs: replacement, reduction, and refinement [1]. These principles state that if it is necessary to use animals in experiments, researchers should make every effort to replace them with non-sentient alternatives, reduce their numbers to a minimum, and refine experiments and housing conditions to minimize pain and distress as much as possible. Thus, the 3Rs concept serves both as a framework designed to minimize animal use and suffering (harm to the animal) and as a means to support high-quality science and translation (benefit to society) [2].

This Special Issue compiles the latest research results and advances relevant to the 3Rs. A total of 23 papers have been published: 12 research articles, 1 commentary, 2 communications, 1 concept paper, 5 reviews, and 2 systematic reviews. The contributions are listed below.

Most of the published articles and communications have focused on the third R: refinement. In terms of husbandry, it has been observed that male CD1 mice raised together with environmental enrichment in well-ventilated cages showed fewer signs of stress (1). Conversely, single-housed mice exhibited changes in the immune–endocrine system (2). Concerning experimental processes, the use of clicker training improved compliance in the catwalk test (3), and acclimation and saphenous vein puncture for blood collection reduced stress in C57Bl/6J mice (4). Two articles explore the use of imaging tools to reduce animal numbers and improve their welfare, employing Positron Emission Tomography (PET) to track animals throughout their lives (5) and camera-based respiration monitoring, which reduces animal handling (6). In the same vein, a gelatin-based voluntary ingestion protocol is proposed to administer drugs (7). Pérez-Martin et al. (8) describe a refined stereotaxic neurosurgery technique for long-term intracerebroventricular device implantation in rodents. A score sheet is proposed to evaluate the animal welfare of the type 2 diabetes rat model induced by streptozotocin following fructose consumption (9). Two papers focus on replacement: the use of organoids to evaluate cellular therapies (10) and a new in vitro assay to determine the biological activity of insulins (11). Peruga and collaborators (12) ponder whether current animal models are useful in researching how female hormones influence orthodontic biomechanics. The fourth R, rehoming, has been a positive experience using golden hamsters after their use in SARS-CoV-2 vaccine research (13).

In their commentary, Verderio et al. (14) provide an overview of the current status of the 3Rs and emphasize the need for bioinformaticians to achieve high standards of animal research. The review articles of this Special Issue have focused on the importance of animal models in biomedical research (15), the most widely used techniques to implement the 3Rs in experimental liver research (16), the adverse impacts of sex bias on science and animal welfare (17), the gaps and challenges in primate pain management (18), and ultrasound-guided surgery as a refinement tool in oncology research (19). Regarding systematic reviews, one focuses on the possible causes and solutions to aggression between grouped male mice (20), and the other summarizes published advances in the refinement protocols made by European Union-based research groups in the last 10 years (21). Finally, in his concept paper, David B. Morton (22) proposes a mathematical model to analyze

Citation: Azkona, G. Implementing the 3Rs in Laboratory Animal Research—From Theory to Practice. *Animals* **2023**, *13*, 3063. https://doi.org/10.3390/ani13193063

Received: 27 September 2023
Accepted: 28 September 2023
Published: 29 September 2023

Copyright: © 2023 by the author. Licensee MDPI, Basel, Switzerland. This article is an open access article distributed under the terms and conditions of the Creative Commons Attribution (CC BY) license (https://creativecommons.org/licenses/by/4.0/).

measurement data to determine the degree of harm (or severity) incurred by an animal during research. Likewise, De Vleeschauwer et al. (23) developed a severity classification for all procedures performed in two Belgian academic biomedical institutions.

Overall, this Special Issue presents a range of perspectives on current research in implementing the 3Rs, from practical applications to theoretical frameworks, all with the shared aim of enhancing the welfare of laboratory animals.

List of Contributions

1. Lee, G.-H.; Kim, K.; Jo, W. Stress Evaluation of Mouse Husbandry Environments for Improving Laboratory Animal Welfare. *Animals* **2023**, *13*, 249. https://doi.org/10.3390/ani13020249.
2. Ortega-Saez, I.; Díez-Solinska, A.; Grífols, R.; Martí, C.; Zamora, C.; Muñoz-Culla, M.; Vegas, O.; Azkona, G. Individualized Housing Modifies the Immune–Endocrine System in CD1 Adult Male Mice. *Animals* **2023**, *13*, 1026. https://doi.org/10.3390/ani13061026.
3. Dickmann, J.; Gonzalez-Uarquin, F.; Reichel, S.; Pichl, D.; Radyushkin, K.; Baumgart, J.; Baumgart, N. Clicker Training Mice for Improved Compliance in the Catwalk Test. *Animals* **2022**, *12*, 3545. https://doi.org/10.3390/ani12243545.
4. Marin, N.; Moragon, A.; Gil, D.; Garcia-Garcia, F.; Bisbal, V. Acclimation and Blood Sampling: Effects on Stress Markers in C57Bl/6J Mice. *Animals* **2023**, *13*, 2816. https://doi.org/10.3390/ani13182816.
5. Palumbo, G.; Kunze, L.H.; Oos, R.; Wind-Mark, K.; Lindner, S.; von Ungern-Sternberg, B.; Bartenstein, P.; Ziegler, S.; Brendel, M. Longitudinal Studies on Alzheimer Disease Mouse Models with Multiple Tracer PET/CT: Application of Reduction and Refinement Principles in Daily Practice to Safeguard Animal Welfare during Progressive Aging. *Animals* **2023**, *13*, 1812. https://doi.org/10.3390/ani13111812.
6. Breuer, L.; Mösch, L.; Kunczik, J.; Buchecker, V.; Potschka, H.; Czaplik, M.; Pereira, C.B. Camera-Based Respiration Monitoring of Unconstrained Rodents. *Animals* **2023**, *13*, 1901. https://doi.org/10.3390/ani13121901.
7. Ruvira, S.; Rodríguez-Rodríguez, P.; Cañas, S.; Ramiro-Cortijo, D.; Aguilera, Y.; Muñoz-Valverde, D.; Arribas, S.M. Evaluation of Parameters Which Influence Voluntary Ingestion of Supplements in Rats. *Animals* **2023**, *13*, 1827. https://doi.org/10.3390/ani13111827.
8. Pérez-Martín, E.; Coto-Vilcapoma, A.; Castilla-Silgado, J.; Rodríguez-Cañón, M.; Prado, C.; Álvarez, G.; Álvarez-Vega, M.A.; Fernández-García, B.; Menéndez-González, M.; Tomás-Zapico, C. Refining Stereotaxic Neurosurgery Techniques and Welfare Assessment for Long-Term Intracerebroventricular Device Implantation in Rodents. *Animals* **2023**, *13*, 2627. https://doi.org/10.3390/ani13162627.
9. Silva-Reis, R.; Faustino-Rocha, A.I.; Silva, J.; Valada, A.; Azevedo, T.; Anjos, L.; Gonçalves, L.; Pinto, M.d.L.; Ferreira, R.; Silva, A.M.S.; et al. Studying and Analyzing Humane Endpoints in the Fructose-Fed and Streptozotocin-Injected Rat Model of Diabetes. *Animals* **2023**, *13*, 1397. https://doi.org/10.3390/ani13081397.
10. García-Delgado, A.B.; Campos-Cuerva, R.; Rosell-Valle, C.; Martin-López, M.; Casado, C.; Ferrari, D.; Márquez-Rivas, J.; Sánchez-Pernaute, R.; Fernández-Muñoz, B. Brain Organoids to Evaluate Cellular Therapies. *Animals* **2022**, *12*, 3150. https://doi.org/10.3390/ani12223150.
11. Rüggeberg, S.; Wanglin, A.; Demirel, Ö.; Hack, R.; Niederhaus, B.; Bidlingmaier, B.; Blumrich, M.; Usener, D. Progress towards the Replacement of the Rabbit Blood Sugar Test for the Quantitative Determination of the Biological Activity of Insulins (USP <121>) with an In Vitro Assay. *Animals* **2023**, *13*, 2953. https://doi.org/10.3390/ani13182953.
12. Peruga, M.; Kawala, B.; Sarul, M.; Kotowicz, J.; Lis, J. Are Currently Selected Laboratory Animals Useful in the Research of How Female Hormones Influence Orthodontic Biomechanics? *Animals* **2023**, *13*, 629. https://doi.org/10.3390/ani13040629.

13. Štrbenc, M.; Kuhar, U.; Lainšček, D.; Orehek, S.; Slavec, B.; Krapež, U.; Malovrh, T.; Majdič, G. Rehoming and Other Refinements and Replacement in Procedures Using Golden Hamsters in SARS-CoV-2 Vaccine Research. *Animals* **2023**, *13*, 2616. https://doi.org/10.3390/ani13162616.
14. Verderio, P.; Lecchi, M.; Ciniselli, C.M.; Shishmani, B.; Apolone, G.; Manenti, G. 3Rs Principle and Legislative Decrees to Achieve High Standard of Animal Research. *Animals* **2023**, *13*, 277. https://doi.org/10.3390/ani13020277.
15. Domínguez-Oliva, A.; Hernández-Ávalos, I.; Martínez-Burnes, J.; Olmos-Hernández, A.; Verduzco-Mendoza, A.; Mota-Rojas, D. The Importance of Animal Models in Biomedical Research: Current Insights and Applications. *Animals* **2023**, *13*, 1223. https://doi.org/10.3390/ani13071223.
16. Martinez-Lopez, S.; Angel-Gomis, E.; Sanchez-Ardid, E.; Pastor-Campos, A.; Picó, J.; Gomez-Hurtado, I. The 3Rs in Experimental Liver Disease. *Animals* **2023**, *13*, 2357. https://doi.org/10.3390/ani13142357.
17. Nunamaker, E.A.; Turner, P.V. Unmasking the Adverse Impacts of Sex Bias on Science and Research Animal Welfare. *Animals* **2023**, *13*, 2792. https://doi.org/10.3390/ani13172792.
18. Paterson, E.A.; Turner, P.V. Challenges with Assessing and Treating Pain in Research Primates: A Focused Survey and Literature Review. *Animals* **2022**, *12*, 2304. https://doi.org/10.3390/ani12172304.
19. Camara Serrano, J.A. Ultrasound Guided Surgery as a Refinement Tool in Oncology Research. *Animals* **2022**, *12*, 3445. https://doi.org/10.3390/ani12233445.
20. Weber, E.M.; Zidar, J.; Ewaldsson, B.; Askevik, K.; Udén, E.; Svensk, E.; Törnqvist, E. Aggression in Group-Housed Male Mice: A Systematic Review. *Animals* **2023**, *13*, 143. https://doi.org/10.3390/ani13010143.
21. Díez-Solinska, A.; Vegas, O.; Azkona, G. Refinement in the European Union: A Systematic Review. *Animals* **2022**, *12*, 3263. https://doi.org/10.3390/ani12233263.
22. Morton, D.B. A Model Framework for the Estimation of Animal 'Suffering': Its Use in Predicting and Retrospectively Assessing the Impact of Experiments on Animals. *Animals* **2023**, *13*, 800. https://doi.org/10.3390/ani13050800.
23. De Vleeschauwer, S.; Lambaerts, K.; Hernot, S.; Debusschere, K. Severity Classification of Laboratory Animal Procedures in Two Belgian Academic Institutions. *Animals* **2023**, *13*, 2581. https://doi.org/10.3390/ani13162581.

Acknowledgments: I would like to express my gratitude to the authors who have contributed their papers to this Special Issue and the reviewers for their invaluable recommendations. I am also thankful to the *Animals* Editorial Office for granting me this opportunity and for their unwavering assistance in organizing and managing this Special Issue.

Conflicts of Interest: The author declares no conflict of interest.

References

1. Russel, W.; Burch, R. *The Principles of Humane Experimental Technique*; Methuen: London, UK, 1959.
2. Graham, M.L.; Prescott, M.J. The multifactorial role of the 3Rs in shifting the harm-benefit analysis in animal models of disease. *Eur. J. Pharmacol.* **2015**, *759*, 19–29. [CrossRef] [PubMed]

Disclaimer/Publisher's Note: The statements, opinions and data contained in all publications are solely those of the individual author(s) and contributor(s) and not of MDPI and/or the editor(s). MDPI and/or the editor(s) disclaim responsibility for any injury to people or property resulting from any ideas, methods, instructions or products referred to in the content.

animals MDPI

Commentary

3Rs Principle and Legislative Decrees to Achieve High Standard of Animal Research

Paolo Verderio [1,*,†], Mara Lecchi [1,†], Chiara Maura Ciniselli [1], Bjorn Shishmani [1], Giovanni Apolone [2] and Giacomo Manenti [3]

1 Unit of Bioinformatics and Biostatistics, Fondazione IRCCS Istituto Nazionale dei Tumori, 20133 Milan, Italy
2 Scientific Directorate, Fondazione IRCCS Istituto Nazionale dei Tumori, 20133 Milan, Italy
3 Unit of Animal Health and Welfare, Fondazione IRCCS Istituto Nazionale dei Tumori, 20133 Milan, Italy
* Correspondence: paolo.verderio@istitutotumori.mi.it; Tel.: +39-02-2390-3201
† These authors contributed equally to this work.

Simple Summary: The 3Rs principle refers to three concepts: Replacement, Reduction and Refinement. These principles should be taken into consideration during the planning and execution of experiments by trying to replace the animal model with an alternative model (if possible), reduce the number of animals by adopting proper and efficient statistical designs, and refine and improve the experimental conditions. The first application of this principle in Europe was reported in the document of the Council Directive of 24 November 1986 (86/609/EEC), and then developed in the updated Directive 2010/63/EU of the European Parliament and of the Council of 22 September 2010. Here, we discuss the perspectives of the 3Rs principle, with a particular focus on the concept of Reduction.

Abstract: Animal experimentation is a vast ecosystem that tries to make different issues such as legislative, ethical and scientific coexist. Research in animal experimentation has made many strides thanks to the 3Rs principle and the attached legislative decrees, but for this very reason, it needs to be evenly implemented both among the countries that have adhered to the decrees and among the team members who design and execute the experimental practice. In this article, we emphasize the importance of the 3Rs principle's application, with a particular focus on the concept of Reduction and related key aspects that can best be handled with the contribution of experts from different fields.

Keywords: 3Rs principle; replacement; reduction; refinement; in vivo experiments; animal welfare

Citation: Verderio, P.; Lecchi, M.; Ciniselli, C.M.; Shishmani, B.; Apolone, G.; Manenti, G. 3Rs Principle and Legislative Decrees to Achieve High Standard of Animal Research. *Animals* **2023**, *13*, 277. https://doi.org/10.3390/ani13020277

Academic Editor: Garikoitz Azkona

Received: 6 December 2022
Revised: 9 January 2023
Accepted: 11 January 2023
Published: 13 January 2023

Copyright: © 2023 by the authors. Licensee MDPI, Basel, Switzerland. This article is an open access article distributed under the terms and conditions of the Creative Commons Attribution (CC BY) license (https://creativecommons.org/licenses/by/4.0/).

1. Introduction

Although about 60 years have passed since its publication, the 3Rs principle has not lost its relevance; indeed, the ideas put forward by Russell and Burch [1] are still current and continue to represent a flexible and valuable tool to guarantee both the welfare of the animals used in laboratories and the quality of the data collected. The 3Rs principle refers to three concepts: Replacement, Reduction and Refinement. During the planning phase and the execution of an in vivo experiment, the researcher should follow this principle as much as possible, trying to (i) replace, where possible, the animal model with an alternative model, (ii) reduce as much as possible the number of animals used by adopting experimental designs that are as efficient as possible and (iii) refine and improve the experimental conditions where animals are involved.

The first application of this principle in Europe was reported in the document of the Council Directive of 24 November 1986 (86/609/EEC) [2], and developed in the updated Directive 2010/63/EU of the European Parliament and of the Council of 22 September 2010 [3], which describes the issues related to the protection of animals used for scientific purposes and regulates the use of animal models in the European Union. All member states have to transpose the Directive 2010/63/EU in its legislation, and Italy must specifically follow the Legislative Decree 26/2014 [4].

Here, we discuss the perspectives of the 3Rs principle with a particular focus on the concept of Reduction and key aspects related to in vivo study implementation, such as blinding and randomization, together with statistical–methodological issues related to experimental design. It is beyond the scope of this communication to report an exhaustive description of the available statistical designs, although proper references will be provided throughout the text.

2. The 3Rs Principle

2.1. Replacement

The concept of Replacement refers to the possibility of replacing the animal model with an alternative one, thanks to the use of non-sentient material [1]. Russel and Burch [1] described a number of alternative methods, such as plants, microorganisms, bio-chemical systems and non-living physicists, and distinguished between relative and absolute replacing techniques [5]. Over time, this concept has evolved into partial replacement (i.e., relative replacement) and full replacement (i.e., absolute replacement). Partial replacement implies the use of another species characterized by a relatively less complex nervous system than the original one and that currently are not considered capable of experiencing suffering [6], such as invertebrates and immature forms of vertebrates or the use of primary cells (and tissues) taken from animals killed solely for this purpose (i.e., not having been used in a scientific procedure that causes suffering). Conversely, as regards to full replacement, in 2005, Buchanan-Smith [7] declared that the 3Rs "still remain the best approach to alternatives" and defined the concept of Replacement as "the set of procedures that completely eliminate the use of some animals". In this case, the animal model is completely eliminated by the use of an alternative method such as tissues and cells, established cell lines and mathematical and computer models.

Alternative Methods

Currently, alternative methods to the animal model include the use of a variety of bioengineered in vitro and ex vivo systems, including organoids, scaffold-based 3D models and microfluidic systems to resemble and capture aspects of human physiology that were unthinkable until few years ago with a good level of realism [8]. The main feature of all these methods relies on the opportunity to address advanced question due to their more intact cytoarchitectural structures and microphysiological processes as well their multiple interacting cell types and multiorgan ones. Although these models can be interpretable by adopting conventional techniques available to pathologists, efforts for their characterization, validation and standardization are needed to confirm their adequacy for the specific question and their fully applicability [9].

For example, concerning bioprinted microtissues, some key aspects should be taken into account, including the selection of cell types and growth media as well as the composition of extracellular scaffolds that support a defined (2D or 3D) architecture [9]. Similarly, mini-organs are designed to reflect the structure and key functions of a complete organ, with a 2D or 3D tissue architecture depending on the nature of the model fabrication process [9]. A special emphasis regards organoids, which are 3D structures organized in a specific spatial pattern characterized by an organ-like architecture resembling the organ of interest and including the biological functions of those tissues without the influence of other organs and systems of the whole organism [9,10]. As recently reported by Kim et al. [11], different cancer types have already been cultured as cancer organoid models. Although cancer organoids represent a powerful and potential in vitro system for drug screening and for predicting the best therapeutic options for individual patients, additional research is needed in order to standardize the set-up of entire organoid protocols.

Already in 2013, Lancaster and colleagues [12] described so-called cerebral organoids to recapitulate the human brain tissue. Then, in 2018, Madhavan et al. [13] reported the generation of oligocortical spheroids to study the myelination of the developing central nervous system. Koike et al., in 2019 [14], developed hepato-biliary-pancreatic organoids as a model for the study of complex human endoderm organogenesis. Chen and colleagues [15] provided in their review a complete picture of the recent advancements in fabricating vascularized tissue and organs, including novel strategies and materials, and their applications. They also explored the limitations of vascularized tissue engineering and some of the promising future directions this technology may bring. Moreover, Costa et al., in 2017 [16], presented a method to produce microfluidic chips containing miniaturized vascular structures to mimic the architectures of blood flow patterns in arterial thrombosis. In the oncological area, the availability of a microfluidic system that recapitulates the physiological and pathological characteristics of human tissues and organs could, for example, open a new, interesting prospective in mimicking the tumor microenvironment (TME) and also in recapitulating the interplays between cancer and the immune system [17]. Further alternative approaches to animal use were presented by Marchesi and colleagues a few years ago [18]; they created an in vitro network representing a simplified model of the nervous and cardiovascular systems' crosstalk. Last year, Barra et al. [19], using bioreactors, simulated the anatomical–physiological complexity of the blood–brain barrier in vitro, potentially contributing to improving the management of neurodegenerative diseases in accordance with the 3R principles. A final note concerns the organ-on-chip (OcC) platforms, a recent innovation for advanced in vitro modeling, and 3D bioprinting [20]. Despite their potential, significant challenges remain and efforts are needed in validating these technologies for biomedical research applications, also in regard to multiorgan models [20].

2.2. Reduction

The principle of Reduction refers to the reduction in the number of experimental units used in an experimental protocol to obtain relevant and robust results [1]. However, this does not imply a mere reduction in the number of experimental units but rather a correct planning of the experiment through the use of experimental designs suited to the objectives and the statistical nature of the variables under investigation. This presupposes the employment of the statistical theory of "Experimental Design" including issues related to the design, methods to control source of bias, and sample size determination, and therefore, it may require the involvement of experts in (bio)statistics.

Reduction can occur, for example, by using results deriving from previous studies or pilot studies planned to characterize the variables under investigation and the variability associated with them, adopting the most efficient experimental design and a sample size appropriate for the goal. Another option to reduce the number of experimental units is to harmonize the methodologies used by various laboratories/research institutions in order to promote the sharing of positive results as well as negative results. This would be of help in reducing the replication of similar experiments and/or negative results. Three levels of reduction can be classified as follows [21]:

(i) Intra-experimental reduction: This kind of reduction concerns the number of animals within each individual experiment or individual protocol within a more complex experiment. By improving the statistical design, carrying out pilot studies and possibly through a retrospective analysis of previously obtained data, it is possible to optimize the minimum number of animals to be used in the experiment in question;
(ii) Supra-experimental reduction: Here, the focus is to disseminate, as much as possible, the concept of reduction among researchers involved at various levels in animal experimentation. This can be achieved through the implementation of ad hoc courses and meetings on experimental designs or statistical methods;
(iii) Extra-experimental reduction: The reduction in this case is obtained through the evolution of experimental practice at an international level, through harmonization of the rules related to experimentation and the development of new research and testing strategies.

By focusing on the intra-experimental level, the reduction in the number of animals used in experiment derives from correct planning of the experimental design and the related statistical analysis. The use of too many EUs, besides being unethical, would lead to a waste of effort and resources in terms of money and time. At the same time, the use of an insufficient number of animals could lead to the loss of a significant result due to the lack of statistical power. Even in this case, there would be a waste of resources. The planning of the entire research flow allows one to obtain the required information and answer(s) to the original scientific question(s) with satisfactory statistical power and optimized sample size.

Study Design

Before starting the experiment, the researcher must postulate a clear hypothesis to test and must already know how the data will be analyzed in order to avoid the unnecessary use of animals in accordance with the 3Rs. The experimenter's question, the type of variables under consideration and, consequently, the statistical test to be applied, as well as the size of the hypothesized effect and the levels of error that one is willing to accept, will influence the design of the experiment and the number of animals necessary for its management. Figure 1 summarizes the main statistical–methodological factors to be considered in planning an experiment.

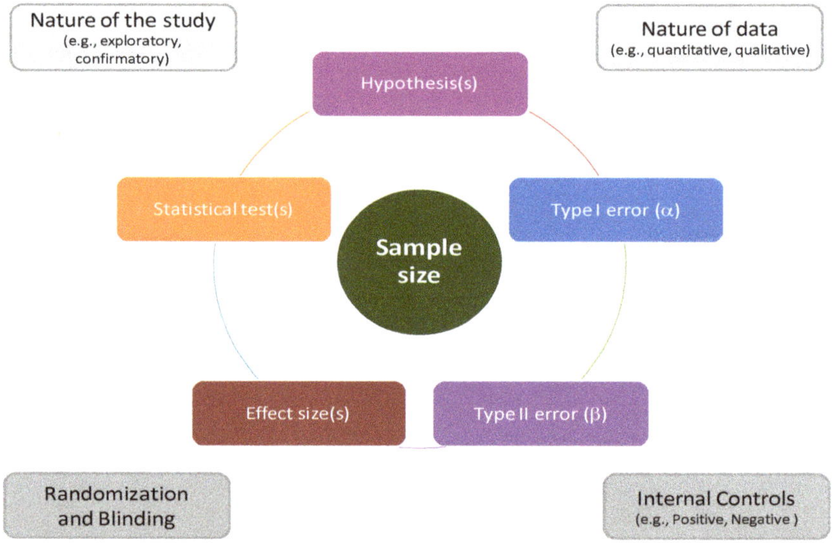

Figure 1. Key statistical–methodological factors to be considered in planning an experiment.

First, the design of the experiment must take into account the nature of the study, i.e., exploratory or confirmatory (Figure 1, upper part). Exploratory studies usually involve the investigation of multiple hypotheses, possibly evaluating multiple objectives [22]. These studies (also called hypothesis-generating studies) are usually used in the initial stages of research to develop new hypotheses, which can be formally tested later. Confirmatory studies, on the other hand, are designed to verify the validity of a specific hypothesis developed a priori. In this case, the statistical analysis should only be focused on the specific hypothesis, and if more statistical tests are performed, adjustments for multiple comparisons (i.e., Bonferroni correction or false discovery rate) should be adopted to reduce the risk of false positive results [22,23].

In addition to the nature of the study, the type of the involved variable(s) should be clearly defined. Quantitative data are measured on a continuous numerical scale (e.g., weight, tumor volume) or on a discrete one (e.g., number of injections, number of metastasis), whereas qualitative data can be measured on a nominal (e.g., genotype) or ordinal scale (e.g., severity score). A special case of nominal scale is represented by the binary scale, in which only two levels are available (e.g., presence/absence of a response). According to the nature of the variable(s), specific statistical models should be adopted.

Once the above-mentioned factors are defined, the next steps are represented by the formal sample size estimation (Figure 1, middle part). Briefly, researchers have to define the hypothesis system, i.e., the null (H_0) and alternative (H_1) hypotheses; the latter can be one-sided or two-sided. In parallel, the level of Type I and II errors should be defined: the first one represents the probability of rejecting H_0 when it is true (i.e., false positives results—α level), whereas the second one represents the probability of not rejecting H_0 when it is false (i.e., false negative results, β level). The complement of Type II error is called statistical power ($\gamma = 1 - \beta$). The last two components that should be defined—according to the nature of the variable(s)—are the statistical test to be implemented and the corresponding effect size, i.e., the smallest difference that we want to detect at the chosen α level. All these inputs are fundamental ingredients for a formal estimation of the sample size.

By moving towards the implementation of the experiments (Figure 1, lower part), it could be useful to evaluate the possibility of including appropriate experimental controls. Specifically, negative controls ensure that an unknown variable does not influence the experimental outcome; for example, animals can be treated with placebo compared to an active treatment or simulated surgery versus surgery. This could defend the researchers against false positive results. Positive controls, on the other hand, ensure that the experiment is actually able to detect the expected effect. Failure to respond in these controls could imply, for example, an experimental bias and ultimately research costs spent without obtaining clear evidence.

Finally, during study implementation, it is important to consider two key aspects: randomization and blinding [23]. The first one equalizes both measured and unmeasured confounders across treatment groups, isolating the experimental treatment as the only difference between them [24]. Moreover, it ensures that other factors, except the treatment/experimental factor under investigation, do not affect the outcome. If the outcomes of the treatment group and control group show differences, this will be the only difference between groups, leading to the conclusion that the observed difference is treatment-induced. Not only a lack of randomization but also errors related to this process are common [25]. Indeed, as summarized in Chusyd DE et al. [26], some authors report non-random allocation as random, or in other studies, animals are randomly allocated to an experimental group together but the data are analyzed as if they were randomized individually, without taking into account the intra-group correlation and ignoring the clustering. This could address misleading conclusions, finding a treatment effect when there is actually no evidence. A full awareness of the identification of the experimental unit and of the related randomization are fundamental steps to avoid recurring mistakes. As previously mentioned, randomization is the process of assigning experimental units to treatments, independently of the pre-randomization characteristics of those units, both observed and unobserved, that could confound the outcome [26]. According to Lazic et al. [27], the experimental units have to meet some conditions: (i) they must be independently randomized to the treatment conditions, (ii) they must not influence each other, especially the outcomes of interest, and (iii) the treatment must be independently applied to each EU. Notably, when animals are caged or housed in the same treatment group (i.e., not independent of each other or they could influence each other's outcomes), the cage and not the single animal has to be considered as the experimental unit. In such a scenario, proper adjustments must be made during sample size estimation, for example, by considering the intra-class correlation (ICC) [28] or by considering the cage as an experimental unit. Especially in this case, where the sample size could be different from the number of animals [29], the involvement of biostatisticians also during the data analysis phase could be suitable in order to apply the most appropriate methods able to ensure robust and replicable results [25]. To prevent any errors, statistical support from the initial phases of the experimental planning is, in any case, strongly recommended.

As mentioned, another key aspect is blinding, which consists of removing as many prejudices of researchers as possible in evaluating the study's measures. It implies, for example, the blindness of the researchers who perform the measurements to the treatment/condition under investigation, as well as the use of anonymized codes. Unfortunately, it is not always possible to apply it, for example, when different types of drug administration (e.g., subcutaneous, oral, intra-muscular) are under investigation. In these cases, it could be useful to involve independent technicians/researchers in the different phases of the experiments (e.g., one for treatment administration and another one for outcome's measurement) or blind the technicians/researchers regarding some details (e.g., dose of the drug) in order to limit as much bias as possible due to unintentional prejudices regarding the treatment/experimental condition of interest.

Once all these elements are defined, and according to the objectives of the study, it will be possible to design the experiment and plan to collect the data in the appropriate way in order to answer the biological question. Inadequate experimental designs may produce biased or inconclusive answers or lack generalizability [30,31]. Without any purpose of providing a complete and exhaustive examination of all the available designs, we briefly summarize the most common ones.

In a Completely Randomized design, experimental units (e.g., mice) are assigned to different treatments at random. In this way, any significant differences between conditions can be fairly attributed to the treatment of interest by ignoring the nuisance factors that could affect treatment conditions. Instead, when a nuisance factor (e.g., reagents batch) may influence the experimental response but is not of interest and is known and controllable, experimental units with a similar nuisance factor should be grouped into blocks, leading to a Randomized Complete Block design. In this design, each treatment is randomly assigned to one experimental unit in each block. Moreover, when the interest is focused on the effect of many discrete factors (e.g., presence/absence of a treatment or different levels of a substance) on the quantitative measurement, it could be useful to implement a Factorial design. In such a design, it is possible to investigate the impact of changes of two or more factors by considering all possible combinations of the levels of each factor and also if the effect of one factor depends on the level of another (i.e., interaction). Other types of designs are the Hierarchical Nested ones that are characterized from EUs sampled multiple times—typically to obtain a more accurate measure of the EU's response—or the Repeated Measures design, in which repeated measurements are made on each experimental unit according to a factor of interest. Finally, if assumptions under the Cross-over design are met, each EU could receive multiple treatments with a wash-out period between exposures and outcome measurement. Detailed descriptions of such designs as well as other types of designs available in the literature are reported in Sorzano COS et al. [23].

By taking into consideration all of the above mentioned issues, it is clear that proper planning of the experiment is essential for all subsequent phases, as mistakes during study planning could lead to irreversible consequences that can potentially invalidate the entire experiment. Furthermore, a clear reporting of all the phases of the experiment, from the design to randomization and analysis, with sufficient details about the methods used, is fundamental to make the research reproducible and to adequately evaluate it. This could be helpful to understand exactly each step and possibly adjust the final statistical analysis, for example, if data were not opportunely analyzed according to the randomization scheme adopted.

2.3. *Refinement*

Russell and Burch indicated that refining the experimental procedures involves not only looking after the animal welfare during the experiment, but also improving the animal's quality of life during all the procedures it is subjected to during its life in captivity [1]. Starting from this concept, Buchanan-Smith [7] proposed a new definition of refinement: "any approach which avoids or minimises the actual or potential pain, distress and other adverse effects experienced at any time during the life of the animals involved, and which enhances their well-being". It is important to notice that well-being is not simply the absence of discomfort, but implies a necessary, active and continuous effort for the improvement of the experimental animal's state. It should be considered that distress may manifest both behaviorally (e.g., overt escape behaviors, approach–avoidance preferences) and physiologically (e.g., movement, vocalization, changes in electroencephalographic activity, heart rate, sympathetic nervous system activity, hypothalamic–pituitary axis activity) [32].

The use of environmental enrichment, which is now explicitly contemplated by the new European Directive 2010/63 [3], is a proven way to provide the animal with greater control of the environment and stimulate the manifestation of behaviors that are inherent to the ecology and etiology of the species used [33]. Regarding the use of environmental enrichments, it is generally thought that the living conditions of animals in captivity are better if they are given the opportunity to express behaviors observed in natural conditions. However, while this is a significant point of view, this is not always the case. In fact, considering that part of the behavioral repertoire of a species can be modified by contingent environmental conditions, and bearing in mind their behavioral flexibility, it follows that animals in captivity can be different from wild ones that live in the environment of origin. The behavioral needs of a captive specimen can therefore be expected to be somewhat different from those of a wild animal. Environmental enrichment must therefore be tailored to each individual situation in order for it to be truly effective. Housing conditions, for example, can significantly improve the welfare of research animals such as rodents, including the provision of shelters and nesting material without neglecting the rather obvious free access to water and food [34]. In addition, all potentially distressing factors should be minimized, such as the noise of ventilation systems. The environment temperature should also be adapted to the physiology of the rodents. If these behavioral needs are not met, the animals can suffer psychophysical stress, which can also compromise the outcome of the experiment. Therefore, improvement strategies do not only improve the welfare of the animals used in research, but also improve the quality and reproducibility of scientific evidence in terms of physical, physiological and ethological needs. Accordingly, enrichment programs must be (i) financially sustainable, (ii) shared with researchers, (iii) established taking into account the time commitment that their implementation requires from the animal facility staff and the fact that they themselves must not interfere with the routine management of animals, as well as (iv) ensure the safety of workers and (v) be monitored with behavioral observations to keep the health status of housed animals under control [35]. Table 1 schematizes the main animal shelters and environmental enrichments.

Table 1. Animal shelters and environmental enrichments.

Type of Improvements	Description
Hygienic	• The environment must be free of insects and parasites and easy to clean and disinfect. In the environment, there must be a rest area, equipped with litter or other soft, dry and clean material. • The surfaces on which the animals move must be as dry and clean as possible, so as to ensure the comfort of the animal during movement, prevent slipping and avoid the onset of foot injuries.
Environmental	• The temperatures in the housing areas must be set according to the species, the age of the animals and the density of the same in order to avoid situations of heat or cold stress. • Ventilation serves to ensure high air quality, low dust and helps to keep the environment at a constant temperature and humidity as much as possible. Finally, air quality should be measured regularly, evaluating the levels of carbon dioxide and oxygen present. • Too low humidity causes pathologies in rodents (for example, ringtail) and excessive heat loss. Too high humidity, on the other hand, could favor the development of ammonia in the premises and in the cages. Relative humidity must therefore be managed according to the ambient temperature.

Table 1. *Cont.*

Type of Improvements	Description
Housing	• The animal should be able to organize the space according to its ethological needs; for this reason, the environment should be equipped with material for the construction of nests, adequate litter, perches or platforms. It is useful to populate the environment with objects that allow the animal to carry out the exploratory behavior of the species. • The environment must provide complex stimuli that engage the animal in the search for the most appropriate responses, expanding the possibility of expression of his behavioral repertoire. An inadequate housing environment could not guarantee the animal's well-being and cause psychophysical discomfort (e.g., induce the development of behavioral atypia, stereotypes, and excessive aggression).
Sensory	If the experimental protocol does not allow the interaction with conspecifics, it is essential that they can at least enjoy indirect social contact (i.e., visual, auditory and olfactory). This support develops positive effects on the cognitive abilities of animals, reduces stress during experimental procedures and ensures better interaction between animals and staff or researchers.
Social	Animals need social interaction with conspecifics and therefore should always be housed in pairs or in groups that are as stable and as harmonious as possible, but can also involve negative and stressful aspects such as the aggression of dominant subjects over subordinates.

General Health, Well-Being and Anxiety-Like Behavior Evaluation

To assess the general health and well-being of mice and minimize their distress, in accordance with the Refinement principle, it is necessary to measure physiological and behavioral indicators. Over the past three decades, numerous tests has been developed to assess compulsive-like behaviors. In 2001, Roughan and Flecknell [36] studied the possibility that objective behavioral analyses may be used to develop an objective scheme for pain scoring in rats following laparotomy. Starting from 150 behaviors, they selected a set of 16 behaviors that had the greatest value in discriminating treated and control groups.

The Nest Building Test is an important test for assessing the general behavior integrity and well-being of laboratory mice [34,37]. Nest building is a behavior that mice perform for comfort, thermoregulation and for housing their pups; therefore, altered nest building behavior may suggest reduced mouse welfare due to several factors (e.g., thermal stress, general malaise, amount of aggression present within the cage). For example, Hess et al. [38] evaluated the use of naturalistic nesting materials to investigate the potential improvement of the nest quality, concluding that the use of a more naturalistic nesting material allows mice to build more naturalistic nests.

In parallel, the elevated maze test (EPM) and open field test [37,39] can be adopted to test anxiety behavior in mice and evaluate the effectiveness of anxiolytic drugs in neurobiological anxiety research. Briefly, the EPM consists of four arms; two of the opposite arms are walled and the remaining two are open. The amount of time mice spend in the mural arms compared to the open arms in a defined short period provides a measure of anxiety or fear. The test allows researchers to gather information on post-traumatic stress disorder and other conditions characterized by anxious behavior and could be used to screen for new compounds for anxiolytic properties. This test has been used, for example, by Ataka et al. [40] to evaluate the anxiety-like behavior in mice subjected to chronic psychological stress (cPS) and the effects of cPS on the interaction between bone marrow-derived microglia and neurons. Similarly, the Open Field Test measures anxiety-like behavior and locomotor activity [34,37]. The test, which is also used in neurobiological studies, allows researchers to evaluate the basis of anxiety and screening for novel targets and anxiolytic compounds, in addition to the general health and well-being of an animal. Briefly, the test implies the use of a camera to monitor the movement of the animal in and around the peripheral and central areas of a 42 × 42 × 42 cm polyvinyl chloride box. Changes in locomotion may be indicative of altered neurological processes and may therefore reflect abnormal brain function. This test was one of those used by Gouveia and L. Hurst [41] to assess the impact of handling on stress and anxiety in laboratory animals.

To complement standard behavioral tests, Home Cage-Monitoring Systems (HCMS) could also be used to assess the general locomotor activity levels and animals' anxiety as well as investigate the pain response [42]. This method, which allows the prolonged and unbiased observations of spontaneous behavior, has been used by Roughan et al. [43] and Radaelli et al. [44]. Roughan and colleagues [43] evaluated the precision of HCS relative to an experienced human observer in differentiating between the pre- and postoperative behavior of groups of mice undergoing anesthesia and administered different doses of an analgesic; they provided evidence about the powerful role of HCS as a tool for investigating pain responses and analgesic effects following various different types of surgery and other potentially painful conditions in mice, and eventually other rodent species. Radaelli and colleagues [44] adopted the HCS as a complement tool to measure the effect of a brightly lit enclosed chamber (R&L) on mice, concluding that R&L lowered normal walking frequency and likely posed a risk of low-grade neuro-inflammation.

Besides the aforementioned behavior-based methods, physiological methods have been introduced to objectively evaluate the animal experience. Mayer and colleagues [32] reviewed and discussed the approaches for evaluating stress in animals using physiological methods, with emphasis on the transition between the conscious and unconscious states.

3. Directive 2010/63/EU—Protection of Animals Used for Scientific Purposes

The use of animal models according to the 3Rs principle in the European Union is regulated by the Directive 2010/63/EU [3]. It is made up of 66 articles and 8 annexes aiming to ensure and adapt scientific and technological progress in the European legislation, which until then was represented by the Council Directive of 24 November 1986 (86/609/EEC [2]). Each state must transpose the Directive 2010/63/EU on the protection of animals used for scientific purposes, and Italy, specifically, must adhere to Legislative Decree 26/2014 [4]. In the 24 years that have passed between the two directives, the acquaintances in the laboratory animal science sector have expanded significantly; it turned out that an update also at the legislative level is therefore essential. In fact, the directive is not a series of rules that favor or prohibit the use of animals in research, but that protect animals used in research laboratories. The primary purpose of the directive is to provide specifications that are as detailed as possible in order to reduce the disparity between member states on methodologies that guarantee the welfare of laboratory animals, making the European landscape more uniform. "Animal Welfare" is in fact a value of the European Union, as described in Article 13 of the Treaty on the Functioning of the European Union [45].

One of the main updates introduced in this European directive with respect to the previous Directive 86/609/EEC is the evaluation of suffering. According to Annex VIII, "The severity of the procedure is determined based on the level of pain, suffering, distress or prolonged damage to which the individual animal is presumably subjected during the procedure itself" [3]. Section I of Annex VIII establishes and defines four categories into which procedures using experimental animals can be divided, as schematized in Table 2.

The assignment of the severity category is based on the risk of the most severe effects, once all the appropriate refinement techniques have been applied. In assigning a procedure to a particular category, the type of procedure itself and other factors such as those summarized in Table 3 are taken into account.

Article 26 of European Regulation 63/2010 provides another strong innovation, "that each breeder, supplier and user sets up an Animal-Welfare Body (AWB)". Each AWB is composed of at least one person in charge of the welfare and care of animals, a veterinarian and in the case of a structure authorized for experimentation, one scientific member as the guarantor of scientific quality. The figure of the biostatistician within the AWB is still a discussed matter [46]; however, it is crucial for verifying that the study design is compatible with the Reduction principle of the 3Rs.

Table 2. Level of pain.

Level of Pain	Description
Non-awakening	Procedures conducted entirely under general anesthesia from which the animal cannot regain consciousness.
Mild	Procedures that are likely to cause mild and short-lived pain, suffering or distress, as well as procedures that do not cause a significant deterioration in the welfare or general condition of the animals.
Moderate	Animal procedures that are likely to cause moderate and short-term pain, suffering or distress, or mild and long-lasting pain, suffering or distress, as well as procedures that are likely to cause moderate deterioration of the welfare or general condition of the animals.
Severe	Procedures on animals that are likely to cause intense pain, suffering or distress, or moderate and long-lasting pain, suffering or distress, as well as procedures that are likely to cause serious deterioration in the welfare or general condition of the animals.

Table 3. Factors related to the procedure.

Factors
Nature of the pain, suffering, distress or prolonged harm caused by the Procedure (in all its elements) and relative intensity
Duration, frequency and multiplicity of the techniques used
Cumulative suffering under the procedure
Impediment of natural behavior, due among other things to limitations of the rules in Housing, breeding and care
Type of species and genotype
Maturity, age and sex of the animal
Training experience of the animal with respect to the procedure
Actual severity of the previous procedures, if the animal is intended to be reused
Methods used to reduce or eliminate pain, suffering and distress, including the Improvement of housing, breeding and care conditions

4. Discussion

The widest possible dissemination of information related to the 3Rs is essential if Russell and Burch's aspiration is to be realized for animal research in the most humane way possible. Although in this commentary we mainly focus on the Reduction principle and the importance of correct study planning and sizing with the most proper experiential design, improvements in both the Replacement and Refinement principles should be considered.

As aforementioned, we are seeing rapid progress in the acceptance of organs-on-a-chip, or, more generally, in the engineering of in vitro models as alternative methods to fulfill the Replacement principle. Increasing evidence supports the potential and utility of these models, for example, for drug/toxicological screening studies [47]. One of the first implementation measures at a legislative level—Commission Implementing Regulation (EU) 2021/1709—concerns the replacement of live animals for the detection of paralytic shellfish poisoning toxins [48]. Similarly, regarding the Refinement principle, improvements to all phases of animal welfare, from housing to analgesia protocols and methods of drug administration, contribute to reduce the suffering of laboratory animals [49]. It is clear that the Reduction principle implies the use of the minimum number of laboratory animals to achieve scientific objectives through the adoption of appropriate study design and statistical methods. Thus, each of the three Rs plays a crucial role in experimental achievement, and they could interact positively or negatively. Indeed, some approaches could simultaneously enhance more than one objective of the 3Rs, or, in other cases, generate contrary effects on two different objectives of the 3Rs. Some examples of positive and negative interactions between the three principles are reported by de Boo et al. [50] and those related to management issues. The harmonization of protocols interacts positively with Replacement and Reduction, the use of in vitro models acts on both Replacement and Refinement, and the implementation of training programs could be viewed as a powerful action point for Reduction and Refinement. On the other hand, as stated by de Boo et al. [50], the implementation of non-animal methods, which require the comparison with the corresponding original in vivo model, could create a negative interaction between the Reduction and the Replacement principles.

In general, in order to conduct useful research, the goal one wants to achieve and the kind of value the experiment possesses are the first aspects to be clearly defined, along with the choice of the most appropriate animal species to reach the most qualitative outcome. Regarding the latter aspect, as reported by Azkona et al. [47], the selection of the species for an animal model should be made to completely resemble the disease completely and to allow a translation of results to humans. Although the choice may be influenced by practical constraints or unscientific reasons, the selection of a species for an animal model requires a multidisciplinary team of specialists who take into account not only financial feasibility, but also biological characteristics, available imaging and molecular techniques, results of previous experiments and even ethical issues for a given species [47].

It should be mentioned that it is possible that discrepancies from the initial experimental project may occur during animal experiment due to some inconveniences. The onset of difficult-to-solve organizational and managerial problems during the implementation phase can lead the researchers to introduce alterations to the original experimental design that affect the subsequent data analysis. In order to avoid invalidating the entire experiment or continuing it with non-robust results, biostatistical support should be required to modify and possibly re-plan an experimental project. This also highlights the importance of a continuous interaction between scientists, managers of facilities and technical personnel, a practice that is a fundamental feature which has become known as the "culture of care". This embodies the commitment to improve animal welfare, scientific quality, personal care and transparency for all interested stakeholders.

Finally, it is important to emphasize that just as the randomization methods, limitations and potential sources of bias of the studies should be declared in the published manuscript following the accepted guidelines that define the standards of reporting the results of animal research (i.e., Animal Research: Reporting of In Vivo Experiments, ARRIVE) [51], the same details should be adequately cited in the research project submitted to the AWB, in order to have—during study planning—a complete picture of all the implementation phases of the study. Because animal design is a delicate ecosystem, all components and scientists must cooperate with each other. In this view, training/education programs, as well as data/protocols sharing are aspects that should be continuously supported. The EU Reference Laboratory for alternatives to animal testing contributes to the development and testing new animal-free methods to be applied in an integrated safety assessment of chemicals, as well as provides informatics tools and databases to support this [52]. It also promotes the dissemination of information and sharing of knowledge on the 3Rs. Moreover, the "Refinement Database" [53] that collects newly published scientific contributions related to improving or safeguarding the welfare of animals used in research could be a useful initiative in this direction.

5. Conclusions

All experimental projects come with trade-offs. It is an ability to conduct animal testing while ensuring the maximum probability of success, maintaining high standards of scientific rigor and accepting practical limits and safeguarding ethics, a skill that constantly requires development. This research field is currently struggling with a high amount of low-power studies, which generate little reliable information. Therefore, a figure such as a biostatistician is needed, able to move and create links in distant fields ranging from the sampling of the population to the ethical aspects of research, for the purpose of valid statistical design, management, analysis and interpretation.

The objectives of harmonization between EU member states, hoped for by the directive, are still a long way off. There are still many gaps to fill before the high standard of animal welfare set by the Union can be achieved [54], and the effort of all the member states is required not only for animal welfare, but for better science and economic reasons [55].

Author Contributions: Conceptualization, P.V. and G.M.; methodology, M.L., C.M.C. and B.S.; data curation, M.L. and C.M.C.; writing—original draft preparation, M.L., C.M.C. and B.S.; writing—review and editing, P.V., G.A. and G.M. All authors have read and agreed to the published version of the manuscript.

Funding: This research received no external funding.

Institutional Review Board Statement: Not applicable.

Informed Consent Statement: Not applicable.

Data Availability Statement: Not applicable.

Acknowledgments: We thank Manuela Gariboldi for her help in deepening the replacement techniques.

Conflicts of Interest: The authors declare no conflict of interest.

References

1. Russell, W.M.S.; Burch, R.L. *The Principles of Humane Experimental Technique*; Methuen & Co, Ltd.: London, UK, 1959.
2. Council Directive 86/609/EEC of 24 November 1986 on the Approximation of Laws, Regulations and Administrative Provisions of the Member States Regarding the Protection of Animals Used for Experimental and Other Scientific Purposes (EEC) No 86/609. OJ L358/0001-0028. Available online: https://eur-lex.europa.eu/legal-content/EN/TXT/PDF/?uri=CELEX:31986L0609&from=EN (accessed on 2 September 2022).
3. Directive 2010/63/EU of the European Parliament and of the Council of 22 September 2010 on the Protection of Animals Used for Scientific Purposes. OJ L276/33-79. Available online: https://eur-lex.europa.eu/legal-content/EN/TXT/PDF/?uri=CELEX:32010L0063&from=EN (accessed on 6 October 2022).
4. Legislative Decree 26/2014—Implementation of Directive 2010/63/EU on the Protection of Animals Used for Scientific Purposes; GU n. 61 of March 14, 2014. Available online: https://www.gazzettaufficiale.it/eli/id/2014/03/14/14G00036/sg (accessed on 2 September 2022).
5. Russell, W.M.S. The increase of humanity in experimentation: Replacement, reduction and refinement. *Lab. Anim. Bur. Collect. Pap.* **1957**, *6*, 23–25.
6. National Centre for the Replacement Refinement & Reduction of Animal Research (NC3RS). Available online: https://nc3rs.org.uk/who-we-are/3rs (accessed on 6 October 2022).
7. Buchanan-Smith, H.M.; Rennie, A.E.; Vitale, A.; Pollo, S.; Prescott, M.J.; Morton, D.B. Harmonising the definition of refinement. *Anim. Welf.* **2005**, *14*, 379–384. [CrossRef]
8. Taebnia, N.; Römling, U.; Lauschke, V.M. In Vitro and Ex Vivo Modeling of Enteric Bacterial Infections. *Gut Microbes* **2023**, *15*, 2158034. [CrossRef] [PubMed]
9. Gan, J.; Bolon, B.; Van Vleet, T.; Wood, C. Chapter 24—Alternative Models in Biomedical Research: In Silico, In Vitro, Ex Vivo, and Nontraditional In Vivo Approaches. In *Haschek and Rousseaux's Handbook of Toxicologic Pathology*, 4th ed.; Haschek, W.M., Rousseaux, C.G., Wallig, M.A., Bolon, B., Eds.; Academic Press: Cambridge, MA, USA, 2022; pp. 925–966. [CrossRef]
10. Krysko, D.V.; Demuynck, R.; Efimova, I.; Naessens, F.; Krysko, O.; Catanzaro, E. In Vitro Veritas: From 2D Cultures to Organ-on-a-Chip Models to Study Immunogenic Cell Death in the Tumor Microenvironment. *Cells* **2022**, *11*, 3705. [CrossRef] [PubMed]
11. Kim, J.; Koo, B.-K.; Knoblich, J.A. Human Organoids: Model Systems for Human Biology and Medicine. *Nat. Rev. Mol. Cell Biol.* **2020**, *21*, 571–584. [CrossRef] [PubMed]
12. Lancaster, M.A.; Renner, M.; Martin, C.-A.; Wenzel, D.; Bicknell, L.S.; Hurles, M.E.; Homfray, T.; Penninger, J.M.; Jackson, A.P.; Knoblich, J.A. Cerebral Organoids Model Human Brain Development and Microcephaly. *Nature* **2013**, *501*, 373–379. [CrossRef] [PubMed]
13. Madhavan, M.; Nevin, Z.S.; Shick, H.E.; Garrison, E.; Clarkson-Paredes, C.; Karl, M.; Clayton, B.L.L.; Factor, D.C.; Allan, K.C.; Barbar, L.; et al. Induction of Myelinating Oligodendrocytes in Human Cortical Spheroids. *Nat. Methods* **2018**, *15*, 700–706. [CrossRef]
14. Koike, H.; Iwasawa, K.; Ouchi, R.; Maezawa, M.; Giesbrecht, K.; Saiki, N.; Ferguson, A.; Kimura, M.; Thompson, W.L.; Wells, J.M.; et al. Modelling Human Hepato-Biliary-Pancreatic Organogenesis from the Foregut–Midgut Boundary. *Nature* **2019**, *574*, 112–116. [CrossRef]
15. Chen, E.P.; Toksoy, Z.; Davis, B.A.; Geibel, J.P. 3D Bioprinting of Vascularized Tissues for in Vitro and in Vivo Applications. *Front. Bioeng. Biotechnol.* **2021**, *9*, 664188. [CrossRef]
16. Costa, P.F.; Albers, H.J.; Linssen, J.E.A.; Middelkamp, H.H.T.; van der Hout, L.; Passier, R.; van den Berg, A.; Malda, J.; van der Meer, A.D. Mimicking Arterial Thrombosis in a 3D-Printed Microfluidic in Vitro Vascular Model Based on Computed Tomography Angiography Data. *Lab Chip* **2017**, *17*, 2785–2792. [CrossRef]
17. Li, Y.; Fan, H.; Ding, J.; Xu, J.; Liu, C.; Wang, H. Microfluidic Devices: The Application in TME Modeling and the Potential in Immunotherapy Optimization. *Front. Genet.* **2022**, *13*, 969723. [CrossRef]

18. Marchesi, N.; Barbieri, A.; Fahmideh, F.; Govoni, S.; Ghidoni, A.; Parati, G.; Vanoli, E.; Pascale, A.; Calvillo, L. Use of Dual-Flow Bioreactor to Develop a Simplified Model of Nervous-Cardiovascular Systems Crosstalk: A Preliminary Assessment. *PLoS ONE* **2020**, *15*, e0242627. [CrossRef] [PubMed]
19. Barra, T.; Falanga, A.; Bellavita, R.; Laforgia, V.; Prisco, M.; Galdiero, S.; Valiante, S. GH625-Liposomes Deliver PACAP through a Dynamic in Vitro Model of the Blood-Brain Barrier. *Front. Physiol.* **2022**, *13*, 932099. [CrossRef] [PubMed]
20. Chliara, M.A.; Elezoglou, S.; Zergioti, I. Bioprinting on Organ-on-Chip: Development and Applications. *Biosensors* **2022**, *12*, 1135. [CrossRef]
21. de Boo, J.; Hendriksen, C. Reduction Strategies in Animal Research: A Review of Scientific Approaches at the Intra-Experimental, Supra-Experimental and Extra-Experimental Levels. *Altern. Lab. Anim.* **2005**, *33*, 369–377. [CrossRef] [PubMed]
22. Kimmelman, J.; Mogil, J.S.; Dirnagl, U. Distinguishing between Exploratory and Confirmatory Preclinical Research Will Improve Translation. *PLoS Biol.* **2014**, *12*, e1001863. [CrossRef]
23. Sorzano, C.O.S.; Parkinson, M. Statistical Experiment Design for Animal Research. 2019. Available online: http://www.hu.usp.br/wp-content/uploads/sites/195/2019/09/Statistical-experiment-design-for-animal-research.pdf (accessed on 16 May 2022).
24. Muhlhausler, B.S.; Bloomfield, F.H.; Gillman, M.W. Whole Animal Experiments Should Be More like Human Randomized Controlled Trials. *PLoS Biol.* **2013**, *11*, e1001481. [CrossRef]
25. Vorland, C.J.; Brown, A.W.; Dawson, J.A.; Dickinson, S.L.; Golzarri-Arroyo, L.; Hannon, B.A.; Heo, M.; Heymsfield, S.B.; Jayawardene, W.P.; Kahathuduwa, C.N.; et al. Errors in the Implementation, Analysis, and Reporting of Randomization within Obesity and Nutrition Research: A Guide to Their Avoidance. *Int. J. Obes.* **2021**, *45*, 2335–2346. [CrossRef]
26. Chusyd, D.E.; Austad, S.N.; Brown, A.W.; Chen, X.; Dickinson, S.L.; Ejima, K.; Fluharty, D.; Golzarri-Arroyo, L.; Holden, R.; Jamshidi-Naeini, J.; et al. From Model Organisms to Humans, the Opportunity for More Rigor in Methodologic and Statistical Analysis, Design, and Interpretation of Aging and Senescence Research. *J. Gerontol. Ser. A* **2022**, *77*, 2155–2164. [CrossRef]
27. Lazic, S.E.; Clarke-Williams, C.J.; Munafò, M.R. What Exactly Is N in Cell Culture and Animal Experiments? *PLoS Biol.* **2018**, *16*, e2005282. [CrossRef]
28. Basson, A.R.; LaSalla, A.; Lam, G.; Kulpins, D.; Moen, E.L.; Sundrud, M.S.; Miyoshi, J.; Ilic, S.; Theriault, B.R.; Cominelli, F.; et al. Artificial Microbiome Heterogeneity Spurs Six Practical Action Themes and Examples to Increase Study Power-Driven Reproducibility. *Sci. Rep.* **2020**, *10*, 5039. [CrossRef]
29. Gaskill, B.N.; Garner, J.P. Power to the People: Power, Negative Results and Sample Size. *J. Am. Assoc. Lab. Anim. Sci.* **2020**, *59*, 9–16. [CrossRef]
30. Karp, N.A.; Fry, D. What Is the Optimum Design for My Animal Experiment? *BMJ Open Sci.* **2021**, *5*, e100126. [CrossRef]
31. Ding, Y.; Lin, H.M. Experimental Design and Data Analysis of In Vivo Fluorescence Imaging Studies. In *In Vivo Fluorescence Imaging. Methods in Molecular Biology*; Bai, M., Ed.; Humana Press: New York, NY, USA, 2016; Volume 1444. [CrossRef]
32. Meyer, R.E. Physiologic Measures of Animal Stress during Transitional States of Consciousness. *Animals* **2015**, *5*, 702–716. [CrossRef] [PubMed]
33. Olsson, I.A.S.; Dahlborn, K. Improving Housing Conditions for Laboratory Mice: A Review of Environmental Enrichment. *Lab. Anim.* **2002**, *36*, 243–270. [CrossRef] [PubMed]
34. Gaskill, B.N.; Karas, A.Z.; Garner, J.P.; Pritchett-Corning, K.R. Nest building as an indicator of health and welfare in laboratory mice. *J. Vis. Exp.* **2013**, *82*, 51012. [CrossRef] [PubMed]
35. Knopp, K.L.; Stenfors, C.; Baastrup, C.; Bannon, A.W.; Calvo, M.; Caspani, O.; Currie, G.; Finnerup, N.B.; Huang, W.; Kennedy, J.D.; et al. Experimental Design and Reporting Standards for Improving the Internal Validity of Pre-Clinical Studies in the Field of Pain: Consensus of the IMI-Europain Consortium. *Scand. J. Pain* **2015**, *7*, 58–70. [CrossRef]
36. Roughan, J.V.; Flecknell, P.A. Behavioural Effects of Laparotomy and Analgesic Effects of Ketoprofen and Carprofen in Rats. *Pain* **2001**, *90*, 65–74. [CrossRef]
37. Guest, P.C.; Rahmoune, H. *Pre-Clinical Models, Techniques and Protocols*; Guest, P.C., Ed.; Springer: New York, NY, USA, 2019. [CrossRef]
38. Hess, S.E.; Rohr, S.; Dufour, B.D.; Gaskill, B.N.; Pajor, E.A.; Garner, J.P. Home Improvement: C57BL/6J Mice given More Naturalistic Nesting Materials Build Better Nests. *J. Am. Assoc. Lab. Anim. Sci.* **2008**, *47*, 25–31. [PubMed]
39. Eltokhi, A.; Rappold, G.; Sprengel, R. Distinct Phenotypes of Shank2 Mouse Models Reflect Neuropsychiatric Spectrum Disorders of Human Patients With SHANK2 Variants. *Front. Mol. Neurosci.* **2018**, *11*, 240. [CrossRef]
40. Ataka, K.; Asakawa, A.; Nagaishi, K.; Kaimoto, K.; Sawada, A.; Hayakawa, Y.; Tatezawa, R.; Inui, A.; Fujimiya, M. Bone Marrow-Derived Microglia Infiltrate into the Paraventricular Nucleus of Chronic Psychological Stress-Loaded Mice. *PLoS ONE* **2013**, *8*, e81744. [CrossRef] [PubMed]
41. Gouveia, K.; Hurst, J.L. Improving the Practicality of Using Non-Aversive Handling Methods to Reduce Background Stress and Anxiety in Laboratory Mice. *Sci. Rep.* **2019**, *9*, 20305. [CrossRef]
42. Grieco, F.; Bernstein, B.J.; Biemans, B.; Bikovski, L.; Burnett, C.J.; Cushman, J.D.; van Dam, E.A.; Fry, S.A.; Richmond-Hacham, B.; Homberg, J.R.; et al. Measuring Behavior in the Home Cage: Study Design, Applications, Challenges, and Perspectives. *Front. Behav. Neurosci.* **2021**, *15*, 735387. [CrossRef] [PubMed]
43. Roughan, J.V.; Wright-Williams, S.L.; Flecknell, P.A. Automated Analysis of Postoperative Behaviour: Assessment of HomeCageScan as a Novel Method to Rapidly Identify Pain and Analgesic Effects in Mice. *Lab. Anim.* **2009**, *43*, 17–26. [CrossRef] [PubMed]

44. Redaelli, V.; Bosi, A.; Luzi, F.; Cappella, P.; Zerbi, P.; Ludwig, N.; Di Lernia, D.; Roughan, J.V.; Porcu, L.; Soranna, D.; et al. Neuroinflammation, Body Temperature and Behavioural Changes in CD1 Male Mice Undergoing Acute Restraint Stress: An Exploratory Study. *PLoS ONE* **2021**, *16*, e0259938. [CrossRef]
45. European Union. Consolidated Version of the Treaty on the Functioning of the European Union, 13 December 2007, 2008/C 115/01. OJ C 115/47. Available online: https://www.refworld.org/docid/4b17a07e2.html (accessed on 6 December 2022).
46. Grignaschi, G.; Redaelli, V.; Luzi, F.; Fornasier, M. The Bodies in Charge of Animal Welfare: What They Do and What They Could Do? *Front. Physiol.* **2018**, *9*, 391. [CrossRef]
47. Azkona, G.; Sanchez-Pernaute, R. Mice in Translational Neuroscience: What R We Doing? *Prog. Neurobiol.* **2022**, *217*, 102330. [CrossRef]
48. Commission Implementing Regulation (EU) 2021/1709 of 23 September 2021 Amending Implementing Regulation (EU) 2019/627 as Regards Uniform Practical Arrangements for the Performance of Official Controls on Products of Animal Origin. Available online: https://eur-lex.europa.eu/eli/reg_impl/2021/1709/oj (accessed on 5 January 2023).
49. Díez-Solinska, A.; Vegas, O.; Azkona, G. Refinement in the European Union: A Systematic Review. *Animals* **2022**, *12*, 3263. [CrossRef]
50. de Boo, M.J.; Rennie, A.E.; Buchanan-Smith, H.M.; Hendriksen, C.F.M. The interplay between replacement, reduction and refinement: Considerations where the Three Rs interact. *Anim. Welf.* **2005**, *14*, 327–332. [CrossRef]
51. Percie du Sert, N.; Hurst, V.; Ahluwalia, A.; Alam, S.; Avey, M.T.; Baker, M.; Browne, W.J.; Clark, A.; Cuthill, I.C.; Dirnagl, U.; et al. The ARRIVE Guidelines 2.0: Updated Guidelines for Reporting Animal Research. *J. Cereb. Blood Flow Metab.* **2020**, *40*, 1769–1777. [CrossRef]
52. Available online: https://joint-research-centre.ec.europa.eu/eu-reference-laboratory-alternatives-animal-testing-eurl-ecvam_en (accessed on 5 January 2023).
53. Animal Welfare Institute. 2022. Available online: https://awionline.org/ (accessed on 4 January 2023).
54. Barletta, M. La Sperimentazione Animale in Europa: L'incidenza Della Direttiva 2010/63 UE Tra Gli Stati Membri E Il Suo Ruolo Nello Sviluppo Del Vaccino Anti COVID-19. *BioLaw J. Riv. Di. BioDiritto* **2021**, *2*, 41–59.
55. Busquet, F.; Kleensang, A.; Rovida, C.; Herrmann, K.; Leist, M.; Hartung, T. New European Union Statistics on Laboratory Animal Use—What Really Counts! *ALTEX* **2020**, *37*, 167–186. [CrossRef] [PubMed]

Disclaimer/Publisher's Note: The statements, opinions and data contained in all publications are solely those of the individual author(s) and contributor(s) and not of MDPI and/or the editor(s). MDPI and/or the editor(s) disclaim responsibility for any injury to people or property resulting from any ideas, methods, instructions or products referred to in the content.

animals

MDPI

Review

The Importance of Animal Models in Biomedical Research: Current Insights and Applications

Adriana Domínguez-Oliva [1], Ismael Hernández-Ávalos [2], Julio Martínez-Burnes [3], Adriana Olmos-Hernández [4], Antonio Verduzco-Mendoza [4] and Daniel Mota-Rojas [5,*]

1 Master's Program in Agricultural and Livestock Sciences [Maestría en Ciencias Agropecuarias], Xochimilco Campus, Universidad Autónoma Metropolitana (UAM), Mexico City 04960, Mexico
2 Clinical Pharmacology and Veterinary Anesthesia, Facultad de Estudios Superiores Cuautitlán, Universidad Nacional Autónoma de México (UNAM), Cuautitlán 54714, Mexico
3 Facultad de Medicina Veterinaria y Zootecnia, Universidad Autónoma de Tamaulipas, Victoria City 87000, Mexico
4 Division of Biotechnology—Bioterio and Experimental Surgery, Instituto Nacional de Rehabilitación-Luis, Guillermo Ibarra Ibarra (INR-LGII), Mexico City 14389, Mexico
5 Neurophysiology, Behavior and Animal Welfare Assessment, DPAA, Universidad Autónoma Metropolitana (UAM), Mexico City 04960, Mexico
* Correspondence: dmota100@yahoo.com.mx

Simple Summary: The present review highlights and examines the importance of animal models in relevant topics concerning current human and animal health. Over the past five years, different animal species have been used to study pandemics, such as the 2019 Coronavirus, diabetes, and obesity. Through murine, primate, porcine, and even aquatic models (e.g., zebrafish), several neurological, behavioral, cardiovascular, and oncological disorders are being understood while developing new therapeutic approaches. Nematodes and arthropods are some of the new alternatives for biomedical science; however, regardless of the species, many animal research studies show the vital role of animal models in advancing biomedical research.

Abstract: Animal research is considered a key element in advance of biomedical science. Although its use is controversial and raises ethical challenges, the contribution of animal models in medicine is essential for understanding the physiopathology and novel treatment alternatives for several animal and human diseases. Current pandemics' pathology, such as the 2019 Coronavirus disease, has been studied in primate, rodent, and porcine models to recognize infection routes and develop therapeutic protocols. Worldwide issues such as diabetes, obesity, neurological disorders, pain, rehabilitation medicine, and surgical techniques require studying the process in different animal species before testing them on humans. Due to their relevance, this article aims to discuss the importance of animal models in diverse lines of biomedical research by analyzing the contributions of the various species utilized in science over the past five years about key topics concerning human and animal health.

Keywords: translational research; animal research; laboratory animals; rodents; primates; pigs; zebrafish; nematodes

Citation: Domínguez-Oliva, A.; Hernández-Ávalos, I.; Martínez-Burnes, J.; Olmos-Hernández, A.; Verduzco-Mendoza, A.; Mota-Rojas, D. The Importance of Animal Models in Biomedical Research: Current Insights and Applications. *Animals* **2023**, *13*, 1223. https://doi.org/10.3390/ani13071223

Academic Editor: Garikoitz Azkona

Received: 21 February 2023
Revised: 19 March 2023
Accepted: 30 March 2023
Published: 31 March 2023

Copyright: © 2023 by the authors. Licensee MDPI, Basel, Switzerland. This article is an open access article distributed under the terms and conditions of the Creative Commons Attribution (CC BY) license (https://creativecommons.org/licenses/by/4.0/).

1. Introduction

The use of animals in scientific research is controversial [1]. However, the transformation of medicine from an art to a science can be mainly attributed to using a wide range of animal models [2], selected according to their functional and genetic characteristics for specific research lines [3]. Animal models contribute significantly to the advance of biomedical science through their meaningful contributions to our growing understanding of pathological and biological processes [4]. Moreover, they enable the development and testing of drugs, vaccines, and surgical techniques applicable to human and veterinary medicine [5].

The term "animal model" comes from the Latin *animae* (alma or spirit) and the word model, which means to imitate or be similar to [6]. Animal models are based on the principle of comparative medicine [7] as instruments that can replicate physiological and pathological processes [8]. The species is selected according to each project's objective and hypothesis [3] but also considers biological, anatomical, functional, and genetic similarities to humans or other animals [6]. Today, most of the species utilized in biomedical research are rodents [9], as they are deemed ideal models for studying pathologies that affect human populations due to their physiological homology [10], which allows them to be employed to further our understanding of such processes as sepsis, obesity, cancer, organ transplants, and biological development, among many others [11,12].

The species used in experimentation are not limited to small mammals. Rhesus monkeys (*Macaca mulata*) are utilized to study high-priority diseases such as the pandemic caused by the severe, acute respiratory syndrome type 2 coronavirus (SARS-CoV-2) [13]. Domestic pigs (*Sus scrofa*) are crucial for organ transplant medicine and immune therapies [14]. New species, including some invertebrates such as fruit flies (*Drosophila melanogaster*), are used to study neurological disorders such as epilepsy [15], nematodes such as *Caenorhabditis elegans* to study obesity [16], and aquatic models, such as the zebrafish (*Danio rerio*), to treat metabolic disorders, including diabetes [17].

The broad range of species used in research has brought exponential advances in medicine, especially with the introduction of genetically modified (transgenic) animals [18] and the implementation of supporting technologies such as nanotechnology and artificial intelligence [19]. In light of this, this article aims to discuss the importance of animal models in diverse lines of biomedical research by analyzing the contributions of the various species utilized in science over the past five years concerning key topics of human and animal health.

2. Search Methodology

The literature search was performed in the Web of Science, Scopus, and PubMed. Keywords related to the use of animal models applied to current research priorities were searched to select the relevant articles, for example, "emerging infectious disease", "diabetes and obesity", "neurodegenerative diseases", "pain therapies", "surgical techniques", "cancer models", and "alternative animal models". The search was limited to articles published in English in the last five years (2019–2023) and related to human and non-human medicine and therapeutics.

3. A Review of Animal Experimentation

Animal models are essential for several biomedical research fields such as cancer biology and therapeutics, neuroscience, pharmacology and toxicology, neurobiology of diseases, endocrinology, public health, palliative medicine, also, in studies in human and animal biology and for the discovery and testing of new drugs, vaccines, and other biologicals (e.g., antibodies, hormones) whose validation requires preclinical studies in animals [6,20]. Currently, these models address current research priorities, considered as those imposing major global threats to human and animal health. These include diseases that have afflicted humankind or increased exponentially in recent years such as SARS-CoV-2, different types of cancer and their therapy, cardiovascular diseases, metabolic and neurodegenerative disorders, and experimental refinement of surgical techniques to treat these issues [21]. The models may involve complete animals or only particular cells, tissues, organs, genes, or other agents that reproduce pathological processes (Figure 1) [8,22]. Species include rats, mice, guinea pigs, dogs, rabbits, birds, ruminants (cows, sheep), horses, fish, frogs, monkeys, cats, reptiles, squid, crabs, bees, chimpanzees, hamsters, sea slugs, pigs, nematodes (roundworm), fruit flies, and protozoans, among others [7].

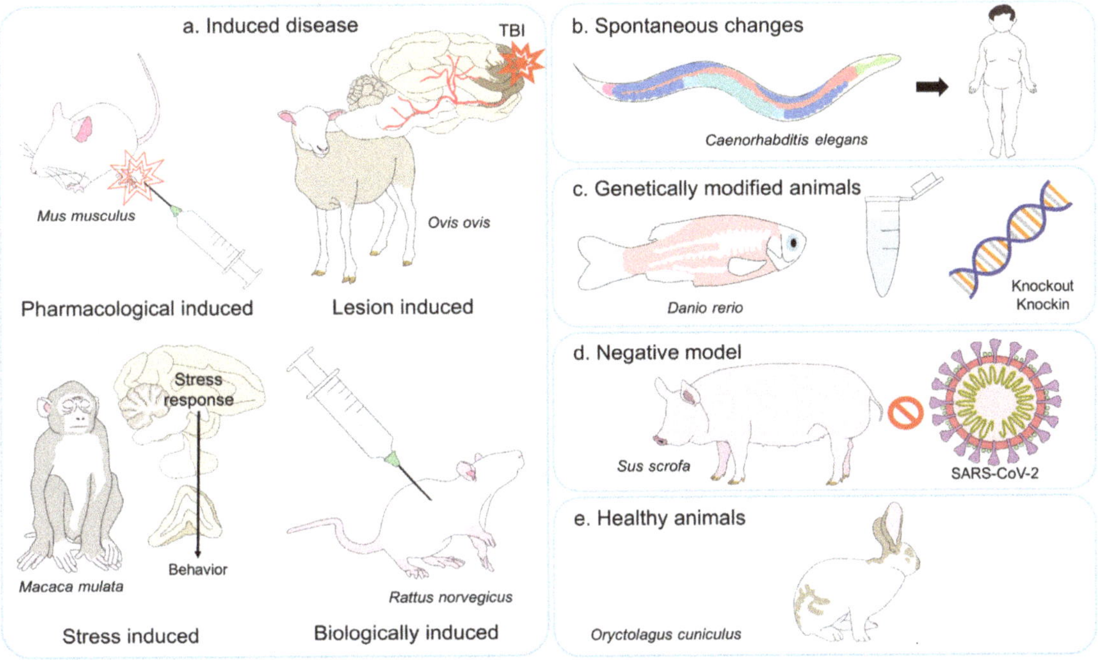

Figure 1. Classification of various animal models. The animals used in science can be divided into five broad types. (**a**) The main ones are models in which animals are induced to present a pathology similar to one that affects humans or other animals by administering drugs or other biologicals, inflicting injuries, or subjecting them to stress or other environmental conditions. In contrast, models based on spontaneous changes (**b**) include animals where the normal course of their life predisposes them to develop a specific disease. (**c**) Genetically-modified test subjects are animals with knockin or knockout genes or proteins. In contrast to using healthy animals (**e**), negative models (**d**) employ individuals that are not susceptible to certain diseases but serve to evaluate susceptibility to a specific pathology. TBI: traumatic brain injury.

The importance of animals in medical science is reflected, for example, in the percentage of Nobel Prizes studies in Physiology or Medicine using animal models (90%) [5]. From 1901 to 2020, two-thirds of those awards (186 of 222 projects [7]) employed animal models to understand pathogenic mechanisms, metabolic diseases, diagnostic and therapeutic procedures, develop vaccines, or test the efficacy of novel drugs [22]. At least 144 species used in those animal-based studies were mammals, and 42% were rodents [7]. Dogs were the first animal model used in metabolic research on gastric secretions [23] and for discovering insulin [24]. To date, rodents are the predominant species in research (Table 1) [9]. However, non-mammal species are trending, and the number of animals depends on the country and its legal regulation regarding the use and reporting of animals in research. Moreover, in some countries, there is no official annual report on animal research (e.g., South America), and not every country counts the same animals (e.g., the United States does not consider rats, mice, fish, birds, amphibians, reptiles, and cephalopods, they are not covered by the Animal Welfare Act). Although it might differ, Tables 1 and 2 show an overview of the use of animals according to species in some countries and a summary of the reported statistics worldwide.

Table 1. Overview of the number of animals used in research, according to the species.

Country	Year	Specie	Percentage (%)	Reference
European Union	2019	Rodents	61.9	[25]
		Fish	24.6	
		Birds	6.2	
		Amphibians	0.5	
		Cephalopods	0.2	
		Dogs	0.1	
		Non-human primates	0.07	
		Other mammals	6.5	
Canada	2020	Birds	50.0	[26]
		Rodents	24.5	
		Fish	11.7	
		Cattle	11.3	
		Amphibians	1.1	
		Pigs	0.4	
		Dogs	0.2	
		Non-human primates	0.1	
		Reptiles	0.1	
		Other animals	0.5	
United Kingdom [1]	2021	Mice	68.2	[27]
		Fish	12.9	
		Rats	6.5	
		Birds	8	
		Dogs	0.14	
		Non-human primates	0.09	
		Cats	0.01	
		Other animals	3.3	
United States [2]	2019	Guinea pigs	23	[28]
		Rabbits	18	
		Hamsters	12	
		Non-human primates	9	
		Dogs	7	
		Pigs	6	
		Cats	2	
		Sheep	2	
		Other species	21	
South Korea	2017	Rodents	91.8	[29]
		Fish	3.3	
		Birds	2.3	
		Rabbits	1	
		Non-human primates	0.08	
		Amphibians	0.07	
		Other species	1.21	
		Total	3,085,259	

[1] Excluding Northern Ireland; [2] Rats, mice, fish, birds, amphibians, reptiles, and cephalopods are not included.

Table 2. Approximate of the number of animals used in research worldwide between 2019–2020.

Country	Number of Animals	References
United States	20,000,000–24,000,000	[30–32]
China	16,000,000	
Japan	11,000,000	
European Union	9,400,000	
Australia	6,700,000	
Canada	5,067,778	
South Korea	4,141,433	
United Kingdom	3,300,000	
Norway	2,282,710	
Germany	2,151,805	
France	1,865,403	
Spain	761,012	
Mexico	685,315	
Switzerland	556,107	
Belgium	437,275	
New Zealand	240,000	

Several Nobel Prizes have been awarded for animal research, and the increasing number of animal models in different countries demonstrates these studies' importance for scientific advancement [7]. However, just as necessary, their use also entails ethical challenges that require surveillance through laws, norms, guides, and strict bioethical committees to monitor the use and care of laboratory animals based on the principles of the 3Rs [33]. In this regard, for 50 years, the National Center for the Replacement, Reduction, and Refinement of Animals in Research (NC3Rs) has promoted Russel and Burch's initiative of the 3Rs to reduce, replace, and refine procedures to improve the conditions of animals used in experimental protocols [34].

These norms differ from one nation to the next. However, one guide recognized internationally is ARRIVE (Animal Research: Reporting of in vivo Experiments), developed in 2010 to improve the in vivo experiments description to increase the reproducibility of results, refine the stages of study design, and clearly report the methods so they can be repeated and tested [35]. A second guide is PREPARE (Planning Research and Experimental Procedures on Animals), which seeks to determine and guarantee quality control in animal studies [36]. Today, for any experimental protocol requiring animals, proposals such as the Animal Study Registry (ASR) help researchers thoroughly plan their study design, methods, and statistical analyses to ensure transparency and reproducibility in their results [37]. Additionally, it is essential to mention that Ethic Committees must approve current experimental protocols within each institute to promote an appropriate use and care for animals in research.

Animal models certainly provide valuable information on the nature of diseases [38]. However, it is important to remember that inter-species limitations exist in anatomy, metabolism, physiology, and genetics [39], so a single preclinical model cannot represent all aspects of pathogenesis due to differences in resistance or susceptibility [38]. Currently, many animals used in biomedical studies undergo some genetic modification, such as transgenesis or the utilization of knockout or knockin genes, to visualize specific changes that would take years to develop under normal conditions [40]. Therefore, the selection of the animals depends on the specific research field; through their use, researchers develop scientific knowledge focused on human and veterinary medicine.

4. Animal Models and Their Application in Distinct Fields of Current Biomedical Science

4.1. Emerging Infectious Diseases

The SARS-CoV-2 virus is the etiologic agent of the coronavirus 2019 disease (COVID-19) [41]. This disease has claimed the lives of over 6.3 million people worldwide since

2019 [42,43]. The lack of knowledge of this virus and its rapid propagation at the onset of the pandemic made it essential to determine its physiopathology and identify therapeutic agents and vaccines that could mitigate its threatening consequences. These fundamental issues were solved using in vivo assays that replicated the virus in animals to untangle its pathogenesis, the immune response, and the adverse effects that might result from the vaccines and therapies proposed before testing in humans and their release to the public [41,44].

The choice of an animal model that would allow researchers to observe the histopathological, radiological, or immune changes that the virus caused required that the test animals be susceptible to lung tissue damage and capable of developing an inflammatory process [45]. Potential species included nonhuman primates, ferrets, rats, mice, Syrian hamsters, lagomorphs, minks, cats, camelids, and even zebrafish [46].

The transgenic mice can express the human angiotensin-converting enzyme II (hACE2), a functional receptor for the SARS-CoV-2 virus that mimics clinical signs observed in humans [47]. Sun et al.'s [48] research with 4.5–30-week-old transgenic mice successfully replicated the virus after intranasal and intragastric inoculation. It led to the discovery of viral loads in the lung, trachea, brain, and feces. Those authors also detected an immune and inflammatory response due to the presence of interleukins (IL). Adult mice showed more lesions in the alveolar epithelial cells, focal pulmonary hemorrhage, and more significant apoptosis of macrophages. Those findings concurred with human reports showing that COVID-19 affected older adults more severely, with the over-65 population representing 80% of all hospitalizations and a 23-fold greater risk of mortality. Reports emphasized clinical signs, such as respiratory distress and cytokine release syndromes [49]. Studies with Syrian hamsters found that while the virus is lung-tropic and infects the respiratory tract by binding to the ACE2 cell surface in the alveoli, causing pneumonia in 67% of the animals, the gastrointestinal signs reported in humans are due to viral replication and dissemination in enterocytes [50].

One animal model that shares multiple similarities with humans for the physiopathology of the SARS-CoV-2 virus is based on Rhesus macaques, African green monkeys (*Chlorocebus aethiops*), and crab-eating macaques (*Cynomolgus macaques*) [51]. The latter has been utilized to replicate the infection conditions in young (males and females of 3–9 years) and old-aged animals (23–29 years-old females). After intranasal and intratracheal viral inoculations, researchers found that nasal swabs (peak viral load of 10^6 copies/μL) had higher viral loads than pharynx and rectal ones (a maximum of 10^4 copies/μL). Additionally, viruses from nasal and pharynx samples were detected for longer periods in elderly monkeys [52]. This relation between age and disease mortality was also reported in Rhesus monkeys. Comparative studies of three nonhuman primates (three 3–5 years and two 15 years old macaques) infected intratracheally revealed that the viral replication detected by nasopharyngeal and anal swabs was persistently detected from 3 days post-infection (dpi) to 11 dpi in elderly animals. In older macaques, 104–107.5 copies/mL were also detected (while young individuals had approximately 104 copies/mL), often accompanied by the development of diffuse severe interstitial pneumonia [53].

The reinfection processes prevalent in human populations were replicated in studies with *C. aethiops*. Infection in six animals caused signs such as fever (50%), hypercapnia (66%), 2–7-fold increases in C-reactive protein concentrations (100%), and coagulopathy (100%) were recorded. That research proved that anal, oral, and nasal swabs could detect viral loads up to 15 dpi [44]. These findings are similar to those from other works with *M. mulata*, where viral RNA was found in swabs from the nose, pharynx, and anus, with amounts increasing up to 3 dpi (in an approximate range of 4–7 copies/mL) [53]. These nonhuman primate models undoubtedly contributed significantly to our understanding of the pathogenicity of COVID-19 and the physiological bases for implementing preventive and diagnostic measures and treatment.

Another important aspect of using animals is that they helped understand the transmission of the virus to other domestic species and showed that pets could acquire the

SARS-CoV-2 virus through contact with an infected human. However, there is no evidence of active pet-to-human transmission [54]. Studies with dogs, pigs, chickens, and ducks showed they were not susceptible to COVID-19 infection due to low viral replication [55]. Identifying susceptible species made it possible to choose appropriate models for developing and testing vaccines [55]. Ferrets, Syrian hamsters, rabbits, transgenic mice [47], and cats were all found to be susceptible, the latter even vulnerable to airborne transmission with the development of clinical signs such as hair loss and pulmonary alterations similar to those seen in humans [56,57]. Apart from domestic cats, wild felines (tigers, lions, pumas, snow leopards) [58] have been reported to show infections by this virus. Kang et al. [59], who reported the first Delta variant (SARS-CoV-2 Delta) case in three domestic cats with COVID-19-positive owners in China, insist that transmission to pets is a topic of concern due to their possible role as silent intermediate hosts.

4.2. Endocrinology and Metabolic Pathologies

Obesity is a public health problem affecting over 600 million people worldwide [60]. Obesity and its associated metabolic syndromes have consequences such as knee osteoarthritis, a disease prevalent in approximately 60% of the overweight population [61], but this is also associated with cancer, cardiovascular disease, hypertension, coronary artery disease, stroke, sleep apnea, asthma, gallstones, steatohepatitis, and dyslipidemia. Over one-third of the world's overweight or obese population is at risk of developing type 2 diabetes mellitus [23]. Using rodent models, researchers have determined that one element that promotes the development of type 2 diabetes mellitus is adipose tissue inflammation due to insulin resistance and excess fat mass [62]. The increase in the presentation of these comorbidities has led to the use of animal models to test new, improved strategies for reducing the incidence of this disease.

The role of the different types of adipose tissue in humans and animals is a crucial line of research that has developed with the use of rodents. For example, adipogenesis suppression and the browning of white adipose tissue (WAT) [63] have been suggested as strategies for preventing obesity [60]. The browning process creates a brown adipose-like tissue (BAT) that can participate in thermogenesis by transforming caloric intake into heat [64]. Since this is part of a central nervous system response to cold, certain medications and exercise can trigger browning as has been observed in obese and lean rats subjected to high-intensity training. In C57BL/6J mice, the transformation of beige adipocytes into WAT can be promoted with diets complemented with resveratrol for 16 weeks, as this induces a change in the intestinal microbiota in treated animals ($p < 0.01$) (increasing microorganisms of the genera *Bacteroides*, *Lachnospiraceae*, *Blautia*, *Lachnoclostridium*, and *Parabacteroides*, among others) that modulates lipid metabolism and has anti-inflammatory properties and anti-obesity effects [65].

The importance of physical activity in treating these conditions has been demonstrated in experiments with 48 Sprague-Dawley male rats, where aerobic exercise for 12 weeks combined with prebiotic fiber supplementation prevented knee joint damage, dyslipidemia, endotoxemia and normalized the effects of insulin resistance ($p < 0.001$) [61]. Studies with these supplements as part of a therapeutic protocol in Wistar rats, administered in presentations such as yogurt, have shown that supplementation with 5% of yogurt reduces levels of oxidative stress (significant decreases in NO levels, $p < 0.05$), and had fewer amounts of inflammatory cell infiltration and collagen deposits in the liver ($p < 0.05$) when compared to animals fed high-fat diets. According to these studies, this supplement could be a potential human therapeutic option [66].

Studies of the human genome have identified hundreds of genetic variants associated with obesity and opened the way to examining these genes in species such as *C. elegans*, a nematode capable of storing fat in the form of lipid droplets inside hypodermal and intestinal cells. *C. elegans* has 14 genes that promote diet-induced obesity and three that prevent it [67]. Those genes are now recognized as potential targets for anti-obesity treatment. Ke et al. [68] found that the knockdown of 23 fat-storing not only reduced excessive fat accumulation but also improved the health and lifespan of this species ($p < 0.05$). The inhibitory effect of flavonoids such as butein on lipogenesis in *C. elegans* succeeded in reducing triglyceride levels by up to 27% without altering food intake or energy expenditure, an effect due to the downregulation of proteins involved in lipid metabolism [69]. Likewise, the appetite suppressant effect of administering vegetable extracts from the *Lentinus strigosus* mushroom (300 and 1000 μg/mL) to *C. elegans* functioned as a natural means of preventing obesity [70]. Studies of this kind allow researchers to address obesity as a complex pathology affected by diverse factors: diet, physical activity, developmental stage, age, genes, and environmental interaction [67].

Another animal species considered a promising model for studying metabolic syndromes is the zebrafish (*D. rerio*). This species has genetic homology with humans, so through genetic mutation, chemical induction, and changes in diet, they can be used to study hyperglycemia, obesity, diabetes, and hypertriglyceridemia [71]. Pigs, meanwhile, share similarities with humans in terms of organ size, lifespan, anatomy, physiology, and metabolic profile [40]. A study of obesity in Iberian pigs showed the pathogenesis of chronic kidney disease caused by overweight and obesity. Although the administration of high-fat diets did not generate diabetes in those pigs by day 100, analyses revealed hypercholesterolemia (142 ± 27 mg/dl), hypertriglyceridemia (75 ± 43), insulin resistance, and glomerular hyperfiltration [72]. These effects also occur in humans [73] and have been studied in obese male mice and ovariectomized females [74].

The domestic dog has been postulated as a valuable model for studying chronic morbidities brought on by environmental conditions since they share morbidity and mortality factors with humans. In this field, Hoffman et al. [75] reported that comorbidities behind chronic conditions such as obesity, arthritis, hypothyroidism, and diabetes reported in humans were also present in 73,835 canines and that those dogs showed a positive association between age and the number of morbidities ($p < 0.001$). Other studies have revealed that obesity in dogs (137/198) is closely linked to the alimentary habits of their owners, finding that the 79.8% of dogs from overweight owners (114 persons) were obese ($p < 0.001$) [76]. Therefore, studies of these animals could provide information on disease interaction.

4.3. Cancer in Biomedicine

According to the World Health Organization [77] and the National Cancer Institute [77,78], the most common types of cancer in humans in 2020 were breast (2.26 million cases), lung (2.21 million), colorectal (1.93 million), prostate (1.41 million), skin (1.20 million), and stomach (1.09 million). These cancers cause 10 million deaths per year. Projections for 2022 estimate that around 1,918,030 new cancer cases will be diagnosed in the United States, with 350 cancer-induced deaths per day, making this disease a primary cause of mortality [79]. The pathogeny of these cancers and testing new treatment options is another field that extensively uses animal models. Over 95% of studies use rats and mice to inject cancer cell lines subcutaneously, study the primary cancer lesion, and follow its growth before excising tumors [80,81]. However, one disadvantage of this subcutaneous tumor model, is that injections in athymic nude mice may not accurately represent the interaction among tumor cells, local stroma, and the tumor's microenvironment, depending on its precise location [82]. Contrarily, orthotopic murine models have been shown to replicate the tumor microenvironment –including metastasis– when inoculated in the original anatomical site of the tumor. In female BALB7c mice, inoculation of mammary cancer cell line 4T1 as a fat pad tumor model showed that 50% of the animals had metastasis to the ovaries, spleen, liver, and sternum. However, when compared to a heterotopic model, orthotopic tumors were smaller (1993.7 ± 197.15 mm^3 vs. 1078.4 ± 300.26 mm^3, $p < 0.05$) and had a significantly lower percentage of infiltrating cells ($p < 0.05$) [83]. Moreover, these orthotropic models, together with in vivo optical metabolic imaging, are proposed as an approach to studying how, for example, the fatty acid uptake by breast cancer cells increases accordingly to tumor aggressiveness and metastatic process ($p < 0.05$) [84] Attacking this complication in tumor development is the principal objective of anticancer therapies, since most deaths from prostate cancer, for example, are due to metastasis into bone structures [80].

Koosha et al. [85] used diosmetin, an anti-tumorigenic, in colon cancer xenografts in 24 male nude mice. Results showed that tumor volume in the group treated with 100 mg/kg of diosmetin was significantly smaller than in the untreated group (264 ± 238.3 vs. 1428 ± 459.6 mm^3, $p < 0.01$). Promisingly, the drug did not produce toxicity even when administered at high doses. Studies of this kind show that laboratory animals allow researchers to test new drugs and better understand disease development but also aid in determining non-toxic doses that can be applied to humans or animals. Using these models as translational media for studying cancer has also revealed the importance of identifying the pain that animals may experience. Pain assessment is important in in human medicine and laboratory animal welfare. In this regard, recognizing degrees of cancer-induced bone pain has been studied by observing behavioral changes in rats and mice, where innate behaviors, such as burrowing, are reduced 9 days after inoculation when compared to control groups ($p < 0.05$) as a result of the nociception associated with the degree of severity of cancer due to reduced bone density [86].

The fact that the canine and human genomes share a high degree of similarity (75%) and that the risks of death due to neoplastic, congenital, and metabolic diseases are comparable means that the dog is an ideal translational model for studying human morbidity and mortality [75,87]. For example, the percentage of neoplasia is similar between dogs and humans (27.4 vs. 25.3%). However, because the types of cancer that affect each species correlate only marginally (Spearman rank $p = 0.661$) [75], dogs have been replaced in many preclinical studies by genetically-modified pigs [87].

Another novel anticancer strategy involves managing nerve-tumor interaction [88] since tumor-specific denervation can suppress neoplasia growth [89]. A study by Kamiya et al. [90] with female Balb/c-nu mice and the use of xenografts in Hras128 rats in a model of chemically-induced breast cancer showed that sympathetic stimulation of the nerves in tumors accelerated cancer growth but that parasympathetic stimulation reduced growth and downregulated the expression of programmed death. In contrast, in the case of late-stage colorectal cancer, parasympathetic denervation via vagotomy and atropine administration in 150 male Wistar rats reduced the incidence of tumors and their weight and volume after eight weeks, as well as cell proliferation, angiogenesis, and regulated expression of the nerve growth factor [89].

These neural anticancer therapies in humans and animals indicate that while sympathetic nerves show cancer-promoting effects in prostate and breast cancer, and melanoma cases, the parasympathetic/vagal nerves are believed to trigger both reactions. For example, vagal nerves can promote prostate, gastric, and colorectal cancers, but suppress breast and pancreatic cancers, due to β-adrenergic and muscarinic effects that modify the behavior of cancer cells, angiogenesis, tumor-associated macrophages, and antitumor immunity [88]. The axonogenesis process in species such as mice, linked to the development of metastasis in breast cancer, showed through immunofluorescence that nerve twigs tend to be sympathetic-like, with no expression of parasympathetic fibers [91].

In addition to the support of laboratory techniques such as immunofluorescence, non-invasive diagnostic methods are a priority in oncology. In immunocompetent genetically-engineered mouse models, Kirkpatrick et al. [92] utilized nanosensors with urine tests to detect protease activity in diverse types of cancer, including lung cancer, achieving 100% specificity and 81% sensitivity. In this way, monitoring with nanosensors and clinical assays in animals has demonstrated that this technique can be an option for conducting accurate, radiation-free diagnostic tests.

Nanoparticles and their application, together with in vivo imaging, can help to test novel luminescent particles and assess their tissue penetration to improve cancer therapy [93]. In vivo imaging enables us to understand tumor growth-related processes such as oxidative mitochondrial metabolism in mouse models with cell lung cancer [94]. Likewise, in a mouse model of brain tumor –glioblastoma– under general anesthesia, modified in vivo optical imaging (Surface enhanced spatially offset Raman scattering) covers the inability of conventional techniques that rely on subcutaneous inoculation of cancerous cells because they cannot read deep tissues [95]. These techniques are the basis for imaging-guided phototherapies that are a current research field to find agents capable of inducing tumor cell apoptosis, such as photodynamic y and photothermal therapy [96].

4.4. Pharmacology and Therapeutics

Parallel to the advances in our knowledge of the physiopathology of diverse conditions, developing and testing new therapeutic options is another field destined for animal models. Algology is a science in constant actualization to provide new and efficient drugs to prevent the consequences of pain by reducing the number and severity of secondary effects in both human and veterinary patients [97,98]. Adequate models are needed to evaluate analgesic efficacy accurately. In the case of treatments for open wounds, Parra et al. [99] applied carprofen (5 mg/kg) and buprenorphine (0.1 mg/kg) to the left hind paw of Sprague Dawley rats of both sexes using a punch biopsy to assess analgesia in an open wound model. Using four behavioral tests associated with aspects of nociception, mechanical and thermal stimulation, guarding behavior, and the weight-bearing test, they found that carprofen promoted recovery of the thermal response to basal levels after just 2 h. The same rat species were utilized to test the renal and gastrointestinal safety of non-steroidal anti-inflammatory drugs (NSAIDs) such as ibuprofen by administering single and multiple oral doses to pediatric patients. Furthermore, the necropsies performed on pigs of different ages (8-week-old and 6-to-7-months-old) in the study by Millecam et al. [100] revealed no severe lesions in the stomach after multiple doses of ibuprofen at 5 mg/kg. However, significant histological score differences ($p < 0.025$) were observed in the duodenum (1.38 vs. 4) and jejunum (3.63 vs. 1.25) between the experimental and control group. Additionally, an increased clearance time for the drug after multiple doses was found, an effect similar to reports in human pediatric patients.

Due to the adverse effects that NSAIDs can generate, especially for treating chronic afflictions such as arthritis and cancer, opioids are another therapeutic option [101]. However, since the long-term use of these drugs is also associated with complications, research has begun to new concepts and explore directions. The opioid-free anesthesia technique was introduced to prevent tolerance and hyperalgesia and reduce the use of these drugs in the postoperative period. This method uses agents such as alpha-2-agonists, ketamine, and local analgesics with distinct action mechanisms in multimodal analgesia [102–104]. Other new opioid-based pharmacological options are transdermal patches impregnated with morphine-like compounds. In 6–12-week-old C57BL/6JJmsSlc mice, patches synthesized with two new opioids (new-opioids 1 and 2, N1 at 3 mg/kg; N2 at 10 mg/kg) showed the same analgesic efficacy as morphine at 3 mg/kg. The effect remained constant, even under repeated administration (in contrast to fentanyl), and the cutaneous trans-permeability rate was greater, at 1.71 ± 0.35 and 3.94 ± 1.36 μg/cm/h [105]. The administration of opioid nanoparticles has also been suggested to prevent opioid tolerance and reduce the severity of adverse effects. Leucine-enkephalin hydrochloride-based nanoparticles with a size of 100–200 nm have been tested in male Sprague Dawley rats by applying them intranasally, reaching the brain directly. After dosing, high concentrations were found in the olfactory bulb and cerebrum between the first 60 min (approximately 80 ng/g and 160 ng/g, respectively), while plasma concentrations were not detected at any evaluation time ($p < 0.0001$). This prevents the side effects of drug transit through peripheral pathways [106].

Techniques based on local anesthesia temporarily relieve pain by inhibiting nerve impulse transmission. However, when used to complement multimodal analgesia protocols, they can be associated with neurotoxicity in both human and veterinary patients [107]. Administration via polymer-based encapsulation is a new strategy designed to prevent toxicity and permit the prolonged release of the active ingredient to give a long-term analgesic effect for up to seven days [107]. A ketamine-polymer-based drug was applied transdermally to Wistar rats to determine its analgesic effects [98]. Results of the tail-flick test and readings from an analgesiometer led them to determine a significant analgesic effect ($p < 0.01$) maintained for 24 h with a peak effect at 8 h and a response time on the test 5.72 s vs. a basal time of 2.44 s. The compound did not produce irritation when tested on rabbit skin. It prevented the secondary effects of intravenous, nasal, or oral administration, so it is a potential option for treating neuropathic pain [108].

4.5. Experimental Surgical Tecniques

In addition to developing novel drugs, advances in surgical technology and techniques have opened fields in microsurgery in human and animal medicine since the 1900s when Carrel and Guthrie performed the first transplants in dogs [109]. Later, in 1950–1960, Buncke and Schultz tested the first microsurgery techniques using models of digital amputations and reimplantation in Rhesus monkeys, performing vascular microsurgery to restore circulatory connections successfully [110]. Anastomosis of 1-mm blood vessels in the ears of adult rabbits by reimplantation was the first demonstration of microsurgery in reconstructive medicine [111].

Today, rodents are considered models for reimplanting extremities and restoring blood vessels because their vascularization is homologous to the human finger [112]. For example, developing heterotrophic osteomyocutaneus flap transplant protocols in Lewis rats furthered our understanding of the mechanisms and pathways involved in the immune response underlying tissue transplant rejection [113]. Likewise, in an experiment with five syngeneic mice and allografts—using a donor-supplied aorta and inferior vena cava—end-to-end anastomosis of those structures showed a 74% success rate as a technique for hind limb transplants [114]. In another study, Tee et al. [115] performed grafts of engineered cardiac muscle flaps in the epicardium of 8 rats. The flaps were transplanted by microsurgery to resolve one of the first limitations: failed vascular anastomosis. Those researchers performed successful end-to-end anastomosis of the carotid artery and jugular vein by placing the flap on the epicardium, achieving a survival rate of 75% during 4 weeks post-surgery, with viable cardiomyocytes and vascular connections between the flap and the epicardium by week 10 [115]. These techniques, tested first in animals, were later used with human patients with coronary artery disease caused by diseases such as squamous cell carcinoma, with a 96% survival rate of the flap in individuals subjected to neck and head surgery [116].

Another advance in biomedicine achieved thanks to experimental work with animal species such as pigs are based on animal-to-human organ transplants. On 7 January 2022, Bartley Griffith's team performed the first heart transplant from a genetically-modified pig to a 57-year-old human patient with terminal heart disease [117]. Although the patient's condition who received that xenotransplant deteriorated two months after surgery, and he died, the procedure set an important precedent. It showed the need to continue research on genetically-engineered animal organs and immunosuppressor drugs since the immune response and organ rejection are still the leading causes of transplant failure, especially when the organs come from other animal species [118].

Due to the physiological similarity between nonhuman primates and humans, procedures for organ transplants are often tested in those species. Over seven years, Lee et al. [119] performed 22 xenotransplants using hearts from transgenic pigs eliminating alpha-galactosidase transferase knockout or expression of the regulatory proteins CD46, CD39, or CD73 in Cynomolgus monkeys (*Macaca fascicularis*). Results showed that survival of the grafts was significantly higher in hearts with double or triple genetic manipulation (11.63 ± 11.29 days vs. 30.83 ± 20.34 days, $p = 0.03$). This is similar to the report by Cui et al. [120] on triple knockout cells from pigs (that do not express any of the three carbohydrate xenoantigens). The complement-dependent cytotoxicity response and the amount of anti-pig IgG/IgM immunoglobulins (Ig) were evaluated in serum from 72 specific pathogen-free (SPF) baboons and in human serum. Results for humans and old-world monkeys showed similar antibody binding, but the cytotoxicity measured in IgM and IgG was lower in the humans ($p < 0.05$ vs. $p < 0.01$).

Observations on the immunosuppressor response to compounds such as anti-thymocyte globulin (20 mg/kg) and rituximab (20 mg/kg) demonstrate that, in addition to the use of transgenic animals, a strict immunosuppressor regimen is a critical element in allotransplants [119]. In this regard, drugs injected in nanoparticles such as mycophenolate mofetil allow low-water soluble compounds to be combined with other compounds and administered as solid lipid nanoparticles to improve their absorption and release by as much as 68% in acid media [121].

In this field, sustained release options such as nanoparticle-anchoring hydrogel scaffolds of the immunosuppressant tacrolimus allowed the localized release of the drug with tissue regeneration in nude female mice or those of the BALB/c line that were given the drug in the hind limb. Those combinations allowed the sustained release of 77% of the drug, without toxicity, within 28 days at <100 ng/mL [122]. Thus, refining these drugs in the future will make it possible to reduce the cases of organ rejection due to the immune response. This finding is significant because their benefits are not accompanied by systemic toxicity, complications, or dose reduction without pharmacological efficacy [123].

4.6. Neurosciences

The field of neuroscience includes surgical and therapeutic procedures involving the central nervous system and conducts studies focused on specific diseases or pathologies of that system. With the discovery of neurological sequelae in COVID-19-infected patients, animal models have allowed researchers to observe the effects that the SARS-CoV-2 virus generates in sporadic cases, including epileptic seizures and encephalitis with a mortality rate of approximately 5.3% [124].

Estimates suggest that approximately 42 million people worldwide suffer brain injuries annually and that 80% of cases are classified as traumatic brain injury (TBI). Animal models based on rodent species are being used to improve our understanding of the physiopathology of TBI [125], though authors such as Vink [126] caution that neuroanatomical differences in the mouse's lissencephalic brain can generate biomechanical responses distinct from those in humans. Moreover, the replication of trauma may be greater in rodents since traumatisms in these animals tend to generate focal instead of diffuse lesions [127]. Grovola et al. [128] used male Yucatán miniature pigs to analyze neurological dysfunction in animals with mild traumatism 1-year postevent. They found a persistent neuroimmune response in animals with morphological changes to the microglia, with increased branches and junctions per cell ($p = 0.026$ and $p = 0.045$, respectively). In other research, models of medullar lesions are widely utilized with species such as rats, which are particularly important because between 236 and 1009 per million humans annually suffer a spinal cord injury [129]. Although this species is the one most often employed to replicate medullar damage, Filipp et al. [129] affirm that between-species differences (quadrupeds, bipeds) must be considered when evaluating the neuroplasticity of the spinal neurons.

Epilepsy is one of the most common neurological conditions, affecting over 50 million people worldwide [130] and 0.6–0.75% of the domestic canine population [131]. Recent studies of the physiopathology of this disorder and the testing of anti-seizure drugs have used fruit flies (*D. melanogaster*) because they manifest seizure-like behavior and share 70% of their genes with humans [15]. The use of the endocannabinoid anandamide (at 2, 20, and 200 µg/mL) in *Drosophilas* prevented induced seizures ($p < 0.0001$). This led to the discovery that the action mechanism of their metabolites is not linked to the cannabinoid receptors but, instead, to transient potential receptors (TRP). This makes the fruit fly a suitable medium for studying this type of drug [132].

Despite its nature and supposed organic simplicity, *Drosophila* has been used to understand the neurobiological bases of processes still considered mysteries by biology, such as sleep, plasticity, and memory [133]. After studying 12,000 exemplars of *D. melanogaster*, Toda et al. [134] reported the existence of the "nemuri" gene, a peptide with antimicrobial properties that favors sleep and helps these flies survive the infection. This suggests that its function could be linked to the immune competence of the sleep process in animals and humans. The association of sleep with long-term memory, known as post-learning sleep, was studied by Lei et al. [135], who found a neural circuit that excites the mushroom body neurons and a connection to the fan-shaped ventral neurons that promotes post-learning sleep during courtship. This finding underlined the association between the longer learning experience and the reinforcement of long-term memory, mechanisms sometimes found in mammals.

Neuroscience techniques applied to species such as nonhuman primates and transgenic models of those species have recently been proposed as useful for studying human evolution and the cerebral functioning of people with autism disorders and neurodegenerative diseases such as Alzheimer's [136]. In humans, Alzheimer's disease is considered the most common neurodegenerative disease accounting for around 80% of cases of dementia worldwide [137]. It is widely recognized that mitochondrial dysfunction is an event that precedes the onset of Alzheimer's, and this has been studied in two lines of mice (APPswe/PSEN1 ΔE9 and C57BL/6J). There, the alteration of mitochondrial homeostasis and increased mitochondrial calcium levels caused damage and neuronal death ($p < 0.0001$) due to deposits of amyloid plaques. Recognition of this physiopathology helped scientists establish the goal of preventing this process as a novel therapeutic approach [138].

Another neurogenerative disease, Parkinson's, has been studied primarily with murine models [139]. Recently, however, researchers recognized that the zebrafish shares more neuroanatomical traits with humans and that mutations of the PARK7 gene in adult fish were associated with the development of Parkinson's in humans [140,141]. Exposure of zebrafish larvae to neurotoxins that act directly on the dopaminergic neurons constitutes a method to mimic the phenotype of Parkinson's disease. Specifically, the MPP+ neurotoxin affected the locomotor function (total distance and velocity) of fish, reducing its performance by 80% and 85%, respectively ($p < 0.001$). Furthermore, no systemic effects were observed, presenting a condition similar to Parkinson's [142].

Palliative treatments to control movement disorders such as dystonia, Huntington's, and Parkinson's disease have also been tested in zebrafish [143]. Treatment of Parkinsonian embryos with substances such as rosmarinic acid (RA) prevents the loss of dopaminergic neurons due to neurotoxicity. This acid has been proposed as a neuroprotector and antioxidant that reduces locomotor deficits measured, for example, by increasing the swimming distance in zebrafish treated with RA at concentrations of 10 or 100 μm (approximately 130 to 150 cm, $p < 0.01$) [144]. Similarly, it has been suggested that herbal medicines based on Tongtian oral liquid have neuroprotective and antioxidant properties. The administration of Tongtian to zebrafish prevented neurotoxicity and the degeneration of dopaminergic neurons ($p < 0.01$ when compared to non-treated fish) while reducing larval behavioral impairment measured as improvements in the total distance (peak distance around 180 cm) and velocity (peak values around 3.5 cm/s) ($p < 0.001$) [145].

Aquatic models are also utilized to study other neurodevelopmental problems, such as autism spectrum disorder in zebrafish and Medaka fish (*Oryzas celebensis*) [146]. Chen et al. [147] found that prenatal exposure to valproic acid (at 5 and 50 μM) in AB lines of zebrafish produced embryos and larvae with signs similar to those seen in autistic humans, including hyperactivity, manifested in a greater frequency of tail-bending, greater distances traveled after touching of the dorsal tail ($p < 0.001$, $p < 0.05$), increased swimming speed under both light and dark conditions, and deficient social interaction, anxiety, and macrocephaly, all as consequences of neuronal cerebral cell proliferation. In a separate study, when applied to 28 neonate rat pups, this acid generated oxidative stress in the cerebellar hemispheres and reduced the count and nuclear size of the Purkinje cells [148]. These findings appeared, as well, in the brains of children with this condition. In the case of rats, administering grape seed extract served as a neuroprotector thanks to its antioxidant effect.

Referring to neurodegenerative disorders, a key strategy is to improve symptomatology through physiotherapy and rehabilitation protocols, another line of research that has increased in importance due to the prevalence of neurological conditions that can affect the quality of life of both humans and animals.

4.7. Physiotherapy and Rehabilitation

Because the number of neurodegenerative and traumatic diseases in humans and animals has been increasing in recent years, one of the main options for these cases is developing and implementing physiotherapy techniques. For example, stimulation of the lateral cerebellar nucleus with low-intensity ultrasound is a non-invasive therapy for reducing the consequences of cerebrovascular accidents in mice after induced ischemic stroke. In those test animals, functional asymmetry of the brain was restored, and pathological electrical cerebral delta activity was reduced, leading to improved performance on the beam-walking test [149].

In cases of osteoarthritis, for example, transcutaneous electrical nerve stimulation techniques (TENS) in physiotherapy protocols utilized in male Sprague Dawley rats with induced pain showed that when applied to the knee joint for 20 min a day for two weeks, TENS reduced the expression of c-fos ($p < 0.05$) (a biomarker of pain) on the day following the intervention (7302.80 $\pm$ 152.40% vs. 5074.50 $\pm$ 199.50%) in all the test animals that, in addition to TENS, did exercise on a treadmill (7333.40 $\pm$ 156.70% vs. 2790.00 $\pm$ 111.88%) [150]. In canine patients, functional neurorehabilitation after Hansen type I intervertebral disc surgery has been tested using a technique with bases similar to TENS called transcutaneous electrical spinal cord stimulation (TESCS). Combined with pharmacological treatment (4-AP) for 90 days, this approach restored ambulation in 88% of 16 animals thanks to the so-called multimodal neurorehabilitation protocol in a study by Martins et al. [151].

In human medicine, TENS has been used with patients with knee osteoarthritis. It improved performance on the stair-climbing test by 0.41 s [152] and reduced pain in individuals with head and neck cancer who had received radiation and developed oral mucositis with the pain. In those patients, 30 min of high-frequency TENS functioned as a non-pharmacological intervention that reduced pain levels at rest by approximately 3.0 from visit #1 to visit #3, as measured by the McGill Pain Questionnaire. However, this approach did not show results for controlling functional pain [153]. Pain reduction allowed the patients to exercise the limb and prevent the loss of mass, muscular strength, and joint instability with some cartilage recovery.

Electroacupuncture is a similar technique used to control chronic inflammatory pain. The action mechanisms of this technique have been studied in murine models after administering the complete Freund's adjuvant to the hind paw. In those animals, electroacupuncture produced analgesia by attenuating neuronal signaling in the dorsal ganglia of the spinal cord, the anterior cingulate cortex, and neurons of the somatosensorial cortex. This suggests that the analgesia generated affects cortical pain pathways and means that the somatosensorial and anterior cingulate cortices may be potential therapeutic targets for developing new options for pain management [154], one of the principal objectives of rehabilitative medicine in humans and animals.

5. New Models and Strategies Applied in Animal Research

The use of poorly developed or unconventional species is expanding to other areas of biomedicine. For example, the zebrafish is used to study anomalies in limbs and craniofacial regions [155]. In those fish, Bergen et al. [156] found 604 genes associated with processes of the formation, mineralization, and regeneration of scales, which demonstrated that those structures are reminiscent of bone. Mutations of these genes in humans generate bone mineralization disease. This suggests that scales could be a model for studying the pathogenesis of skeletal diseases, calcification, and matrix formation [156]. In another fish species –Medaka, the Japanese rice fish (*Oryzias latipes*)– researchers found that the electrocardiogram pattern was more similar to that of humans than those of rats and mice. This led authors such as Yonekura et al. [157] to use it as a model for testing cardiovascular therapies and the response of action potentials to verapamil, which causes bradycardia, an effect also seen in humans [158].

In addition to the use of mammals such as domesticated dogs as models for research on urinary pathologies due to their anatomical and physiological similarity to humans [159], the diverse species that have been incorporated into biomedical science include protozoans, platyhelminths, planarians, cnidarians, bivalve mollusks, gastropods, cephalopods, annelids such as the tardigrades, and arthropods such as hexapods, crustaceans, arachnids, and various insects in studies in broad fields of investigation [160]. In dermo-cosmetology, extraction of hyaluronic acid from mollusks such as *Mytilus galloprovincialis* and *Crassostrea gigas* to treat wounds in Wistar rats accelerated the processes of wound repair and re-epithelization, allowing lesions to heal completely within 15 days of treatment, in contrast to the results attained with commercial healing creams [161]. Another application of a cephalopod (*Octopus vulguris*) is in reconstructive medicine due to its capacity to regenerate nerves and adjacent tissues such as muscle and blood vessels. Despite these technical advances s in medical research, additional studies are required to determine markers, antibodies, and imaging techniques designed to take advantage of those species [162].

Non-animal alternatives such as cell cultures, 3D tissue cultures or organs-on-chips, mathematical models, stem cells, bioprinting, in silico tests, and advanced computer simulations have been increasing in recent years [163]. In leading research countries and regions such as the United States, United Kingdom, China, Germany, Japan, Canada, and Australia, among others [164], there has been a particular interest in replacing animal models with another methods. This is promoted by ethical pressures, the 3Rs initiative, and official instances such as the National Institute of Health [165]. An example of this is the new US law sponsored by the Food and Drug Administration (FDA), which states that drugs no longer require animal testing before human clinical trials [166]. Another example could be Canada and the statistics regarding the number of rats and fish used as animal models from 2019 to 2020. In 2019, rats and fish went from 3.9% and 19.9% to 2.6% and 11.7%, respectively [26,167].

When mentioning tissue engineering, the so-called "organoids"—transplantable tissues created by engineering—have raised expectations for replacing animals, resolving specific bioethical issues by making the study of pathologies and drug testing more specialized [168]. Protocols for head and neck squamous carcinoma have been published, using patient-derived organoids to study therapeutic agents and their drug sensitivity [169]. However, as materials that depend on in vitro handling and do not come from organisms that provide blood flow or the biochemical conditions of a live individual, their development and clinical application require further advances, not only in medicine but also in applicable biotechnologies [168]. Current trials aim to establish the vascularization of organoids, such as in human brains [170] or kidney organoids., In vitro culturing under millifluidic chips and endothelial cells is an alternative to creating vascular networks that need future studies but can be an option to research nephropathies [171]. Complex vascular networks made with mesodermal progenitor cells by Wörsdörfer et al. [172] replicated the ultrastructure of a blood vessel in tumor organoids with endothelial cell junctions, luminal caveolae, microvesicles, and antiangiogenic responsiveness to stimuli. Moreover, 3D bioprinting of organoids derived from stem cells (e.g., ectoderm, mesoderm, and endoderm) is another alternative to replicate developmental diseases in the brain, skin, kidney, heart, intestine, lung, and liver [173]. Those biotechnological advances include approaches in which animal models are accompanied by artificial intelligence [174].

The support that robotics and artificial intelligence provide to the advance of science has improved the technologies involved in techniques of robot-assisted, minimally-invasive surgery [175]. Recently, machine learning techniques have been used with animal models to help diagnose or identify specific behavioral or physiological changes in species. In this regard, models of Parkinson's disease in zebrafish have used video recording to teach the machine to differentiate between a movement disorder and a parkinsonian fish, a technique that may apply to cases of motor diseases in humans [140]. Deep learning algorithms, a type of machine learning, are another approach to the future of biomedical science, particularly in diagnosing a wide range of diseases. Based on CT images, it has been tested in hepatocellular carcinomas [176] and COVID-19 diagnosis, showing 85.2% accuracy a specificity and sensitivity of 88 and 87%, respectively [177]. A similar accuracy percentage (91%) was also obtained when testing deep learning to identify genetic syndromes according to facial features [178]. In veterinary medicine, automatizing facial recognition to assess pain is a current approach applied in cats, with an accuracy above 72% [179]. These applications suggest that new diagnostic tools might not require animal models. Nonetheless, implementing these technologies depends on their ability to simulate the physiology of a live organism, especially humans, to improve the replicability of results [180].

The replicability of animal models in preclinical protocols depends on their internal and external validity for transposing results to humans. However, the complexity of some human conditions and the physiological differences among species have led authors such as Pound and Ritskes-Hoitinga [181] to recommend focusing on techniques and technologies prioritizing human research. However, it is important to remember that experimentation with human subjects involves many serious ethical and legal controversies such as those surrounding experimentation with nonhuman primates [182]. One ethical way to deal with this topic consists in establishing and following norms and guidelines such as the 3R principles that promote the rational and humane use of laboratory species [33].

In summary, important advances in human and veterinary medicine have been mainly achieved thanks to animal species that allow us to improve our understanding of the etiology, pathology, physiology, and toxicology of diverse conditions that affect both humans and nonhuman animals [5]. However, using these species requires evaluating ethical considerations, existing limitations, the options available, earlier studies, and, above all, focusing on the welfare of laboratory species to fully recognize their enormous contributions to science.

6. Conclusions

Animal models—including a broad diversity of species of vertebrates and invertebrates—are a key element for experimental research aimed at replicating human and animal pathologies. Over the past five years, significant advances regarding worldwide priority diseases such as COVID-19, breast cancer, diabetes, obesity, and Parkinson, among others, were made in species such as nonhuman primates, rodents, lagomorphs, dogs, pigs, and even invertebrates such as zebrafish and nematodes. Moreover, before human clinical trials, novel therapeutic drugs, diagnostic techniques, and surgical procedures such as flaps or organ transplants have also been refined in animals.

These examples show the importance of using animals in biomedical research to study emerging or poorly understood human and animal diseases, and development of novel therapeutic options, including nanoparticles and in vivo techniques. Although animals will remain an essential element of science in the near future, due to their remarkable contributions, the ethical aspect of animal experimentation is significant.

The ethical pressure and the application of initiatives to reduce and replace the number of animals used in experimental protocols is leading to new strategies such as genetic engineering, artificial intelligence, organs-on-chips, mathematical models, bioprinting of organs, and advanced machine learning technologies. This multimodal approach is considered the best option for addressing the ethical dilemmas raised by using laboratory animals while emphasizing their valuable contributions to human and animal medicine.

Author Contributions: All the authors contributed to the conceptualization, writing, reading, and approval of the final manuscript. All authors have read and agreed to the published version of the manuscript.

Funding: This research received no external funding.

Institutional Review Board Statement: Not applicable.

Informed Consent Statement: Not applicable.

Data Availability Statement: Not applicable.

Conflicts of Interest: The authors declare no conflict of interest.

References

1. Davies, G.; Gorman, R.; Greenhough, B.; Hobson-West, P.; Kirk, R.G.W.; Message, R.; Myelnikov, D.; Palmer, A.; Roe, E.; Ashall, V.; et al. Animal research nexus: A new approach to the connections between science, health and animal welfare. *Med. Humanit.* **2020**, *46*, 499–511. [CrossRef] [PubMed]
2. Hobson-West, P.; Davies, A. Societal Sentience: Constructions of the public in animal research policy and practice. *Sci. Technol. Hum. Values* **2018**, *43*, 671–693. [CrossRef] [PubMed]
3. Robinson, N.B.; Krieger, K.; Khan, F.M.; Huffman, W.; Chang, M.; Naik, A.; Yongle, R.; Hameed, I.; Krieger, K.; Girardi, L.N.; et al. The current state of animal models in research: A review. *Int. J. Surg.* **2019**, *72*, 9–13. [CrossRef]
4. Lee, K.H.; Lee, D.W.; Kang, B.C. The 'R' principles in laboratory animal experiments. *Lab. Anim. Res.* **2020**, *36*, 45. [CrossRef]
5. Andersen, M.L.; Winter, L.M.F. Animal models in biological and biomedical research—Experimental and ethical concerns. *An. Acad. Bras. Cienc.* **2019**, *91*, e20170238. [CrossRef]
6. Juárez-Portilla, C.; Zepeda-Hernández, R.C.; Sánchez-Salcedo, J.A.; Flores-Muñoz, M.; López-Franco, Ó.; Cortés-Sol, A. El uso de los animales en la investigación y en la enseñanza: Lineamientos y directrices para su manejo. *Rev. Eduscientia. Divulg. Cienc. Educ.* **2019**, *2*, 4–19.
7. Jota Baptista, C.V.; Faustino-Rocha, A.I.; Oliveira, P.A. Animal models in pharmacology: A brief history awarding the nobel prizes for physiology or medicine. *Pharmacology* **2021**, *106*, 356–368. [CrossRef] [PubMed]
8. Swearengen, J.R. Choosing the right animal model for infectious disease research. *Anim. Model. Exp. Med.* **2018**, *1*, 100–108. [CrossRef]
9. Makowska, I.J.; Weary, D.M. A good life for laboratory rodents? *ILAR J.* **2020**, *60*, 373–388. [CrossRef]
10. Carbone, L. Estimating mouse and rat use in American laboratories by extrapolation from Animal Welfare Act-regulated species. *Sci. Rep.* **2021**, *11*, 493. [CrossRef]
11. Smith, J.R.; Bolton, E.R.; Dwinell, M.R. The rat: A model used in biomedical research. In *Rat Genomics*; Hayman, T., Smith, J.R., Dwinell, M.R., Shimoyama, M., Eds.; Springer: Hertfordshire, UK, 2019; pp. 1–41.
12. Joyce, C.; Scallan, C.D.; Mateo, R.; Belshe, R.B.; Tucker, S.N.; Moore, A.C. Orally administered adenoviral-based vaccine induces respiratory mucosal memory and protection against RSV infection in cotton rats. *Vaccine* **2018**, *36*, 4265–4277. [CrossRef] [PubMed]
13. Munster, V.J.; Feldmann, F.; Williamson, B.N.; van Doremalen, N.; Pérez-Pérez, L.; Schulz, J.; Meade-White, K.; Okumura, A.; Callison, J.; Brumbaugh, B.; et al. Respiratory disease in rhesus macaques inoculated with SARS-CoV-2. *Nature* **2020**, *585*, 268–272. [CrossRef] [PubMed]
14. Mariscal, A.; Caldarone, L.; Tikkanen, J.; Nakajima, D.; Chen, M.; Yeung, J.; Cypel, M.; Liu, M.; Keshavjee, S. Pig lung transplant survival model. *Nat. Protoc.* **2018**, *13*, 1814–1828. [CrossRef] [PubMed]
15. Lasko, P.; Lüthy, K. Investigating rare and ultrarare epilepsy syndromes with Drosophila models. *Fac. Rev.* **2021**, *10*, 10. [CrossRef]
16. Babac, B.D.; Milton Dulay, R.R.; Arwen Calpito, R.S.; Domingo, M.A.; Grace Macamos, M.M.; Mangabat, A.R.; Zhyra Zoberiaga, N.L. In-vitro activity of ethanolic extract of Lentinus strigosus mycelia in N2 wild strain Caenorhabditis elegans-An animal model for obesity and its chemical composition. *J. Appl. Biol. Biotech.* **2021**, *9*, 41–46. [CrossRef]
17. Zang, L.; Maddison, L.A.; Chen, W. Zebrafish as a Model for Obesity and Diabetes. *Front. Cell Dev. Biol.* **2018**, *6*, 91. [CrossRef]
18. Fernandes, M.R.; Pedroso, A.R. Animal experimentation: A look into ethics, welfare and alternative methods. *Rev. Assoc. Med. Bras.* **2017**, *63*, 923–928. [CrossRef]
19. Adir, O.; Poley, M.; Chen, G.; Froim, S.; Krinsky, N.; Shklover, J.; Shainsky-Roitman, J.; Lammers, T.; Schroeder, A. Integrating artificial intelligence and nanotechnology for precision cancer medicine. *Adv. Mater.* **2020**, *32*, 1901989. [CrossRef]
20. Bale, T.L.; Abel, T.; Akil, H.; Carlezon, W.A.; Moghaddam, B.; Nestler, E.J.; Ressler, K.J.; Thompson, S.M. The critical importance of basic animal research for neuropsychiatric disorders. *Neuropsychopharmacology* **2019**, *44*, 1349–1353. [CrossRef]
21. Zeggini, E.; Baumann, M.; Götz, M.; Herzig, S.; Hrabe de Angelis, M.; Tschöp, M.H. Biomedical research goes viral: Dangers and opportunities. *Cell* **2020**, *181*, 1189–1193. [CrossRef]
22. Okechukwu, I.B. Introductory Chapter: Animal models for human diseases, a major contributor to modern medicine. In *Experimental Animal Models of Human Diseases—An Effective Therapeutic Strategy*; Batholomew, I., Ed.; InTech: London, UK, 2018; pp. 3–10.
23. Kleinert, M.; Clemmensen, C.; Hofmann, S.M.; Moore, M.C.; Renner, S.; Woods, S.C.; Huypens, P.; Beckers, J.; de Angelis, M.H.; Schürmann, A.; et al. Animal models of obesity and diabetes mellitus. *Nat. Rev. Endocrinol.* **2018**, *14*, 140–162. [CrossRef]
24. Wright, J.R. Frederick Banting's actual great idea: The role of fetal bovine islets in the discovery of insulin. *Islets* **2021**, *13*, 121–133. [CrossRef]
25. Understanding Animal Research. EU-Wide Animal Research Statistics. Available online: https://www.understandinganimalresearch.org.uk/news/eu-wide-animal-research-statistics-2019 (accessed on 9 March 2023).
26. Canadian Council on Animal Care. CCAC Animal Data Report 2020. Available online: https://speakingofresearch.files.wordpress.com/2021/11/canada-2020.pdf (accessed on 9 March 2020).
27. Speaking of Research. UK Animal Research Statistics. Available online: https://speakingofresearch.com/facts/uk-statistics/ (accessed on 9 March 2023).
28. Speaking of Research. US animal Research Statistics. Available online: https://speakingofresearch.com/facts/statistics/ (accessed on 9 March 2023).

29. Speaking of Research. Rise in Animal Research in South Korea in 2017. Available online: https://speakingofresearch.com/2018/04/12/rise-in-animal-research-in-south-korea-in-2017/ (accessed on 9 March 2023).
30. Statista. Annual Number of Animals Used in Research and Testing in Selected Countries Worldwide as of 2020. Available online: https://www.statista.com/statistics/639954/animals-used-in-research-experiments-worldwide/ (accessed on 9 March 2023).
31. Speaking of Research. Worldwide Animal Research Statistics. Available online: https://speakingofresearch.com/facts/animal-research-statistics/ (accessed on 9 March 2023).
32. Gil, G.; Rodríguez, X. Experimentos con Animales en México: 20 Años de Caos y Riesgo. Available online: https://aristeguinoticias.com/2405/mexico/__trashed-25/ (accessed on 9 March 2023).
33. Hubrecht, R.C. The 3Rs and humane experimental technique: Implementing change. *Animals* **2019**, *9*, 754. [CrossRef] [PubMed]
34. Marquardt, N.; Feja, M.; Hünigen, H.; Plendl, J.; Menken, L.; Fink, H.; Bert, B. Euthanasia of laboratory mice: Are isoflurane and sevoflurane real alternatives to carbon dioxide? *PLoS ONE* **2018**, *13*, e0203793. [CrossRef] [PubMed]
35. Percie du Sert, N.; Hurst, V.; Ahluwalia, A.; Alam, S.; Avey, M.T.; Baker, M.; Browne, W.J.; Clark, A.; Cuthill, I.C.; Dirnagl, U.; et al. The ARRIVE guidelines 2.0: Updated guidelines for reporting animal research. *J. Cereb. Blood Flow Metab.* **2020**, *40*, 1769–1777. [CrossRef]
36. Smith, A.J.; Clutton, R.E.; Lilley, E.; Hansen, K.E.A.; Brattelid, T. PREPARE: Guidelines for planning animal research and testing. *Lab. Anim.* **2018**, *52*, 135–141. [CrossRef] [PubMed]
37. Bert, B.; Heinl, C.; Chmielewska, J.; Schwarz, F.; Grune, B.; Hensel, A.; Greiner, M.; Schönfelder, G. Refining animal research: The Animal Study Registry. *PLoS Biol.* **2019**, *17*, e3000463. [CrossRef]
38. Singh, A.; Gupta, U. Animal models of tuberculosis: Lesson learnt. *Indian J. Med. Res.* **2018**, *147*, 456. [CrossRef]
39. Kim, T.-W.; Che, J.-H.; Yun, J.-W. Use of stem cells as alternative methods to animal experimentation in predictive toxicology. *Regul. Toxicol. Pharmacol.* **2019**, *105*, 15–29. [CrossRef]
40. Yang, H.; Wu, Z. Genome editing of pigs for agriculture and biomedicine. *Front. Genet.* **2018**, *9*, 360. [CrossRef]
41. Muñoz-Fontela, C.; Dowling, W.E.; Funnell, S.G.P.; Gsell, P.-S.; Riveros-Balta, A.X.; Albrecht, R.A.; Andersen, H.; Baric, R.S.; Carroll, M.W.; Cavaleri, M.; et al. Animal models for COVID-19. *Nature* **2020**, *586*, 509–515. [CrossRef] [PubMed]
42. Dissanayake, H.A.; de Silva, N.L.; Sumanatilleke, M.; de Silva, S.D.N.; Gamage, K.K.K.; Dematapitiya, C.; Kuruppu, D.C.; Ranasinghe, P.; Pathmanathan, S.; Katulanda, P. Prognostic and therapeutic role of vitamin D in COVID-19: Systematic review and meta-analysis. *J. Clin. Endocrinol. Metab.* **2022**, *107*, 1484–1502. [CrossRef] [PubMed]
43. World Health Organization (WHO). WHO Coronavirus (COVID-19) Dashboard. Available online: https://covid19.who.int/ (accessed on 9 June 2022).
44. Woolsey, C.; Borisevich, V.; Prasad, A.N.; Agans, K.N.; Deer, D.J.; Dobias, N.S.; Heymann, J.C.; Foster, S.L.; Levine, C.B.; Medina, L.; et al. Establishment of an African green monkey model for COVID-19 and protection against re-infection. *Nat. Immunol.* **2021**, *22*, 86–98. [CrossRef] [PubMed]
45. Cleary, S.J.; Pitchford, S.C.; Amison, R.T.; Carrington, R.; Robaina Cabrera, C.L.; Magnen, M.; Looney, M.R.; Gray, E.; Page, C.P. Animal models of mechanisms of SARS-CoV-2 infection and COVID-19 pathology. *Br. J. Pharmacol.* **2020**, *177*, 4851–4865. [CrossRef]
46. Da Costa, C.B.P.; Cruz, A.C.D.M.; Penha, J.C.Q.; Castro, H.C.; Da Cunha, L.E.R.; Ratcliffe, N.A.; Cisne, R.; Martins, F.J. Using in vivo animal models for studying SARS-CoV-2. *Expert Opin. Drug Discov.* **2022**, *17*, 121–137. [CrossRef]
47. Younes, S.; Younes, N.; Shurrab, F.; Nasrallah, G.K. Severe acute respiratory syndrome coronavirus-2 natural animal reservoirs and experimental models: Systematic review. *Rev. Med. Virol.* **2021**, *31*, e2196. [CrossRef]
48. Sun, S.-H.; Chen, Q.; Gu, H.-J.; Yang, G.; Wang, Y.-X.; Huang, X.-Y.; Liu, S.-S.; Zhang, N.-N.; Li, X.-F.; Xiong, R.; et al. A mouse model of SARS-CoV-2 infection and pathogenesis. *Cell Host Microbe* **2020**, *28*, 124–133.e4. [CrossRef]
49. Mueller, A.L.; McNamara, M.S.; Sinclair, D.A. Why does COVID-19 disproportionately affect older people? *Aging* **2020**, *12*, 9959–9981. [CrossRef]
50. Devaux, C.A.; Lagier, J.-C.; Raoult, D. New insights into the physiopathology of COVID-19: SARS-CoV-2-associated gastrointestinal illness. *Front. Med.* **2021**, *8*, 640073. [CrossRef]
51. Bi, Z.; Hong, W.; Yang, J.; Lu, S.; Peng, X. Animal models for SARS-CoV-2 infection and pathology. *MedComm* **2021**, *2*, 548–568. [CrossRef]
52. Urano, E.; Okamura, T.; Ono, C.; Ueno, S.; Nagata, S.; Kamada, H.; Higuchi, M.; Furukawa, M.; Kamitani, W.; Matsuura, Y.; et al. COVID-19 cynomolgus macaque model reflecting human COVID-19 pathological conditions. *Proc. Natl. Acad. Sci. USA* **2021**, *118*, e2104847118. [CrossRef] [PubMed]
53. Yu, P.; Qi, F.; Xu, Y.; Li, F.; Liu, P.; Liu, J.; Bao, L.; Deng, W.; Gao, H.; Xiang, Z.; et al. Age-related rhesus macaque models of COVID-19. *Anim. Model. Exp. Med.* **2020**, *3*, 93–97. [CrossRef] [PubMed]
54. Csiszar, A.; Jakab, F.; Valencak, T.G.; Lanszki, Z.; Tóth, G.E.; Kemenesi, G.; Tarantini, S.; Fazekas-Pongor, V.; Ungvari, Z. Companion animals likely do not spread COVID-19 but may get infected themselves. *GeroScience* **2020**, *42*, 1229–1236. [CrossRef] [PubMed]
55. Kumar, R.; Harilal, S.; Al-Sehemi, A.G.; Pannipara, M.; Behl, T.; Mathew, G.E.; Mathew, B. COVID-19 and domestic animals: Exploring the species barrier crossing, zoonotic and reverse zoonotic transmission of SARS-CoV-2. *Curr. Pharm. Des.* **2021**, *27*, 1194–1201. [CrossRef]

56. Shi, J.; Wen, Z.; Zhong, G.; Yang, H.; Wang, C.; Huang, B.; Liu, R.; He, X.; Shuai, L.; Sun, Z.; et al. Susceptibility of ferrets, cats, dogs, and other domesticated animals to SARS–coronavirus 2. *Science* **2020**, *368*, 1016–1020. [CrossRef]
57. Hosie, M.J.; Hofmann-Lehmann, R.; Hartmann, K.; Egberink, H.; Truyen, U.; Addie, D.D.; Belák, S.; Boucraut-Baralon, C.; Frymus, T.; Lloret, A.; et al. Anthropogenic infection of cats during the 2020 COVID-19 pandemic. *Viruses* **2021**, *13*, 185. [CrossRef]
58. Pramod, R.K.; Nair, A.V.; Tambare, P.K.; Chauhan, K.; Kumar, T.V.; Rajan, R.A.; Mani, B.M.; Asaf, M.; Pandey, A.K. Reverse zoonosis of coronavirus disease-19: Present status and the control by one health approach. *Vet. World* **2021**, *14*, 2817–2826. [CrossRef]
59. Kang, K.; Chen, Q.; Gao, Y.; Yu, K. Detection of SARS-CoV-2 B.1.617.2 (Delta) variant in three cats owned by a confirmed COVID-19 patient in Harbin, China. *Vet. Med. Sci.* **2022**, *8*, 945–946. [CrossRef]
60. da Costa Rodrigues, K.C.; Pereira, R.M.; de Campos, T.D.P.; de Moura, R.F.; da Silva, A.S.R.; Cintra, D.E.; Ropelle, E.R.; Pauli, J.R.; de Araújo, M.B.; de Moura, L.P. The role of physical exercise to improve the browning of white adipose tissue via POMC neurons. *Front. Cell. Neurosci.* **2018**, *12*, 88. [CrossRef]
61. Rios, J.L.; Bomhof, M.R.; Reimer, R.A.; Hart, D.A.; Collins, K.H.; Herzog, W. Protective effect of prebiotic and exercise intervention on knee health in a rat model of diet-induced obesity. *Sci. Rep.* **2019**, *9*, 3893. [CrossRef]
62. Burhans, M.S.; Hagman, D.K.; Kuzma, J.N.; Schmidt, K.A.; Kratz, M. Contribution of adipose tissue inflammation to the development of type 2 diabetes mellitus. *Compr. Physiol.* **2018**, *9*, 1–58. [CrossRef] [PubMed]
63. Khalafi, M.; Mohebbi, H.; Symonds, M.E.; Karimi, P.; Akbari, A.; Tabari, E.; Faridnia, M.; Moghaddami, K. The impact of moderate-intensity continuous or high-intensity interval training on adipogenesis and browning of subcutaneous adipose tissue in obese male rats. *Nutrients* **2020**, *12*, 925. [CrossRef]
64. Lee, H.E.; Yang, G.; Han, S.-H.; Lee, J.-H.; An, T.-J.; Jang, J.-K.; Lee, J.Y. Anti-obesity potential of Glycyrrhiza uralensis and licochalcone A through induction of adipocyte browning. *Biochem. Biophys. Res. Commun.* **2018**, *503*, 2117–2123. [CrossRef]
65. Wang, P.; Li, D.; Ke, W.; Liang, D.; Hu, X.; Chen, F. Resveratrol-induced gut microbiota reduces obesity in high-fat diet-fed mice. *Int. J. Obes.* **2020**, *44*, 213–225. [CrossRef] [PubMed]
66. Lasker, S.; Rahman, M.M.; Parvez, F.; Zamila, M.; Miah, P.; Nahar, K.; Kabir, F.; Sharmin, S.B.; Subhan, N.; Ahsan, G.U.; et al. High-fat diet-induced metabolic syndrome and oxidative stress in obese rats are ameliorated by yogurt supplementation. *Sci. Rep.* **2019**, *9*, 20026. [CrossRef] [PubMed]
67. Bouyanfif, A.; Jayarathne, S.; Koboziev, I.; Moustaid-Moussa, N. The nematode Caenorhabditis elegans as a model organism to study metabolic effects of ω-3 polyunsaturated fatty acids in obesity. *Adv. Nutr.* **2019**, *10*, 165–178. [CrossRef] [PubMed]
68. Ke, W.; Reed, J.N.; Yang, C.; Higgason, N.; Rayyan, L.; Wählby, C.; Carpenter, A.E.; Civelek, M.; O'Rourke, E.J. Genes in human obesity loci are causal obesity genes in C. elegans. *PLoS Genet.* **2021**, *17*, e1009736. [CrossRef]
69. Farias-Pereira, R.; Zhang, Z.; Park, C.; Kim, D.; Kim, K.; Park, Y. Butein inhibits lipogenesis in *Caenorhabditis elegans*. *BioFactors* **2020**, *46*, 777–787. [CrossRef]
70. Yusela Kris, A.; Rich Milton, D.; Sofronio, K. Effect of Lentinus strigosus extract on the food intake and locomotion of N2 wild strain Caenorhabditis elegans as model for obesity. *J. Appl. Pharm. Sci.* **2020**, *10*, 23–28. [CrossRef]
71. Benchoula, K.; Khatib, A.; Jaffar, A.; Ahmed, Q.U.; Sulaiman, W.M.A.W.; Wahab, R.A.; El-Seedi, H.R. The promise of zebrafish as a model of metabolic syndrome. *Exp. Anim.* **2019**, *68*, 407–416. [CrossRef]
72. Rodríguez, R.R.; González-Bulnes, A.; Garcia-Contreras, C.; Elena Rodriguez-Rodriguez, A.; Astiz, S.; Vazquez-Gomez, M.; Luis Pesantez, J.; Isabel, B.; Salido-Ruiz, E.; González, J.; et al. The Iberian pig fed with high-fat diet: A model of renal disease in obesity and metabolic syndrome. *Int. J. Obes.* **2020**, *44*, 457–465. [CrossRef] [PubMed]
73. Chagnac, A.; Zingerman, B.; Rozen-Zvi, B.; Herman-Edelstein, M. Consequences of glomerular hyperfiltration: The role of physical forces in the pathogenesis of chronic kidney disease in diabetes and obesity. *Nephron* **2019**, *143*, 38–42. [CrossRef] [PubMed]
74. Rodríguez-Rodríguez, A.E.; Donate-Correa, J.; Luis-Lima, S.; Díaz-Martín, L.; Rodríguez-González, C.; Pérez-Pérez, J.A.; Acosta-González, N.G.; Fumero, C.; Navarro-Díaz, M.; López-Álvarez, D.; et al. Obesity and metabolic syndrome induce hyperfiltration, glomerulomegaly, and albuminuria in obese ovariectomized female mice and obese male mice. *Menopause* **2021**, *28*, 1296–1306. [CrossRef]
75. Hoffman, J.M.; Creevy, K.E.; Franks, A.; O'Neill, D.G.; Promislow, D.E.L. The companion dog as a model for human aging and mortality. *Aging Cell* **2018**, *17*, e12737. [CrossRef]
76. Suarez, L.; Bautista-Castaño, I.; Peña Romera, C.; Montoya-Alonso, J.A.; Corbera, J.A. Is dog owner obesity a risk factor for canine obesity? a "one-health" study on human–animal interaction in a region with a high prevalence of obesity. *Vet. Sci.* **2022**, *9*, 243. [CrossRef]
77. World Health Organization (WHO). Cancer. Available online: https://www.who.int/news-room/fact-sheets/detail/cancer (accessed on 9 March 2023).
78. National Cancer Institute. Common Cancer Types. Available online: https://www.cancer.gov/types/common-cancers (accessed on 9 March 2023).
79. Siegel, R.L.; Miller, K.D.; Fuchs, H.E.; Jemal, A. Cancer statistics, 2022. *CA Cancer J. Clin.* **2022**, *72*, 7–33. [CrossRef] [PubMed]
80. Berish, R.B.; Ali, A.N.; Telmer, P.G.; Ronald, J.A.; Leong, H.S. Translational models of prostate cancer bone metastasis. *Nat. Rev. Urol.* **2018**, *15*, 403–421. [CrossRef]

81. Onaciu, A.; Munteanu, R.; Munteanu, V.C.; Gulei, D.; Raduly, L.; Feder, R.-I.; Pirlog, R.; Atanasov, A.G.; Korban, S.S.; Irimie, A.; et al. Spontaneous and induced animal models for cancer research. *Diagnostics* **2020**, *10*, 660. [CrossRef]
82. Valcourt, D.M.; Kapadia, C.H.; Scully, M.A.; Dang, M.N.; Day, E.S. Best practices for preclinical in vivo testing of cancer nanomedicines. *Adv. Healthc. Mater.* **2020**, *9*, 2000110. [CrossRef]
83. Santana-Krímskaya, S.E.; Kawas, J.R.; Zarate-Triviño, D.G.; Ramos-Zayas, Y.; Rodríguez-Padilla, C.; Franco-Molina, M.A. Orthotopic and heterotopic triple negative breast cancer preclinical murine models: A tumor microenvironment comparative. *Res. Vet. Sci.* **2022**, *152*, 364–371. [CrossRef]
84. Madonna, M.C.; Duer, J.E.; Lee, J.V.; Williams, J.; Avsaroglu, B.; Zhu, C.; Deutsch, R.; Wang, R.; Crouch, B.T.; Hirschey, M.D.; et al. In vivo optical metabolic imaging of long-chain fatty acid uptake in orthotopic models of triple-negative breast cancer. *Cancers* **2021**, *13*, 148. [CrossRef] [PubMed]
85. Koosha, S.; Mohamed, Z.; Sinniah, A.; Alshawsh, M.A. Evaluation of anti-tumorigenic effects of diosmetin against human colon cancer xenografts in athymic nude mice. *Molecules* **2019**, *24*, 2522. [CrossRef]
86. Sliepen, S.H.J.; Diaz-del Castillo, M.; Korioth, J.; Olsen, R.B.; Appel, C.K.; Christoph, T.; Heegaard, A.-M.; Rutten, K. Cancer-induced bone pain impairs burrowing behaviour in mouse and rat. *In Vivo* **2019**, *33*, 1125–1132. [CrossRef] [PubMed]
87. Overgaard, N.H.; Fan, T.M.; Schachtschneider, K.M.; Principe, D.R.; Schook, L.B.; Jungersen, G. Of Mice, Dogs, Pigs, and Men: Choosing the Appropriate Model for Immuno-Oncology Research. *ILAR J.* **2018**, *59*, 247–262. [CrossRef] [PubMed]
88. Kamiya, A.; Hiyama, T.; Fujimura, A.; Yoshikawa, S. Sympathetic and parasympathetic innervation in cancer: Therapeutic implications. *Clin. Auton. Res.* **2021**, *31*, 165–178. [CrossRef]
89. Sadighparvar, S.; Darband, S.G.; Ghaderi-Pakdel, F.; Mihanfar, A.; Majidinia, M. Parasympathetic, but not sympathetic denervation, suppressed colorectal cancer progression. *Eur. J. Pharmacol.* **2021**, *913*, 174626. [CrossRef]
90. Kamiya, A.; Hayama, Y.; Kato, S.; Shimomura, A.; Shimomura, T.; Irie, K.; Kaneko, R.; Yanagawa, Y.; Kobayashi, K.; Ochiya, T. Genetic manipulation of autonomic nerve fiber innervation and activity and its effect on breast cancer progression. *Nat. Neurosci.* **2019**, *22*, 1289–1305. [CrossRef]
91. Grelet, S.; Fréreux, C.; Obellianne, C.; Noguchi, K.; Howley, B.V.; Dalton, A.C.; Howe, P.H. TGFβ-induced expression of long noncoding lincRNA Platr18 controls breast cancer axonogenesis. *Life Sci. Alliance* **2022**, *5*, e202101261. [CrossRef]
92. Kirkpatrick, J.D.; Warren, A.D.; Soleimany, A.P.; Westcott, P.M.K.; Voog, J.C.; Martin-Alonso, C.; Fleming, H.E.; Tammela, T.; Jacks, T.; Bhatia, S.N. Urinary detection of lung cancer in mice via noninvasive pulmonary protease profiling. *Sci. Transl. Med.* **2020**, *12*, eaaw0262. [CrossRef]
93. Xu, X.; An, H.; Zhang, D.; Tao, H.; Dou, Y.; Li, X.; Huang, J.; Zhang, J. A self-illuminating nanoparticle for inflammation imaging and cancer therapy. *Sci. Adv.* **2019**, *5*, eaat2953. [CrossRef]
94. Momcilovic, M.; Jones, A.; Bailey, S.T.; Waldmann, C.M.; Li, R.; Lee, J.T.; Abdelhady, G.; Gomez, A.; Holloway, T.; Schmid, E.; et al. In vivo imaging of mitochondrial membrane potential in non-small-cell lung cancer. *Nature* **2019**, *575*, 380–384. [CrossRef]
95. Nicolson, F.; Andreiuk, B.; Andreou, C.; Hsu, H.-T.; Rudder, S.; Kircher, M.F. Non-invasive in vivo imaging of cancer using surface-enhanced spatially offset Raman Spectroscopy (SESORS). *Theranostics* **2019**, *9*, 5899–5913. [CrossRef] [PubMed]
96. Chen, L.; Zuo, W.; Xiao, Z.; Jin, Q.; Liu, J.; Wu, L.; Liu, N.; Zhu, X. A carrier-free metal-coordinated dual-photosensitizers nanotheranostic with glutathione-depletion for fluorescence/photoacoustic imaging-guided tumor phototherapy. *J. Colloid Interface Sci.* **2021**, *600*, 243–255. [CrossRef]
97. Hernández-Avalos, I.; Mota-Rojas, D.; Mendoza-Flores, J.E.; Casas-Alvarado, A.; Flores-Padilla, K.; Miranda-Cortes, A.E.; Torres-Bernal, F.; Gómez-Prado, J.; Mora-Medina, P. Nociceptive pain and anxiety in equines: Physiological and behavioral alterations. *Vet. World* **2021**, *14*, 2984–2995. [CrossRef] [PubMed]
98. Singh, S.; Kumar, A.; Mittal, G. Ketamine-polymer based drug delivery system for prolonged analgesia: Recent advances, challenges and future prospects. *Expert Opin. Drug Deliv.* **2021**, *18*, 1117–1130. [CrossRef] [PubMed]
99. Parra, S.; Thanawala, V.J.; Rege, A.; Giles, H. A novel excisional wound pain model for evaluation of analgesics in rats. *Korean J. Pain* **2021**, *34*, 165–175. [CrossRef] [PubMed]
100. Millecam, J.; van Bergen, T.; Schauvliege, S.; Antonissen, G.; Martens, A.; Chiers, K.; Gehring, R.; Gasthuys, E.; Vande Walle, J.; Croubels, S.; et al. Developmental pharmacokinetics and safety of ibuprofen and its enantiomers in the conventional pig as potential pediatric animal model. *Front. Pharmacol.* **2019**, *10*, 505. [CrossRef]
101. Domínguez-Oliva, A.; Casas-Alvarado, A.; Miranda-Cortés, A.E.; Hernández-Avalos, I. Clinical pharmacology of tramadol and tapentadol, and their therapeutic efficacy in different models of acute and chronic pain in dogs and cats. *J. Adv. Vet. Anim. Res.* **2021**, *8*, 404. [CrossRef]
102. Mulier, J.P. Is opioid-free general anesthesia for breast and gynecological surgery a viable option? *Curr. Opin. Anaesthesiol.* **2019**, *32*, 257–262. [CrossRef]
103. Cicirelli, V.; Burgio, M.; Lacalandra, G.M.; Aiudi, G.G. Local and regional anaesthetic techniques in canine ovariectomy: A review of the literature and technique description. *Animals* **2022**, *12*, 1920. [CrossRef]
104. Miranda-Cortés, A.E.; Ruiz-García, A.G.; Olivera-Ayub, A.E.; Garza-Malacara, G.; Ruiz-Cervantes, J.G.; Toscano-Zapien, J.A.; Hernández-Avalos, I. Cardiorespiratory effects of epidurally administered ketamine or lidocaine in dogs undergoing ovariohysterectomy surgery: A comparative study. *Iran. J. Vet. Res.* **2020**, *21*, 92–96. [PubMed]

105. Komatsu, A.; Miyano, K.; Nakayama, D.; Mizobuchi, Y.; Uezono, E.; Ohshima, K.; Karasawa, Y.; Kuroda, Y.; Nonaka, M.; Yamaguchi, K.; et al. Novel opioid analgesics for the development of transdermal opioid patches that possess morphine-like pharmacological profiles rather than fentanyl: Possible opioid switching alternatives among patch formula. *Anesth. Analg.* **2022**, *134*, 1082–1093. [CrossRef]
106. Godfrey, L.; Iannitelli, A.; Garrett, N.L.; Moger, J.; Imbert, I.; King, T.; Porreca, F.; Soundararajan, R.; Lalatsa, A.; Schätzlein, A.G.; et al. Nanoparticulate peptide delivery exclusively to the brain produces tolerance free analgesia. *J. Control. Release* **2018**, *270*, 135–144. [CrossRef]
107. Wang, B.; Wang, S.; Zhang, Q.; Deng, Y.; Li, X.; Peng, L.; Zuo, X.; Piao, M.; Kuang, X.; Sheng, S.; et al. Recent advances in polymer-based drug delivery systems for local anesthetics. *Acta Biomater.* **2019**, *96*, 55–67. [CrossRef] [PubMed]
108. Wang, R.; Gan, J.; Li, R.; Duan, J.; Zhou, J.; Lv, M.; Qi, R. Controlled delivery of ketamine from reduced graphene oxide hydrogel for neuropathic pain: In vitro and in vivo studies. *J. Drug Deliv. Sci. Technol.* **2020**, *60*, 101964. [CrossRef]
109. Kania, K.; Chang, D.K.; Abu-Ghname, A.; Reece, E.M.; Chu, C.K.; Maricevich, M.; Buchanan, E.P.; Winocour, S. Microsurgery training in plastic surgery. *Plast. Reconstr. Surg.—Glob. Open* **2020**, *8*, e2898. [CrossRef]
110. Ozols, D.; Zarins, J.; Petersons, A. Novel technique for toe-to-hand transplantation: The fourth-toe as an alternative option for toe-to-hand transplantation for pediatric patients. *Tech. Hand Up. Extrem. Surg.* **2019**, *23*, 74–80. [CrossRef]
111. Dvořák, Z.; Stupka, I. Atypical replantation and reconstruction of frozen ear. *Medicine* **2020**, *99*, e20068. [CrossRef]
112. Khachatryan, A.; Tevosyan, A.; Novoselskiy, D.; Arakelyan, G.; Yushkevich, A.; Nazaretovich Nazarian, D. Experimental reimplantation models. In *Microsurgery Manual for Medical Students and Residents*; Springer International Publishing: Cham, Switzerland, 2021; pp. 139–144.
113. Goutard, M.; Randolph, M.A.; Taveau, C.B.; Lupon, E.; Lantieri, L.; Uygun, K.; Cetrulo, C.L.; Lellouch, A.G. Partial heterotopic hindlimb transplantation model in rats. *J. Vis. Exp.* **2021**, *172*, e62586. [CrossRef]
114. Vernon, R.; Wang, J.; Song, M.; Wilson, N.; Moris, D.; Cendales, L. Vascularized Composite Allotransplantation: A Functional Hind Limb Model in Mice. *J. Surg. Res.* **2020**, *250*, 119–124. [CrossRef]
115. Tee, R.; Morrison, W.A.; Dilley, R.J. A novel microsurgical rodent model for the transplantation of engineered cardiac muscle flap. *Microsurgery* **2018**, *38*, 544–552. [CrossRef]
116. Scaglioni, M.; Giovanoli, P.; Scaglioni, M.F.; Yang, J.C. Microsurgical head and neck reconstruction in patients with coronary artery disease: A perioperative assessment algorithm. *Microsurgery* **2019**, *39*, 290–296. [CrossRef] [PubMed]
117. Kotz, D. UM Medicine Performs Historic Xenotransplantation. Available online: https://www.umaryland.edu/news/archived-news/january-2022/um-medicine-performs-historic-xenotransplantation.php (accessed on 7 June 2022).
118. Wang, W.; He, W.; Ruan, Y.; Geng, Q. First pig-to-human heart transplantation. *Innovation* **2022**, *3*, 100223. [CrossRef]
119. Lee, S.J.; Kim, J.S.; Chee, H.K.; Yun, I.J.; Park, K.S.; Yang, H.S.; Park, J.H. Seven years of experiences of preclinical experiments of xeno-heart transplantation of pig to non-human primate (Cynomolgus Monkey). *Transplant. Proc.* **2018**, *50*, 1167–1171. [CrossRef]
120. Cui, Y.; Yamamoto, T.; Raza, S.S.; Morsi, M.; Nguyen, H.Q.; Ayares, D.; Cooper, D.K.C.; Hara, H. Evidence for GTKO/β4GalNT2KO pigs as the preferred organ-source for old world nonhuman primates as a preclinical model of xenotransplantation. *Transplant. Direct.* **2020**, *6*, e590. [CrossRef] [PubMed]
121. Iqbal, A.; Zaman, M.; Wahab Amjad, M.; Adnan, S.; Abdul Ghafoor Raja, M.; Haider Rizvi, S.F.; Mustafa, M.W.; Farooq, U.; Abbas, G.; Shah, S. Solid lipid nanoparticles of mycophenolate mofetil: An attempt to control the release of an immunosuppressant. *Int. J. Nanomed.* **2020**, *15*, 5603–5612. [CrossRef] [PubMed]
122. Li, R.; Liang, J.; He, Y.; Qin, J.; He, H.; Lee, S.; Pang, Z.; Wang, J. Sustained release of immunosuppressant by nanoparticle-anchoring hydrogel scaffold improved the survival of transplanted stem cells and tissue regeneration. *Theranostics* **2018**, *8*, 878–893. [CrossRef]
123. Xie, H.; Zhu, H.; Zhou, K.; Wan, J.; Zhang, L.; Yang, Z.; Zhou, L.; Chen, X.; Xu, X.; Zheng, S.; et al. Target-oriented delivery of self-assembled immunosuppressant cocktails prolongs allogeneic orthotopic liver transplant survival. *J. Control. Release* **2020**, *328*, 237–250. [CrossRef]
124. Natoli, S.; Oliveira, V.; Calabresi, P.; Maia, L.F.; Pisani, A. Does SARS-Cov-2 invade the brain? Translational lessons from animal models. *Eur. J. Neurol.* **2020**, *27*, 1764–1773. [CrossRef]
125. Badaut, J.; Adami, A.; Huang, L.; Obenaus, A. Noninvasive magnetic resonance imaging stratifies injury severity in a rodent model of male juvenile traumatic brain injury. *J. Neurosci. Res.* **2019**, *98*, 129–140. [CrossRef]
126. Vink, R. Large animal models of traumatic brain injury. *J. Neurosci. Res.* **2018**, *96*, 527–535. [CrossRef]
127. Ackermans, N.L.; Varghese, M.; Wicinski, B.; Torres, J.; De Gasperi, R.; Pryor, D.; Elder, G.A.; Gama Sosa, M.A.; Reidenberg, J.S.; Williams, T.M.; et al. Unconventional animal models for traumatic brain injury and chronic traumatic encephalopathy. *J. Neurosci. Res.* **2021**, *99*, 2463–2477. [CrossRef] [PubMed]
128. Grovola, M.R.; Paleologos, N.; Brown, D.P.; Tran, N.; Wofford, K.L.; Harris, J.P.; Browne, K.D.; Shewokis, P.A.; Wolf, J.A.; Cullen, D.K.; et al. Diverse changes in microglia morphology and axonal pathology during the course of 1 year after mild traumatic brain injury in pigs. *Brain Pathol.* **2021**, *31*, e12953. [CrossRef]
129. Filipp, M.; Travis, B.; Henry, S.; Idzikowski, E.; Magnuson, S.; Loh, M.; Hellenbrand, D.; Hanna, A. Differences in neuroplasticity after spinal cord injury in varying animal models and humans. *Neural Regen. Res.* **2019**, *14*, 7. [CrossRef]

130. Perucca, P.; Bahlo, M.; Berkovic, S.F. The genetics of epilepsy. *Annu. Rev. Genom. Hum. Genet.* **2020**, *21*, 205–230. [CrossRef] [PubMed]
131. Hobbs, S.L.; Law, T.H.; Volk, H.A.; Younis, C.; Casey, R.A.; Packer, R.M.A. Impact of canine epilepsy on judgement and attention biases. *Sci. Rep.* **2020**, *10*, 17719. [CrossRef] [PubMed]
132. Jacobs, J.A.; Sehgal, A. Anandamide metabolites protect against seizures through the TRP channel water witch in Drosophila melanogaster. *Cell Rep.* **2020**, *31*, 107710. [CrossRef]
133. Dissel, S. Drosophila as a model to study the relationship between sleep, plasticity, and memory. *Front. Physiol.* **2020**, *11*, 533. [CrossRef]
134. Toda, H.; Williams, J.A.; Gulledge, M.; Sehgal, A. A sleep-inducing gene, nemuri, links sleep and immune function in Drosophila. *Science* **2019**, *363*, 509–515. [CrossRef]
135. Lei, Z.; Henderson, K.; Keleman, K. A neural circuit linking learning and sleep in Drosophila long-term memory. *Nat. Commun.* **2022**, *13*, 609. [CrossRef]
136. Shi, L.; Su, B. A transgenic monkey model for the study of human brain evolution. *Zool. Res.* **2019**, *40*, 236–238. [CrossRef] [PubMed]
137. Hoffe, B.; Holahan, M.R. The use of pigs as a translational model for studying neurodegenerative diseases. *Front. Physiol.* **2019**, *10*, 838. [CrossRef]
138. Calvo-Rodriguez, M.; Hou, S.S.; Snyder, A.C.; Kharitonova, E.K.; Russ, A.N.; Das, S.; Fan, Z.; Muzikansky, A.; Garcia-Alloza, M.; Serrano-Pozo, A.; et al. Increased mitochondrial calcium levels associated with neuronal death in a mouse model of Alzheimer's disease. *Nat. Commun.* **2020**, *11*, 2146. [CrossRef]
139. Kin, K.; Yasuhara, T.; Kameda, M.; Date, I. Animal models for parkinson's disease research: Trends in the 2000s. *Int. J. Mol. Sci.* **2019**, *20*, 5402. [CrossRef] [PubMed]
140. Hughes, G.L.; Lones, M.A.; Bedder, M.; Currie, P.D.; Smith, S.L.; Pownall, M.E. Machine learning discriminates a movement disorder in a zebrafish model of Parkinson's disease. *Dis. Model. Mech.* **2020**, *13*, dmm045815. [CrossRef] [PubMed]
141. Barnhill, L.M.; Murata, H.; Bronstein, J.M. Studying the pathophysiology of Parkinson's disease using zebrafish. *Biomedicines* **2020**, *8*, 197. [CrossRef]
142. Kalyn, M.; Hua, K.; Mohd Noor, S.; Wong, C.E.D.; Ekker, M. Comprehensive analysis of neurotoxin-induced ablation of dopaminergic neurons in zebrafish larvae. *Biomedicines* **2019**, *8*, 1. [CrossRef]
143. Vaz, R.L.; Outeiro, T.F.; Ferreira, J.J. Zebrafish as an animal model for drug discovery in parkinson's disease and other movement disorders: A systematic review. *Front. Neurol.* **2018**, *9*, 347. [CrossRef]
144. Zhao, Y.; Han, Y.; Wang, Z.; Chen, T.; Qian, H.; He, J.; Li, J.; Han, B.; Wang, T. Rosmarinic acid protects against 1-methyl-4-phenyl-1,2,3,6-tetrahydropyridine-induced dopaminergic neurotoxicity in zebrafish embryos. *Toxicol. Vitr.* **2020**, *65*, 104823. [CrossRef]
145. Dongjie, S.; Rajendran, R.S.; Xia, Q.; She, G.; Tu, P.; Zhang, Y.; Liu, K. Neuroprotective effects of Tongtian oral liquid, a Traditional Chinese Medicine in the Parkinson's disease-induced zebrafish model. *Biomed. Pharmacother.* **2022**, *148*, 112706. [CrossRef]
146. Kodera, K.; Matsui, H. Zebrafish, medaka and turquoise killifish for understanding human neurodegenerative/neurodevelopmental disorders. *Int. J. Mol. Sci.* **2022**, *23*, 1399. [CrossRef] [PubMed]
147. Chen, J.; Lei, L.; Tian, L.; Hou, F.; Roper, C.; Ge, X.; Zhao, Y.; Chen, Y.; Dong, Q.; Tanguay, R.L.; et al. Developmental and behavioral alterations in zebrafish embryonically exposed to valproic acid (VPA): An aquatic model for autism. *Neurotoxicol. Teratol.* **2018**, *66*, 8–16. [CrossRef] [PubMed]
148. Arafat, E.A.; Shabaan, D.A. The possible neuroprotective role of grape seed extract on the histopathological changes of the cerebellar cortex of rats prenatally exposed to Valproic Acid: Animal model of autism. *Acta Histochem.* **2019**, *121*, 841–851. [CrossRef]
149. Baek, H.; Sariev, A.; Lee, S.; Dong, S.-Y.; Royer, S.; Kim, H. Deep cerebellar low-intensity focused ultrasound stimulation restores interhemispheric balance after ischemic stroke in mice. *IEEE Trans. Neural Syst. Rehabil. Eng.* **2020**, *28*, 2073–2079. [CrossRef]
150. Chen, C.-D.; Kim, S.-K.; HwangBo, G. Effect of Acute Phase Pain Control Using TENS on Pain Relief in Knee Osteoarthritis in a Rat Model. *J. Korean Soc. Phys. Med.* **2021**, *16*, 15–20. [CrossRef]
151. Martins, Â.; Gouveia, D.; Cardoso, A.; Carvalho, C.; Silva, C.; Coelho, T.; Gamboa, Ó.; Ferreira, A. Functional neurorehabilitation in dogs with an incomplete recovery 3 months following intervertebral disc surgery: A case series. *Animals* **2021**, *11*, 2442. [CrossRef] [PubMed]
152. Iijima, H.; Eguchi, R.; Shimoura, K.; Yamada, K.; Aoyama, T.; Takahashi, M. Transcutaneous electrical nerve stimulation improves stair climbing capacity in people with knee osteoarthritis. *Sci. Rep.* **2020**, *10*, 7294. [CrossRef]
153. Lee, J.E.; Anderson, C.M.; Perkhounkova, Y.; Sleeuwenhoek, B.M.; Louison, R.R. Transcutaneous electrical nerve stimulation reduces resting pain in head and neck cancer patients. *Cancer Nurs.* **2019**, *42*, 218–228. [CrossRef]
154. Hsiao, I.-H.; Liao, H.-Y.; Lin, Y. Optogenetic modulation of electroacupuncture analgesia in a mouse inflammatory pain model. *Sci. Rep.* **2022**, *12*, 9067. [CrossRef]
155. Truong, B.T.; Artinger, K.B. The power of zebrafish models for understanding the co-occurrence of craniofacial and limb disorders. *Genesis* **2021**, *59*, e23407. [CrossRef]

156. Bergen, D.J.M.; Tong, Q.; Shukla, A.; Newham, E.; Zethof, J.; Lundberg, M.; Ryan, R.; Youlten, S.E.; Frysz, M.; Croucher, P.I.; et al. Regenerating zebrafish scales express a subset of evolutionary conserved genes involved in human skeletal disease. *BMC Biol.* **2022**, *20*, 21. [CrossRef]
157. Yonekura, M.; Kondoh, N.; Han, C.; Toyama, Y.; Ohba, T.; Ono, K.; Itagaki, S.; Tomita, H.; Murakami, M. Medaka as a model for ECG analysis and the effect of verapamil. *J. Pharmacol. Sci.* **2018**, *137*, 55–60. [CrossRef]
158. Marzęda, P.; Drozd, M.; Tchórz, M.; Kisiel, K. Importance of early intervention in verapamil overdose—Case Report and antidotes review. *J. Pre-Clin. Clin. Res.* **2021**, *15*, 142–147. [CrossRef]
159. Ruetten, H.; Vezina, C.M. Relevance of dog as an animal model for urologic diseases. In *Progress in Molecular Biology and Translational Science*; Tao, Y.-X., Ed.; Academic Press: Oxford, UK, 2022; pp. 35–65.
160. Gajski, G.; Žegura, B.; Ladeira, C.; Pourrut, B.; Del Bo', C.; Novak, M.; Sramkova, M.; Milić, M.; Gutzkow, K.B.; Costa, S.; et al. The comet assay in animal models: From bugs to whales—(Part 1 Invertebrates). *Mutat. Res. Mutat. Res.* **2019**, *779*, 82–113. [CrossRef]
161. Bouriga, N.; Mili, S.; Bahri, W.R.; Jammeli, B.; Ben-Attia, M.; Quignard, J.-P.; Trabelsi, M. Skin Wound Healing Potential and Antioxidant Effect of Hyaluronic Acid Extracted from Mytilus galloprovincialis and Crassostrea gigas. *Pharm. Chem. J.* **2022**, *56*, 381–386. [CrossRef]
162. Imperadore, P.; Uckermann, O.; Galli, R.; Steiner, G.; Kirsch, M.; Fiorito, G. Nerve regeneration in the cephalopod mollusc Octopus vulgaris: Label-free multiphoton microscopy as a tool for investigation. *J. R. Soc. Interface* **2018**, *15*, 20170889. [CrossRef] [PubMed]
163. NIH When Are Alternatives to Animals Used in Research? Available online: https://grants.nih.gov/grants/policy/air/alternatives (accessed on 10 March 2023).
164. Conroy, G. These Are the 10 Best Countries for Life Sciences Research. Available online: https://www.nature.com/nature-index/news-blog/ten-best-countries-life-sciences-research-rankings (accessed on 10 March 2023).
165. Van Norman, G.A. Limitations of animal studies for predicting toxicity in clinical trials. *JACC Basic Transl. Sci.* **2019**, *4*, 845–854. [CrossRef]
166. Hernandez, J. The FDA No Longer Requires All Drugs to Be Tested on Animals before Human Trials. Available online: https://www.npr.org/2023/01/12/1148529799/fda-animal-testing-pharmaceuticals-drug-development (accessed on 10 March 2023).
167. Canadian Council on Animal Care. CCAC Animal Data Report 2019. Available online: https://speakingofresearch.files.wordpress.com/2021/08/canada-2019.pdf (accessed on 10 March 2023).
168. Xinaris, C. Organoids for replacement therapy: Expectations, limitations and reality. *Curr. Opin. Organ Transplant.* **2019**, *24*, 555–561. [CrossRef] [PubMed]
169. Driehuis, E.; Kretzschmar, K.; Clevers, H. Establishment of patient-derived cancer organoids for drug-screening applications. *Nat. Protoc.* **2020**, *15*, 3380–3409. [CrossRef]
170. Matsui, T.K.; Tsuru, Y.; Hasegawa, K.; Kuwako, K. Vascularization of human brain organoids. *Stem Cells* **2021**, *39*, 1017–1024. [CrossRef]
171. Homan, K.A.; Gupta, N.; Kroll, K.T.; Kolesky, D.B.; Skylar-Scott, M.; Miyoshi, T.; Mau, D.; Valerius, M.T.; Ferrante, T.; Bonventre, J.V.; et al. Flow-enhanced vascularization and maturation of kidney organoids in vitro. *Nat. Methods* **2019**, *16*, 255–262. [CrossRef] [PubMed]
172. Wörsdörfer, P.; Dalda, N.; Kern, A.; Krüger, S.; Wagner, N.; Kwok, C.K.; Henke, E.; Ergün, S. Generation of complex human organoid models including vascular networks by incorporation of mesodermal progenitor cells. *Sci. Rep.* **2019**, *9*, 15663. [CrossRef] [PubMed]
173. Rawal, P.; Tripathi, D.M.; Ramakrishna, S.; Kaur, S. Prospects for 3D bioprinting of organoids. *Bio-Des. Manuf.* **2021**, *4*, 627–640. [CrossRef]
174. Briceño, J. Artificial intelligence and organ transplantation: Challenges and expectations. *Curr. Opin. Organ Transpl.* **2020**, *25*, 393–398. [CrossRef]
175. Fattahi Sani, M.; Ascione, R.; Dogramadzi, S. Mapping surgeons hand/finger movements to surgical tool motion during conventional microsurgery using machine learning. *J. Med. Robot. Res.* **2021**, *6*, 2150004. [CrossRef]
176. Azer, S.A. Deep learning with convolutional neural networks for identification of liver masses and hepatocellular carcinoma: A systematic review. *World J. Gastrointest Oncol.* **2019**, *11*, 1218–1230. [CrossRef]
177. Wang, S.; Kang, B.; Ma, J.; Zeng, X.; Xiao, M.; Guo, J.; Cai, M.; Yang, J.; Li, Y.; Meng, X.; et al. A deep learning algorithm using CT images to screen for Corona virus disease (COVID-19). *Eur. Radiol.* **2021**, *31*, 6096–6104. [CrossRef]
178. Gurovich, Y.; Hanani, Y.; Bar, O.; Nadav, G.; Fleischer, N.; Gelbman, D.; Basel-Salmon, L.; Krawitz, P.M.; Kamphausen, S.B.; Zenker, M.; et al. Identifying facial phenotypes of genetic disorders using deep learning. *Nat. Med.* **2019**, *25*, 60–64. [CrossRef]
179. Feighelstein, M.; Shimshoni, I.; Finka, L.R.; Luna, S.P.L.; Mills, D.S.; Zamansky, A. Automated recognition of pain in cats. *Sci. Rep.* **2022**, *12*, 9575. [CrossRef]
180. Herrmann, K. Beyond the 3Rs: Expanding the use of human-relevant replacement methods in biomedical research. *ALTEX* **2019**, *36*, 343–352. [CrossRef]

181. Pound, P.; Ritskes-Hoitinga, M. Is it possible to overcome issues of external validity in preclinical animal research? Why most animal models are bound to fail. *J. Transl. Med.* **2018**, *16*, 304. [CrossRef] [PubMed]
182. Carvalho, C.; Gaspar, A.; Knight, A.; Vicente, L. Ethical and scientific pitfalls concerning laboratory research with non-human primates, and possible solutions. *Animals* **2018**, *9*, 12. [CrossRef] [PubMed]

Disclaimer/Publisher's Note: The statements, opinions and data contained in all publications are solely those of the individual author(s) and contributor(s) and not of MDPI and/or the editor(s). MDPI and/or the editor(s) disclaim responsibility for any injury to people or property resulting from any ideas, methods, instructions or products referred to in the content.

animals

MDPI

Review

The 3Rs in Experimental Liver Disease

Sebastian Martinez-Lopez [1,2,†], Enrique Angel-Gomis [1,2,†], Elisabet Sanchez-Ardid [3,4], Alberto Pastor-Campos [5], Joanna Picó [1] and Isabel Gomez-Hurtado [1,2,3,*]

1 Instituto ISABIAL, Hospital General Universitario Dr. Balmis, 03010 Alicante, Spain; sebastian.martinez@goumh.umh.es (S.M.-L.); enrique.angel@alu.umh.es (E.A.-G.); joanna_pc@hotmail.com (J.P.)
2 Departamento de Medicina Clínica, Universidad Miguel Hernández, 03550 Sant Joan, Spain
3 CIBERehd, Instituto de Salud Carlos III, 28220 Madrid, Spain; esanchezar@santpau.cat
4 Servicio de Patología Digestiva, Institut de Recerca IIB-Sant Pau, Hospital de Santa Creu i Sant Pau, 08025 Barcelona, Spain
5 Oficina de Investigación Responsable, Universidad Miguel Hernández, 03202 Elche, Spain; albertopastor@umh.es
* Correspondence: gomezhurtado_isa@gva.es; Tel.: +34-965-913-928
† These authors contributed equally to this work.

Simple Summary: This article provides a review of recent studies that explore the application of the 3Rs in experimental animal models of liver disease, comparing different models for a correct search for replacement, refinement, and reduction methods. However, the limitations of each of the replacement techniques are identified, which highlights that although the use of animal models is still necessary, their number can be reduced and adjusted to the scientific question of interest, always taking into account their welfare and using alternative techniques to answer more specific questions.

Abstract: Patients with cirrhosis present multiple physiological and immunological alterations that play a very important role in the development of clinically relevant secondary complications to the disease. Experimentation in animal models is essential to understand the pathogenesis of human diseases and, considering the high prevalence of liver disease worldwide, to understand the pathophysiology of disease progression and the molecular pathways involved, due to the complexity of the liver as an organ and its relationship with the rest of the organism. However, today there is a growing awareness about the sensitivity and suffering of animals, causing opposition to animal research among a minority in society and some scientists, but also about the attention to the welfare of laboratory animals since this has been built into regulations in most nations that conduct animal research. In 1959, Russell and Burch published the book "The Principles of Humane Experimental Technique", proposing that in those experiments where animals were necessary, everything possible should be done to try to replace them with non-sentient alternatives, to reduce to a minimum their number, and to refine experiments that are essential so that they caused the least amount of pain and distress. In this review, a comprehensive summary of the most widely used techniques to replace, reduce, and refine in experimental liver research is offered, to assess the advantages and weaknesses of available experimental liver disease models for researchers who are planning to perform animal studies in the near future.

Keywords: replacement; refinement; reduction; liver; research; disease

Citation: Martinez-Lopez, S.; Angel-Gomis, E.; Sanchez-Ardid, E.; Pastor-Campos, A.; Picó, J.; Gomez-Hurtado, I. The 3Rs in Experimental Liver Disease. *Animals* **2023**, *13*, 2357. https://doi.org/10.3390/ani13142357

Academic Editor: Garikoitz Azkona

Received: 14 June 2023
Revised: 16 July 2023
Accepted: 17 July 2023
Published: 19 July 2023

Copyright: © 2023 by the authors. Licensee MDPI, Basel, Switzerland. This article is an open access article distributed under the terms and conditions of the Creative Commons Attribution (CC BY) license (https://creativecommons.org/licenses/by/4.0/).

1. Introduction

The liver is a very important organ in organic homeostasis, with different functions, such as maintaining plasma glucose and ammonia levels, drug detoxification, bile synthesis, and storage and processing of key nutrients [1]. The hepatic response to insults like alcohol, infections, drugs and toxins, cancer, obesity and metabolic syndrome, genetic diseases, and autoimmune conditions is liver fibrosis, like wound healing [2]. Cirrhosis represents the end-stage of any chronic liver disease when liver acini are substituted by nodules [3] characterized by vasodilatation and generalized hypotension, related to portal hypertension, mainly due to increased intrahepatic resistance [4]. Patients with decompensated liver cirrhosis have a poor prognosis [5] and increased mortality, most often attributed to direct complications resulting from the loss of liver function, portal hypertension, and the development of hepatocellular carcinoma and, as such, has been estimated to account for one million deaths worldwide per year [6].

Experimental animal models have been used extensively to understand the underlying mechanisms of human disease, particularly liver cirrhosis. One of the principal reasons for their use is that animals and humans have many anatomical and biological similarities [7]. Throughout the 20th century, there have been great advances in biomedical sciences (invention of antibiotics, new methods of diagnosis and treatment of diseases, surgical techniques, the improvement of vaccination) thanks mainly to animal models, whose use in recent decades has increased exponentially, saving millions of lives and significantly increasing life expectancy. Currently, in Europe, the most used animal species according to the last report of EU Commission (Commission Staff Working Document) are rodents (mouse: 48.9% and rat: 8.4%), domestic fowl (5.3%), rabbits (4.3%), zebrafish (3.5%), guinea pig (1.4%), amphibians, cephalopods and reptiles (0.5%), dogs, cats, and non-human primate models (0.2%). Nevertheless, rodents are the most used for various reasons: Their high resistance to successive inbreeding, decreasing genetic variability between individual animals, their rapid reproduction rate, their small size and ease of handling, and their low cost in terms of accommodation and maintenance. Besides, mice are the species of choice used for geneti engineering and associated basic research. Notably, it is crucial to acknowledge as well that mice are not humans, and depending on the process under study, key differences between human and murine biology may affect our results [8,9], thus, being aware of them would allow us to better implement the 3Rs in our research (Figure 1).

Animal models of liver disease can be induced by different approaches [10]: (a) Oral or intraperitoneal (IP) administration of chemical compounds causing a direct injury and inflammatory reaction in the hepatocytes (CCl_4 [11], thioacetamide [12], dimethylnitrosamine [13], dioxin [14], sodium arsenate [15], and ethanol [16]), (b) special diet causing non-alcoholic fatty liver disease (NAFLD) and cirrhosis, such as choline-deficient, L-amino acid-defined, methionine-deficient diet [17–19], and high-fat diet [20], (c) surgery, like bile duct ligation (BDL) [21], leading to cholestasis, infiltration of inflammatory cells in the portal area and liver fibrosis, or (d) fibrosis induced by deposition of immune complexes in the portal area and around the central vein area (concanavalin A [22] and xenogenic serum [23]). The resulting liver disease is usually presented as a painless clinical picture, with jaundice (especially in cholestatic models like BDL) and brightly colored urine. Jaundice is easily identifiable in albino strains by yellowing fur or ears, but in pigmented animals (black or agouti strains), it is not always immediate to detect. The color change in urine is also not easy to detect and requires the use of metabolic cages or using white bedding. In models of liver damage, early detection of the appearance of cirrhosis or of any color change in the urine could serve as endpoint criteria, but it is sometimes difficult to ascertain despite daily monitoring of the animals. Additionally, in advanced cirrhosis, ascites may appear in the animal (excessive accumulation of fluid in the abdomen). In most animal models, the onset of ascites is used as the endpoint to prevent animals from experiencing pain, stress, or discomfort [24].

In 1959, Russell and Burch [25] developed what was the first idea of good practice in animal research (that, over time, has been expanded with other Good Research Practice guidelines not specific to animals): The concept of the 3Rs (reduce, replace, and refine) to reduce the number of animals used, to look for an alternative for total or if not partial replacement, and refine the techniques used to minimize the pain or suffering of animals as much as possible. However, scientists have not stopped there, and new proposals have been developed based on Carol Newton's three Ss (3Ss) [26], adapting them to what is called "the Three Cs" (3Cs) [27]: Full science, objective criteria, and culture of care. Nowadays, the pillar of any research that is carried out, even before starting a study, is to find the best way to do it, trying, whenever possible, not to use experimental animals. However, today, studies on experimental animals are still irreplaceable. It seems, however, that researchers are at the beginning of a new stage, and investigators are closer to replacement thanks to industrial development and research in the search for new biomaterials, computational development and its greater accessibility, and the use of artificial intelligence (AI), which will very probably change in a few years the vision of science as the community has seen it up to now. Meanwhile, governments, as well as the research community itself, have committed to work together for a responsible science where those studies that require animals will focus on an Integral Science: Proper scientific methodology, honesty, adherence to regulations; Objective Criteria: Retrospective evaluation harm-benefit analysis (3Rs: reduction, refinement, and replacement); and Culture of Care: bioethics, animal welfare, and responsibility.

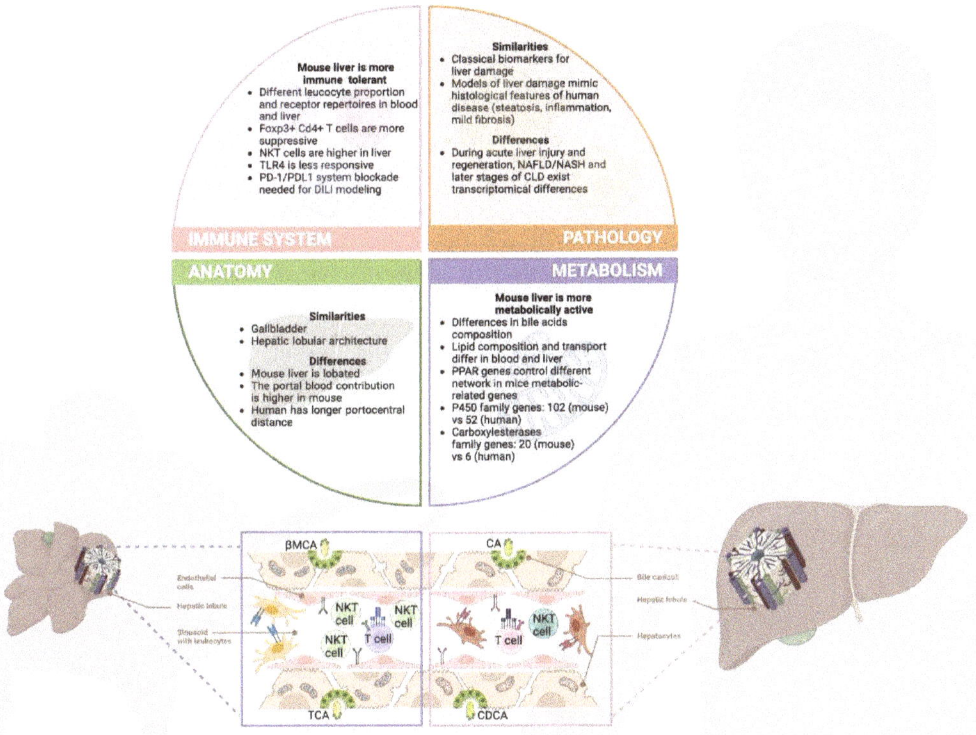

Figure 1. Relevant human and mouse differences in liver physiopathology. Although mice remain one of the best approximations to the human systems, key differences in liver biology need to be considered while designing projects in order to avoid failed experiments and inconclusive results, thus meeting the 3Rs principles. With respect to liver anatomy, albeit the external appearance differs between mice and humans, the functional microscopic lobule architecture is quite conserved in both [28]. This aspect translates into similar phenotypic manifestations during liver pathology (steatosis, inflammation, and fibrosis), although the symptomatology of advanced human disease is difficult to mimic in mice [29]. This might be attributed to the differences in rodents' local immune system and metabolism. Regarding liver immunology, mice are considered more tolerant, with a reduced response to bacterial products and potent regulatory lymphocytes compared to humans [30,31]. Metabolism is considered to be more active in murine models, being exemplified in

the expansion of metabolic enzyme genes and their respective increased production [32,33]. Remarkably, these distinctions are further exemplified in omics analyses that highlight partially divergent genomic responses to liver injury and disease progression [34,35], ultimately indicating that improved models, such as humanized mice, could reduce the gap between the two species [36]. This image was created using Biorender platform.

This review aims to help researchers, mainly those who work with experimental models of liver damage, to be aware that sometimes the animal model can be replaced by alternative techniques and the importance of the 3Rs in the day-to-day of their work with experimental animals.

2. The 3RS from an Institutional Point of View

Advances in scientific research have prompted the search for alternative methods in animal models of liver damage, with the aim of promoting their replacement, reduction, and refinement. Institutionally, university ethics committees and their research compliance offices play a key role in promoting and overseeing these approaches.

European legislation (Directive 2010/63/EU of the European parliament and of the council) is clear in the 3R's approach: When there is a validated alternative method that replaces, reduces, or refines the use of animals, its use is mandatory. Existing validated replacement methods are the main reason for an unfavorable evaluation of an animal research project. However, not only in Europe, in the United States, the Animal Welfare Act (AWA) and its associated regulations [37] serve as the primary framework for overseeing animal research. According to these regulations, research protocols involving animals must undergo review and approval by Institutional Animal Care and Use Committees (IACUCs). The IACUCs play a crucial role in ensuring that the principal investigator has thoroughly considered alternative methods for proposed research activities involving animals. Similarly, in Japan, the Act on Welfare and Management of Animals [38] establishes the framework for animal research oversight. This regulation emphasizes the importance of considering the appropriate use of animals in research and encourages the exploration of alternative methods that minimize the reliance on animal use. It also highlights the goal of minimizing the number of animals involved in research whenever possible.

The replacement of traditional animal models implies the use of non-invasive techniques, such as cell cultures and in vitro models, which allow the mechanisms of liver damage to be studied more accurately and ethically. These in silico and ex vivo alternatives, based on the use of cell lines or isolated tissues, considerably reduce the need for animal experimentation and offer relevant results for understanding liver pathophysiology.

The reduction focuses on minimizing the number of animals used in experiments and optimizing research protocols to obtain the maximum information with the least possible impact. All kinds of institutional ethics review boards for animal research projects and offices for research compliance encourage the implementation of strategies that make it possible to achieve scientific objectives with a smaller number of animals through a rigorous experimental design and the application of appropriate statistical techniques. Current international regulations (European Union, USA, and Japan) do not require a biostatistician to be part of an ethics committee, but the existence of this figure should be mandatory as it is key to correctly applying the r for reduction.

Refinement involves improving experimental conditions to minimize pain, stress, and suffering in animals used in models of liver damage. Ethics committees and research compliance offices promote the adoption of measures that guarantee animal welfare, such as the use of adequate anesthesia, constant monitoring of vital signs, and the implementation of non-invasive sample collection techniques. Animal welfare is a crucial concern and is currently the best attended R worldwide since current legislation does regulate the existence of the attending veterinarian (USA research facilities and their IACUCs) [37], designated veterinarian (European Union breeders, suppliers and users of animals for research) (Directive 2010/63/EU), the person responsible for the welfare and care of the animals (Animal-Welfare Bodies in the European Union) (Directive 2010/63/EU), and the official in Charge of Animal Welfare (Japan) [38].

In summary, research institutions and research compliance offices establish various types of review boards to oversee animal research projects. These boards, with the help of workers with very specific professional profiles, play an essential role in promoting and supervising alternative methods in animal models of liver injury. Through the promotion of replacement, reduction, and refinement, it seeks to guarantee the scientific and ethical integrity of research while protecting animal welfare. These approaches contribute to moving towards a future where the reliance on animal experimentation in this crucial scientific field is significantly reduced and, possibly in the longer term, completely replaced.

3. Animal Models in Liver Disease

The animal species used for animal models of cirrhosis depends on the main objectives of the study [33], but the most commonly used are rodents, especially mice and rats. Mice are mainly used if the primary goal of the study is organ harvesting to assess liver fibrosis and inflammation, and where complex surgical interventions are not required. In mice, due to their small size, surgical interventions are more complicated. Although a high degree of technical expertise is always necessary, in mice, an ocular microscope is sometimes needed and is not always available. Thus, in surgery models, rats are used more often since reproducibility is much higher, and mortality is lower [33].

Within each species, there are also differences between strains. In mice, for example, there are interstrain differences in the development of diet-induced NAFLD between C57BL/6, BALB/c, and C3H/HeN mice [39,40]. There are gender-specific differences too, for example, in the development of steatosis in rats [41]. Male animals are generally used to eliminate potential confounding factors, such as the complex female hormonal status, but it has detrimental consequences for women's health [42], so the legislation ultimately requires detailed experiments performed on both sexes. Therefore, each model has specific advantages and disadvantages and must, therefore, be chosen according to the research questions to be answered.

4. 3RS in Experimental Models of Liver Disease

4.1. Replacement

Since the publication of Rusell and Burch's guide for reduction of animal pain in 1959 [25], the application of less aggressive handling and experimentation techniques to animal research in the form of the 3Rs principles has permeated every aspect of scientific research, including its legislation [43]. These principles have their major (European) legislative acknowledgment in Directive 2010/63/EU, which states that "An experiment shall not be performed if another satisfactory scientific method of obtaining the result sought, not entailing the use of an animal, is reasonably and practicably available".

Owing to the variability regarding etiology and pathophysiology of liver diseases, a plethora of animal models have been developed over the years in order to adequately study these pathologies [29]. In addition, a wide variety of in vitro and in silico approaches have concurrently been engineered with the aim of replacing animal use in hepatic research (Table 1). Those replacement alternatives will be explored in this section.

4.1.1. 2D In Vitro Liver Models: Monocultures and Cocultures

In general, the most widely used method for studying cell biology is the traditional 2D monolayer culture, where isolated cells are seeded in an appropriate culture medium over a flat, stiff polystyrene surface. In the field of liver fibrosis, Hepatic Stellate Cells (HSC) cultures have been usually used for high-throughput screening of fibrosis-related compounds and potential treatments [44,45]. Classic in vitro HSC cultures are based on stellate cells transdifferentiation, transitioning from a quiescent, high-lipid, healthy liver phenotype cells to a myofibroblast-like, activated phenotype, typical of a diseased liver [3,44]. 2D hepatocyte monocultures (commonly hepatocellular carcinoma-derived cell lines) have also been extensively used in different liver pathology studies, such as NAFLD. In these models, steatosis is induced in cultured cells by the addition of a mixture of free fatty acids to the culture medium [46].

However, the aforementioned strategies both fail to recapitulate the complex intercellular interactions characteristic of hepatopathies, the genomic and proteomic changes that cells undergo, and their repercussion in cytokine and extracellular matrix (ECM) production. Despite their utility, classic monocultures are of limited use for studying liver diseases, as these strategies do not consider intercellular interactions between cell types. Bearing that in mind, coculture systems have been used in order to explore communication between liver cells, investigating the intense crosstalk established by means of cytokines, growth factors, chemokines, reactive oxygen species, and plasma proteins, therefore, allowing researchers to measure changes in gene/protein expression and functional/physiological effects [3,44].

Even though this approach is unable to differentiate between the effects of one cell type over the other, it identifies the overall effect of cell interaction. If the experimental design involves coculture of non-adherent and adherent cells, both populations can be cultured together, washing the non-adherent one before sample analysis, thus measuring specific adherent cell changes. Following this strategy, transwell inserts offer the possibility of establishing indirect cocultures, physically separating both cell types but allowing secreted factors exchange. A halfway point between coculture and paracrine cellular interaction pathways involves the stimulation of a cell type with the conditioned medium of another cell culture, containing all soluble factors, exosomes, microparticles, and cytokines produced [47–49].

Collagen overexpression and its accumulation, forming liver scar tissue and fibrous septa, are the most characteristic events of the disease observable in ECM, being a hallmark of the progression of chronic liver disease. ECM proteins and their interaction with the cells condition cellular behaviors, such as cell morphology and gene/protein expression, by means of the activation of multiple signaling pathways [50,51]. Interestingly, research has shown differences in the stiffness of the surface the cells are grown into activate those crucial signaling pathways. For instance, primary HSC cultures seeded on supports of different stiffness adopt different phenotypes depending on it (12 kPa, extremely cirrhotic rat liver; 0.4 kPa, healthy rat liver). These values obviously contrast with the approximately 10,000 kPa stiffness of a regular culture dish, bringing to light the inaccuracy of the most common in vitro supports in which experiments are carried out [45,52]. In vivo, HSCs are subjected to different mechanical/pressure forces depending on their location (portal ducts are stiffer than pericentral regions, etc.) [44,53], thus showing a heterogeneous profile that is very difficult to imitate in vitro. In addition, elegant research has demonstrated that mechanical stress increases TGF-β mRNA and protein expression levels, inducing an epithelial-mesenchymal transition in HSCs [54,55].

Taken together, these studies show that reductionist 2D monoculture does not fully address the complex multicellular processes and mechanical heterogeneity that shape the healthy fibrotic liver, being unable to accurately recapitulate the physiological interactions in which cytokines, nutrients, oxygen, and pressure shape cell behavior.

4.1.2. 3D In Vitro Liver Models

It is clear that highly accurate in vitro liver studies require a precise recreation of the liver microenvironment, taking into account inter-cell interactions, paracrine signaling, and secreted mediators. Hence, 3D in vitro models represent the next logical step towards mimicking in vivo conditions as closely as possible, allowing researchers to construct complex microscale ultra-structures for cultures. When achieved, these 3D strategies can provide a fast, high-throughput, accurate liver culture model for hepatic research [56] (Table 1).

- Cell Stacking: The most conceptually simple strategy for three-dimensional cell engineering uses monolayer cultured cells in the "height" dimension. Cell sheet stacking undertakes this approach by using temperature-responsive culture dishes coated with polymers (poly(*N*-isopropylacrylamide) (PNIPAAm)) that change their hydrophilicity, hence the adherence of the cells, depending on the temperature. This allows the cells to be harvested when the hydrophilic conditions are met (20 °C) without any damage to membrane proteins. The preservation of intercellular interaction allows the stacking of several cell sheets and the reproduction of a more complex 3D structure [57,58]. In hepatic research, the combination of both parenchymal and non-parenchymal liver cell sheets has emulated some liver functions, such as albumin and urea synthesis. However, the absence of oxygen supply to the inner zone of the 3D-engineered tissue due to the lack of vascularization conditions limits cell viability to a short period of time that ends in an ischemic event [59,60].

- Spheroids as a 3D in vitro hepatic system: Spheroids are cellular clusters that can include one or more cell types and that typically acquire a spheroidal structure in the presence or absence of a matrix, recapitulating the hepatic in vivo 3D environment, being the 3D spheroid system applied to hepatocyte culture one of the first models of this technique developed [61]. Hepatocyte spheroids promote intercellular and ECM–cell interactions, the development of metabolites, gas, and nutrient gradients across the spheric-like structure and have been demonstrated to imitate and maintain liver-specific functions [62]. This approach has probed its utility in the study of drug pharmacokinetics and toxicity tests, making feasible its application for the determination of an appropriate drug dose [63]. Current hepatic 3D spheroid-assembling methods have evolved [64,65], giving rise to a wide variety of protocols, being some of the most used the micromolds of non-adherent materials [66], rotation of cultured cells on 96-well non-adherent plates [61], stirred bioreactors [67], and methods that depend on gravitational aggregation of cells using hanging drop cultures. The availability of commercial solutions for hanging drop culture systems (ED Biomatrix. Ann Arbor, MI, USA; InSphero. Schlieren, Switzerland) has eased an initial standardization strategy for the technique [68]. A variable that clearly limits the standardization of spheroids includes their non-homogeneous size, which directly influences experiments' reproducibility [63]. Several technologies have investigated the generation of standard homogeneous spheroids in order to overcome this limitation by means of micromold-based spheroids [69], or by inoculating primary hepatocytes into pre-cultured 3D spheroidal nylon-scaffolds of non-parenchymal cells producing size-consistent particles with hepato-specific functions [70]. Even though there are some commercial solutions for spheroid assembly, their cost and difficulty make them unattractive for their implantation in most laboratories. On top of that, the clustering systems previously defined share several common limitations with monolayer cultures, such as the absence of a standardized culture scaffold stiffness and topography [71].
- 3D matrix/scaffold-based cultures: The use of 3D scaffolds for 3D culture of hepatic cell populations is a frequently used system just by itself or in combination with other technologies (such as its application in homogeneous spheroid generation explained in the previous section). Culture matrixes can be developed out of synthetic (scaffolds synthetic polymers [72]) or biological materials biological sources (naturally derived hydrogels: Alginates, celluloses, or polyethylene [73]). Synthetic materials are extremely flexible and can be easily fabricated, but their microarchitecture and biomechanical behavior, key properties for proper cell culture, are not adequate. The current strategy to overcome these limits is based on the development of 3D scaffolds from biological ECM-derived materials, such as basement membrane or type-I collagen. These ECM-derived gels are characterized by the preservation of cell–ECM interactions. Despite that, these biomaterials also have several limitations, mainly the heterogeneity of mechanical properties between different batches of gels and the incapacity of resembling physiological liver 3D microarchitecture [74]. In those conditions, cultured cells encapsulate randomly in ECM-derived gel scaffolds, randomly self-organizing into not natural liver microstructures and also not reproducing and/or

maintaining specific liver functions over time due to the absence of tissue-specificity of these biomaterials [44].

- 3D bioprinting of liver tissue: 3D bioprinting includes a group of techniques for the fabrication of three-dimensional, cell-packed, liver-like structures with a determined geometry by the manipulation of a combination of biological materials, biochemicals, and living cells, usually called "bioinks" [74]. Based on their handling, the most common printing technologies can be classified into three different categories: Droplet-based, photocuring, and extrusion-based bioprinting [45,74].
    1. Droplet-based bioprinting also known as inkjet-based bioprinting, was the first 3D bioprinting technique developed. Originally proposed by adapting a common inkjet printer to work with bioink-loaded cartridges, current droplet-based bioprinters use an actuator to deposit droplets of biomaterial onto a substrate to generate a 3D structure [75]. In addition to their simplicity and commercial availability, droplet-based bioprinters still remain the first choice for applications that require the printing of a high accuracy pattern, being able to even generate single-cell droplets of bioink, thus building constructs with a similar resolution to cellular dimension [76]. However, their popularity is being progressively reduced due to the overstress that is generated in the cells by the thermal and piezoelectric actuators that form the printer [77].
    2. Photocuring bioprinting: Currently known as laser-based bioprinting due to the implementation of this lighting technology to treat photo cross-linkable bioinks, this approach can generate highly accurate 3D structures with a resolution of up to 50 μm [45]. Laser-based bioprinting is especially recommended for high-fidelity applications at low resolutions (i.e., vascular network ultrastructure printing [78]). Photocuring bioprinters can effectively mimic the lobular vascular network from a technical, structural point of view. However, the use of this technology strongly limits the variety of bioinks available due to the necessity of photo-crosslinking properties or the use of functionalizing reagents [78]. In combination with their low throughput, the cell-unfriendliness of the method, and the absence of commercially available systems, laser-based bioprinting is not a standardized 3D modeling method yet [45].
    3. Extrusion-based bioprinting: It is based on the extrusion and deposition of a bioink filament on the printing surface [77]. Depending on the mechanical approach by which the biomaterial is extruded, three different subcategories are identified: Pneumatic extrusion (the most widespread, controls the flow of ink and pressure through an air compressor), extrusion with a syringe pump (controls precisely the extruded volume of bioink), and screw extrusion (the least popular due to the enormous stress that screw rotation places on the cells) [74]. Extrusion-based bioprinting is currently the most popular technology for liver tissues. The main reason is that it overcomes the cellular stress and damage generated by droplet-based approaches [77]. As stated before, there is a wide variety of bioprinting technologies, with their respective market approach, features, price, and technology. Conversely, the design, standardization and optimization of tissue-specific bioinks is still a challenging task in this field [79]. The materials used for bioprinting need to be cell-and printing-process compatible and provide the mechanical and functional properties that mimic liver tissue, thus allowing maintenance of hepato-specific cell behavior in long-term cultures [46]. Private companies have developed some commercially available solutions for bioprinting materials, such as Insphero®®, whose extrusion-based 3D spheroids of hepatocytes, HSCs, and Kupffer cells were printed in liver ECM-hydrogels. Another example is the Novogel 2.0 from Organovo®®, which has been used to extrusion-bioprint HSCs/HUVEC cocultures allowing long-term maintenance of cells [45]. Nevertheless, the use of proprietary printing techniques dramatically

increases costs, and the need for a complicated experimental setup restricts its accessibility to the research community.

- Decellularized Liver scaffolds: Decellularization of liver scaffolds is an elegant strategy to overcome the macro- and microstructure-related issues derived from previously mentioned 3D culture techniques, where liver microanatomy could not be fully reproduced. Decellularized liver scaffold aims to provide a 3D structure that has the optimal conditions for hepatic cell culture survival and proliferation while preserving the intrinsic organ vascular network [80]. This approach is based on eliminating parenchymal and non-parenchymal populations of the liver by perfusing the organ with decellularizing solutions (i.e., detergents, delipidating enzymes, and nucleic acid-degrading enzymes) while preserving three-dimensional ECM structure [45]. Decellularized livers can then be used as 3D ECM bio-scaffolds for recellularization and tissue engineering [44]. A key advantage of decellularization-recellularization protocols lies in the preservation of liver-specific ECM macro- and microstructure and crucial bioactive molecules for ECM-cell interactions, thus assuring the necessary stimuli for hepatic cells engraftment, survival, and physiological behavior, consequently allowing to overcome other 3D platforms' limitations [80]. Some drawbacks of these culture models include the fact that decellularizating perfusion protocols must be tissue-, disease-, and species-specific, mainly due to the significant variations of cell populations abundance and proportions and ECM protein composition that characterize healthy and diseased liver and could affect the recapitulation of 3D environment during cell elimination [81]. Another limitation of the strategy is the limited availability of human liver tissue for the optimization of the model (and the technical complication of human liver perfusion, forcing researchers to adopt alternative tissue in vitro incubation methods). Hence, most liver decellularization studies have been performed in rats, even though successful decellularizations have been reported of whole sheep and pig livers [80].
- Precision cut liver slices (PCLSs): PCLSs are a versatile, ex vivo model that maintains complete liver architecture and populations (including infiltrating immune cells), therefore, avoiding the limitations of previous models when it comes to macro- and microarchitecture mimicking, as it is a genuine slice of tissue [82]. Among their main benefits, their reproducibility, low cost, and high overall cell viability should be highlighted, allowing researchers to perform experiments in cultured PCLSs for 5–7 days (although this relatively short time could be considered a drawback of the system) [82]. In addition, the combination of PCLSs with microfluidic devices has boosted the in vitro evaluation of multi-organ interaction and culture technology [45]. Nevertheless, access to fresh human liver tissue (the most valuable for research) is one of the pivotal limitations of PCLS: The scarcity of healthy human liver biopsies in combination with the coordination between surgeons, clinician teams, and researchers strongly limits its use [82]. Another drawback is the lack of biopsies on healthy tissue, forcing researchers to work with distal portions of the diseased liver that are not actually healthy. The same situation happens with diseased tissue, as research usually needs a clearly defined background to proceed and identify interesting outcomes. However, harmed liver biopsies are extracted from patients undergoing different treatments/therapies and might have a combined diseased background with metabolic, alcohol/drug intake, cancer, or genetic components [83]. Bearing that in mind, the main alternative for liver slice obtention are animal organs, which, unfortunately, have shown different proteomic profiles and outcomes, depending on the species [44]. Lastly, a clear disadvantage of PCLSs independently of human or animal sources is that slice preparation inevitably generates two cut-injured surfaces that undergo regeneration processes that result in HSCs activation [45].
- Liver-on-chip: As stated before, liver cells in vivo are put under constant and diverse stimuli. Organs-on-Chip, specifically Liver-on-Chip, are microfluidic devices that in vitro recapitulate the physiological properties of tissue microenvironment. Liver-

on-Chip specifically considers fluid flows and 3D architecture of the hepatic sinusoid, allowing for 3D cell culture with a greater degree of in vivo-like environmental cues, highlighting fluidic shear stress as one of the key features of healthy and diseased liver 3D culture [84,85]. Liver-on-Chip devices can be classified in the following categories: Gravity-driven perfusion platforms, Pump-driven perfusion platforms, and 3D mass culture systems [85].

1. Gravity-driven perfusion platforms: They are the simplest version of these microfluidic devices, but also the ones that less resemble physiological conditions, relying on gravity to perfuse cultured cells. This approach eliminates the necessity of tubing, pump components, and power supply dependence, hence, flow rate, nutrient supply, and shear stress are not constant over time [86]. In this device, the tilting of the chip forces culture media to flow from one reservoir to another through microfluidic channels where liver cells are cocultured. The fluidic conductions are engineered so that the cells experience a unidirectional flow of media (by returning media with different connections to the original reservoir). Despite some new designs that try to avoid flow-related issues (bi-directional, recirculating organ-on-chip), the ease of use and resilience against bubbles make simple, gravity-pumped organs-on-chip a valuable and affordable option [86,87].
2. Pump-driven perfusion platforms: They constitute a higher level of physiological similarity than the previous ones, in which peristaltic/syringe pumps deliver a high enough, consistent flow rate, thus assuring a better grade of nutrient supply and shear stress [88]. These systems allow researchers to design far more complex chips to study specific biological aspects of cells in culture, even combining 2D perfusion systems with a gradient generator to imitate the metabolic zonation of the liver [89] or the development of membrane-based, pump-driven devices for the coculture of primary human hepatocytes with hepatic non-parenchymal cells, successfully resembling albumin and urea secretion in comparison with classic culture techniques [90].
3. 3D mass culture systems: Designed to promote cell proliferation and aggregation into three-dimensional, self- or engineered-organized structures (following similar principles as those explained in the spheroids section of this paper) while putting cells under the flow conditions previously described. One of the best examples in liver disease is the device developed by Lee et al. [91]: A mimetic sinusoid-like structure formed by a wide inner channel for cell culture and an outer channel for culture medium flow. The microfluidic arrangement is connected by perpendicular channels that allow a continuously perfused 3D culture with limited shear stress that highly imitates the physical distribution of sinusoidal endothelium and even LSECs fenestrae. Unfortunately, the cost and complexity of these microfluidic systems, in combination with the absence of popular, commercial options, clearly limit the applicability of the technology for investigation. The vast majority of liver-on-chip and microfluidic organ-on-chip devices are developed by the research groups themselves, compromising their reproducibility [46]. In addition, the complex and diverse etiology of most liver diseases, in which not only the liver, but many other players, such as adipose tissue and gut, take part, bring the necessity of studying the interactions between multiple organs in these already complex devices. Even though multi-organ systems and body-on-chip approaches are starting to be a reality, these highly complicated microfluidic culture strategies are still more a field under research rather than a real, affordable, and accessible alternative to non-specialized research teams [92].

- Organoids: Organoids, as defined by Lancaster and Knoblich in 2014 [93] and Huch and Koo in 2015 [94], are an "in vitro 3D cellular cluster derived from tissue-resident stem/progenitor cells, embryonic stem cells, or induced pluripotential stem cells capable of self-renewal and self-organization that recapitulate the functionality of the tissue of origin." From this general definition, liver organoids arise as three-dimensional culture systems developed by embedding hepatic stem, progenitor, or tissue resident cells in hydrogel matrixes that resemble ECM properties [46]. The beginning of organoid culture starts with the isolation of the aforementioned stem/progenitor/pluripotent cells from embryonic or adult tissues, which then are cultured and stimulated in media with growth factors and support matrixes that imitate endogenous, physiological signals that give rise to liver tissue during embryonic patterning [95]. Organoids completely outplay 2D and even some 3D culture models (Table 1) when it comes to (patho)physiological modeling, cell differentiation, interaction, and migration. In addition, their high genomic stability and easy culture expansion make them a highly suitable model for long-term storage and high-throughput compound screening [96]. Even though many researchers have defined specific protocols and schemes for organoid generation, this technology is still extremely proprietary. The long, time-consuming protocols and particular growth factors, stimulant agents, and molecule cocktails required for each step of organoid generation keep this 3D culture method as a not widespread option [97]. However, no matter the strategy followed, the process of liver organoid formation (which obviously mimics physiological liver organogenesis) considers three key factors: Scaffolding/matrix material, signaling for cell differentiation, and starting cell types. Firstly, ECM stiffness, origin (biological or artificial), and composition have a great impact on the cell behavior and its metabolism, as previously commented. In organoid generation protocol, commercial Matrigel (a biological ECM extracted from Engelberth–Holm–Swarm (EHS) mouse sarcoma) is the most popular scaffolding material for liver organoids [98]. Although biological-source hydrogels better imitate the structure and microstructure of liver ECM while also being naturally embedded with a milieu of proteins, growth factors, cytokines, and signaling molecules that biologically prime organoids' cells, their lack of a defined composition and consistency limit the reproducibility of assays and large-scale, commercial production [99]. Furthermore, the murine origin of matrigel restricts its applicability to clinical or human-based research processes. Secondly, cell signaling differentiation stimuli must be carefully implemented in culture media to prime liver development pathways and self-organization. Finally, the starting cell types play a pivotal role in the whole liver organoid formation process by directly influencing their evolution. Monoculture or coculture of different starting cells (primary/adult stem cells, embryonic stem cells, induced pluripotent stem cells, etc.) follow different differentiation paths, therefore requiring specific treatment [100]. Benefits notwithstanding, liver organoids remain an uncommon culture strategy for basic research groups. The time-consuming, low efficiency, and poor control of morphology and composition, in combination with the tendency to express immature fetal markers or differentiated cells, are obstacles that basic researchers have to overcome. The scarcity of primary liver tissue that conditions organoid development to immortalized or tumoral cell lines, a lack of exposure of liver organoids to gut microbiome, and the absence of optimized differentiation protocol on the industrial scale are the main issues that keep this technology as a promising but not yet applicable technology at mass, commercial production, or clinical applications [100,101].

4.1.3. Computational Models

Computational models and in silico approaches have acquired great importance along with the evolution of computers and the rising relevance of data science, which allows researchers to investigate large databases in a high-throughput, fast manner. In silico strategies have been developed to study food–drug interactions, liver metabolism modeling, and fibrotic process simulation that recapitulate collagen deposition and inflammatory processes in central venous regions [102–104]. However, Drug-Induced Liver Injury (DILI) research stands out as one of the most benefitted areas when it comes to computational models. Chemical transforming and clearing via first-pass metabolism and detoxification reactions is the main task of the liver. Due to these metabolic functions, the withdrawal of many pharmaceutical products is related to the modification of chemical species carried out by the aforementioned hepatic metabolism. Therefore, it is easy to understand the broad interest and potential in developing in silico tools for preclinical and even basic computational testing of hepatotoxic effects, which allow researchers to avoid in vitro and in vivo unsuccessful experiments [105].

Different approaches have been designed to mathematically reproduce liver physiology and its modifications by xenobiotic interactions, disease progression, and other pathologies. Knowledge-based prediction algorithms are computer programs that can be trained with input and output data to determine the probability of a biological outcome or event. Then, by integrating real input cases, properly trained knowledge-based prediction algorithms can predict potential endpoints of hepatotoxicity or liver damage by preliminary properties of certain molecules, drugs, or substances in development [106]. A similar strategy is used by cheminformatics-based models. In this case, in silico models benefit from the structure–function relationships established for chemical and biological molecules, in a procedure named Quantitative Structure-Activity Relationship (QSAR). QSAR models relate molecular descriptors (molecular weight, number of carbon atoms, fingerprints, molecular patterns, etc.) with their biological activity and then determine possible outcomes [107,108]. Bioactivity-based models can improve structure-based technology by integrating biological data, such as gene expression profiles, biomarkers (mitochondrial damage, oxidative stress, etc.), or toxicologic information, to better predict the required endpoints [109,110]. Alternatively, expert knowledge approaches can use QSAR methods and code explicit decision rules that identify molecular patterns as related with a determined outcome/pathology [111].

Despite the fast improvement experienced by computational algorithms and models in recent times, their major drawback is the heavy dependence on the quality of training data. Even with the invaluable help of expert hepatologists and the implementation of explicit decision rules in models based on their knowledge, bigger, high-quality datasets (obtained from in vitro or in vivo experiments) are required for algorithm training [112]. In addition, the intricate and complex metabolic and biological pathways involved in liver physiology and the relatively limited knowledge of many of them keep computational models as a still-under-development technology [113].

Table 1. Comparison of advantages and disadvantages of popular replacement strategies in hepatic research.

	Advantages	Disadvantages	Key References
2D Mono-/cocultures	Easy, reproducible handling High throughput	Intercellular interactions Macro-, microenvironment Mechanical heterogeneity of liver Reductionist view of disease	[44]
Cell stacking	3D environment Preservation of intercellular proteins	Short incubation time Absence of oxygen supply Ischemic events in long-term cultures	[59,60]
Spheroids	Intercellular and ECM interactions Metabolites, gas and nutrient gradients across spheroid	Non-standardized size Difficult reproducibility of experiments Elevated price	[61–63]
3D scaffold-based cultures	Allows better spheroid standardization ECM-cell interactions	Heterogeneous mechanical properties between scaffold batches Possibility of random, non-liver organizations of cells	[72,74]
3D Bioprinting	High versatility thanks to multiple bioprinting techniques	High cost of proprietary reactives Lack of ECM interactions	[45,75,78]

Table 1. *Cont.*

	Advantages	Disadvantages	Key References
Decellularized liver scaffolds	Maintenance of liver macro- and microstructure ECM interactions Preservation of organ vascular network	Highly specific decellularization protocols with difficult reproducibility Limited source of human liver tissue	[75,80,81]
Precision cut liver slices	Complete maintenance of liver architecture and cell populations High reproducibility Low cost Combination with other technologies (e.g., liver-on-chip)	Scarring process at cut-injured surfaces Short culture time Limited source of human liver tissue Lack of healthy liver samples Precise coordination between clinicians and researchers required	[82]
Liver-on-chip	Good recapitulation of physiological conditions Shear stress, nutrients supply and waste clearance Metabolic zonation emulation	High cost of technology Absence of commercial/popular solutions Limited reproducibility Restricted diffusion due to complex experimental design and infrastructure	[85–91]
Organoids	Excellent (patho)physiological modelling Good cell differentiation, interaction and migration Culture expansion and high genomic stability	Non-standardized generation protocols Long, time consuming engineering process Highly complex experimental design and infrastructure Matrix-related limitations (murine Matrigel, batch effect, etc.)	[46,96,97,100,101]
Computational models	Exceptionally high throughput Technology under constant improvement Positive benefit/cost ratio Excellent coupling with further validations	Limited to existent physiological/expert knowledge Requires in vitro/in vivo experiments for validation Heavy dependence on training data quality	[102,104,106,111]

4.2. Refinement

Refinement in animal research refers to the process of changing experimental procedures or husbandry practices in order to minimize any potential for pain, suffering, distress, or harm to the animals involved. Since its original definition by Russell and Burch [25], the concept of refinement has evolved, not only including the time in which animals are being used but also focusing on alleviating any other adverse effects that animals may experience during their lifetime [114]. In that sense, the collective effort from researchers and authorities nowadays is directed towards reducing discomfort in animals as well as improving their quality of life, while maintaining the scientific rigor in experimental procedures.

In liver disease research, many aspects of the steps involved are susceptible to refinement. From less stressful handling animal techniques to modern methods of non-invasive imaging, several improvements have emerged to fulfill this purpose:

4.2.1. Animal Husbandry and Monitoring

In recent years, there has been a growing recognition of the specific environmental needs of the different species used in research, as it can affect the wellbeing of the animals and the reproducibility of the results [115,116]. In the specific case of mice, as they are considered social animals, they should be housed in stable groups (European Commission Recommendation 2007/526/EC), which, in principle, would allow the animals to express their natural behavior. It is also true that group housing benefits humans for practical and economic reasons, given our needs for standardization and ease of sanitization [116]. In that sense, male housing is one of the aspects of husbandry still subjected to fine-tuning nowadays [117,118]. The social structures which free-living mice exhibit most typically in nature (one male, several females, and their offspring) cannot be replicated in the laboratory for practical reasons [119,120]. As such, males housed in groups display territorial behaviors, giving rise to social stress, violence, pain, or even death [118]. When this aggression happens, it may induce variability in the results and the appearance of artifacts detrimental to chronic liver disease research, such as wound-associated infections [121]. Spontaneous bacterial peritonitis, for example, is a characteristic and severe infection in the progression of advanced liver disease which constitutes a subject of study on its own [11,122]. Thus, keeping mice free from other infectious diseases would benefit the scientific validity of the results in this area.

It is also worth noting that lack of social interaction in the alternative of single housing for males could be as detrimental as the distress generated by the common aggressive behaviors displayed when caged in groups [117,123–125]. In consequence, it is crucial to assess which housing method is more suited to the needs of our experiments while having the welfare of the animals as a priority. Depending on the strain or the model investigators are working with, for instance, they could consider individual housing if avoiding the injurious fighting in aggressive strains [126–128] outweighs the stress associated with the lack of social interactions. Conversely, if it is needed to avoid the effects of isolation, reducing the animals per cage [129,130], as well as forming groups from littermates when possible [131], have been suggested as methods to ameliorate aggression. Other feasible strategies to implement are related to housing management. It is known that cage cleaning, although essential for the health of the animals, disrupts odor cues involved in the establishment of the social hierarchy of the group, promoting, as a consequence, aggression [132,133]. Accordingly, transferring nesting material from the old cage to the new one has shown to reduce aggression and stress [133]. Moreover, it is widely recognized that other adverse stimuli during handling, such as mouse identification by ear-based methods like ear-notching, can be painful and aversive for mice [134]. Although reliable alternatives, like tail tattooing, have emerged, ear-notching remains one of the most used methods due to its simplicity, permanence, and the possibility of using the removed tissue for genotyping [135]. Additionally, there is still controversy regarding the suitability and welfare cost of both, being tail tattooing also stressful and painful to mice [135,136]. Alternatively, if identification is meant for short-term purposes, using non-toxic fur dyes is considered a refined alternative to the methods discussed [137]. On the other hand, tail-based restraining techniques, albeit a common practice, are considered aversive as well, making the alternatives of cupping and tunnel handling [138] a great refinement method to reduce aggression [139]. As a matter of fact, cupping and tunnel handling have been shown to ameliorate the behavioral signs of stress compared to tail-based methods, even when full restraining is unavoidable for the procedure [138,140]. In the model of CCL_4-induced cirrhosis [141], for example, an adequate immobilization of the animal is crucial during oral gavage of the compound to avoid the complications associated with this dosing route [142]. Thus, efficient non-aversive handling [143], together with the refined technique of three-finger scruffing during restraining, would greatly improve the welfare of the animals throughout gavage-based protocols [140,144].

Another successful method that has been shown to improve animals' quality of life is environmental enrichment [117,145,146]. Although its definition is still not clearly defined, it can be referred to as any practice that provides the animal with means to express, at least partially, natural behaviors, such as exploring and foraging, thereby promoting their wellness [147]. The basic approach, now contemplated explicitly by the current codes of practice (2007/526/EC), usually consists in the addition of nesting material to the cage, as well as material meant for hiding and climbing [148]. Conversely, researchers should be cautious with additional increases in the complexity of the enrichment as it has shown mixed results [147]. Running wheels, for instance, have been proven to increase competitive behavior leading to aggression and stress [149,150]. As valuable items are susceptible to monopolization, especially under aggressive strains, complex structural enrichment should be tailored to each project individually.

To assess whether the measures mentioned above have an impact on the well-being of laboratory animals, several tools have emerged in order to monitor their health. Simple 'Cage side' behavioral indicators have been suggested as a helpful guide to examine their general condition routinely [118]. How mice build their nest and sleep or whether they are feeding enough can give us important hints about their comfort levels. When it comes to pain, more than a decade ago, the so-called "grimace scales" were proposed as a standardized coding system for mouse facial expressions triggered by noxious stimuli [151]. To this date, the system has been adapted for numerous species [152] and has been widely recognized as a potent tool for fast evaluation of suffering. Although in mice, it has mostly been used for scoring pain retrospectively, it offers the potential to support animal welfare by allowing researchers to intervene more precisely when needed [152]. Regarding anxiety, more sensitive tests, such as the elevated maze test and open field test, can be used to identify this adverse effect in mice when greater levels of stress are expected [153,154]. If further thorough approaches are needed, Home Cage-Monitoring stands out as a non-bias system for long-term monitoring of animals subjected to painful procedures and their response to analgesia [155]. While its implementation may pose some practical challenges, this powerful tool has been recommended as a complement to standard behavioral tests and to refine our management of pain, as well as to facilitate the implementation of humane endpoints [155,156]. Interestingly, the latter have been historically interpreted as a time point to perform euthanasia [157]. While it may be correct in many cases, recent research has focused on refining the concept itself by introducing non-euthanasia interventions when the endpoint is met [158]. An extended period with analgesia or facilitating access to food and water for mice after a painful and invalidating procedure would prevent further decline in their state, avoiding premature euthanasia and the use of additional animals [158]. If death is finally unavoidable, refined methods have also emerged as alternatives to sacrifice by carbon dioxide inhalation [159]. Despite being the most commonly used practice these days, doubts about its humanness still linger, making the options of anesthetic overdose or anesthetizing prior to cervical dislocation preferable in order to avoid additional distress in the animals' final moments [157].

4.2.2. Refinement in Liver Research

Once an animal is included in a specific protocol of liver research, there are still several aspects to improve from their experience welfare-wise. As mentioned, it is undeniable that any in-vivo experimentation on complex and severe diseases would prompt adverse effects on animals ranging from mild discomfort to long-lasting and intense pain, depending on the model. Surprisingly, in the case of mice and rats, this aspect is considered to be insufficiently explored and poorly tackled [160,161]. Several circumstances are thought to contribute to this phenomenon: Insufficient evidence-based information to guide effective treatments, concerns about unwanted interaction of analgesics in the experimental outcomes or lack of specifications, and underreporting analgesia in projects and published articles stand out among others [161]. Although it may not be realistic to expect that all pain is going to be successfully managed, simple strategies, such as anticipation of pain, multimodal analgesia, or a close evaluation of suffering and response to therapy post-intervention, as mentioned, should significantly foster welfare [160,162,163].

- Surgery models: In the BDL model, the surgical model of chronic liver disease, it is needless to say how critical the role of the surgeon is in order to avoid unnecessary damage and the consequent pain during the procedure. Until the second half of the XX century, for instance, severe infections or bleeding were common in animals subjected to this model [164,165]. This was likely caused by the challenges intrinsic to rodent surgery. In contrast to bigger species where there is a dedicated surgical space and the surgeon is usually assisted, rodent surgery is routinely performed alone and outside a sterile environment. Thus, the responsible for performing the surgery must induce, maintain, and recover the animal from anesthesia while ensuring crucial aseptic conditions, especially when operating the abdominal cavity [166,167]. Unfortunately, this rather common setup is still far from the standards for bigger species, indicating an important room for the refinement of surgical procedures in rodents. Luckily, microsurgery has advanced at a rapid pace during the last two decades, optimizing the protocols for BDL in terms of a reduction of vascular structures and bile duct manipulation, improvements in aseptic techniques, and optimization of the postoperative mice, which resulted in an overall reduction of tissue damage [168]. Earlier studies also implemented the use of prophylactic antibiotics to reduce the risk of infection and early mortality [168]. Nevertheless, with the increasing importance of intestinal microbiota in the gut-liver axis [169] and the aforementioned spontaneous infections during cirrhosis progression [122], the application of antibiotics has been limited to highly specific circumstances. Remarkably, nowadays, BDL is considered as a low-complication, highly reproducible model for cholestatic disease in mice, albeit it is skill-intensive [167]. Additional refinements have been made in the model of partial BDL, in which only certain lobes are affected by the cholestasis, implying that the unaffected lobes can be used as internal controls while also reducing systemic distress [170]. Despite these innovations, pain is an integral part of the BDL model. Not only the laparotomy but especially the biliary pressure and the inflammation caused by the ligation of the duct, are a source of significant chronic pain and distress [171]. This pain is usually associated with notable levels of morbidity and mortality [30,172] and is considered traditionally neglected as well [173]. In that sense, local lidocaine-based anesthesia, either topically or injected in combination with bupivacaine, could provide multimodal analgesia at the incision site during laparotomy [174] when used along inhaled anesthetics such as isoflurane [160,172]. Alternatively, the ketamine-xylazine cocktail can be used as IP anesthetic, although tailoring the dose is necessary when applied to mice with chronic liver disease [160,175]. If this cocktail is used, preventive butorphanol or metamizole have been explored as a pain-anticipation strategy, showing positive results in the control with analgesia perioperatively [176]. In order to address further postoperative pain, buprenorphine has been the systemic analgesic of choice since the inception of the BDL model, albeit its insufficient use [173,177]. Interestingly, the multimodal combination of buprenorphine and carprofen outperformed buprenorphine alone controlling pain in postoperative mice [163], which could represent a refined analgesic protocol in BDL. Finally, when external pain mitigation is not enough to alleviate discomfort in this model, given its aggressiveness, early humane endpoints have been suggested to reduce excessive suffering in these mice [178].

- Substance Administration: Regarding the CCL_4-induced chronic liver disease model, it is considered less severe than the BDL model [171] but not exempt from distress nor mortality [144,179]. Notably, depending on the route of administration of the toxin, the progression of liver damage can be more aggressive, leading to key differences between experimental designs, in turn, leaving room for optimization and refinement [141]. Intraperitoneal administration is the most common treatment route for CCL_4 [180] (also for dimethylnitrosamine [13] or dioxin [14]), which, unlike thioacetamide [12], cannot be administered in drinking water due to their chemical and safety attributes [29]. This administration, spanning from 6 to 12 weeks depending on the protocol, renders a reproducible fibrosis that does not fully progress into cirrhosis in most cases [141]. Additionally, the process requires an experienced handler although the injection is considered technically simpler than other methods, in order to avoid organ puncture, bleeding, and infections [142]. Conversely, when advanced stages of the disease are needed, oral gavage CCL_4 represents a viable alternative [181] despite being considered to induce greater rates of mortality [29]. While earlier studies showed molecules, such as isothiocyanate or dihydrocollidine, to be safer when inducing cirrhosis orally [180], recent ones demonstrate that refined dosing protocols, in conjunction with careful handling during gavage, not only reduced mortality rates but also led to a full development of cirrhosis with portal hypertension which is difficult to replicate in rodents [182]. Remarkably, this latter study suggests that misplaced installation of CCL_4 might be responsible for most of the deaths during the protocol, which, in turn, implies that further improvements during oral gavage would benefit the model [182]. Indeed, this route of administration is deemed technically challenging. Besides aspiration, gastrointestinal tract irritation, or even rupture in the worst scenarios, are some of the adverse effects associated with oral gavage [142]. In addition to learning a proper technique, utilizing softer intragastric probes would reduce gastrointestinal irritation [183]. When dealing with struggling individuals, habituation to the process might reduce resistance and stress. As a matter of fact, it has been shown that handling and restraint could be sufficient as an acclimatization protocol replacing the sham gavage treatment [184]. Further strategies to reduce discomfort during the intragastrical administration include the coating of the probes with sweet solutions when the model allows it [185] or, in extreme cases, the use of mild inhaled anesthesia for sensitive or aggressive animals [186].

- Imaging techniques: As mentioned, one of the outcomes of both, the BDL and CCl4 model, is the induction of a fibrotic process [3]. As such, the evaluation of fibrous deposition in the mouse liver plays a central role in these protocols. In order to assess the effect of genetic models or drugs on fibrogenesis, for instance, the liver is usually studied post-mortem with different molecular and histological techniques. Consequently, if the research is aimed at understanding a dynamic process, several time points along the protocol are usually scheduled with their corresponding higher number of mice enrolled [187]. Attempts to implement liver biopsy in mice, the gold standard in humans, have been made, striving for a reduction of the animals used in these cases [188]. Still, the concerns about its safety in rodents and its welfare implications make non-invasive approaches more suitable for following fibrosis progression [189]. Computed tomography [190] and magnetic resonance imaging [191] have been used to study liver processes, such as regeneration or metastasis growth. Nevertheless, the complex machinery required for these methodologies has prevented the spread of its use. Ultrasound-based approaches, on the other hand, are considered simpler, low-cost, and efficient, being widely used in human cirrhosis diagnosis [189]. In particular, transient elastography has been recognized as a potent tool to assess liver stiffness, which is proportional to the grade of fibrosis [192]. In fact, recent advances in the methodology have dealt with the technical difficulties related to the size of the mice, facilitating further longitudinal studies and promoting refinement and reduction in liver fibrosis research [193].
- Blood collection: Another excellent refined system to follow degenerative diseases of the liver is through the use of biomarkers present in blood and other biological fluids [194,195]. Traditional evaluation of liver function is based on the serum levels of hepatic enzymes, such as alanine aminotransferase (ALT), aspartate aminotransferase (AST), or alkaline phosphatase (ALP), as readout of hepatocyte damage. Nevertheless, additional proteins are gaining importance as prognostic markers of survival like Albumin or Liver-type fatty acid binding protein [196]. Moreover, MicroRNAs have been shown to be reliable markers of disease progression, especially miR-122, which not only is liver-specific but also strongly correlates with necroinflammatory activity in the liver and fibrosis progression [197]. Recent strategies based on omic technologies have also contributed to the identification of novel molecular markers fined-tuned to specific liver injury models [198,199], such as Keratin-18 or Glutamate dehydrogenase, ultimately increasing the set of tools available to follow disease progression. This increasing pool of markers is thought to have better sensitivity and specificity compared with the classical ALT and AST, giving rise to the use of combinatory panels of makers, which helps to accurately detect and predict risk during liver disease progression in a less invasive and humane manner [194,200]. When it comes to collecting the blood from which most of liver-relevant biomarkers are assessed, there are also key differ-

ences to keep in mind between the existing approaches. In terms of animal welfare and experimental outcomes, not only the collection technique alters the values of crucial parameters, such as ALT and AST [201], but as they require more manipulation or are more invasive, they also induce greater levels of distress in the animals. Although intracardiac puncture or collection from the vena cava render higher volumes of blood, these are usually employed secondary to other interventions, such as organ harvesting, hence refinement is only applicable to suffering prevention during the procedure through a proper anesthesia and analgesia. On the contrary, with the advances in bioanalytical techniques, novel systems that require as little as 10–50 μL to function have emerged, promoting the development of microsampling techniques and favoring the implementation of longitudinal studies [202]. There is increasing evidence that microsampling techniques provide similar results to traditional approaches [203,204]. As such, they have been employed to perform several extractions from the same mouse, enabling the study of dynamic processes, like pharmacokinetics or toxicokinetics, from fewer mice, fostering reduction along refinement [204,205]. Regarding the welfare of this approach, it is true that taking several samples from one animal could be distressing, nevertheless, skilled handlers with the appropriate route of collection would maintain stress at a minimum [203,205]. For instance, facial vein puncture or sublingual puncture produce lower levels of stress when compared to other techniques like retrobulbar sinus puncture [206,207]. The latter is so aggressive that it requires the use anesthesia, its use is proposed as terminal, and even its use is discouraged in some countries [208]. Other routes that have been shown to induce acceptable levels of distress when applied to serial sampling are the lateral tail vein [205] and the saphenous vein [204]. Although there is still research to be done given the novelty of this field, the use of serial microsampling has a promising niche in the study of DILI [209].

- Genetic models: Finally, in the intersection between reduction and refinement, we can find the use of genetically modified animals as potential tools to refine liver studies ranging from drug metabolism to liver fibrosis, among other pathologies [18,210]. These models, most commonly generated in mice, allow researchers to mimic and study human liver diseases without the need for complex surgeries or special diets in many of the cases, thereby reducing manipulation and the consequent additional distress in the experimental animal. For example, in the BALB/c.*Mdr2*$^{-/-}$ model, mice lack a canicular phospholipid transporter leading to an imbalance of phosphatidylcholine transport to the bile, which causes sclerosing cholangitis and liver injury. These animals develop a spontaneous progressive chronic biliary liver disease [187] that has been used to study biliary fibrosis progression without the need to generate the traditional BDL model through surgery [211]. Other interesting examples can be found for liver fibrosis or NAFLD elsewhere [212,213]. Even though these models have been proven useful as a result of similarities found between human and murine genetics and physiology, it is crucial to acknowledge that the remaining differences still constitute a burden in translational research (Figure 1). In particular, this aspect has affected drug development the most, having over 90% of the molecules that showed positive results after preclinical studies discarded in later stages of the clinical trials because of different levels of toxicity [214]. Remarkably, differences at the level of liver metabolism are at the core of the majority of these incompatibilities (Figure 1). About 50% of the molecules that caused DILI in humans showed no toxicity in the animal models [215]. To tackle this phenomenon, genetically humanized mice, which express human xenobiotic receptors, drug-metabolizing enzymes, and transporters, were first developed with positive albeit limited results [216]. Consequently, in order to mimic better the physiology and microenvironment of the human organ, mice with a chimeric liver, in which the hepatocytes have been replaced by human ones, were established to replace drug metabolisms accurately in vivo [214,216]. Interestingly, these humanized models have been successfully used for the evaluation of novel therapies, substituting more complex organisms like primates [217], or for the study of human liver viral infections [218]. Although the benefits of these models are unparalleled, there are several caveats still to address. The repopulation of the liver by the human hepatocytes is not perfect (>85%), which needs to be considered in the pharmacokinetic studies. Additionally, for the human hepatocytes to survive in the mice, different levels of immunodeficiency, depending on the specific model, must be generated to avoid rejection [18,214]. The implications of this lack of a complete immune response in humanized mice are also to be accounted for when modeling human liver pathologies since innate and adaptive immunity are involved in most of them [18]. Regardless of this pitfall, which could potentially be circumvented with the infusion of human immune cells as well [219], several attempts with promising results have been made to model chronic liver diseases, such as NAFLD [220] or liver fibrosis [221]. Whilst this is an incipient field, and the numerous variables and components of humanized mice make them extremely complex, they are a powerful tool for biomedical research, contributing to the refinement of the models by mimicking human disease better, while contributing to reduction by limiting the number of non-humanized mice used in preclinical studies.

4.3. Reduction

Since 3Rs principles were established in the fifties, the definition of reduction has been gradually settled due to its complexity [43]. Most definitions focus on reduction at the experimental or research project level, but certain reduction strategies may take an indirect approach by altering the research methodology or by modifying factors that are not directly associated with scientific procedures [222]. Currently, definitions from the scientific community globally refer to the reduction in the number of experimental units in the most efficient manner with the purpose of producing relevant and robust results to answer the scientific question, provided that all possible replacement strategies are considered [43,222,223]. However, as it is described in Russel and Burch's book, the main aim of the 3Rs is to diminish or remove inhumanity in the treatment of laboratory animals [25], and therefore, some of the alternative strategies in animal testing cannot be defined separately as one unique principle [224]. In this regard, a balance between lowering the number of animals, acquiring solid evidence, and avoiding unnecessary injury to individuals is urgently needed [225]. Therefore, the avoidance of animal re-use is necessary when the cumulative impact of pain and suffering outweighs the utilization of new animals, based on the EU Directive 86/609/EEC [222].

The reduction principle can be subdivided into three approaches [222]: Intra-experimental Reduction, Supra-experimental Reduction, and Extra-experimental Reduction [25,222]. Intra-experimental reduction is the approach most frequently used, and it focuses on the strategies related with individual experiments in order to elaborate a proper experimental design [25,222]. By contrast, Supra-experimental approach is independent of the experimental research context, and it is related to conditions and settings in which animal experiments are performed and periodically overlap with refinement [222]. Last but not least, the Extra-experimental approach of reduction is not related to animal experiments and consists of studying the advancement of experimental methods on a global scal, achieved by implementing standardized guidelines for harmonization, as well as by introducing novel research and testing strategies [25,222], which frequently include not only replacement strategies but refinement as well. Taking all these approaches into account entails a better comprehension of reduction definition. The current review focuses on reduction strategies separately related to each approach.

- The experimental design: These days, experts in animal research claim for the necessity of including a new 6Rs approach (including robustness, registration, and reporting) since the majority of scientists are mostly concerned merely with animal welfare when the 3Rs principles are applied [226]. In this sense, the UK National Centre for the Replacement, Refinement and Reduction of Animals in Research (NC3Rs) has recently created the new NC3Rs Experimental Design Assistant (EDA) that promotes the new 3Rs principles for scientific value, facilitating both the advancement of robust study protocols and enabling the inclusion of timestamps in the resultant protocols, gaining reproducibility and reducing variance [226]. Consequently, during the preliminary experimental phase, scientists should receive counsel from bio-statistical experts for the purpose of methodological scheduling the experimental design since it is of utmost importance and will define the research outcomes [25,222,224]. It is well-reported that during the nineties, over 60 percent of the published papers exhibited statistical errors and this fact led to an increased number of animals required for answering the scientific hypotheses [224]. Some of the solutions given by animal researchers were not only doing comprehensive research in the literature on the matter but also an improvement of scientists' training regarding high technological statistics and avoiding duplicate tests sharing positive as negative results with the scientific community [222,224]. As for the experimental design, the key elements to consider are: The hypothesis, objectives, types of variables (quantitative or qualitative), controls needed (positive or negative), the nature of the study (exploratory or confirmatory), the effect size of a given treatment, types of errors, the levels of bearable errors, and the statistical tests, among others [223]. Moreover, during experiment execution, blinding and randomization are two more crucial factors to bear in mind [223]. Regarding blinding, since in some instances it is inevitable, some measures, in particular using anonymous codes for identification and changing independent personnel for treatment administration, are suggested so as to reduce researchers' prejudices about an experimental condition [223]. As for randomization, randomized block experimental designs have risen in importance as an alternative to completely randomized designs since they are considered more powerful and convenient and, in comparison, a reduced animal number is required. These designs are focused on reducing systematic errors by creating separated blocks composed of an experimental unit for a given treatment, in which an animal cage, not a single animal, is exposed to a drug combination, dosage, etc., in order to find a treatment effect [223,227]. This concept includes possible different variables between the EU that can affect the statistical outcomes because of possible independent-treatment effects from matched ages and weight, up to refinement conditions, such as the cage position on the rack and stress and treatment duration of each block [223,227]. Taking all of these delimited variables into consideration, in addition, using a proper estimator of the sample size per group of individuals results in powerful statistical evidence with a sufficient number of animals [223]. In order to attain the estimated size of the number of animals, certain resources serve as supportive assistance, including the online assistance organization, the NC3Rs EDA's website, and the widely utilized Gpower 3.1 software of Düsseldorf's University.

- Data quality in the study design: Contemporary approaches to reduction focus on experimental design mostly for creating reproducible and quality data with fewer animals. Considerable apprehension has been raised in recent years regarding reproducibility, as evidenced by reviews, which was able to corroborate results in merely six out of fifty-three preclinical cancer trials [222,228] (Rt. Some of the reasons were included inappropriate cell lines and mouse models [222]. To a lesser extent, some experts have already considered that microbiota by itself can represent a source of variation in terms of immunological response in rodent models, and although, unfortunately, it is inevitable, it is suggested to be considered both in the first steps of the experimental design and analyzing the results in comparison with standardized studies in different laboratories on the same model [229,230]. Another aspect to take into account in the reduction in animal experimentation is the sex as a biological variable (SABV). It is undeniable that for over a century, males have been preferred as experimental animals due to the possible effects of female hormones, and this fact could have potentially introduced bias and compromised the representativeness of the obtained results [231]. This approach can affect the sample size and the reduction principle in different aspects. Firstly, it has been discussed that when sex does not have a clear impact on the treatment, only a small number of additional animals are needed in order to have an adequate number of males and females per treatment. By contrast, the possible interactions of sex with treatment's outcomes exponentially increase the required sample size [231]. In pilot and exploratory studies, the use of a single sex is generally accepted because of a reduction of the sample size since their main purpose is to discover potential effects and not the sex impact. However, large cohort studies could have resulted in a waste of animals and other resources if treatment effects are sex-dependent [231]. In this sense, according to the hypothesis and mouse model, governing authorities should enforce absolute transparency in relation to the inclusion and exclusion of animals, particularly concerning the selection of animal gender [231]. Secondly, the use of a single sex entails a misuse of female littermates in the breeding. In this sense, there exists an evident issue of notable concern in breeding facilities, which often exhibit animals in stock generating a surplus that surpasses ten percent of the overall bred population, and this can be addressed by facilitating information on that stock in repositories aiming to achieve a more optimal balance in the production and consumption of experimental animals [222]. Further aspects to be mindful of are the training and educational level, which are mandatory in the vast majority of European countries when it comes to animal experimentation. Outcomes' variability can be disrupted when accredited but non-experienced researchers perform experiments with animals, creating stress to animals regarding refinement conditions such as handling [222]. Furthermore, pharmaceutical companies incorporate the Good Laboratory Practice (GLP) and Good Manufacturing Practice (GMP), which both contribute to reduction by limiting the occurrence of dubious results and the necessity for re-testing, as high-quality and dependable data are employed, with clearly defined protocols in standard operating procedures (SOPs) [222].

- Use of up-to-date techniques: In contemporary times, novel technologies, as independent approaches within experimental design, have surfaced and gained recognition within the scientific community, aligning with the objective of implementing the reduction principle. During the past two decades, select fields of biomedical research, such as studies from pharmacological companies focusing on toxicity assessments, production, and vaccine testing, were developed in order to decrease the used animal number [222]. Notably, in the context of toxicity testing, the implementation of standardized protocols based on reversed toxicology has facilitated the direct evaluation of chemical exposure levels, deviating from the conventional four-phase approach. Consequently, this paradigm shift has led to a diminished reliance on animal models. Another notable example pertains to the meticulous control of vaccine batch production, wherein emphasis is placed on biochemical and physicochemical tests instead of evaluating the final product. Noteworthy historical instances include the replacement of a considerable amount of animals with a unique egg yolk for polyclonal antibody production and the utilization of recombinant DNA in hormone production, particularly during quality assessment procedures [222]. Additionally, due to the high social pressure on replacement of experimental animal models, the cosmetic industry has largely replaced their usage for in vitro techniques in mutagenicity, phototoxicity, and skin corrosion [232]. In vitro techniques related with the replacement are specified in the replacement section. Some of the most remarkable novelty techniques related to reduction are the use of longitudinal imaging and multi-omics technology, which boost the information collected per animal [193,233,234]. Another innovative method related with genetic expression is the well-known CRISPR-Cas9 system, which although not yet standardized in practice, could prevent the creation of undesired gender from breeding in the near future [234].
- Reduction principle in liver research: First and foremost, precisely selecting a liver damage model according to the scientific question is imperative for researchers in order to fulfill the appropriate study design and reduce the number of animals with utmost

effort. Additionally, this choice must be thoroughly chosen alongside a plenitude of factors, such as rodent strain and gender since the abundant diversity regarding clinical and histopathological features of the disease is already acknowledged [29]. Concerning liver injury models in rodents, over the years, as it has been described, the scientific community has specifically tailored different models in accordance with the stages of hepatic damage [11]. Nevertheless, although they are handy for disease study, some of these models, such as ALD models, cannot reproduce the features of human disease. What is more, NAFLD/NASH models tend to recreate the disease, but they are not equally reproducible and count with a high disparity between them (methionine- and choline-deficient diet, high fat diet, Western diet, etc., because of the heterogenous diet composition, mice strains and gender are usually reported [11] (Figure 1).

As for gender's part, it is universally accepted to use males in experimental NAFLD models since more severe liver histology changes have been appreciated in HFD and MCD diet in comparison with females [11,235]. Currently, the fact that female and male livers are metabolically different due to related genes with sex-specific effects on hepatic metabolism and because of the protective role of estrogen in liver damage has been addressed [235,236]. For instance, there is substantial evidence related to the innate immune system that demonstrates that innate immune cells (Kupffer cells and liver neutrophils) from males lead to the promotion of liver inflammation and fibrosis, meanwhile macrophages from females exhibit a tolerant profile by anti-fibrotic tendencies [235]. In fact, not only is it generally accepted that males are more susceptible to liver tumor development than females, but also a higher tendency of NAFLD-HCC in men can be noticed, whose effect can be justified by IL-6 production of Kupffer cells and the inhibitory role of estrogens [235]. Nonetheless, it has been reported that female C57BL/6 mice subjected to a high-fructose diet exhibited comparable liver steatosis to males despite experiencing greater hepatic inflammation, and what is more, they tend to lose this protective effect upon undergoing ovariectomy [236]. In this sense, it has also been reported that during a menopausal stage, both female animals and women present similar clinical outcomes in liver disease [236]. Taking all this into consideration, the majority of studies accept sex and age as independent variables. However, researchers opened a heated debate in this sense: Therapeutic targets and treatment responses should be evaluated upon hormone effects in preclinical to epidemiological studies and clinical trials in the female population to accomplish personalized medicine [236]. This debate has the potential to result in the inclusion of a larger number of animals initially but fewer in the long run, as the effectiveness of the drug should also be tested in female patients during clinical trials.

As for advanced chronic liver disease (CLD) models, it is undeniable that CCl4 has emerged as a widely utilized experimental model for liver fibrosis induction [11,141]. This model might be believed to be standardized, in spite of that, the reality is quite different since nearly all laboratories show protocols with variations in terms of length of treatment, doses, and administration methods [141]. This lack of standardization of SOPs affects the outcome of experimentation and increases the number of animals used.

Therefore, a correct choice of mouse strain to be used is important when it comes to reducing the number of animals used, considering the main objective of our study. In CCl_4 models, BALB/c inbred mice are more susceptible to fibrosis development in comparison with C57BL/6 and FVB/N to an even lesser extent. Despite this fact, C57BL/6 are preferably used because of their availability in genetic modifications [141]. Moreover, fibrogenesis' extent can be modified by the frequency and duration of the treatment. In this sense, a scenario like human stage 3 fibrosis can be reproduced in C57BL/6, with three times per week during four weeks or twice per week during six weeks of CCl_4 IP [141]. However, this administration technique is generally argued because of the lack of fibrosis' advanced stage obtained in C57BL/6 [11]. Additionally, it is highly crucial to establish the endpoint of the experiment after the CCl_4 treatment since immediate proinflammatory states should be observed 24–48 h after the last dose, while for settled fibrosis and cirrhosis states, tissue harvest must be performed after 2–4 and 6–8 weeks, respectively [141]. Recently, some authors have reported a combined model of CCl4 and HFD feeding for the development not only of advanced stages of fibrosis but also of HCC, recreating human disease's stages according to adapted treatment in weeks [11,131]. In this regard, a clear hypothesis upon a proper study design should be engaged with a specific animal model.

5. Conclusions

Experimental disease models are still necessary to be able to understand the pathophysiology of each of them and to be able to investigate the development of new therapeutic targets. In liver diseases, animal models are used to understand the molecular pathways involved and to develop possible treatments in preclinical studies. As has been described, there are numerous complementary and replacement methods to reduce animal testing. Unfortunately, currently, they are still not enough to recapitulate the complexity of the liver as an organ and its relationship with the rest of the organism through the circulatory system. In replacement models, in addition to these limitations, the need for special equipment and the enormous costs must also be considered. Furthermore, there is a wide variety of causes in the field of liver cirrhosis, with molecular pathways involved in its evolution still unknown. Therefore, animal models seem essential to study the clinical entity of cirrhosis until equivalent in vitro methods become available. Moreover, the cellular components used in most of these in vitro methods come from animals, so animal donors would be required in any case, although it is true that researchers would be applying the principle of reduction by reducing the number of animals used. Animal replacement in the field of liver diseases is only the beginning, with the concepts of reduction and refinement of techniques gaining special importance as long as the use of animals is necessary.

Author Contributions: Conceptualization, I.G.-H., S.M.-L. and E.A.-G.; Writing—original draft preparation, S.M.-L., E.A.-G. and J.P.; Writing—review and editing, I.G.-H., S.M.-L., E.A.-G., J.P., A.P.-C. and E.S.-A.; Supervision, I.G.-H. All authors have read and agreed to the published version of the manuscript.

Funding: This research has been partially funded by Ministry of Sciences, Innovation and Universities, Ma-drid, Spain, grant number PID2019-107036RB-I00; and by Institute of Health Carlos III (ISCIII), grant number PMP21/00082 cofunded by European Union—Next GenerationEU and Recovery, Transformation and Resilience Plan.

Institutional Review Board Statement: Not applicable.

Informed Consent Statement: Not applicable.

Data Availability Statement: Not applicable.

Conflicts of Interest: The authors declare no conflict of interest.

References

1. Driskill, J.H.; Pan, D. The Hippo Pathway in Liver Homeostasis and Pathophysiology. *Annu. Rev. Pathol.* **2021**, *16*, 299–322. [CrossRef] [PubMed]
2. Schuppan, D.; Afdhal, N.H. Liver cirrhosis. *Lancet* **2008**, *371*, 838–851. [CrossRef] [PubMed]
3. Yanguas, S.C.; Cogliati, B.; Willebrords, J.; Maes, M.; Colle, I.; van den Bossche, B.; de Oliveira, C.; Andraus, W.; Alves, V.A.F.; Leclercq, I.; et al. Experimental models of liver fibrosis. *Arch. Toxicol.* **2016**, *90*, 1025–1048. [CrossRef] [PubMed]
4. Bosch, J.; Pizcueta, P.; Feu, F.; Fernandez, M.; Garcia-Pagan, J.C. Pathophysiology of portal hypertension. *Gastroenterol. Clin. N. Am.* **1992**, *21*, 1–14. [CrossRef]
5. Bataller, R.; Brenner, D.A. Liver fibrosis. *J. Clin. Investig.* **2005**, *115*, 209–218. [CrossRef]
6. Mokdad, A.A.; Lopez, A.D.; Shahraz, S.; Lozano, R.; Mokdad, A.H.; Stanaway, J.; Murray, C.J.; Naghavi, M. Liver cirrhosis mortality in 187 countries between 1980 and 2010: A systematic analysis. *BMC Med.* **2014**, *12*, 145. [CrossRef]
7. Franco, N.H. Animal Experiments in Biomedical Research: A Historical Perspective. *Animals* **2013**, *3*, 238–273. [CrossRef]
8. The Human Microbiome Project Consortium. Structure, function and diversity of the healthy human microbiome. *Nature* **2012**, *486*, 207–214. [CrossRef]
9. Mestas, J.; Hughes, C.C. Of mice and not men: Differences between mouse and human immunology. *J. Immunol.* **2004**, *172*, 2731–2738. [CrossRef]

10. Zhou, W.C.; Zhang, Q.B.; Qiao, L. Pathogenesis of liver cirrhosis. *World J. Gastroenterol.* **2014**, *20*, 7312–7324. [CrossRef]
11. Runyon, B.A.; Sugano, S.; Kanel, G.; Mellencamp, M.A. A rodent model of cirrhosis, ascites, and bacterial peritonitis. *Gastroenterology* **1991**, *100*, 489–493. [CrossRef] [PubMed]
12. Karantonis, H.C.; Gribilas, G.; Stamoulis, I.; Giaginis, C.; Spiliopoulou, C.; Kouraklis, G.; Demopoulos, C.; Theocharis, S.E. Platelet-activating factor involvement in thioacetamide-induced experimental liver fibrosis and cirrhosis. *Dig. Dis. Sci.* **2010**, *55*, 276–284. [CrossRef] [PubMed]
13. Lee, M.F.; Liu, M.L.; Cheng, A.C.; Tsai, M.L.; Ho, C.T.; Liou, W.S.; Pan, M.H. Pterostilbene inhibits dimethylnitrosamine-induced liver fibrosis in rats. *Food Chem.* **2013**, *138*, 802–807. [CrossRef] [PubMed]
14. Pierre, S.; Chevallier, A.; Teixeira-Clerc, F.; Ambolet-Camoit, A.; Bui, L.C.; Bats, A.S.; Fournet, J.C.; Fernandez-Salguero, P.; Aggerbeck, M.; Lotersztajn, S.; et al. Aryl hydrocarbon receptor-dependent induction of liver fibrosis by dioxin. *Toxicol. Sci.* **2014**, *137*, 114–124. [CrossRef] [PubMed]
15. Wu, J.; Cheng, M.L.; Li, L.; Li, C.X.; Jiang, L.; Zhang, Y.; Ou, B. Model establishment of liver fibrosis in oral arsenic solution exposed mice. *Zhonghua Yi Xue Za Zhi* **2009**, *89*, 1455–1459. [PubMed]
16. Wang, J.Y.; Guo, J.S.; Li, H.; Liu, S.L.; Zern, M.A. Inhibitory effect of glycyrrhizin on NF-kappaB binding activity in CCl4- plus ethanol-induced liver cirrhosis in rats. *Liver* **1998**, *18*, 180–185. [CrossRef]
17. Koca, S.S.; Bahcecioglu, I.H.; Poyrazoglu, O.K.; Ozercan, I.H.; Sahin, K.; Ustundag, B. The treatment with antibody of TNF-alpha reduces the inflammation, necrosis and fibrosis in the non-alcoholic steatohepatitis induced by methionine- and choline-deficient diet. *Inflammation* **2008**, *31*, 91–98. [CrossRef]
18. Kawaguchi, K.; Sakaida, I.; Tsuchiya, M.; Omori, K.; Takami, T.; Okita, K. Pioglitazone prevents hepatic steatosis, fibrosis, and enzyme-altered lesions in rat liver cirrhosis induced by a choline-deficient L-amino acid-defined diet. *Biochem. Biophys. Res. Commun.* **2004**, *315*, 187–195. [CrossRef]
19. Ibanez, P.; Solis, N.; Pizarro, M.; Aguayo, G.; Duarte, I.; Miquel, J.F.; Accatino, L.; Arrese, M. Effect of losartan on early liver fibrosis development in a rat model of nonalcoholic steatohepatitis. *J. Gastroenterol. Hepatol.* **2007**, *22*, 846–851. [CrossRef]
20. Tan, T.C.; Crawford, D.H.; Jaskowski, L.A.; Subramaniam, V.N.; Clouston, A.D.; Crane, D.I.; Bridle, K.R.; Anderson, G.J.; Fletcher, L.M. Excess iron modulates endoplasmic reticulum stress-associated pathways in a mouse model of alcohol and high-fat diet-induced liver injury. *Lab. Investig.* **2013**, *93*, 1295–1312. [CrossRef]
21. Yang, L.; Kwon, J.; Popov, Y.; Gajdos, G.B.; Ordog, T.; Brekken, R.A.; Mukhopadhyay, D.; Schuppan, D.; Bi, Y.; Simonetto, D.; et al. Vascular endothelial growth factor promotes fibrosis resolution and repair in mice. *Gastroenterology* **2014**, *146*, 1339–1350.e1. [CrossRef] [PubMed]
22. Liang, J.; Zhang, B.; Shen, R.W.; Liu, J.B.; Gao, M.H.; Li, Y.; Li, Y.Y.; Zhang, W. Preventive effect of halofuginone on concanavalin A-induced liver fibrosis. *PLoS ONE* **2013**, *8*, e82232. [CrossRef] [PubMed]
23. Wang, B.E. Animals with liver fibrosis induced by albumin immunization. *Zhonghua Yi Xue Za Zhi* **1989**, *69*, 503–505, 536. [PubMed]
24. Königshofer, P.; Brusilovskaya, K.; Schwabl, P.; Reiberger, T. Animal models of portal hypertension. *Biochim. Biophys. Acta (BBA)-Mol. Basis Dis.* **2019**, *1865*, 1019–1030. [CrossRef] [PubMed]
25. Russell, W.M.S.; Burch, R.L. *The Principles of Humane Experimental Technique*; Methuen: London, UK, 1959.
26. Smith, A.J.; Hawkins, P. Good Science, Good Sense and Good Sensibilities: The Three Ss of Carol Newton. *Animals* **2016**, *6*, 70. [CrossRef]
27. Téllez Ballesteros, E.E.; Vanda Cantón, B. Las tres Ces como ampliación de las tres Erres para una praxis en la investigación biomédica. *Rev. Bioética Derecho* **2021**, *51*, 123–139. [CrossRef]
28. Kruepunga, N.; Hakvoort, T.B.M.; Hikspoors, J.; Kohler, S.E.; Lamers, W.H. Anatomy of rodent and human livers: What are the differences? *Biochim. Biophys. Acta Mol. Basis Dis.* **2019**, *1865*, 869–878. [CrossRef]
29. Nevzorova, Y.A.; Boyer-Diaz, Z.; Cubero, F.J.; Gracia-Sancho, J. Animal models for liver disease—A practical approach for translational research. *J. Hepatol.* **2020**, *73*, 423–440. [CrossRef]
30. Zschaler, J.; Schlorke, D.; Arnhold, J. Differences in innate immune response between man and mouse. *Crit. Rev. Immunol.* **2014**, *34*, 433–454. [CrossRef]
31. Warren, H.S.; Fitting, C.; Hoff, E.; Adib-Conquy, M.; Beasley-Topliffe, L.; Tesini, B.; Liang, X.; Valentine, C.; Hellman, J.; Hayden, D.; et al. Resilience to bacterial infection: Difference between species could be due to proteins in serum. *J. Infect. Dis.* **2010**, *201*, 223–232. [CrossRef]
32. Nelson, D.R.; Zeldin, D.C.; Hoffman, S.M.; Maltais, L.J.; Wain, H.M.; Nebert, D.W. Comparison of cytochrome P450 (CYP) genes from the mouse and human genomes, including nomenclature recommendations for genes, pseudogenes and alternative-splice variants. *Pharmacogenetics* **2004**, *14*, 1–18. [CrossRef] [PubMed]
33. Lian, J.; Nelson, R.; Lehner, R. Carboxylesterases in lipid metabolism: From mouse to human. *Protein Cell* **2018**, *9*, 178–195. [CrossRef] [PubMed]
34. Campos, G.; Schmidt-Heck, W.; De Smedt, J.; Widera, A.; Ghallab, A.; Putter, L.; Gonzalez, D.; Edlund, K.; Cadenas, C.; Marchan, R.; et al. Inflammation-associated suppression of metabolic gene networks in acute and chronic liver disease. *Arch. Toxicol.* **2020**, *94*, 205–217. [CrossRef] [PubMed]
35. Jiang, C.; Li, P.; Ruan, X.; Ma, Y.; Kawai, K.; Suemizu, H.; Cao, H. Comparative Transcriptomics Analyses in Livers of Mice, Humans, and Humanized Mice Define Human-Specific Gene Networks. *Cells* **2020**, *9*, 2566. [CrossRef]

36. Luo, Y.; Lu, H.; Peng, D.; Ruan, X.; Eugene Chen, Y.; Guo, Y. Liver-humanized mice: A translational strategy to study metabolic disorders. *J. Cell Physiol.* **2022**, *237*, 489–506. [CrossRef]
37. USDA Animal & Plant Health Inspection Service. *USDA Animal Care: Animal Welfare Act and Animal Welfare Regulations*; USDA APHIS: Riverdale, MD, USA, 2022; p. 265.
38. *Act on Welfare and Management of Animals, Japan*; Ministry of Environment: Tokyo, Japan, 2014.
39. Ihara, M.; Ueno, A.; Tsuda, Y.; Takahashi, H.; Yamazaki, T.; Kudo, M.; Doi, Y.; Murai, O.; Tsukamoto, Y. A case of repeated occlusion in the common iliac artery due to an unexpected stent deformation. *Cardiovasc. Interv. Ther.* **2015**, *30*, 162–167. [CrossRef]
40. Farrell, G.C.; Mridha, A.R.; Yeh, M.M.; Arsov, T.; Van Rooyen, D.M.; Brooling, J.; Nguyen, T.; Heydet, D.; Delghingaro-Augusto, V.; Nolan, C.J.; et al. Strain dependence of diet-induced NASH and liver fibrosis in obese mice is linked to diabetes and inflammatory phenotype. *Liver Int.* **2014**, *34*, 1084–1093. [CrossRef]
41. Stoppeler, S.; Palmes, D.; Fehr, M.; Holzen, J.P.; Zibert, A.; Siaj, R.; Schmidt, H.H.; Spiegel, H.U.; Bahde, R. Gender and strain-specific differences in the development of steatosis in rats. *Lab. Anim.* **2013**, *47*, 43–52. [CrossRef]
42. Plevkova, J.; Brozmanova, M.; Harsanyiova, J.; Sterusky, M.; Honetschlager, J.; Buday, T. Various aspects of sex and gender bias in biomedical research. *Physiol. Res.* **2020**, *69*, S367. [CrossRef]
43. Díaz, L.; Zambrano, E.; Flores, M.E.; Contreras, M.; Crispín, J.C.; Alemán, G.; Bravo, C.; Armenta, A.; Valdés, V.J.; Tovar, A. Ethical considerations in animal research: The principle of 3R's. *Rev. Investig. Clin.* **2021**, *73*, 199–209. [CrossRef]
44. Mazza, G.; Al-Akkad, W.; Rombouts, K. Engineering in vitro models of hepatofibrogenesis. *Adv. Drug Deliv. Rev.* **2017**, *121*, 147–157. [CrossRef] [PubMed]
45. van Grunsven, L.A. 3D in vitro models of liver fibrosis. *Adv. Drug Deliv. Rev.* **2017**, *121*, 133–146. [CrossRef] [PubMed]
46. Ramos, M.J.; Bandiera, L.; Menolascina, F.; Fallowfield, J.A. In vitro models for non-alcoholic fatty liver disease: Emerging platforms and their applications. *iScience* **2022**, *25*, 103549. [CrossRef] [PubMed]
47. Heymann, F.; Tacke, F. Immunology in the liver—From homeostasis to disease. *Nat. Rev. Gastroenterol. Hepatol.* **2016**, *13*, 88–110. [CrossRef]
48. Weiskirchen, R.; Tacke, F. Cellular and molecular functions of hepatic stellate cells in inflammatory responses and liver immunology. *Hepatobiliary Surg. Nutr.* **2014**, *3*, 344–363. [CrossRef]
49. Friedman, S.L.; Arthur, M.J. Activation of cultured rat hepatic lipocytes by Kupffer cell conditioned medium. Direct enhancement of matrix synthesis and stimulation of cell proliferation via induction of platelet-derived growth factor receptors. *J. Clin. Investig.* **1989**, *84*, 1780–1785. [CrossRef]
50. Chen, L.; Brigstock, D.R. Integrins and heparan sulfate proteoglycans on hepatic stellate cells (HSC) are novel receptors for HSC-derived exosomes. *FEBS Lett.* **2016**, *590*, 4263–4274. [CrossRef]
51. Heino, J. The collagen receptor integrins have distinct ligand recognition and signaling functions. *Matrix Biol.* **2000**, *19*, 319–323. [CrossRef]
52. Iwasaki, A.; Sakai, K.; Moriya, K.; Sasaki, T.; Keene, D.R.; Akhtar, R.; Miyazono, T.; Yasumura, S.; Watanabe, M.; Morishita, S.; et al. Molecular Mechanism Responsible for Fibronectin-controlled Alterations in Matrix Stiffness in Advanced Chronic Liver Fibrogenesis. *J. Biol. Chem.* **2016**, *291*, 72–88. [CrossRef]
53. Caliari, S.R.; Perepelyuk, M.; Cosgrove, B.D.; Tsai, S.J.; Lee, G.Y.; Mauck, R.L.; Wells, R.G.; Burdick, J.A. Stiffening hydrogels for investigating the dynamics of hepatic stellate cell mechanotransduction during myofibroblast activation. *Sci. Rep.* **2016**, *6*, 21387. [CrossRef]
54. O'Connor, J.W.; Gomez, E.W. Biomechanics of TGFbeta-induced epithelial-mesenchymal transition: Implications for fibrosis and cancer. *Clin. Transl. Med.* **2014**, *3*, 23. [CrossRef] [PubMed]
55. Zhubanchaliyev, A.; Temirbekuly, A.; Kongrtay, K.; Wanshura, L.C.; Kunz, J. Targeting Mechanotransduction at the Transcriptional Level: YAP and BRD4 Are Novel Therapeutic Targets for the Reversal of Liver Fibrosis. *Front. Pharmacol.* **2016**, *7*, 462. [CrossRef]
56. Fitzgerald, K.A.; Malhotra, M.; Curtin, C.M.; FJ, O.B.; CM, O.D. Life in 3D is never flat: 3D models to optimise drug delivery. *J. Control Release* **2015**, *215*, 39–54. [CrossRef] [PubMed]
57. Sasagawa, T.; Shimizu, T.; Yamato, M.; Okano, T. Endothelial colony-forming cells for preparing prevascular three-dimensional cell-dense tissues using cell-sheet engineering. *J. Tissue Eng. Regen. Med.* **2016**, *10*, 739–747. [CrossRef] [PubMed]
58. Tsuda, Y.; Kikuchi, A.; Yamato, M.; Nakao, A.; Sakurai, Y.; Umezu, M.; Okano, T. The use of patterned dual thermoresponsive surfaces for the collective recovery as co-cultured cell sheets. *Biomaterials* **2005**, *26*, 1885–1893. [CrossRef]
59. Kim, K.; Utoh, R.; Ohashi, K.; Kikuchi, T.; Okano, T. Fabrication of functional 3D hepatic tissues with polarized hepatocytes by stacking endothelial cell sheets in vitro. *J. Tissue Eng. Regen. Med.* **2017**, *11*, 2071–2080. [CrossRef]
60. Kim, K.; Ohashi, K.; Utoh, R.; Kano, K.; Okano, T. Preserved liver-specific functions of hepatocytes in 3D co-culture with endothelial cell sheets. *Biomaterials* **2012**, *33*, 1406–1413. [CrossRef]
61. Cox, C.R.; Lynch, S.; Goldring, C.; Sharma, P. Current Perspective: 3D Spheroid Models Utilizing Human-Based Cells for Investigating Metabolism-Dependent Drug-Induced Liver Injury. *Front. Med. Technol.* **2020**, *2*, 611913. [CrossRef]
62. Donato, M.T.; Gallego-Ferrer, G.; Tolosa, L. In Vitro Models for Studying Chronic Drug-Induced Liver Injury. *Int. J. Mol. Sci.* **2022**, *23*, 11428. [CrossRef]
63. Drewitz, M.; Helbling, M.; Fried, N.; Bieri, M.; Moritz, W.; Lichtenberg, J.; Kelm, J.M. Towards automated production and drug sensitivity testing using scaffold-free spherical tumor microtissues. *Biotechnol. J.* **2011**, *6*, 1488–1496. [CrossRef]

64. Thomas, R.J.; Bennett, A.; Thomson, B.; Shakesheff, K.M. Hepatic stellate cells on poly(DL-lactic acid) surfaces control the formation of 3D hepatocyte co-culture aggregates in vitro. *Eur. Cell Mater.* **2006**, *11*, 16–26; discussion 26. [CrossRef]
65. Wong, S.F.; No, D.Y.; Choi, Y.Y.; Kim, D.S.; Chung, B.G.; Lee, S.H. Concave microwell based size-controllable hepatosphere as a three-dimensional liver tissue model. *Biomaterials* **2011**, *32*, 8087–8096. [CrossRef] [PubMed]
66. Gevaert, E.; Dolle, L.; Billiet, T.; Dubruel, P.; van Grunsven, L.; van Apeldoorn, A.; Cornelissen, R. High throughput micro-well generation of hepatocyte micro-aggregates for tissue engineering. *PLoS ONE* **2014**, *9*, e105171. [CrossRef] [PubMed]
67. Tostoes, R.M.; Leite, S.B.; Serra, M.; Jensen, J.; Bjorquist, P.; Carrondo, M.J.; Brito, C.; Alves, P.M. Human liver cell spheroids in extended perfusion bioreactor culture for repeated-dose drug testing. *Hepatology* **2012**, *55*, 1227–1236. [CrossRef] [PubMed]
68. Foty, R. A simple hanging drop cell culture protocol for generation of 3D spheroids. *J. Vis. Exp.* **2011**, e2720. [CrossRef]
69. Jun, Y.; Kang, A.R.; Lee, J.S.; Jeong, G.S.; Ju, J.; Lee, D.Y.; Lee, S.H. 3D co-culturing model of primary pancreatic islets and hepatocytes in hybrid spheroid to overcome pancreatic cell shortage. *Biomaterials* **2013**, *34*, 3784–3794. [CrossRef]
70. Kostadinova, R.; Boess, F.; Applegate, D.; Suter, L.; Weiser, T.; Singer, T.; Naughton, B.; Roth, A. A long-term three dimensional liver co-culture system for improved prediction of clinically relevant drug-induced hepatotoxicity. *Toxicol. Appl. Pharmacol.* **2013**, *268*, 1–16. [CrossRef]
71. Hakkinen, K.M.; Harunaga, J.S.; Doyle, A.D.; Yamada, K.M. Direct comparisons of the morphology, migration, cell adhesions, and actin cytoskeleton of fibroblasts in four different three-dimensional extracellular matrices. *Tissue Eng. Part. A* **2011**, *17*, 713–724. [CrossRef]
72. Tsang, V.L.; Bhatia, S.N. Fabrication of three-dimensional tissues. *Adv. Biochem. Eng. Biotechnol.* **2007**, *103*, 189–205. [CrossRef]
73. Celiz, A.D.; Smith, J.G.; Langer, R.; Anderson, D.G.; Winkler, D.A.; Barrett, D.A.; Davies, M.C.; Young, L.E.; Denning, C.; Alexander, M.R. Materials for stem cell factories of the future. *Nat. Mater.* **2014**, *13*, 570–579. [CrossRef]
74. Gu, Z.; Fu, J.; Lin, H.; He, Y. Development of 3D bioprinting: From printing methods to biomedical applications. *Asian J. Pharm. Sci.* **2020**, *15*, 529–557. [CrossRef] [PubMed]
75. Rider, P.; Kacarevic, Z.P.; Alkildani, S.; Retnasingh, S.; Schnettler, R.; Barbeck, M. Additive Manufacturing for Guided Bone Regeneration: A Perspective for Alveolar Ridge Augmentation. *Int. J. Mol. Sci.* **2018**, *19*, 3308. [CrossRef] [PubMed]
76. Zhou, X.; Wu, H.; Wen, H.; Zheng, B. Advances in Single-Cell Printing. *Micromachines* **2022**, *13*, 80. [CrossRef] [PubMed]
77. Guagliano, G.; Volpini, C.; Briatico-Vangosa, F.; Cornaglia, A.I.; Visai, L.; Petrini, P. Toward 3D-Bioprinted Models of the Liver to Boost Drug Development. *Macromol. Biosci.* **2022**, *22*, e2200264. [CrossRef] [PubMed]
78. Xu, H.Q.; Liu, J.C.; Zhang, Z.Y.; Xu, C.X. A review on cell damage, viability, and functionality during 3D bioprinting. *Mil. Med. Res.* **2022**, *9*, 70. [CrossRef] [PubMed]
79. Dey, M.; Ozbolat, I.T. 3D bioprinting of cells, tissues and organs. *Sci. Rep.* **2020**, *10*, 14023. [CrossRef]
80. Rossi, E.A.; Quintanilha, L.F.; Nonaka, C.K.V.; Souza, B.S.F. Advances in Hepatic Tissue Bioengineering with Decellularized Liver Bioscaffold. *Stem Cells Int.* **2019**, *2019*, 2693189. [CrossRef]
81. Caralt, M.; Uzarski, J.S.; Iacob, S.; Obergfell, K.P.; Berg, N.; Bijonowski, B.M.; Kiefer, K.M.; Ward, H.H.; Wandinger-Ness, A.; Miller, W.M.; et al. Optimization and critical evaluation of decellularization strategies to develop renal extracellular matrix scaffolds as biological templates for organ engineering and transplantation. *Am. J. Transplant.* **2015**, *15*, 64–75. [CrossRef]
82. Palma, E.; Doornebal, E.J.; Chokshi, S. Precision-cut liver slices: A versatile tool to advance liver research. *Hepatol. Int.* **2019**, *13*, 51–57. [CrossRef]
83. Zimmermann, M.; Armeanu, S.; Smirnow, I.; Kupka, S.; Wagner, S.; Wehrmann, M.; Rots, M.G.; Groothuis, G.M.; Weiss, T.S.; Konigsrainer, A.; et al. Human precision-cut liver tumor slices as a tumor patient-individual predictive test system for oncolytic measles vaccine viruses. *Int. J. Oncol.* **2009**, *34*, 1247–1256.
84. Polidoro, M.A.; Ferrari, E.; Marzorati, S.; Lleo, A.; Rasponi, M. Experimental liver models: From cell culture techniques to microfluidic organs-on-chip. *Liver Int.* **2021**, *41*, 1744–1761. [CrossRef]
85. Dalsbecker, P.; Beck Adiels, C.; Goksor, M. Liver-on-a-chip devices: The pros and cons of complexity. *Am. J. Physiol. Gastrointest. Liver Physiol.* **2022**, *323*, G188–G204. [CrossRef] [PubMed]
86. Esch, M.B.; Ueno, H.; Applegate, D.R.; Shuler, M.L. Modular, pumpless body-on-a-chip platform for the co-culture of GI tract epithelium and 3D primary liver tissue. *Lab. Chip* **2016**, *16*, 2719–2729. [CrossRef]
87. Busek, M.; Aizenshtadt, A.; Koch, T.; Frank, A.; Delon, L.; Martinez, M.A.; Golovin, A.; Dumas, C.; Stokowiec, J.; Gruenzner, S.; et al. Pump-less, recirculating organ-on-a-chip (rOoC) platform. *Lab. Chip* **2023**, *23*, 591–608. [CrossRef]
88. Giobbe, G.G.; Michielin, F.; Luni, C.; Giulitti, S.; Martewicz, S.; Dupont, S.; Floreani, A.; Elvassore, N. Functional differentiation of human pluripotent stem cells on a chip. *Nat. Methods* **2015**, *12*, 637–640. [CrossRef] [PubMed]
89. Kang, Y.B.; Eo, J.; Mert, S.; Yarmush, M.L.; Usta, O.B. Metabolic patterning on a chip: Towards in vitro liver zonation of primary rat and human hepatocytes. *Sci. Rep.* **2018**, *8*, 8951. [CrossRef] [PubMed]
90. Prodanov, L.; Jindal, R.; Bale, S.S.; Hegde, M.; McCarty, W.J.; Golberg, I.; Bhushan, A.; Yarmush, M.L.; Usta, O.B. Long-term maintenance of a microfluidic 3D human liver sinusoid. *Biotechnol. Bioeng.* **2016**, *113*, 241–246. [CrossRef]
91. Lee, P.J.; Hung, P.J.; Lee, L.P. An artificial liver sinusoid with a microfluidic endothelial-like barrier for primary hepatocyte culture. *Biotechnol. Bioeng.* **2007**, *97*, 1340–1346. [CrossRef]
92. Slaughter, V.L.; Rumsey, J.W.; Boone, R.; Malik, D.; Cai, Y.; Sriram, N.N.; Long, C.J.; McAleer, C.W.; Lambert, S.; Shuler, M.L.; et al. Validation of an adipose-liver human-on-a-chip model of NAFLD for preclinical therapeutic efficacy evaluation. *Sci. Rep.* **2021**, *11*, 13159. [CrossRef]

93. Lancaster, M.A.; Knoblich, J.A. Organogenesis in a dish: Modeling development and disease using organoid technologies. *Science* **2014**, *345*, 1247125. [CrossRef]
94. Huch, M.; Koo, B.K. Modeling mouse and human development using organoid cultures. *Development* **2015**, *142*, 3113–3125. [CrossRef]
95. Prior, N.; Inacio, P.; Huch, M. Liver organoids: From basic research to therapeutic applications. *Gut* **2019**, *68*, 2228–2237. [CrossRef]
96. Soret, P.A.; Magusto, J.; Housset, C.; Gautheron, J. In Vitro and In Vivo Models of Non-Alcoholic Fatty Liver Disease: A Critical Appraisal. *J. Clin. Med.* **2020**, *10*, 36. [CrossRef]
97. Tang, X.-Y.; Wu, S.; Wang, D.; Chu, C.; Hong, Y.; Tao, M.; Hu, H.; Xu, M.; Guo, X.; Liu, Y. Human organoids in basic research and clinical applications. *Signal Transduct. Target. Ther.* **2022**, *7*, 168. [CrossRef]
98. Hughes, C.S.; Postovit, L.M.; Lajoie, G.A. Matrigel: A complex protein mixture required for optimal growth of cell culture. *Proteomics* **2010**, *10*, 1886–1890. [CrossRef]
99. Giobbe, G.G.; Crowley, C.; Luni, C.; Campinoti, S.; Khedr, M.; Kretzschmar, K.; De Santis, M.M.; Zambaiti, E.; Michielin, F.; Meran, L.; et al. Extracellular matrix hydrogel derived from decellularized tissues enables endodermal organoid culture. *Nat. Commun.* **2019**, *10*, 5658. [CrossRef]
100. Zhu, X.; Zhang, B.; He, Y.; Bao, J. Liver Organoids: Formation Strategies and Biomedical Applications. *Tissue Eng. Regen. Med.* **2021**, *18*, 573–585. [CrossRef]
101. Harrison, S.P.; Baumgarten, S.F.; Verma, R.; Lunov, O.; Dejneka, A.; Sullivan, G.J. Liver Organoids: Recent Developments, Limitations and Potential. *Front. Med.* **2021**, *8*, 574047. [CrossRef]
102. Cvitanovic Tomas, T.; Moskon, M.; Mraz, M.; Rozman, D. Computational Modelling of Liver Metabolism and its Applications in Research and the Clinics. *Acta Chim. Slov.* **2018**, *65*, 253–265. [CrossRef]
103. Guttman, Y.; Kerem, Z. Computer-Aided (In Silico) Modeling of Cytochrome P450-Mediated Food-Drug Interactions (FDI). *Int. J. Mol. Sci.* **2022**, *23*, 8498. [CrossRef]
104. Yoshizawa, M.; Sugimoto, M.; Tanaka, M.; Sakai, Y.; Nishikawa, M. Computational simulation of liver fibrosis dynamics. *Sci. Rep.* **2022**, *12*, 14112. [CrossRef] [PubMed]
105. Weaver, R.J.; Valentin, J.P. Today's Challenges to De-Risk and Predict Drug Safety in Human "Mind-the-Gap". *Toxicol. Sci.* **2019**, *167*, 307–321. [CrossRef] [PubMed]
106. Przybylak, K.R.; Cronin, M.T. In silico models for drug-induced liver injury—Current status. *Expert. Opin. Drug Metab. Toxicol.* **2012**, *8*, 201–217. [CrossRef] [PubMed]
107. Vall, A.; Sabnis, Y.; Shi, J.; Class, R.; Hochreiter, S.; Klambauer, G. The Promise of AI for DILI Prediction. *Front. Artif. Intell.* **2021**, *4*, 638410. [CrossRef] [PubMed]
108. Choi, R.Y.; Coyner, A.S.; Kalpathy-Cramer, J.; Chiang, M.F.; Campbell, J.P. Introduction to Machine Learning, Neural Networks, and Deep Learning. *Transl. Vis. Sci. Technol.* **2020**, *9*, 14.
109. Lamb, J. The Connectivity Map: A new tool for biomedical research. *Nat. Rev. Cancer* **2007**, *7*, 54–60. [CrossRef]
110. Zhu, X.W.; Sedykh, A.; Liu, S.S. Hybrid in silico models for drug-induced liver injury using chemical descriptors and in vitro cell-imaging information. *J. Appl. Toxicol.* **2014**, *34*, 281–288. [CrossRef]
111. Lin, J.; Li, M.; Mak, W.; Shi, Y.; Zhu, X.; Tang, Z.; He, Q.; Xiang, X. Applications of In Silico Models to Predict Drug-Induced Liver Injury. *Toxics* **2022**, *10*, 788. [CrossRef]
112. Poloznikov, A.; Gazaryan, I.; Shkurnikov, M.; Nikulin, S.; Drapkina, O.; Baranova, A.; Tonevitsky, A. In vitro and in silico liver models: Current trends, challenges and opportunities. *ALTEX* **2018**, *35*, 397–412. [CrossRef]
113. Ellison, C.; Hewitt, M.; Przybylak, K. In Silico Models for Hepatotoxicity. *Methods Mol. Biol.* **2022**, *2425*, 355–392. [CrossRef]
114. Buchanan-Smith, H.M.; Rennie, A.; Vitale, A.; Pollo, S.; Prescott, M.J.; Morton, D.B. Harmonising the definition of refinement. *Anim. Welf.* **2005**, *14*, 379–384. [CrossRef]
115. Baumans, V.; Van Loo, P.L.P. How to improve housing conditions of laboratory animals: The possibilities of environmental refinement. *Vet. J.* **2013**, *195*, 24–32. [CrossRef] [PubMed]
116. Baumans, V. Science-based assessment of animal welfare: Laboratory animals. *Rev. Sci. Tech. (Off. Int. Des Epizoot)* **2005**, *24*, 503–513. [CrossRef]
117. Van Loo, P.L.; Van Zutphen, L.F.; Baumans, V. Male management: Coping with aggression problems in male laboratory mice. *Lab. Anim.* **2003**, *37*, 300–313. [CrossRef]
118. Kappel, S.; Hawkins, P.; Mendl, M.T. To Group or Not to Group? Good Practice for Housing Male Laboratory Mice. *Animals* **2017**, *7*, 88. [CrossRef]
119. Pocock, M.J.; Searle, J.B.; White, P.C.L. Adaptations of animals to commensal habitats: Population dynamics of house mice Mus musculus domesticus on farms. *J. Anim. Ecol.* **2004**, *73*, 878–888. [CrossRef]
120. Latham, N.; Mason, G. From house mouse to mouse house: The behavioural biology of free-living Mus musculus and its implications in the laboratory. *Appl. Anim. Behav. Sci.* **2004**, *86*, 261–289. [CrossRef]
121. Fahey, J.R.; Olekszak, H. An Overview of Typical Infections of Research Mice: Health Monitoring and Prevention of Infection. *Curr. Protoc. Mouse Biol.* **2015**, *5*, 235–245. [CrossRef]
122. Sanchez, E.; Such, J.; Chiva, M.T.; Soriano, G.; Llovet, T.; Merce, J.; Sancho, F.; Munoz, C.; Song, X.Y.; Perez-Mateo, M.; et al. Development of an experimental model of induced bacterial peritonitis in cirrhotic rats with or without ascites. *Am. J. Gastroenterol.* **2007**, *102*, 1230–1236. [CrossRef]

123. Valzelli, L.J.P. The "isolation syndrome" in mice. *Psychopharmacologia* **1973**, *31*, 305–320. [CrossRef]
124. Kudryavtseva, N.N.; Bakshtanovskaya, I.V.; Koryakina, L.A. Social model of depression in mice of C57BL/6J strain. *Pharmacol. Biochem. Behav.* **1991**, *38*, 315–320. [CrossRef] [PubMed]
125. Chourbaji, S.; Zacher, C.; Sanchis-Segura, C.; Spanagel, R.; Gass, P. Social and structural housing conditions influence the development of a depressive-like phenotype in the learned helplessness paradigm in male mice. *Behav. Brain Res.* **2005**, *164*, 100–106. [CrossRef] [PubMed]
126. Guillot, P.V.; Chapouthier, G. Intermale aggression and dark/light preference in ten inbred mouse strains. *Behav. Brain Res.* **1996**, *77*, 211–213. [CrossRef] [PubMed]
127. Lewejohann, L.; Kloke, V.; Heiming, R.S.; Jansen, F.; Kaiser, S.; Schmitt, A.; Lesch, K.P.; Sachser, N. Social status and day-to-day behaviour of male serotonin transporter knockout mice. *Behav. Brain Res.* **2010**, *211*, 220–228. [CrossRef] [PubMed]
128. Rodriguez-Arias, M.; Navarrete, F.; Daza-Losada, M.; Navarro, D.; Aguilar, M.A.; Berbel, P.; Minarro, J.; Manzanares, J. CB1 cannabinoid receptor-mediated aggressive behavior. *Neuropharmacology* **2013**, *75*, 172–180. [CrossRef]
129. Van Loo, P.L.; Mol, J.A.; Koolhaas, J.M.; Van Zutphen, B.F.; Baumans, V. Modulation of aggression in male mice: Influence of group size and cage size. *Physiol. Behav.* **2001**, *72*, 675–683. [CrossRef]
130. Bailoo, J.D.; Murphy, E.; Varholick, J.A.; Novak, J.; Palme, R.; Wurbel, H. Evaluation of the effects of space allowance on measures of animal welfare in laboratory mice. *Sci. Rep.* **2018**, *8*, 713. [CrossRef]
131. Van Loo, P.L.; de Groot, A.C.; Van Zutphen, B.F.; Baumans, V. Do male mice prefer or avoid each other's company? Influence of hierarchy, kinship, and familiarity. *J. Appl. Anim. Welf. Sci.* **2001**, *4*, 91–103. [CrossRef]
132. Gray, S.; Hurst, J.L. The effects of cage cleaning on aggression within groups of male laboratory mice. *Anim. Behav.* **1995**, *49*, 821–826. [CrossRef]
133. Van Loo, P.; Kruitwagen, C.; Van Zutphen, L.; Koolhaas, J.; Baumans, V. Modulation of aggression in male mice: Influence of cage cleaning regime and scent marks. *Anim. Welf.* **2000**, *9*, 281–295. [CrossRef]
134. Weber, E.M.; Dallaire, J.A.; Gaskill, B.N.; Pritchett-Corning, K.R.; Garner, J.P. Aggression in group-housed laboratory mice: Why can't we solve the problem? *Lab. Anim.* **2017**, *46*, 157–161. [CrossRef] [PubMed]
135. Roughan, J.V.; Sevenoaks, T. Welfare and Scientific Considerations of Tattooing and Ear Tagging for Mouse Identification. *J. Am. Assoc. Lab. Anim. Sci.* **2019**, *58*, 142–153. [CrossRef] [PubMed]
136. Gaskill, B.N.; Stottler, A.M.; Garner, J.P.; Winnicker, C.W.; Mulder, G.B.; Pritchett-Corning, K.R. The effect of early life experience, environment, and genetic factors on spontaneous home-cage aggression-related wounding in male C57BL/6 mice. *Lab. Anim.* **2017**, *46*, 176–184. [CrossRef]
137. Burn, C.C.; Mazlan, N.H.B.; Chancellor, N.; Wells, D.J. The Pen Is Milder Than the Blade: Identification Marking Mice Using Ink on the Tail Appears More Humane Than Ear-Punching Even with Local Anaesthetic. *Animals* **2021**, *11*, 1664. [CrossRef] [PubMed]
138. Davies, J.R.; Purawijaya, D.A.; Bartlett, J.M.; Robinson, E.S.J. Impact of Refinements to Handling and Restraint Methods in Mice. *Animals* **2022**, *12*, 2173. [CrossRef] [PubMed]
139. Weber, E.M.; Zidar, J.; Ewaldsson, B.; Askevik, K.; Udén, E.; Svensk, E.; Törnqvist, E. Aggression in group-housed male mice: A systematic review. *Animals* **2022**, *13*, 143. [CrossRef] [PubMed]
140. Henderson, L.J.; Dani, B.; Serrano, E.M.N.; Smulders, T.V.; Roughan, J.V. Benefits of tunnel handling persist after repeated restraint, injection and anaesthesia. *Sci. Rep.* **2020**, *10*, 14562. [CrossRef]
141. Scholten, D.; Trebicka, J.; Liedtke, C.; Weiskirchen, R. The carbon tetrachloride model in mice. *Lab. Anim.* **2015**, *49*, 4–11. [CrossRef]
142. Turner, P.V.; Brabb, T.; Pekow, C.; Vasbinder, M.A. Administration of substances to laboratory animals: Routes of administration and factors to consider. *J. Am. Assoc. Lab. Anim. Sci.* **2011**, *50*, 600–613.
143. Gouveia, K.; Hurst, J.L. Improving the practicality of using non-aversive handling methods to reduce background stress and anxiety in laboratory mice. *Sci. Rep.* **2019**, *9*, 20305. [CrossRef]
144. Labitt, R.N.; Oxford, E.M.; Davis, A.K.; Butler, S.D.; Daugherity, E.K. A Validated Smartphone-based Electrocardiogram Reveals Severe Bradyarrhythmias during Immobilizing Restraint in Mice of Both Sexes and Four Strains. *J. Am. Assoc. Lab. Anim. Sci.* **2021**, *60*, 201–212. [CrossRef] [PubMed]
145. Marashi, V.; Barnekow, A.; Ossendorf, E.; Sachser, N. Effects of different forms of environmental enrichment on behavioral, endocrinological, and immunological parameters in male mice. *Horm. Behav.* **2003**, *43*, 281–292. [CrossRef] [PubMed]
146. Zidar, J.; Weber, E.M.; Ewaldsson, B.; Tjader, S.; Lilja, J.; Mount, J.; Svensson, C.; Svensk, E.; Uden, E.; Tornqvist, A.E. Group and Single Housing of Male Mice: Collected Experiences from Research Facilities in Sweden. *Animals* **2019**, *9*, 1010. [CrossRef] [PubMed]
147. Ratuski, A.S.; Weary, D.M. Environmental Enrichment for Rats and Mice Housed in Laboratories: A Metareview. *Animals* **2022**, *12*, 414. [CrossRef]
148. Olsson, I.A.; Dahlborn, K. Improving housing conditions for laboratory mice: A review of "environmental enrichment". *Lab. Anim.* **2002**, *36*, 243–270. [CrossRef]
149. Howerton, C.L.; Garner, J.P.; Mench, J.A. Effects of a running wheel-igloo enrichment on aggression, hierarchy linearity, and stereotypy in group-housed male CD-1 (ICR) mice. *Appl. Anim. Behav. Sci.* **2008**, *115*, 90–103. [CrossRef]
150. Leduc, R.Y.; Rauw, G.; Baker, G.B.; McDermid, H.E. What Goes Around Can Come Around: An Unexpected Deleterious Effect of Using Mouse Running Wheels for Environmental Enrichment. *J. Am. Assoc. Lab. Anim. Sci.* **2017**, *56*, 194–201.

151. Langford, D.J.; Bailey, A.L.; Chanda, M.L.; Clarke, S.E.; Drummond, T.E.; Echols, S.; Glick, S.; Ingrao, J.; Klassen-Ross, T.; Lacroix-Fralish, M.L.; et al. Coding of facial expressions of pain in the laboratory mouse. *Nat. Methods* **2010**, *7*, 447–449. [CrossRef]
152. Mota-Rojas, D.; Olmos-Hernandez, A.; Verduzco-Mendoza, A.; Hernandez, E.; Martinez-Burnes, J.; Whittaker, A.L. The Utility of Grimace Scales for Practical Pain Assessment in Laboratory Animals. *Animals* **2020**, *10*, 1838. [CrossRef]
153. Ataka, K.; Asakawa, A.; Nagaishi, K.; Kaimoto, K.; Sawada, A.; Hayakawa, Y.; Tatezawa, R.; Inui, A.; Fujimiya, M. Bone marrow-derived microglia infiltrate into the paraventricular nucleus of chronic psychological stress-loaded mice. *PLoS ONE* **2013**, *8*, e81744. [CrossRef]
154. Eltokhi, A.; Rappold, G.; Sprengel, R. Distinct Phenotypes of Shank2 Mouse Models Reflect Neuropsychiatric Spectrum Disorders of Human Patients with SHANK2 Variants. *Front. Mol. Neurosci.* **2018**, *11*, 240. [CrossRef] [PubMed]
155. Roughan, J.V.; Wright-Williams, S.L.; Flecknell, P.A. Automated analysis of postoperative behaviour: Assessment of HomeCageScan as a novel method to rapidly identify pain and analgesic effects in mice. *Lab. Anim.* **2009**, *43*, 17–26. [CrossRef] [PubMed]
156. Grieco, F.; Bernstein, B.J.; Biemans, B.; Bikovski, L.; Burnett, C.J.; Cushman, J.D.; van Dam, E.A.; Fry, S.A.; Richmond-Hacham, B.; Homberg, J.R.; et al. Measuring Behavior in the Home Cage: Study Design, Applications, Challenges, and Perspectives. *Front. Behav. Neurosci.* **2021**, *15*, 735387. [CrossRef] [PubMed]
157. Herrmann, K.; Flecknell, P. The application of humane endpoints and humane killing methods in animal research proposals: A retrospective review. *Altern. Lab. Anim.* **2018**, *46*, 317–333. [CrossRef] [PubMed]
158. Williams, W.O.; Baneux, P. Humane Intervention Points: Refining endpoint terminology to incorporate non-euthanasia intervention options to improve animal welfare and preserve experimental outcomes. *Lab. Anim.* **2022**, *56*, 482–489. [CrossRef]
159. Brunt, M.W.; Amendola, L.; Weary, D.M. Attitudes of laboratory animal professionals and researchers towards carbon dioxide euthanasia for rodents and perceived barriers to change. *Lab. Anim.* **2021**, *55*, 531–539. [CrossRef]
160. Ahmadi-Noorbakhsh, S.; Farajli Abbasi, M.; Ghasemi, M.; Bayat, G.; Davoodian, N.; Sharif-Paghaleh, E.; Poormoosavi, S.M.; Rafizadeh, M.; Maleki, M.; Shirzad-Aski, H.; et al. Anesthesia and analgesia for common research models of adult mice. *Lab. Anim. Res.* **2022**, *38*, 40. [CrossRef]
161. Foley, P.L.; Kendall, L.V.; Turner, P.V. Clinical Management of Pain in Rodents. *Comp. Med.* **2019**, *69*, 468–489. [CrossRef]
162. Penderis, J.; Franklin, R.J. Effects of pre- versus post-anaesthetic buprenorphine on propofol-anaesthetized rats. *Vet. Anaesth. Analg.* **2005**, *32*, 256–260. [CrossRef]
163. Oliver, V.L.; Thurston, S.E.; Lofgren, J.L. Using Cageside Measures to Evaluate Analgesic Efficacy in Mice (Mus musculus) after Surgery. *J. Am. Assoc. Lab. Anim. Sci.* **2018**, *57*, 186–201.
164. Aller, M.A.; Duran, M.; Ortega, L.; Arias, J.L.; Nava, M.P.; Prieto, I.; Arias, J. Comparative study of macro- and microsurgical extrahepatic cholestasis in the rat. *Microsurgery* **2004**, *24*, 442–447. [CrossRef] [PubMed]
165. Symeonidis, A.; Trams, E.G. Morphologic and functional changes in the livers of rats after ligation or excision of the common bile duct. *Am. J. Pathol.* **1957**, *33*, 13–27. [PubMed]
166. Hoogstraten-Miller, S.L.; Brown, P.A. Techniques in aseptic rodent surgery. *Curr. Protoc. Immunol.* **2008**, *82*, 1.12.1–1.12.14. [CrossRef] [PubMed]
167. Tag, C.G.; Sauer-Lehnen, S.; Weiskirchen, S.; Borkham-Kamphorst, E.; Tolba, R.H.; Tacke, F.; Weiskirchen, R. Bile duct ligation in mice: Induction of inflammatory liver injury and fibrosis by obstructive cholestasis. *J. Vis. Exp.* **2015**, e52438. [CrossRef]
168. Aller, M.A.; Arias, N.; Prieto, I.; Agudo, S.; Gilsanz, C.; Lorente, L.; Arias, J.L.; Arias, J. A half century (1961–2011) of applying microsurgery to experimental liver research. *World J. Hepatol.* **2012**, *4*, 199–208. [CrossRef] [PubMed]
169. Albillos, A.; de Gottardi, A.; Rescigno, M. The gut-liver axis in liver disease: Pathophysiological basis for therapy. *J. Hepatol.* **2020**, *72*, 558–577. [CrossRef]
170. Yokota, S.; Ono, Y.; Nakao, T.; Zhang, P.; Michalopoulos, G.K.; Khan, Z. Partial Bile Duct Ligation in the Mouse: A Controlled Model of Localized Obstructive Cholestasis. *J. Vis. Exp.* **2018**, e56930. [CrossRef]
171. Tang, G.; Seume, N.; Hager, C.; Kumstel, S.; Abshagen, K.; Bleich, A.; Vollmar, B.; Talbot, S.R.; Zhang, X.; Zechner, D. Comparing distress of mouse models for liver damage. *Sci. Rep.* **2020**, *10*, 19814. [CrossRef]
172. Georgiev, P.; Jochum, W.; Heinrich, S.; Jang, J.H.; Nocito, A.; Dahm, F.; Clavien, P.A. Characterization of time-related changes after experimental bile duct ligation. *Br. J. Surg.* **2008**, *95*, 646–656. [CrossRef]
173. Secklehner, J.; Richardson, C.A. The reporting of animal welfare details in liver research: A review of studies describing bile duct ligation in mice (2011–2013). *J. Hepatol.* **2015**, *62*, 250–251. [CrossRef]
174. Durst, M.S.; Arras, M.; Palme, R.; Talbot, S.R.; Jirkof, P. Lidocaine and bupivacaine as part of multimodal pain management in a C57BL/6J laparotomy mouse model. *Sci. Rep.* **2021**, *11*, 10918. [CrossRef] [PubMed]
175. Tammam, O.; Taha, A.; El-Sherif, M. Optimization of Xylazine-Ketamine anesthetic dose in mice suffering chronic liver injury. *J. Anesth. Crit. Care* **2019**, *11*, 6–8.
176. Bauer, C.; Schillinger, U.; Brandl, J.; Meyer-Lindenberg, A.; Ott, A.; Baumgartner, C. Comparison of pre-emptive butorphanol or metamizole with ketamine +medetomidine and s-ketamine + medetomidine anaesthesia in improving intraoperative analgesia in mice. *Lab. Anim.* **2019**, *53*, 459–469. [CrossRef] [PubMed]
177. Liles, J.H.; Flecknell, P.A. The influence of buprenorphine or bupivacaine on the post-operative effects of laparotomy and bile-duct ligation in rats. *Lab. Anim.* **1993**, *27*, 374–380. [CrossRef]
178. Zhang, X.; Kumstel, S.; Tang, G.; Talbot, S.R.; Seume, N.; Abshagen, K.; Vollmar, B.; Zechner, D. A rational approach of early humane endpoint determination in a murine model for cholestasis. *ALTEX* **2020**, *37*, 197–207. [CrossRef]

179. Ernst, L.; Zieglowski, L.; Schulz, M.; Moss, M.; Meyer, M.; Weiskirchen, R.; Palme, R.; Hamann, M.; Talbot, S.R.; Tolba, R.H. Severity assessment in mice subjected to carbon tetrachloride. *Sci. Rep.* **2020**, *10*, 15790. [CrossRef]
180. Chang, M.L.; Yeh, C.T.; Chang, P.Y.; Chen, J.C. Comparison of murine cirrhosis models induced by hepatotoxin administration and common bile duct ligation. *World J. Gastroenterol.* **2005**, *11*, 4167–4172. [CrossRef]
181. Gomez-Hurtado, I.; Santacruz, A.; Peiro, G.; Zapater, P.; Gutierrez, A.; Perez-Mateo, M.; Sanz, Y.; Frances, R. Gut microbiota dysbiosis is associated with inflammation and bacterial translocation in mice with CCl4-induced fibrosis. *PLoS ONE* **2011**, *6*, e23037. [CrossRef]
182. Fortea, J.I.; Fernandez-Mena, C.; Puerto, M.; Ripoll, C.; Almagro, J.; Banares, J.; Bellon, J.M.; Banares, R.; Vaquero, J. Comparison of Two Protocols of Carbon Tetrachloride-Induced Cirrhosis in Rats—Improving Yield and Reproducibility. *Sci. Rep.* **2018**, *8*, 9163. [CrossRef]
183. Õkva, K.; Tamoseviciute, E.; Ciziute, A.; Pokk, P.; Ruksenas, O.; Nevalainen, T. Refinements for intragastric gavage in rats. *Scand. J. Lab. Anim. Sci.* **2006**, *33*, 243–252.
184. Karrberg, L.; Andersson, L.; Kastenmayer, R.J.; Ploj, K. Refinement of habituation procedures in diet-induced obese mice. *Lab. Anim.* **2016**, *50*, 397–399. [CrossRef] [PubMed]
185. Hoggatt, A.F.; Hoggatt, J.; Honerlaw, M.; Pelus, L.M. A spoonful of sugar helps the medicine go down: A novel technique to improve oral gavage in mice. *J. Am. Assoc. Lab. Anim. Sci.* **2010**, *49*, 329–334. [PubMed]
186. Jones, C.P.; Boyd, K.L.; Wallace, J.M. Evaluation of Mice Undergoing Serial Oral Gavage While Awake or Anesthetized. *J. Am. Assoc. Lab. Anim. Sci.* **2016**, *55*, 805–810. [PubMed]
187. Ikenaga, N.; Peng, Z.W.; Vaid, K.A.; Liu, S.B.; Yoshida, S.; Sverdlov, D.Y.; Mikels-Vigdal, A.; Smith, V.; Schuppan, D.; Popov, Y.V. Selective targeting of lysyl oxidase-like 2 (LOXL2) suppresses hepatic fibrosis progression and accelerates its reversal. *Gut* **2017**, *66*, 1697–1708. [CrossRef]
188. Kristiansen, M.N.; Veidal, S.S.; Rigbolt, K.T.; Tolbol, K.S.; Roth, J.D.; Jelsing, J.; Vrang, N.; Feigh, M. Obese diet-induced mouse models of nonalcoholic steatohepatitis-tracking disease by liver biopsy. *World J. Hepatol.* **2016**, *8*, 673–684. [CrossRef]
189. Masuzaki, R.; Kanda, T.; Sasaki, R.; Matsumoto, N.; Ogawa, M.; Matsuoka, S.; Karp, S.J.; Moriyama, M. Noninvasive Assessment of Liver Fibrosis: Current and Future Clinical and Molecular Perspectives. *Int. J. Mol. Sci.* **2020**, *21*, 4906. [CrossRef]
190. Will, O.M.; Damm, T.; Campbell, G.M.; von Schonfells, W.; Acil, Y.; Will, M.; Chalaris-Rissmann, A.; Ayna, M.; Drucker, C.; Gluer, C.C. Longitudinal micro-computed tomography monitoring of progressive liver regeneration in a mouse model of partial hepatectomy. *Lab. Anim.* **2017**, *51*, 422–426. [CrossRef]
191. Becker, A.S.; Schneider, M.A.; Wurnig, M.C.; Wagner, M.; Clavien, P.A.; Boss, A. Radiomics of liver MRI predict metastases in mice. *Eur. Radiol. Exp.* **2018**, *2*, 11. [CrossRef]
192. Bastard, C.; Bosisio, M.R.; Chabert, M.; Kalopissis, A.D.; Mahrouf-Yorgov, M.; Gilgenkrantz, H.; Mueller, S.; Sandrin, L. Transient micro-elastography: A novel non-invasive approach to measure liver stiffness in mice. *World J. Gastroenterol.* **2011**, *17*, 968–975. [CrossRef]
193. Czernuszewicz, T.J.; Aji, A.M.; Moore, C.J.; Montgomery, S.A.; Velasco, B.; Torres, G.; Anand, K.S.; Johnson, K.A.; Deal, A.M.; Zukic, D.; et al. Development of a Robotic Shear Wave Elastography System for Noninvasive Staging of Liver Disease in Murine Models. *Hepatol. Commun.* **2022**, *6*, 1827–1839. [CrossRef]
194. Fu, S.; Wu, D.; Jiang, W.; Li, J.; Long, J.; Jia, C.; Zhou, T. Molecular Biomarkers in Drug-Induced Liver Injury: Challenges and Future Perspectives. *Front. Pharmacol.* **2019**, *10*, 1667. [CrossRef] [PubMed]
195. Manna, S.K.; Thompson, M.D.; Gonzalez, F.J. Application of mass spectrometry-based metabolomics in identification of early noninvasive biomarkers of alcohol-induced liver disease using mouse model. *Adv. Exp. Med. Biol.* **2015**, *815*, 217–238. [CrossRef] [PubMed]
196. Iwasa, M.; Shigefuku, R.; Eguchi, A.; Tamai, Y.; Takei, Y. Update on blood-based biomarkers for chronic liver diseases prognosis: Literature review and institutional experience. *JGH Open* **2021**, *5*, 1250–1256. [CrossRef] [PubMed]
197. Hayes, C.N.; Chayama, K. MicroRNAs as Biomarkers for Liver Disease and Hepatocellular Carcinoma. *Int. J. Mol. Sci.* **2016**, *17*, 280. [CrossRef]
198. Yang, B.; Zhang, J.; Sun, L.; Huang, T.; Kong, Y.; Li, L.; Sun, Z.; Yin, M.; Li, X. Mapping Novel Biomarkers of Liver Injury by Tissue Proteomic Analysis. *ACS Omega* **2021**, *6*, 7127–7138. [CrossRef] [PubMed]
199. Dragoi, D.; Benesic, A.; Pichler, G.; Kulak, N.A.; Bartsch, H.S.; Gerbes, A.L. Proteomics Analysis of Monocyte-Derived Hepatocyte-Like Cells Identifies Integrin Beta 3 as a Specific Biomarker for Drug-Induced Liver Injury by Diclofenac. *Front. Pharmacol.* **2018**, *9*, 699. [CrossRef] [PubMed]
200. Vali, Y.; Lee, J.; Boursier, J.; Petta, S.; Wonders, K.; Tiniakos, D.; Bedossa, P.; Geier, A.; Francque, S.; Allison, M.; et al. Biomarkers for staging fibrosis and non-alcoholic steatohepatitis in non-alcoholic fatty liver disease (the LITMUS project): A comparative diagnostic accuracy study. *Lancet Gastroenterol. Hepatol.* **2023**, *8*, 714–725. [CrossRef]
201. Schnell, M.A.; Hardy, C.; Hawley, M.; Propert, K.J.; Wilson, J.M. Effect of blood collection technique in mice on clinical pathology parameters. *Hum. Gene Ther.* **2002**, *13*, 155–161. [CrossRef]
202. Chapman, K.; Burnett, J.; Corvaro, M.; Mitchell, D.; Robinson, S.; Sangster, T.; Sparrow, S.; Spooner, N.; Wilson, A. Reducing pre-clinical blood volumes for toxicokinetics: Toxicologists, pathologists and bioanalysts unite. *Bioanalysis* **2014**, *6*, 2965–2968. [CrossRef]

203. Joyce, A.P.; Wang, M.; Lawrence-Henderson, R.; Filliettaz, C.; Leung, S.S.; Xu, X.; O'Hara, D.M. One mouse, one pharmacokinetic profile: Quantitative whole blood serial sampling for biotherapeutics. *Pharm. Res.* **2014**, *31*, 1823–1833. [CrossRef]
204. Patel, N.J.; Wickremsinhe, E.; Hui, Y.H.; Barr, A.; Masterson, N.; Ruterbories, K.; Weller, J.; Hanes, J.; Kern, T.; Perkins, E. Evaluation and Optimization of Blood Micro-Sampling Methods: Serial Sampling in a Cross-Over Design from an Individual Mouse. *J. Pharm. Pharm. Sci.* **2016**, *19*, 496–510. [CrossRef] [PubMed]
205. Sadler, A.M.; Bailey, S.J. Validation of a refined technique for taking repeated blood samples from juvenile and adult mice. *Lab. Anim.* **2013**, *47*, 316–319. [CrossRef]
206. Harikrishnan, V.S.; Hansen, A.K.; Abelson, K.S.; Sorensen, D.B. A comparison of various methods of blood sampling in mice and rats: Effects on animal welfare. *Lab. Anim.* **2018**, *52*, 253–264. [CrossRef] [PubMed]
207. Hackett, M.J.; Kinderknecht, K.D.; Niemuth, N.A.; Taylor, J.A.; Gibbs, S.T.; Novak, J.; Harbo, S.J. A Factorial Analysis of Drug and Bleeding Effects in Toxicokinetic Studies. *Toxicol. Sci.* **2019**, *170*, 234–246. [CrossRef] [PubMed]
208. Whittaker, A.L.; Barker, T.H. The Impact of Common Recovery Blood Sampling Methods, in Mice (Mus Musculus), on Well-Being and Sample Quality: A Systematic Review. *Animals* **2020**, *10*, 989. [CrossRef]
209. Hattori, N.; Yokoyama, H.; Takumi, A.; Kawaguchi, S.; Yamaguchi, K.; Shimazaki, T.; Shibui, Y.; Saito, K.; Saito, Y. Effect of microsampling (50 muL) on toxicological evaluation of methapyrilene, a hepatotoxic substance, in a collaborative 28-day study in female rats. *J. Toxicol. Sci.* **2023**, *48*, 1–14. [CrossRef]
210. Nagarajan, P.; Mahesh Kumar, M.J.; Venkatesan, R.; Majundar, S.S.; Juyal, R.C. Genetically modified mouse models for the study of nonalcoholic fatty liver disease. *World J. Gastroenterol.* **2012**, *18*, 1141–1153. [CrossRef]
211. Zhou, T.; Kyritsi, K.; Wu, N.; Francis, H.; Yang, Z.; Chen, L.; O'Brien, A.; Kennedy, L.; Ceci, L.; Meadows, V.; et al. Knockdown of vimentin reduces mesenchymal phenotype of cholangiocytes in the Mdr2(-/-) mouse model of primary sclerosing cholangitis (PSC). *eBioMedicine* **2019**, *48*, 130–142. [CrossRef]
212. Bao, Y.L.; Wang, L.; Pan, H.T.; Zhang, T.R.; Chen, Y.H.; Xu, S.J.; Mao, X.L.; Li, S.W. Animal and Organoid Models of Liver Fibrosis. *Front. Physiol.* **2021**, *12*, 666138. [CrossRef]
213. Flessa, C.M.; Nasiri-Ansari, N.; Kyrou, I.; Leca, B.M.; Lianou, M.; Chatzigeorgiou, A.; Kaltsas, G.; Kassi, E.; Randeva, H.S. Genetic and Diet-Induced Animal Models for Non-Alcoholic Fatty Liver Disease (NAFLD) Research. *Int. J. Mol. Sci.* **2022**, *23*, 15791. [CrossRef]
214. Xu, D.; Peltz, G. Can Humanized Mice Predict Drug "Behavior" in Humans? *Annu. Rev. Pharmacol. Toxicol.* **2016**, *56*, 323–338. [CrossRef] [PubMed]
215. Olson, H.; Betton, G.; Robinson, D.; Thomas, K.; Monro, A.; Kolaja, G.; Lilly, P.; Sanders, J.; Sipes, G.; Bracken, W.; et al. Concordance of the toxicity of pharmaceuticals in humans and in animals. *Regul. Toxicol. Pharmacol.* **2000**, *32*, 56–67. [CrossRef] [PubMed]
216. Scheer, N.; Wilson, I.D. A comparison between genetically humanized and chimeric liver humanized mouse models for studies in drug metabolism and toxicity. *Drug Discov. Today* **2016**, *21*, 250–263. [CrossRef] [PubMed]
217. Paulk, N.K.; Pekrun, K.; Zhu, E.; Nygaard, S.; Li, B.; Xu, J.; Chu, K.; Leborgne, C.; Dane, A.P.; Haft, A.; et al. Bioengineered AAV Capsids with Combined High Human Liver Transduction In Vivo and Unique Humoral Seroreactivity. *Mol. Ther.* **2018**, *26*, 289–303. [CrossRef] [PubMed]
218. Ishida, Y.; Chung, T.L.; Imamura, M.; Hiraga, N.; Sen, S.; Yokomichi, H.; Tateno, C.; Canini, L.; Perelson, A.S.; Uprichard, S.L.; et al. Acute hepatitis B virus infection in humanized chimeric mice has multiphasic viral kinetics. *Hepatology* **2018**, *68*, 473–484. [CrossRef]
219. Billerbeck, E.; Mommersteeg, M.C.; Shlomai, A.; Xiao, J.W.; Andrus, L.; Bhatta, A.; Vercauteren, K.; Michailidis, E.; Dorner, M.; Krishnan, A.; et al. Humanized mice efficiently engrafted with fetal hepatoblasts and syngeneic immune cells develop human monocytes and NK cells. *J. Hepatol.* **2016**, *65*, 334–343. [CrossRef]
220. Bissig-Choisat, B.; Alves-Bezerra, M.; Zorman, B.; Ochsner, S.A.; Barzi, M.; Legras, X.; Yang, D.; Borowiak, M.; Dean, A.M.; York, R.B.; et al. A human liver chimeric mouse model for non-alcoholic fatty liver disease. *JHEP Rep.* **2021**, *3*, 100281. [CrossRef]
221. Benten, D.; Kluwe, J.; Wirth, J.W.; Thiele, N.D.; Follenzi, A.; Bhargava, K.K.; Palestro, C.J.; Koepke, M.; Tjandra, R.; Volz, T.; et al. A humanized mouse model of liver fibrosis following expansion of transplanted hepatic stellate cells. *Lab. Investig.* **2018**, *98*, 525–536. [CrossRef]
222. de Boo, J.; Hendriksen, C. Reduction strategies in animal research: A review of scientific approaches at the intra-experimental, supra-experimental and extra-experimental levels. *Altern. Lab. Anim.* **2005**, *33*, 369–377. [CrossRef]
223. Verderio, P.; Lecchi, M.; Ciniselli, C.M.; Shishmani, B.; Apolone, G.; Manenti, G. 3Rs Principle and Legislative Decrees to Achieve High Standard of Animal Research. *Animals* **2023**, *13*, 277. [CrossRef]
224. The Nuffield Council on Bioethics. *The Ethics of Research Involving Animals*; The Nuffield Council on Bioethics: London, UK, 2005; p. 376.
225. Clark, J.M. The 3Rs in research: A contemporary approach to replacement, reduction and refinement. *Br. J. Nutr.* **2018**, *120*, S1–S7. [CrossRef] [PubMed]
226. Strech, D.; Dirnagl, U. 3Rs missing: Animal research without scientific value is unethical. *BMJ Open Sci.* **2019**, *3*, bmjos-2018-000048. [CrossRef] [PubMed]
227. Festing, M.F. Randomized block experimental designs can increase the power and reproducibility of laboratory animal experiments. *ILAR J.* **2014**, *55*, 472–476. [CrossRef]

228. Begley, C.G.; Ellis, L.M. Raise standards for preclinical cancer research. *Nature* **2012**, *483*, 531–533. [CrossRef] [PubMed]
229. De Angelis, I.; Ricceri, L.; Vitale, A. The 3R principle: 60 years taken well. Preface. *Ann. Ist. Super. Sanita* **2019**, *55*, 398–399. [CrossRef] [PubMed]
230. Hansen, A.K.; Hansen, C.H.; Krych, L.; Nielsen, D.S. Impact of the gut microbiota on rodent models of human disease. *World J. Gastroenterol.* **2014**, *20*, 17727–17736. [CrossRef]
231. Bie, P.; Debrabant, B. Gonadal sex and animal experimentation: Perfection vs. 3R principle? *Basic. Clin. Pharmacol. Toxicol.* **2020**, *127*, 111–119. [CrossRef] [PubMed]
232. Kroeger, M. How omics technologies can contribute to the '3R' principles by introducing new strategies in animal testing. *Trends Biotechnol.* **2006**, *24*, 343–346. [CrossRef]
233. Frohlich, E.; Loizou, G.D. Editorial: 3Rs-Strategies for reduction and refinement of animal studies. *Front. Pharmacol.* **2023**, *14*, 1200965. [CrossRef]
234. eBioMedicine. The 3Rs of Animal Research. *eBioMedicine* **2022**, *76*, 103900. [CrossRef]
235. Lonardo, A.; Nascimbeni, F.; Ballestri, S.; Fairweather, D.; Win, S.; Than, T.A.; Abdelmalek, M.F.; Suzuki, A. Sex Differences in Nonalcoholic Fatty Liver Disease: State of the Art and Identification of Research Gaps. *Hepatology* **2019**, *70*, 1457–1469. [CrossRef] [PubMed]
236. Cvitanovic Tomas, T.; Urlep, Z.; Moskon, M.; Mraz, M.; Rozman, D. LiverSex Computational Model: Sexual Aspects in Hepatic Metabolism and Abnormalities. *Front. Physiol.* **2018**, *9*, 360. [CrossRef] [PubMed]

Disclaimer/Publisher's Note: The statements, opinions and data contained in all publications are solely those of the individual author(s) and contributor(s) and not of MDPI and/or the editor(s). MDPI and/or the editor(s) disclaim responsibility for any injury to people or property resulting from any ideas, methods, instructions or products referred to in the content.

animals

MDPI

Review

Unmasking the Adverse Impacts of Sex Bias on Science and Research Animal Welfare

Elizabeth A. Nunamaker [1] and Patricia V. Turner [1,2,*]

[1] Global Animal Welfare and Training, Charles River Laboratories, Wilmington, MA 01887, USA; elizabeth.nunamaker@crl.com
[2] Department of Pathobiology, University of Guelph, Guelph, ON N1G 2W1, Canada
* Correspondence: pvturner@uoguelph.ca

Simple Summary: Sex bias—the use of one sex over the other—is a common practice in biomedical research, typically with over selection of male animals. There are a number of reasons that this practice is common, but it has resulted in dosing errors and unintended side effects in a number of cases in women when products are given without adequate testing in female animals. Sex bias can also result in animal welfare issues given that both sexes are born in approximately equal numbers. Welfare issues include overproduction of the unwanted sex, inadequate ability to recognize and treat pain in female animals, stress associated with differential housing needs based on animal sex, and potential wastage of animals if study results are incorrect or studies need to be redone because only one sex was studied. Even though many government agencies and funding sources now require both sexes to be studied in biomedical research, single-sex studies are still common. More systematic planning and reporting of study details is needed, as well as exploring sex selection technology used in other animal production sectors when single-sex studies are justified, to reduce animal waste.

Abstract: Sex bias in biomedical and natural science research has been prevalent for decades. In many cases, the female estrous cycle was thought to be too complex an issue to model for, and it was thought to be simpler to only use males in studies. At times, particularly when studying efficacy and safety of new therapeutics, this sex bias has resulted in over- and under-medication with associated deleterious side effects in women. Many sex differences have been recognized that are unrelated to hormonal variation occurring during the estrous cycle. Sex bias also creates animal welfare challenges related to animal over-production and wastage, insufficient consideration of welfare (and scientific) impact related to differential housing of male vs female animals within research facilities, and a lack of understanding regarding differential requirements for pain recognition and alleviation in male versus female animals. Although many funding and government agencies require both sexes to be studied in biomedical research, many disparities remain in practice. This requires further enforcement of expectations by the Institutional Animal Care and Use Committee when reviewing protocols, research groups when writing grants, planning studies, and conducting research, and scientific journals and reviewers to ensure that sex bias policies are enforced.

Keywords: sex bias; animal welfare; biomedical research; reproducibility; translatability; drug development

Citation: Nunamaker, E.A.; Turner, P.V. Unmasking the Adverse Impacts of Sex Bias on Science and Research Animal Welfare. *Animals* **2023**, *13*, 2792. https://doi.org/10.3390/ani13172792

Academic Editor: Garikoitz Azkona

Received: 21 July 2023
Revised: 30 August 2023
Accepted: 30 August 2023
Published: 2 September 2023

Copyright: © 2023 by the authors. Licensee MDPI, Basel, Switzerland. This article is an open access article distributed under the terms and conditions of the Creative Commons Attribution (CC BY) license (https://creativecommons.org/licenses/by/4.0/).

1. Introduction

An historical overreliance on male animals in the drug development process has resulted in women taking drugs at inappropriate doses and experiencing side effects rarely reported in men. This has required the FDA to reevaluate safety and efficacy study results to (re)establish differential dosing for men and women after the drugs were approved [1,2]. Many FDA-approved therapeutics have elevated blood levels and longer elimination times in women, as well as being associated with a higher incidence of adverse drug reactions [3].

Had both sexes of animals been used in the preclinical testing phase, the sex-related differences in drug metabolism and clearance may have been identified sooner, and many adverse drug reactions in women may have been avoided [2].

This sex bias in biomedical research has been well documented. In a study across 10 fields of biology, it was found that 80% of the animals used were male [4]. It was further found that, even when both sexes were used, only one-third of studies analyzed the results by sex [4]. While there are many different justifications used to propagate sex bias in biomedical research [5], appropriate experimental design should be used to effectively accommodate both sexes, maintain or increase power, and avoid interpretation errors and associated increased costs [6–8]. Beyond experimental and human health concerns, sex bias also represents an animal welfare concern. Specifically, there is evidence to suggest that sex bias results in overproduction of research animals (see Section 4.1), inadequate pain management (see Section 4.2), and significant animal wastage (see Section 4.4).

The ubiquity of sex bias in biomedical research has developed over time into the current situation. In this paper, we discuss how sex bias developed over the past 100 years and the ongoing efforts to address it. The underlying justifications used, scientific and not, are explored, and the implications for animal welfare are described. Lastly, specific methods to minimize sex bias in biomedical studies are suggested.

2. How Did Sex Bias Develop and How Pervasive Is It?

Sex-based differences in animal studies are not a revolutionary concept. Differences in biological responses have been documented in the literature since the 1930s [9–12]. Early studies were focused on the increased variability in learning behavior of female rats. The perceived higher variability of responses to experimental challenges in females and the differences between the sexes provided a justification for studying only male animals as a means of simplification. This was further compounded by prominent literature in the 1960s that encouraged investigators to keep animal numbers low and reduce experimental variability by specifically using a single sex [13]. By the late 1970s, the literature was flooded with studies demonstrating a sex difference in many physiological systems [14]. Instead of using this as justification to study both sexes, the rationale for studying only males became entrenched in animal-based research studies [15].

An awareness about the limitations of the pervasive bias towards the use of male animals first became apparent in the 1990s [16–18]. Sex bias gained considerable interest between 1997 and 2000, when the US Food and Drug Administration suspended distribution and sales of eight different prescription drugs due to severe adverse effects that were reported in women taking them [19]. Ultimately, the root cause of the suspensions was systemic sex bias in the drug development process resulting in dose recommendation errors. The compounds were initially screened using cell cultures of male origin, preclinical testing was performed in male animals only, and clinical testing was primarily completed in men [19]. These events collectively sparked the interest in studying the sex and gender bias that is still seen today in both biomedical and clinical research.

In 2010, Beery and Zucker conducted a survey of animal use in neuroscience and biomedical research and found that, after 20 years of awareness of sex bias, the practice continued to be pervasive in animal studies. Specifically, they found a male bias in 8 of 10 surveyed disciplines, with single-sex studies of male animals outnumbering those of females, 5.5 to 1 [4]. In response to this demonstrated male bias, the National Institutes of Health began requiring that sex be included as an experimental variable in grant applications [20–22]. Other funding agencies subsequently followed their lead. Despite these efforts, the scientific literature continues to be full of examples of male sex bias in animal research [5], especially in the fields of pain management [23], cardiovascular disease [24,25], diabetes mellitus [25], alcohol-related diseases [26], and development of surgical methods [27]. Potentially equally damaging to scientific rigor and reproducibility, it was also common to not report sex at all during this time, as occurred in upwards of 25% of published animals studies and 76% of cell culture studies [27].

In 2020, when a 10-year follow-up study was conducted on sex reporting [5], there was some evidence of improvement, but sex bias remained pervasive. There was an increase in the proportion of studies that included both sexes, but no change in the proportion of studies that included data analyzed by sex. Most studies continued to fail to provide a rationale for the use of animals of a single sex in their studies. Further, there was also a lack of sex-based analyses, and those that conduct a sex-based analysis relied on misconceptions surrounding the hormonal variability of females. This data suggests that there is still significant work to be done in experimental design and data analysis to include both sexes.

3. Why One Sex May Be Preferred in Research Settings

There are several reasons why one sex may be preferred over the other in a research setting [8]. Some of these have been used historically to propagate the sex bias seen today, while other reasons are legitimate justification for the use of a single sex. Woitowich and colleagues captured and consolidated the justifications found in the literature into six themes: 30% known sex difference or sex effect, 27% increased experimental variability, 13% experimental conditions, 13% limited sample size, 10% inability to sex subjects, and 7% issues with animal husbandry [5]. These are each discussed below.

3.1. Known Sex Differences or Sex Effects on Research

There are diseases and physiologic processes that occur in only one sex. There are many known sex-linked traits and conditions, such as hemophilia A, Duchenne muscular dystrophy, Fragile X syndrome, breast cancer, conception, and in utero fetal development. The disease or physiologic process of interest inherently limits the ability to study the disease in both sexes. However, it is also critical to understand the role of sex in fundamental physiology and diseases processes, making it important to include both sexes to compare and find these sex effects whenever possible [23,28]. Likewise, when studying safety and efficacy of therapeutics, it's important to ensure that there are no sex-based differences in treatment safety or efficacy.

3.2. Increased Experimental Variability

Females have been long excluded from studies due to misconceptions about their estrous (or menstrual) cycle increasing day-to-day experimental variability and because including them results in a need for more research animals. However, the estrous cycle is typically not a variable that contributes significantly to experimental variability [29–34]. Empirical research across multiple rodent species and traits demonstrates that females are not more variable than males, and that for most traits, female estrous cyclicity need not be considered [8]. Even when the estrous cycle is a known or significant variable, experiments can be designed around it, and hormone variation can be incorporated into the design to account for that variability [29–33]. Successful incorporation of hormone variation in study design has previously been documented [30,32,35,36]. It is also noteworthy that individual variation has been documented to be a larger source of variability in behavior than estrous state [37].

Practically speaking, a good approach is to compare males with two or more groups of females where the stage of the ovarian cycle is known. A three-group design in mice or rats, for example, could compare males with females on two specific days of the estrous cycle. Alternatively, a five-group design would compare males with females on each of the four days of the mouse or rat estrous cycle [31]. The later design would allow for detection of a sex difference, specifically isolated to a precise day of the estrous cycle.

3.3. Experimental Condition

The experimental condition may have inherent sex differences or existing biases that make it difficult to include both sexes. Examples of this include situations such as not including females because collecting vaginal smears to control for stage of estrous cycle adds stress to the animal experience or only using adult females for a behavior study because adult males and juveniles of either sex rarely vocalize [5]. Neither of these examples are strong justifications and push-back would be warranted. A more reasonable justification to exclude one sex would be in the case of studying uterine tumors, which cannot occur in the male sex, or prostate cancer, which does not occur in females. The justification for excluding one sex based on experimental condition should be closely scrutinized to ensure that the reasoning is scientifically sound.

3.4. Limited Sample Size

Sample size limitations can make it difficult to analyze data for sex-based effects. These limitations may be due to either having a limited resource (e.g. small population of unique animals) or costs. Potentially increasing the number of animals in a study is a concern due to the associated costs; however, per the Guide for the Care and Use of Laboratory Animals, cost is not an acceptable justification for reduction of animal numbers [38]. A better approach to eliminating the sex bias implications of limited sample size is by using factorial designs to reduce the need for additional animals while including both males and females [8].

3.5. Inability to Sex Subjects

There are times when the sex of the animals or tissues being used is not obvious or is truly unknown. This may be the case when working with embryos or slaughterhouse tissues. However, the sex of these tissues can be determined through the use of various molecular techniques, including polymerase chain reaction (PCR), gas chromatography-mass spectrometry (GC-MS), high performance liquid chromatography/mass spectrometry (HPLC-MS/MS), and enzyme-linked immunosorbent assay (ELISA) on those tissues [39–43].

3.6. Animal Husbandry

Animal husbandry limitations can unknowingly contribute to sex bias in research. This can be due to limited vivarium space, sex-based aggression [44,45], response to husbandry procedures [46], and relative ease of social housing [45] (see Section 4.3 below). To avoid unscheduled breeding and unanticipated offspring, males and females are typically housed in single-sex groups. The ease of social housing can vary by species and by animal age. Aggression between same-sex conspecifics can be a significant welfare concern leading to the use of housing strategies to meet the local or national regulatory requirements, as well as the limitations of the vivarium size (see 4.3 below for a further discussion). An investigator may opt to use one sex during fetal or neonatal development stages for part of a study while using juveniles and adults for other portions of a study with the goal of simplifying the husbandry. Unfortunately, this can create unintended sex bias in a study. Identifying husbandry effects or limitations and effectively preparing for them can prevent this unintended source of sex bias from animal studies.

4. The Impact of Sex Bias on Animal Welfare

There are many examples demonstrating how sex bias in biomedical research can impact animal welfare. In this section, we will highlight how sex bias can result in overproduction of research animals, result in inadequate pain management, ignore underlying differences in male versus female stress responses and physiology, and result in animal wastage and poorly reproducible and translatable results.

4.1. Overproduction of Research Animals: Ethics and Sex Bias

Minimizing animal waste is an important component of reduction, one of the 3R's tenets. There is significant and continued interest within the laboratory animal community in reducing surplus animals produced for biomedical research [47]. It has been estimated that >110 million mice and rats are used in science and education each year, although one recent estimate suggests that >110 million mice and rats are used in research each year in the US alone [48–53]. Conservatively, and on average, at least 30% overproduction exists in a given colony, even with the most efficient breeding methods in commercial settings [54]. For the EU, this results in an estimated 12.5 million animal surplus; for the UK, there is an excess of at least 1.6 million mice and rats produced [55]. Together, these suggest that there could be overproduction of 25 million or more mice and rats worldwide. Some of these unneeded animals may be used for training, harvesting tissues and fluids for subsequent research, or humanely killed and donated or sold for animal feed; however, many can only be incinerated after killing because of strict regulations surrounding disposal of genetically engineered animals worldwide [56]. Some countries have tried to address the ethical concerns created by animal overproduction. In 2022, in an effort to reduce unwanted male dairy calves and male layer chicks produced on farms, German legislation was enacted that makes it illegal to kill surplus production animals without cause, including animals produced for biomedical research [57]. This approach may have the unintended consequence of driving research animal production and, ultimately, biomedical science to countries or regions with less rigorous animal welfare standards in an effort to minimize the significant financial and resource burdens of providing lifetime care to unwanted and unneeded research animals.

There are numerous reasons underlying surplus production of animals for biomedical research; however, sex bias is a significant contributor for smaller species, including rodents and rabbits (see Table 1). This can be exacerbated by age and weight restrictions for a given experiment, animal order, or assay. For example, traditional pertussis vaccine potency tests in mice have required all mice in a given group to vary by no more than 4g in body weight, driving the use of one sex because of significant body weight dimorphism in mice, and resulting in significant ordering wastage due to body weight gain variation between the time when animals arrive in the facility and when they can be studied [58]. Some have suggested using mixed sex groups for vaccine potency and challenge trials; however, this is not practical for the majority of assays in which adult animals are used in studies with a duration of three or more weeks [59] because of the risk of unwanted pregnancies. A better approach when these assays must be conducted in mice is to challenge why such tight weight ranges are required when they don't exist in the human population to whom the vaccines are administered. Broader weight range acceptability would permit animals of both sexes to be used in these assays. Annually, hundreds of thousands of mice are still used for vaccine potency testing, so the impact of this consideration is not insignificant.

Table 1. Primary issues related to overproduction of research rodents (adapted from [60]).

Issue Resulting in Overproduction of Rodents	Definition of Issue
Sex preference	Use of one sex in preference to equal use of both sexes. Remainder of lesser used sex becomes surplus.
Body weight and age requirement	Requirement for a narrow age and/or body weight requirement, such that any animals outside of these requirements become surplus (any time up to study start).
Breeding pressures	Variable demand for animals of multiple strains, for example, changes in week-to-week orders, as well as short notice orders for animals. Creates pressures to have large numbers of animals ready at a given moment.
Timed mating	Inexact procedure such that more animals are mated than are needed. Surplus also created if study is cancelled after mating has occurred.
Poor health status	Animals with clinical disease may need to be euthanized, and more animals are bred to account for these losses. In addition, a large surplus may need to be produced if mouse strains need to be replaced or rederived because of unwanted colony infections (clinical or subclinical).
Part use of litter	Selective use of animals in a litter with remainder as surplus.
Study cancellation	Sudden cancellation of studies such that it may not be possible to reallocate animals to a different study within a given institution.

Table 1. *Cont.*

Issue Resulting in Overproduction of Rodents	Definition of Issue
Genetics of breeding	Proportion of animals may not have the required genotype, and some animals may not be of appropriate quality, e.g., stunted or malformed.
Historical use	Use of specific strains that are no longer in common use because of historical database
Duplication of animal colonies	Duplication of in-house breeding colonies of multiple strains of mice and rats in large institutions
Ineffective management practices	Failure to manage breeding colonies efficiently, for example, by maintaining breeding at low levels when future use is uncertain.

When considering stock and inbred strains, sex bias generally leads to an overproduction of female rats and male mice and rabbits [60,61]. Sex bias is a less of an issue for purpose-bred large animal species, including dogs, primates, and pigs since, within biomedical research, the majority are used for toxicology studies that often require equal numbers of male and female animals [60]. It is more difficult to estimate the impact of sex bias for genetically engineered rodent colonies as there may be a large breeding surplus related to a specifically desired and restricted genotype or adverse phenotype that requires colonies to be maintained largely as heterozygotes [62]. The examples in Figure 1 below demonstrate how sex bias may contribute to overproduction of rodents from inbred, stock, and genetically modified animal colonies. It is important to note that some males and females need to be kept back as replacement breeders or to replace animals unsuitable for study (e.g., with malformations), and these form part of the managed surplus. If the scenario is repeated across many colonies or orders, the overproduction issue becomes significantly magnified.

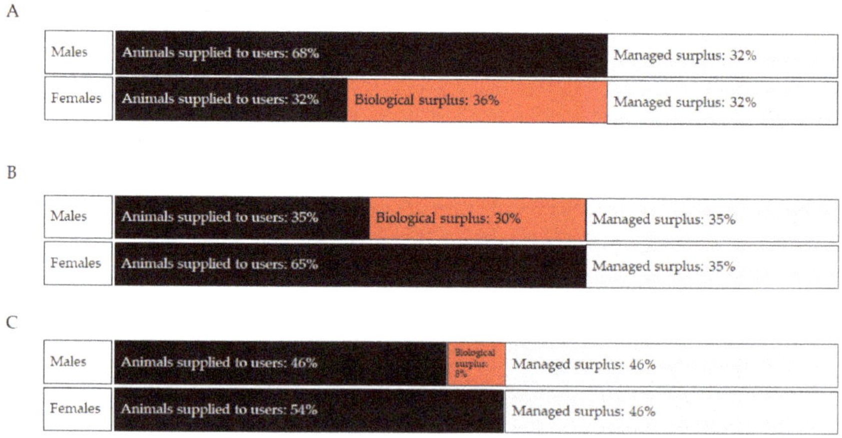

Figure 1. Use and surplus rats and mice by sex. In this example, the animals used for scientific procedures are represented by black bars, the excess animals created but not used (Managed surplus) are represented by grey bars, and the excess animals created but not used due to sex bias (Biological surplus) are represented by red bars. (**A**) Rats used and surplus animals. (**B**) Mice (inbred and outbred) used and surplus animals. (**C**) Mice (genetically modified) used and surplus animals. Adapted from [60].

4.2. Pain Recognition and Mitigation in Laboratory Animals

The pain literature, including how pain is modeled, studied, and mitigated in laboratory animals, is fraught with male bias [63]. This has contributed to challenges in identifying pain and managing it appropriately in research animals.

4.2.1. Pain Response by Sex

It has been recognized for many decades that male and female rodents respond differently to acute pain initiated by standard analgesiometry tests, with female rodents generally demonstrating a lower pain threshold to mechanical, hot thermal, chemical, and inflammatory nociception assays (for a review, see [64]). Potential reasons underlying sex differences in pain processing include the potential modulating effect of ovarian hormones on pain-evoked behaviors in females, with hypersensitivity to pain noted in the proestrus and estrus phases of the estrous cycle, that is likely linked to circulating estrogen and testosterone levels [65]. In addition, there are sex differences in neural mediation of pain, neuroimmune modulation of pain, and genetic mediation of pain, in addition to qualitative sex differences in cognitive, social, and environmental factors that modulate pain (reviewed by [63]). In contrast, for complex pain models, such as chronic inflammation and neuropathic pain, there has been unclear evidence for sex differences in rodents in pain perception [64], whereas in women, complex chronic painful conditions, such as irritable bowel syndrome, migraine, diabetic neuropathy, postoperative pain, and fibromyalgia are more commonly reported in women and last longer with a higher pain intensity [63,65]. The lack of concordance in female rodents may be due to insufficient power in study designs to detect sex differences in addition to a lack of recognition and study of biologically different processes for pain signaling between sexes [63].

Surprisingly, there has been minimal study of sex differences in pain perception for other animal species, including in research, farm, zoologic, and companion animal settings. This may be because of the expense of these models and the difficulty in achieving sufficient sample sizes of a given species and breed, let alone sex, when enrolling veterinary patients in clinical trials. It may also be due to poor recognition in veterinary medicine of species and sex differences in pain processing and response. For example, pain in cats is poorly recognized by many veterinary practitioners compared to pain in dogs, even for the same procedures [66,67]. It has only been relatively recently that neuter procedures for female cats and dogs (i.e., ovariohysterectomy) have been objectively identified to be more painful and require significantly more analgesia than neuter procedures for male cats and dogs (i.e., castration) [68]. Understanding sex differences in pain sensitivity and response in animals are areas requiring more research to support animal welfare.

4.2.2. Pain Mitigation by Sex

Despite decades of research defining sex differences in response to acute and chronic pain, a recent systematic review and meta-analysis examining potential differences in opioid-induced pain relief by sex in humans found inconclusive findings [69]. This is unlikely to be a result of sex having no effect on opioid-induced analgesia and is more likely to be a result of confounding factors. Women typically have higher fat stores compared to men, resulting in a higher apparent volume of distribution whereas men are typically larger and have faster clearance rates for drugs [70]. Differential metabolism of drugs by hepatic cytochrome P450 isoenzymes by sex may also explain differential responses to analgesic drugs. For example, because hepatic expression of Cyp3A4 is higher in women, the effects of some opioids, such as fentanyl, may be reduced compared to men. Conversely, CYP2D6 has higher expression in males, meaning that codeine and other opioids which are preferentially metabolized by CYP2D6 will have a lesser effect in males [71,72]. In mice and rats, males tend to have more body fat than females of the same species, which may skew the effects of lipophilic opioid analgesics oppositely than for humans [73]. Nonsteroidal anti-inflammatory drugs (NSAIDs) and glucocorticoids are known to have different activities and side effects in humans based on sex, likely due to differences in innate and acquired immune system activity and hormonal fluctuations during ovulation in women as well as differences between male and female NSAID pharmacokinetics and pharmacodynamics, but this is poorly characterized [74]. Minimal information is available in the veterinary or laboratory animal medicine literature about differential NSAID activities and sensitivities within a given species by sex. Certainly, to ensure good animal welfare after painful procedures this topic should be prioritized as an area of research to avoid under- and overdosing animals.

4.3. Welfare Impact of Differential Housing by Sex in Biomedical Research

An area that has received insufficient attention is the impact of differential housing of many animal species by sex in research settings (discussed from a different viewpoint in 3.6 above). This is an area that has an important impact on animal welfare as well as reproducibility and translatability of experimental findings. In conventional housing systems used in North America and elsewhere, it is not uncommon for intact, sexually mature males of certain species, including mice, guinea pigs, domestic and mini-pigs, primates, and rabbits to be housed individually because of space restrictions for housing, which generally results in animals being unable to escape agonistic interactions and move away, as would occur under more extensive housing environments. Fighting and wounding can be significant, and males may be routinely housed individually after reaching sexual maturity. In contrast, females of a given species are routinely housed in groups, with offspring, if being held for breeding. For most species, this might somewhat mimic the natural state in which few sexually mature males would be within the same social group; however, these animals would be in constant contact with females and juveniles of the same species rather than living completely solitary lives. These differential housing details are rarely mentioned in published methodology, and yet social housing of vertebrate species is thought to be a critical determinant of individual fitness and health [75]. In addition to inducing states of chronic stress, social isolation may also impact metabolism, biological rhythms, cognition, immunity and inflammation, and oxidative stress and aging in many species, including humans (reviewed by [75]). How this commonly employed differential housing environment impacts male behavior and physiology and whether this approach may skew data when only male animals are used in some types of research are unknown.

Many efforts have been made to try to socially co-house some males of some species to enhance their welfare, but it is impossible to make hard and fast rules for how to do this successfully for all breeds and/or strains of a given species within the constraints of conventional housing [76]. For example, keeping male mice from the same litter together, grouping males prior to sexual maturity, and transferring less heavily soiled nesting material at cage change have been successful methods for keeping some strains of male mice together [77]. Other factors, such as increased cage density and specific strain, were strongly predictive of significant aggression in mice [78], and more work needs to be done to find solutions for compatible housing of male research animals across species [44].

4.4. Sex Bias and Research Animal Waste

The final example of how sex bias and research animal waste can adversely impact animal welfare and the 3Rs relates to animal waste due to poor reproducibility and translatability of research. In a review of >15,000 biomedical research publications in 2014, only 50% of authors reported animal sex, and when reported, sex bias was noted and varied by preclinical model, with strongest male bias in cardiovascular studies and strongest female bias in infectious disease studies [25]. Beyond initial investigative studies, this sex bias in seeking therapeutic targets and new medicines creates a real risk for misinformation given that female animals are not simply scaled down versions of males [79]). Animal models can only be relevant for both male and female humans when both sexes are used. Single-sex studies may result in animal waste if new test articles are inactive in one sex; they can also result in human safety risks if an agent proves to be more potent in one sex, and this is not identified because of single-sex animal studies. For example, calcitonin gene-related peptide (CGRP) antagonists are of interest for the treatment of migraine in humans. Injection of CGRP triggers migraines in people, and when initially modelled in rodents, poor efficacy was noted in the CGRP-migraine model. Subsequently, when researchers returned to the original studies, it was noted that the testing was conducted exclusively in male rats. When female rats were used instead, very significant improvements were seen when CGRP antagonists were given [80]. Given that migraines occur more often in women, this reinforces the need to use both sexes in animal studies to avoid making incorrect generalizations and to avoid wasting animals in studies.

Not only is including both sexes in experimental design important, but reporting either the existence or the lack of sex difference is also critical to minimizing animal waste. There is a misconception that a finding of no sex difference (a negative result) need not be reported. When both sexes are used, it remains common practice to only report when a difference between sexes is identified. In fact, when a sex difference is identified, half of those studies treat it as a major finding and highlight the finding in the title or abstract [81]. Conversely, in the 44% of published studies in which a sex difference was not specifically found, there is no mention of evaluating the data for a sex-based effect. Sex is not uniformly treated or ignored as a biological variable between scientific fields. The sexes are most commonly compared in endocrinology studies (93%) and least often evaluated in neuroscience studies (33%) [81]. As a result of this practice, experiments may later need to be repeated in both sexes due to the missing information. This potentially contributes to significant animal waste.

5. How Can Sex Bias Be Minimized in Biomedical Research?

Increased awareness of the potential harms of sex bias is an important first step in addressing the issue; however, the scientific community needs tangible and practical solutions to help it overcome the pitfalls of sex bias. There are already several tools available to help guide the scientific planning and reporting practices, and learning how to effectively use these tools can help to propagate excellence in study design, data analysis, and reporting behaviors. Funding agencies reinforce these good behaviors and emphasize minimizing sex bias by increasingly requiring that investigators include both sexes in their research proposals or include strong scientific justification for why they are not needed. Once funded studies are completed, accountability for complete and transparent reporting in scientific reports is key to minimizing sex bias in scientific literature. If the scientific community doesn't self-govern in this space, some countries may use legislation to minimize sex bias and the associated animal wastage. On top of study design and transparent reporting practices, technology advancements also may be helpful in minimizing animal wastage when a single sex may be legitimately needed.

5.1. Awareness and Education

When there is awareness that sex bias exists and the associated welfare harms are identified, refined practices can be taught and preemptively employed to prevent sex bias at each step in the scientific process (Table 2). Mindful elimination of sex bias should follow the PREPARE (Planning Research and Experimental Procedures on Animals: Recommendations for Excellence) Guidelines through to reporting following the ARRIVE (Animal Research: Reporting of In Vivo Experiments) Guidelines [82–84].

Table 2. A guide for incorporating sex as a biological variable in animal research studies (adapted from [85]).

Study Aspect	Actionable Items
Literature search	• Perform literature search for sex difference in research area of interest • Critically evaluate the literature for quality of reference to identify sex differences • Understand the strengths and limitations of available literature
Study design	• Use both sexes of animals unless scientifically justifiable • Design studies with sufficient power to determine differences between experimental groups and sexes • Best practice: factorial design with the sample size split between males and females
Data collection	• Collect data from a similar number of animals of both sexes • Collect data on variables that could influence sex differences (e.g., housing differences)

Table 2. *Cont.*

Study Aspect	Actionable Items
Data analysis	• Separate male and female data sets and specifically test for sex differences
Reporting	• Describe study design • Report numbers, ages, and weights of males and females used • Report sex differences tested for and identified (or not)

The path toward minimizing sex bias in animal studies begins with an appropriate and rigorous literature search for sex differences in the targeted area of research interest. Evaluating the resulting search findings is important for developing an understanding of both the strengths and limitations of the current literature and will aid in the development of appropriate design of future studies. There could be known differences between the sexes that may or may not be relevant to the current research question. Critically evaluating the previous supporting work will help to determine if there is an underlying sex difference that needs to be addressed and accommodated. As part of the literature search, the PREPARE Guidelines checklist specifically recommends that researchers: (1) form a clear hypothesis with primary and secondary outcomes; (2) consider the use of systematic reviews; (3) decide upon databases and information specialists to be consulted and construct search terms; (4) assess the relevance of the species to be used, including its biology and suitability to answer the experimental questions with the least suffering and its welfare needs; and (5) assess the reproducibility and translatability of the project. A complete literature search will ultimately help minimize experimental bias, including sex bias, and inappropriate statistical methodology, which are common contributors to poor study design [82].

A well-thought-out study design is key to adequately powering a study to clearly identify and accommodate sex differences [6]. Because many researchers are not trained to do this, it may be important to engage a biostatistician who can assist with the process. Arguably, the most appropriate study design to identify a sex effect is a factorial design [7,8]. This approach reduces the need for additional research subjects while appropriately powering the experiment to identify both the desired experimental effect and any specific difference between the sexes. Factorial design simulations demonstrate that there is no loss of power to detect treatment effects when splitting the sample size across sexes in most scenarios [7]. It may be considered best practice to use a factorial experimental design and split the sample size across both male and female animals.

Once a study has been appropriately designed, using a systematic approach to conduct the experiment will help prevent introducing additional sex bias. One way to accomplish this is to use housing strategies that avoid differential housing of animals based on sex and that account for potential aggression between conspecifics of the same sex [86]. The housing system can introduce sex bias and potentially undermine the most well-designed animal study. As such, there is a need to assess housing effects on research animals and research paradigms to minimize unintended introduction of sex bias into animal studies [44]. Furthermore, blinding observers to the sex of the animals when possible will help to remove any preconceived biases that observer may have.

Using appropriate data analysis methods, including blinding of analysts to animal sex and treatment group, is important for identifying true positive differences between the sexes. It is noteworthy that finding no sex difference is just as significant as the presence of a difference. Similar to the issues with experimental design, many scientists are not trained on best practices for detecting sex-based difference. As a result, it is common for studies to incorrectly claim a sex difference when there is none, and vice versa [81]. As such, there is a need for continuing efforts to train researchers on how to appropriately test for and report sex differences in their data to promote rigor and reproducibility in biomedical research [7,8,81].

The final step in minimizing sex bias is transparent reporting of all aspects of the study. Complete reporting, following the ARRIVE 2.0 Guidelines [83], ensures that a scientist from another institution can accurately recreate the experimental condition and data analysis from the details provided in the manuscript and achieve similar experimental results. Complete and transparent reporting also allows the scientific community to assess study results for sex effects or possible sex bias in the study design or study analysis. Following this entire process from start to finish will help minimize sex bias in animal research and improve reproducibility and translatability of animal-based research.

5.2. Funding Agency Requirements

Funding agencies have recognized that sex is an important biological variable in biomedical research which should be controlled. Many now require researchers to use both sexes in their experiments or clearly justify why they are only using either males or females in their studies. This trend began with the National Institutes of Health [87] announcing a policy aimed at integrating sex as a biological variable (SABV) into biomedical research in 2014, which went into effect in January 2016. The Canadian Institutes of Health Research [88] followed by requiring applicants to specifically integrate sex and gender into experimental design. While the European Commission has had a long-standing policy to question when sex and gender are relevant in the objectives and methodologies of a project, it hasn't included a reporting requirement like the NIH or CIHR. More recently, the UK Research and Innovation Medical Research Council released their new guidance [89] that requires the specification of sex in the experimental design, effective September 2022.

5.3. Accountability in Reporting Practices

There is a long-standing need to improve the reporting of experimental methods and materials [25]. To improve the quality and utility of animal-based study results, it has been previously recommended that journals and funding agencies mandate that reporting of animal studies include complete descriptions of all experimental details [90]. In 2010, a working group sponsored by the National Centre for the Replacement, Refinement and Reduction of Animals in Research (NC3Rs) published the ARRIVE Guidelines [83]. The purpose was to improve transparency in research reporting to help address the reproducibility crisis. Improved reporting of animal sex was included in these guidelines. While many journals have endorsed these guidelines, compliance and enforcement has been poor, and there continues to be incomplete reporting [84].

Building on this need for improved accountability in reporting practices, sex has been specifically identified as a variable to report. The SAGER (Sex and gender equity in research) guidelines were released in 2016 [91]. These guidelines provide a comprehensive approach to reporting sex and gender information in study design, data analysis, results, and interpretation of findings. While the SAGER guidelines are primarily designed for and by scientific writers, they are also useful to reviewers and editors to help ask questions such as whether sex is relevant to the research in question and/or have the authors adequately addressed sex-based effects or justified the absence of such analysis.

In 2020, after 10 years in practice, the ARRIVE guidelines were updated and reorganized to facilitate their use and renamed as ARRIVE 2.0 [84,92]. These guidelines specifically described sex as a property of the sample, and an independent variable that potentially affects the outcome measures. The authors acknowledged that sex effects can be accounted for in the randomization or blocking strategy and that including sex as a variable can increase power, thereby increasing the ability to detect a real effect with fewer animals. The ARRIVE 2.0 guidelines list animal details, including sex, as item 8 of the Essential 10 minimum reporting requirements [84].

While having guidelines and checklists that are endorsed by scientific journals is a good starting point, it requires effort from all members of the scientific community to create a culture of accountability in scientific reporting. Using guidelines such as SAGER and ARRIVE 2.0 makes it easy to report important experimental variables, but consistent use of the guidelines can be difficult without behavior modification by all parties. It isn't sufficient for journals to simply endorse the guidelines; the guidelines must be fully adopted across all roles (editors, reviewers, and authors) and integrated into standard writing practices. Only together, through concerted efforts at the funding agency, institution, and publishing levels will the consideration of sex as a biological variable become standard practice in biomedical research [93].

5.4. Legislation

Using legislation to achieve sex balance of research animals is potentially an extreme approach to achieving the goal, but it is not unheard of. As mentioned previously, current German animal welfare legislation does not allow for killing of animals without a reasonable cause [94]. As a result, there have been criminal complaints filed against at least 15 biomedical research facilities for euthanasia of surplus animals [54]. While criminal charges have not been made to date, this highlights the importance of cooperative engagement and adoption of guidelines by the entire scientific community to voluntarily address sex as a biological variable to avoid such drastic measures.

5.5. Technology

Using a single sex is legitimately needed for some studies. In these cases, thoughtful uses of technology can be paramount to minimizing needless animal waste resulting from overproduction. Using sexed semen is commonplace in some fields of veterinary medicine [95–98]. Sexed semen has been used in the beef and dairy industries since 1989 to minimize production of select sexes with a high level of success [95], and this technology has been expanded to use with pigs, horses, and small ruminants [99–101]. A pilot study in Ireland demonstrated the welfare benefits of using sexed semen to reduce unwanted production of surplus male dairy calves [102]. Although not without some financial costs, use of this technology could occur for common laboratory species to minimize production of the particular unwanted sex and decrease surplus animal creation. Similarly, CRISPR-Cas 9 technology can be used to limit the in utero development of fetuses of a select sex [103]. Production of sex-specific offspring can become a heritable trait, making it easier to continue production of single sex litters in subsequent rounds of breeding. Using these technologies could significantly decrease the number of animals being euthanized due to overproduction, thus improving welfare for research animals. However, careful evaluation of single sex litters for unanticipated effects of genetic manipulation must be conducted and reported as intrauterine position and the sex of adjacent fetuses in utero have well-documented effects on a number of traits and behaviors later in life [104–107].

6. Conclusions

Sex bias in biomedical studies is bad for both scientific advancement and animal welfare. Not only has sex bias led to large financial losses for the pharmaceutical industry and harms for women as patients, but there are also significant welfare impacts for animals. There is currently a need to study both sexes in multiple research disciplines and to recognize when sex may or may not be an experimental variable. Ultimately, it is critically important that sex be considered at each stage of the scientific process. Additionally, using technology to minimize animal waste when a single sex is justified is important to ensure positive animal welfare. Collectively, these actions will improve both reproducibility and translatability to propel scientific discovery and therapeutic success while promoting animal welfare.

Author Contributions: Conceptualization, E.A.N. and P.V.T.; investigation, E.A.N. and P.V.T.; writing—original draft preparation, E.A.N. and P.V.T.; writing—review and editing, E.A.N. and P.V.T. All authors have read and agreed to the published version of the manuscript.

Funding: This work received no external funding.

Institutional Review Board Statement: Not applicable.

Informed Consent Statement: Not applicable.

Data Availability Statement: No new data were created.

Conflicts of Interest: The authors declare no conflict of interest.

References

1. Zakiniaeiz, Y.; Cosgrove, K.P.; Potenza, M.N.; Mazure, C.M. Balance of the Sexes: Addressing Sex Differences in Preclinical Research. *Yale J. Biol. Med.* **2016**, *89*, 255–259. [PubMed]
2. Cahill, L. Equal ≠ the same: Sex differences in the human brain. *Cerebrum* **2014**, *2014*, 5. [PubMed]
3. Zucker, I.; Prendergast, B.J. Sex differences in pharmacokinetics predict adverse drug reactions in women. *Biol. Sex Differ.* **2020**, *11*, 32. [CrossRef] [PubMed]
4. Beery, A.K.; Zucker, I. Sex bias in neuroscience and biomedical research. *Neurosci. Biobehav. Rev.* **2011**, *35*, 565–572. [CrossRef]
5. Woitowich, N.C.; Beery, A.; Woodruff, T. A 10-year follow-up study of sex inclusion in the biological sciences. *eLife* **2020**, *9*, e56344. [CrossRef]
6. Karp, N.A.; Fry, D. What is the optimum design for my animal experiment? *BMJ Open Sci.* **2021**, *5*, e100126. [CrossRef]
7. Phillips, B.; Haschler, T.N.; Karp, N.A. Statistical simulations show that scientists need not increase overall sample size by default when including both sexes in in vivo studies. *PLoS Biol.* **2023**, *21*, e3002129. [CrossRef]
8. Beery, A.K. Inclusion of females does not increase variability in rodent research studies. *Curr. Opin. Behav. Sci.* **2018**, *23*, 143–149. [CrossRef]
9. McNemar, Q.; Stone, C.P. Sex differences in rats on three learning tasks. *J. Comp. Psychol.* **1932**, *14*, 171–180. [CrossRef]
10. Broadhurst, P.L. Determinants of emotionality in the rat. I. Situational factors. *Br. J. Psychol.* **1957**, *48*, 1–12. [CrossRef]
11. Broadhurst, P.L.; Sinha, S.N.; Singh, S.D. The effect of stimulant and depressant drugs on a measure of emotional reactivity in the rat. *J. Genet. Psychol.* **1959**, *95*, 217–226. [CrossRef] [PubMed]
12. Van Liere, E.J.; Stickney, J.C.; Marsh, D.F. Sex Differences in Blood Pressure of Dogs. *Science* **1949**, *109*, 489. [CrossRef] [PubMed]
13. Lewis, J.J. *An Introduction to Pharmacology*; E. & S. Livingstone Ltd.: Edinburgh, UK; London, UK, 1960; p. 49.
14. Beatty, W.W. Gonadal hormones and sex differences in nonreproductive behaviors in rodents: Organizational and activational influences. *Horm. Behav.* **1979**, *12*, 112–163. [CrossRef] [PubMed]
15. Docherty, J.R.; Stanford, S.C.; Panettieri, R.A.; Alexander, S.P.; Cirino, G.; George, C.H.; Hoyer, D.; Izzo, A.A.; Ji, Y.; Lilley, E.; et al. Sex: A change in our guidelines to authors to ensure that this is no longer an ignored experimental variable. *Br. J. Pharmacol.* **2019**, *176*, 4081–4086. [CrossRef]
16. Zucker, I.; Beery, A.K. Males still dominate animal studies. *Nature* **2010**, *465*, 690. [CrossRef] [PubMed]
17. Berkley, K.J. Vive la différence! *Trends Neurosci.* **1992**, *15*, 331–332. [CrossRef] [PubMed]
18. Sechzer, J.A.; Rabinowitz, V.C.; Denmark, F.L.; McGinn, M.F.; Weeks, B.M.; Wilkens, C.L. Sex and Gender Bias in Animal Research and in Clinical Studies of Cancer, Cardiovascular Disease, and Depression. *Ann. N. Y. Acad. Sci.* **1994**, *736*, 21–48. [CrossRef]
19. Hughes, R.N. Sex does matter: Comments on the prevalence of male-only investigations of drug effects on rodent behaviour. *Behav. Pharmacol.* **2007**, *18*, 583–589. [CrossRef]
20. Clayton, J.A. Applying the new SABV (sex as a biological variable) policy to research and clinical care. *Physiol. Behav.* **2018**, *187*, 2–5. [CrossRef]
21. National Institutes of Health. Consideration of Sex as a Biological Variable in NIH-Funded Research. National Institutes of Health Notice Number: NOT-OD-15–102. Available online: https://grants.nih.gov/grants/guide/notice-files/not-od-15-102.html (accessed on 28 February 2023).
22. Clayton, J.A.; Collins, F.S. Policy: NIH to balance sex in cell and animal studies. *Nature* **2014**, *509*, 282–283. [CrossRef]
23. Sadler, K.E.; Mogil, J.S.; Stucky, C.L. Innovations and advances in modelling and measuring pain in animals. *Nat. Rev. Neurosci.* **2022**, *23*, 70–85. [CrossRef] [PubMed]
24. Ramirez, F.D.; Motazedian, P.; Jung, R.G.; Di Santo, P.; MacDonald, Z.; Simard, T.; Clancy, A.A.; Russo, J.J.; Welch, V.; Wells, G.A.; et al. Sex Bias Is Increasingly Prevalent in Preclinical Cardiovascular Research: Implications for Translational Medicine and Health Equity for Women: A Systematic Assessment of Leading Cardiovascular Journals over a 10-Year Period. *Circulation* **2017**, *135*, 625–626. [CrossRef] [PubMed]
25. Flórez-Vargas, O.; Brass, A.; Karystianis, G.; Bramhall, M.; Stevens, R.; Cruickshank, S.; Nenadic, G. Research: Bias in the reporting of sex and age in biomedical research on mouse models. *eLife* **2016**, *5*, e13615. [CrossRef] [PubMed]
26. Li, M.; Qu, Y.; Zhong, J.; Che, Z.; Wang, H.; Xiao, J.; Wang, F.; Xiao, J. Sex bias in alcohol research: A 20-year comparative study. *Front. Neuroendocrinol.* **2021**, *63*, 100939. [CrossRef]
27. Yoon, D.Y.; Mansukhani, N.A.; Stubbs, V.C.; Helenowski, I.B.; Woodruff, T.K.; Kibbe, M.R. Sex bias exists in basic science and translational surgical research. *Surgery* **2014**, *156*, 508–516. [CrossRef]
28. Navarro, K.L.; Huss, M.; Smith, J.C.; Sharp, P.; Marx, J.O.; Pacharinsak, C. Mouse Anesthesia: The Art and Science. *ILAR J.* **2021**, *62*, 238–273. [CrossRef] [PubMed]
29. Mogil, J.S.; Chanda, M.L. The case for the inclusion of female subjects in basic science studies of pain. *Pain* **2005**, *117*, 1–5. [CrossRef]

30. Meziane, H.; Ouagazzal, A.M.; Aubert, L.; Wietrzych, M.; Krezel, W. Estrous cycle effects on behavior of C57BL/6J and BALB/cByJ female mice: Implications for phenotyping strategies. *Genes Brain Behav.* **2007**, *6*, 192–200. [CrossRef]
31. Becker, J.B.; Arnold, A.P.; Berkley, K.J.; Blaustein, J.D.; Eckel, L.A.; Hampson, E.; Herman, J.P.; Marts, S.; Sadee, W.; Steiner, M.; et al. Strategies and methods for research on sex differences in brain and behavior. *Endocrinology* **2005**, *146*, 1650–1673. [CrossRef]
32. Johnson, J.L.; Greaves, L.; Repta, R. Better science with sex and gender: Facilitating the use of a sex and gender-based analysis in health research. *Int. J. Equity Health* **2009**, *8*, 14. [CrossRef]
33. Holdcroft, A. Integrating the dimensions of sex and gender into basic life sciences research: Methodologic and ethical issues. *Gend. Med.* **2007**, *4* (Suppl. B), S64–S74. [CrossRef] [PubMed]
34. Smarr, B.L.; Grant, A.D.; Zucker, I.; Prendergast, B.J.; Kriegsfeld, L.J. Sex differences in variability across timescales in BALB/c mice. *Biol. Sex Differ.* **2017**, *8*, 7. [CrossRef] [PubMed]
35. Bramlett, H.M.; Dietrich, W.D. Neuropathological protection after traumatic brain injury in intact female rats versus males or ovariectomized females. *J. Neurotrauma* **2001**, *18*, 891–900. [CrossRef] [PubMed]
36. Greenspan, J.D.; Craft, R.M.; LeResche, L.; Arendt-Nielsen, L.; Berkley, K.J.; Fillingim, R.; Gold, M.; Holdcroft, A.; Lautenbacher, S.; Mayer, E.A.; et al. Studying sex and gender differences in pain and analgesia: A consensus report. *Pain* **2007**, *132* (Suppl. S1), S26–S45. [CrossRef] [PubMed]
37. Levy, D.R.; Hunter, N.; Lin, S.; Robinson, E.M.; Gillis, W.; Conlin, E.B.; Anyoha, R.; Shansky, R.M.; Datta, S.R. Mouse spontaneous behavior reflects individual variation rather than estrous state. *Curr. Biol.* **2023**, *33*, 1358–1364. [CrossRef]
38. National Research Council of the National Academies. *Guide for the Care and Use of Laboratory Animals*, 8th ed.; The National Academies Press: Washington, DC, USA, 2011.
39. Langen, M.; Peters, U.; Körner, U.; Gissel, C.; Stanislawski, D.; Klein, G. Semiquantitative detection of male pork tissue in meat and meat products by PCR. *Meat Sci.* **2010**, *86*, 821–824. [CrossRef]
40. Gokulakrishnan, P.; Kumar, R.R.; Sharma, B.D.; Mendiratta, S.K.; Malav, O.; Sharma, D. Determination of sex origin of meat and meat products on the DNA basis: A review. *Crit. Rev. Food Sci. Nutr.* **2015**, *55*, 1303–1314. [CrossRef]
41. Peura, T.; Hyttinen, J.M.; Turunen, M.; Jänne, J. Areliable sex determination assay for bovine preimplantation embryos using the polymerase chain reaction. *Theriogenology* **1991**, *35*, 547–555. [CrossRef]
42. Ewen, K.A.; Koopman, P. Mouse germ cell development: From specification to sex determination. *Mol. Cell. Endocrinol.* **2010**, *323*, 76–93. [CrossRef]
43. Mauvais-Jarvis, F.; Arnold, A.P.; Reue, K. A Guide for the Design of Pre-Clinical Studies on Sex Differences in Metabolism. *Cell Metab.* **2017**, *25*, 1216–1230. [CrossRef]
44. Theil, J.H.; Ahloy-Dallaire, J.; Weber, E.M.; Gaskill, B.N.; Pritchett-Corning, K.R.; Felt, S.A.; Garner, J.P. The epidemiology of fighting in group-housed laboratory mice. *Sci. Rep.* **2020**, *10*, 16649. [CrossRef] [PubMed]
45. Jensen, P.; Buitenhuis, B.; Kjaer, J.; Zanella, A.; Mormede, P.; Pizzari, T. Genetics and genomics of animal behaviour and welfare—Challenges and possibilities. *Appl. Anim. Behav. Sci.* **2008**, *113*, 383–403. [CrossRef]
46. Sensini, F.; Inta, D.; Palme, R.; Brandwein, C.; Pfeiffer, N.; Riva, M.A.; Gass, P.; Mallien, A.S. The impact of handling technique and handling frequency on laboratory mouse welfare is sex-specific. *Sci. Rep.* **2020**, *10*, 17281. [CrossRef] [PubMed]
47. Lewejohann, L.; Schwabe, K.; Häger, C.; Jirkof, P. Impulse for animal welfare outside the experiment. *Lab Anim.* **2020**, *54*, 150–158. [CrossRef]
48. Canadian Council for Animal Care. CCAC Animal Data Report 2018. Available online: https://ccac.ca/Documents/AUD/2018-Animal-Data-Report.pdf (accessed on 18 June 2023).
49. Report from the Commission to the European Parliament and the Council: 2019 Report on the Statistics on the Use of Animals for Scientific Purposes in the Member States of the European Union in 2015–2017. Available online: https://op.europa.eu/en/publication-detail/-/publication/04a890d4-47ff-11ea-b81b-01aa75ed71a1 (accessed on 18 June 2023).
50. New South Wales 2018 Animal Use in Research Statistics. Available online: https://www.animalethics.org.au/__data/assets/pdf_file/0011/1200710/nsw-2018-animal-use-in-research-statistics-report.pdf (accessed on 18 June 2023).
51. United Kingdom Home Office. Animals in Science Statistics. Available online: https://www.gov.uk/government/collections/animals-in-science-statistics (accessed on 18 June 2023).
52. Federal Ministry of Food and Agriculture (Germany). Versuchstierdaten 2018: Animals Used under § 7 (2) of the Animal Protection Act by Species. Available online: https://www.bmel.de/SharedDocs/Downloads/DE/_Tiere/Tierschutz/Tierversuche/Versuchstierdaten2018.pdf?__blob=publicationFile&v=1 (accessed on 18 June 2023).
53. Carbone, L. Estimating mouse and rat use in American laboratories by extrapolation from Animal Welfare Act-regulated species. *Sci. Rep.* **2021**, *11*, 493. [CrossRef]
54. Wewetzer, H.; Wagenknecht, T.; Bert, B.; Schönfelder, G. The fate of surplus laboratory animals: Minimizing the production of surplus animals has the greatest potential to reduce the number of laboratory animals. *EMBO Rep.* **2023**, *24*, e56551. [CrossRef]
55. Hose, K.; Nagel-Riedasch, S.; Schenkel, J.; Buch, T. Use surplus laboratory animals as animal feed. *Lab Anim.* **2022**, *51*, 233. [CrossRef] [PubMed]
56. Buch, T.; Davidson, J.; Hose, K.; Jerchow, B.; Nagel-Riedasch, S.; Schenkel, J. Reducing surplus experimental animal generation. *Lab Anim.* **2022**, *56*, 305. [CrossRef]
57. Feldwisch-Drentrup, H. Germany weighs whether culling excess lab animals is a crime. *Science* **2022**, *376*, 567–568. [CrossRef]

58. Food and Drug Administration (FDA). Title 21 CFR 620.4. Code of Federal Regulations 1 April 1996. Chapter I—FDA Administration Subchapter F—BIOLOGICS, Part 620—Additional Standards for Bacterial Products, Subpart A—Pertussis Vaccine, Sec. 620.4—Potency Test. From the U.S. Government Publishing Office. Available online: https://www.govinfo.gov/content/pkg/CFR-1996-title21-vol7/html/CFR-1996-title21-vol7-sec620-4.htm (accessed on 18 June 2023).
59. Jennings, M.; Morton, D.B.; Charton, E.; Cooper, J.; Hendriksen, C.; Martin, S. Application of the Three Rs to challenge assays used in vaccine testing: Tenth report of the BVAAWF/FRAME/RSPCA/UFAW Joint Working Group on Refinement. *Biologicals* **2012**, *38*, 684–695. [CrossRef]
60. LASA. The Production and Disposition of Laboratory Rodents Surplus to the Requirements for Scientific Procedures. A Report of a LASA Taskforce Meeting Held on 12 June 1998. Available online: https://www.lasa.co.uk/PDF/Surplus.pdf (accessed on 13 June 2023).
61. National Research Council (US) Committee on Guidelines for the Use of Animals in Neuroscience and Behavioral Research. Appendix B, Estimating Animal Numbers. In *Guidelines for the Care and Use of Mammals in Neuroscience and Behavioral Research*; National Academies Press: Washington, DC, USA, 2003; pp. 181–190.
62. Robinson, V.; Jennings, M.; Working Group. Refinement and reduction in the production of genetically modified mice: Sixth report of the BVAAWF/FRAME/RSPCA/UFAW Joint Working Group on Refinement. *Altern. Lab. Anim.* **2004**, *32* (Suppl. 1A), 373–375. [CrossRef] [PubMed]
63. Mogil, J.S. Qualitative sex differences in pain processing: Emerging evidence of a biased literature. *Nat. Rev. Neurosci.* **2020**, *21*, 353–365. [CrossRef]
64. Hurley, R.W.; Adams, M.C. Sex, gender, and pain: An overview of a complex field. *Anesth. Analg.* **2008**, *107*, 309–317. [CrossRef]
65. Kaur, S.; Benton, W.L.; Tongkhuya, S.A.; Lopez, C.M.C.; Uphouse, L.; Averitt, D.L. Sex Differences and Estrous Cycle Effects of Peripheral Serotonin-Evoked Rodent Pain Behaviors. *Neuroscience* **2018**, *384*, 87–100. [CrossRef] [PubMed]
66. Dohoo, S.E.; Dohoo, I.R. Postoperative use of analgesics in dogs and cats by Canadian veterinarians. *Can. Vet. J.* **1996**, *37*, 546–551.
67. Hunt, J.R.; Knowles, T.G.; Lascelles, B.D.; Murrell, J.C. Prescription of perioperative analgesics by UK small animal veterinary surgeons in 2013. *Vet. Rec.* **2015**, *176*, 493. [CrossRef]
68. Quarterone, C.; Luna, S.P.L.; Crosignani, N.; de Oliveira, F.A.; Lopes, C.; da Maia Lima, A.F.; de Araújo Aguiar, A.J. Ovariohysterectomy requires more post-operative analgesia than orchiectomy in dogs and cats. *Can. Vet. J.* **2017**, *58*, 1191–1194.
69. Pisanu, C.; Franconi, F.; Gessa, G.L.; Mameli, S.; Pisanu, G.M.; Campesi, I.; Leggio, L.; Agabio, R. Sex differences in the response to opioids for pain relief: A systematic review and meta-analysis. *Pharmacol. Res.* **2019**, *148*, 104447. [CrossRef] [PubMed]
70. Schwartz, J.B. The influence of sex on pharmacokinetics. *Clin. Pharmacokinet.* **2003**, *42*, 107–121. [CrossRef]
71. Franconi, F.; Campesi, I. Pharmacogenomics, pharmacokinetics and pharmacodynamics: Interaction with biological differences between men and women. *Br. J. Pharmacol.* **2014**, *171*, 580–594. [CrossRef]
72. Klein, K.; Zanger, U.M. Pharmacogenomics of Cytochrome P450 3A4: Recent progress toward the "missing heritability" problem. *Front. Genet.* **2013**, *4*, 12. [CrossRef] [PubMed]
73. Smith, J.C. A Review of Strain and Sex Differences in Response to Pain and Analgesia in Mice. *Comp. Med.* **2019**, *69*, 490–500. [CrossRef] [PubMed]
74. Farkouh, A.; Baumgärtel, C.; Gottardi, R.; Hemetsberger, M.; Czejka, M.; Kautzky-Willer, A. Sex-Related Differences in Drugs with Anti-Inflammatory Properties. *J. Clin. Med.* **2021**, *10*, 1441. [CrossRef] [PubMed]
75. Lemonnier, C.; Bize, P.; Boonstra, R.; Dobson, F.S.; Criscuolo, F.; Viblanc, V.A. Effects of the social environment on vertebrate fitness and health in nature: Moving beyond the stress axis. *Horm. Behav.* **2022**, *145*, 105232. [CrossRef]
76. Kappel, S.; Hawkins, P.; Mendl, M.T. To Group or Not to Group? Good Practice for Housing Male Laboratory Mice. *Animals* **2017**, *7*, 88. [CrossRef]
77. Zidar, J.; Weber, E.M.; Ewaldsson, B.; Tjäder, S.; Lilja, J.; Mount, J.; Svensson, C.I.; Svensk, E.; Udén, E.; Törnqvist, E. Group and Single Housing of Male Mice: Collected Experiences from Research Facilities in Sweden. *Animals* **2019**, *9*, 1010. [CrossRef]
78. Lidster, K.; Owen, K.; Browne, W.J.; Prescott, M.J. Cage aggression in group-housed laboratory male mice: An international data crowdsourcing project. *Sci. Rep.* **2019**, *9*, 15211. [CrossRef]
79. Wilson, L.A.B.; Zajitschek, S.R.K.; Lagisz, M.; Mason, J.; Haselimashhadi, H.; Nakagawa, S. Sex differences in allometry for phenotypic traits in mice indicate that females are not scaled males. *Nat. Commun.* **2022**, *13*, 7502. [CrossRef]
80. Avona, A.; Burgos-Vega, C.; Burton, M.D.; Akopian, A.N.; Price, T.J.; Dussor, G. Dural Calcitonin Gene-Related Peptide Produces Female-Specific Responses in Rodent Migraine Models. *J. Neurosci.* **2019**, *39*, 4323–4331. [CrossRef]
81. Garcia-Sifuentes, Y.; Maney, D.L. Reporting and misreporting of sex differences in the biological sciences. *eLife* **2021**, *10*, e70817. [CrossRef]
82. Smith, A.J.; Clutton, R.E.; Lilley, E.; Hansen, K.E.A.; Brattelid, T. PREPARE: Guidelines for planning animal research and testing. *Lab Anim.* **2018**, *52*, 135–141. [CrossRef]
83. Kilkenny, C.; Browne, W.J.; Cuthill, I.C.; Emerson, M.; Altman, D.G. Improving Bioscience Research Reporting: The ARRIVE Guidelines for Reporting Animal Research. *PLoS Biol.* **2010**, *8*, e1000412. [CrossRef] [PubMed]
84. du Sert, N.P.; Hurst, V.; Ahluwalia, A.; Alam, S.; Avey, M.T.; Baker, M.; Browne, W.J.; Clark, A.; Cuthill, I.C.; Dirnagl, U.; et al. The ARRIVE guidelines 2.0: Updated guidelines for reporting animal research. *PLoS Biol.* **2020**, *18*, e3000410. [CrossRef]
85. Denfeld, Q.E.; Lee, C.S.; Habecker, B.A. A primer on incorporating sex as a biological variable into the conduct and reporting of basic and clinical research studies. *Am. J. Physiol. Heart Circ. Physiol.* **2022**, *322*, H350–H354. [CrossRef]

86. Garner, J.P.; Gaskill, B.N.; Weber, E.M.; Ahloy-Dallaire, J.; Pritchett-Corning, K.R. Introducing Therioepistemology: The study of how knowledge is gained from animal research. *Lab Anim.* **2017**, *46*, 103–113. [CrossRef]
87. NIH Policy on Sex as a Biological Variable. Available online: https://orwh.od.nih.gov/sex-gender/nih-policy-sex-biological-variable (accessed on 28 February 2023).
88. Canadian Institutes of Health Research. Does Sex Make a Difference in Preclinical Research? Available online: https://cihr-irsc.gc.ca/e/49346.html (accessed on 28 February 2023).
89. UK Research and Innovation. Sex in Experimental Design. Available online: https://www.ukri.org/councils/mrc/guidance-for-applicants/policies-and-guidance-for-researchers/sex-in-experimental-design/ (accessed on 28 February 2023).
90. Bolon, B. Gender agenda: Sex bias can be justified in animal research. *Nature* **2012**, *466*, 28. [CrossRef]
91. Heidari, S.; Babor, T.F.; De Castro, P.; Tort, S.; Curno, M. Sex and Gender Equity in Research: Rationale for the SAGER guidelines and recommended use. *Res. Integr. Peer Rev.* **2016**, *1*, 2. [CrossRef]
92. du Sert, N.P.; Ahluwalia, A.; Alam, S.; Avey, M.T.; Baker, M.; Browne, W.J.; Clark, A.; Cuthill, I.C.; Dirnagl, U.; Emerson, M.; et al. Reporting animal research: Explanation and elaboration for the ARRIVE guidelines 2.0. *PLoS Biol.* **2020**, *18*, e3000411. [CrossRef]
93. Tannenbaum, C.; Ellis, R.P.; Eyssel, F.; Zou, J.; Schiebinger, L. Sex and gender analysis improves science and engineering. *Nature* **2019**, *575*, 137–146. [CrossRef] [PubMed]
94. Bundesministerium der Justiz. Available online: https://www.gesetze-im-internet.de/tierschg/BJNR012770972.html (accessed on 25 May 2023).
95. Lonergan, P. Review: Historical and futuristic developments in bovine semen technology. *Animal* **2018**, *12*, s4–s18. [CrossRef]
96. Neculai-Valeanu, A.S.; Ariton, A.M. Game-Changing Approaches in Sperm Sex-Sorting: Microfluidics and Nanotechnology. *Animals* **2021**, *11*, 1182. [CrossRef] [PubMed]
97. Wheeler, M.B.; Rutledge, J.J.; Fischer-Brown, A.; VanEtten, T.; Malusky, S.; Beebe, D.J. Application of sexed semen technology to in vitro embryo production in cattle. *Theriogenology* **2006**, *65*, 219–227. [CrossRef]
98. Li, J.; Zhu, S.; He, X.; Sun, R.; He, Q.; Gan, Y.; Liu, S.; Funahashi, H.; Li, Y. Application of a microfluidic sperm sorter to in vitro production of dairy cattle sex-sorted embryos. *Theriogenology* **2016**, *85*, 1211–1218. [CrossRef]
99. Roca, J.; Parrilla, I.; Bolarin, A.; Martinez, E.A.; Rodriguez-Martinez, H. Will AI in pigs become more efficient? *Theriogenology* **2016**, *86*, 187–193. [CrossRef]
100. Zuidema, D.; Kerns, K.; Sutovsky, P. An Exploration of Current and Perspective Semen Analysis and Sperm Selection for Livestock Artificial Insemination. *Animals* **2021**, *11*, 3563. [CrossRef]
101. Squires, E. Current Reproductive Technologies Impacting Equine Embryo Production. *J. Equine Vet. Sci.* **2020**, *89*, 102981. [CrossRef] [PubMed]
102. Balzani, A.; Aparacida Vaz do Amaral, C.; Hanlon, A. A perspective on the use of sexed semen to reduce the number of surplus male dairy calves in Ireland: A pilot study. *Front. Vet. Sci.* **2021**, *7*, 623128. [CrossRef]
103. Douglas, C.; Maciulyte, V.; Zohren, J.; Snell, D.M.; Mahadevaiah, S.K.; Ojarikre, O.A.; Ellis, P.J.I.; Turner, J.M.A. CRISPR-Cas9 effectors facilitate generation of single-sex litters and sex-specific phenotypes. *Nat. Commun.* **2021**, *12*, 6926. [CrossRef] [PubMed]
104. vom Saal, F.S. Sexual differentiation in litter-bearing mammals: Influence of sex of adjacent fetuses in utero. *J. Anim. Sci.* **1989**, *67*, 1824–1840. [CrossRef]
105. Ryan, B.C.; Vandenbergh, J.G. Intrauterine position effects. *Neurosci. Biobehav. Rev.* **2002**, *26*, 665–678. [CrossRef]
106. Clark, M.M.; Galef, B.G., Jr. A gerbil dam's fetal intrauterine position affects the sex ratios of litters she gestates. *Physiol. Behav.* **1995**, *57*, 297–299. [CrossRef] [PubMed]
107. Fishman, R.; Vortman, Y.; Shanas, U.; Koren, L. Cortisol advantage of neighbouring the opposite sex in utero. *R. Soc. Open Sci.* **2018**, *5*, 171636. [CrossRef] [PubMed]

Disclaimer/Publisher's Note: The statements, opinions and data contained in all publications are solely those of the individual author(s) and contributor(s) and not of MDPI and/or the editor(s). MDPI and/or the editor(s) disclaim responsibility for any injury to people or property resulting from any ideas, methods, instructions or products referred to in the content.

animals

MDPI

Article

Are Currently Selected Laboratory Animals Useful in the Research of How Female Hormones Influence Orthodontic Biomechanics?

Małgorzata Peruga [1,*], Beata Kawala [2], Michał Sarul [3], Jakub Kotowicz [4] and Joanna Lis [5]

1 Independent Researcher, 93-410 Łódź, Poland
2 Department of Dentofacial Orthopaedics and Orthodontics, Wroclaw Medical University, 50-376 Wrocław, Poland
3 Department of Integrated Dentistry, Wroclaw Medical University, Krakowska 26, 50-425 Wroclaw, Poland
4 Independent Researcher, 37-700 Przemyśl, Poland
5 Independent Adult Orthodontics Clinic, Department of Dentofacial Orthopaedics and Orthodontics, Wroclaw Medical University, 50-376 Wrocław, Poland
* Correspondence: gosiaperuga@gmail.com

Simple Summary: Animal experiments should be carried out in consultation with veterinarians, who, having clinical knowledge, will help doctors choose the appropriate animal model. Our review of the literature shows how the harmful and unethical duplication of research from other research centers can be. Familiarization with the experimental protocols on an animal model is important each time we want to later extrapolate the obtained results to the target species, i.e., human. This is due to the law on the implementation of the procedures on people. This article focuses on orthodontic teeth movement. While experiments with laboratory animals seem easy, there are many pitfalls. Our goal in this article is to collect data on the maintenance of laboratory animals as models and to critically analyze them based on our literature review.

Abstract: Animal testing was and remains the only method of introducing a certain treatment and medical procedure on humans. On the other hand, animals have their rights resulting from applicable legal acts, including Directive 2010/63/EU and, indirectly, the World Medical Association International Code of Medical Ethics (Helsinki Declaration, 1975, amended 2000). Thus, the question arises whether the credibility of the results of hormonal and orthodontic tests obtained so far and their usefulness for the human population is scientifically justified and worth sacrificing laboratory animals for. Especially that, according to statistical data, about 50% of laboratory animals are euthanized at the conclusion of the experiments. The aim of this article was to determine whether animal experiments are scientifically or morally justified in bringing significant evidence in studies that may validate the influence of changes in the concentration of female hormones secreted by the ovaries in various phases of the menstrual cycle in young patients on the duration of an increased tooth movement rate in orthodontic treatment. Papers reporting the results of the original research into female hormones, either natural or exogeneous ones, likely to alternate the orthodontic tooth movement rate were critically evaluated in terms of animal selection. Thorough analysis supported by veterinary knowledge proved that none of the publications enabled an extrapolation of the results to humans. The evaluation of the relation between the rate of tooth movement upon loading with orthodontic forces and hormones either secreted during the menstrual cycle of women or released from the contraceptives already present in the market, does not require sacrificing laboratory animals.

Keywords: veterinary ethics; qualitative and quantitative research designs; animal experimental model; steroid hormones

Citation: Peruga, M.; Kawala, B.; Sarul, M.; Kotowicz, J.; Lis, J. Are Currently Selected Laboratory Animals Useful in the Research of How Female Hormones Influence Orthodontic Biomechanics? . *Animals* **2023**, *13*, 629. https://doi.org/10.3390/ani13040629

Academic Editor: Melanie L. Graham

Received: 17 November 2022
Revised: 15 January 2023
Accepted: 8 February 2023
Published: 10 February 2023

Copyright: © 2023 by the authors. Licensee MDPI, Basel, Switzerland. This article is an open access article distributed under the terms and conditions of the Creative Commons Attribution (CC BY) license (https://creativecommons.org/licenses/by/4.0/).

1. The Use of Laboratory Animals in Orthodontics

Animal testing was and remains the only method of introducing a certain treatment and medical procedure on humans. The social acceptance of such experiments has been fervently debated since 13th century, when Saint Thomas Aquinas saw animals as useful machines, whilst Saint Francis of Assisi believed them to be humans' smaller brothers. Descartes continued the debate in the 16th century. However, the real rebellion aimed at anti-vivisection was pioneered by a woman, Fanny Matin, who in 1845 married Claude Bernard who was infamous for conducting experiments on stray dogs [1]. Noteworthy, at that time women had no voting rights, therefore Fanny Matin's opposition to vivisection reverberated across the scientific world and turned out to be the driving force behind the shift in the researchers' approach to respecting animals. Female sensitivity was also the catalyst for a mass protest in London. A conflict between Professor William Bayliss, a scientist who experimented on a terrier dog, thus discovering hormones, and whom the majority of the society considered to be a sadist, ended with a bronze monument of a dog being erected in London. The fate of a simple terrier had become the spur for adopting legislation protecting laboratory animals. However, it was not until the 21st century that the world of science changed forever: animals now have their 'protection' under existing legislation, including Directive 2010/63/EU [2], and it is primarily recognized in European Union countries. This directive was transposed into Polish legislation in 2015 [3], and states that animals must be provided with medical and veterinary care and conditions adequate to their health and until their natural death regardless of whether the experiment has ended. However, it must be emphasized that animals subjected to experiments very rarely fully recover, so their "natural" death issue is rather controversial [4]. Furthermore, animals react and metabolize substances differently in comparison to humans. For example, the lethal dose of potassium cyanide for rabbits and mice is, respectively, twice and seven times higher (obviously relative to their body mass) than for humans. In this term, many researchers, such as Ober [5] as well as Azkona [6], believe that in the case of interdisciplinary research work, cooperation between clinicians and scientists and qualified veterinarians is necessary.

Since experimental animals are present throughout all fields of science related to human healthcare, unsurprisingly, they are also involved in the development of dentistry, specifically in the field of orthodontics. The treatment of malocclusions in the case of young adult women, aged 20 to 30, is naturally linked to a menstrual cycle, which constitutes a sequence of recurring fluctuations of hormones that prepare the body for potential pregnancy: estrogens and especially progesterone. These hormones regulate osteoblast activity and the production of collagen, responsible—among other things—for a proper periodontal structure [7–12]. The natural cycle of sex hormones physiological fluctuations is disrupted by ovariectomy or hormonal contraception taken by a patient. Although modern oral contraceptives contain much lower doses of hormones when compared to medication used in the past, they still may damage periodontium, as they contribute shifting a balance between aerobe and anaerobe microorganisms towards the latter [13–15].

Any impairment in the alveolar bone structure immediately affects the rate of orthodontically induced tooth movement. It is therefore quite likely that if loading the teeth is planned in accordance with the menstrual cycle, one obtains the most efficient tooth movement due to the highest "plasticity" of periodontal tissues, which results in shortening the treatment time. Nevertheless, the World Medical Association International Code of Medical Ethics (Helsinki Declaration, 1975, amended 2000) requires that medical support, including orthodontic procedures, be carried out on at least two species, animals, before it can be used in humans [16]. Therefore, since the orthodontic experiment must burden laboratory animals so much, the credibility of the results of hormonal and orthodontic tests obtained so far and their usefulness for the human population should be critically assessed. Especially that, according to statistical data, about 50% of laboratory animals are euthanized at the conclusion of the experiments [17,18].

In this commentary, we aimed to determine whether animal experiments are scientifically or morally justified in bringing significant evidence to studies that may validate the influence of changes in the concentration of female hormones secreted by the ovaries in various phases of the menstrual cycle in young patients on the duration of an increased tooth movement rate in orthodontic treatment.

To this end, we searched PubMed, Elsevier ScienceDirect Journals, and EBSCOhost (Medline) electronic databases for laboratory research performed on animals, concerning orthodontic movement depending on the level of hormones during the menstrual cycle and while using hormonal contraceptives. We included review, clinical, comparative, intervention papers, and unusual cases, using the following keywords: estrogen, progesterone, hormonal contraception, menstrual cycle, heat, orthodontics, and orthodontic treatment. We disregarded the research performed on humans, as well as in vitro or in silico texts. We found 12 studies published between 1997 and 2022, which discussed the related subjects we target in this review.

2. Selected Studies

The impact of female hormones and hormonal contraception on orthodontic movement has been the subject of research in various centers, done on different species of animals. Most tests were done on Wistar rats, i.e., white rats used in pharmacological, toxicological, nutrition, and behavioral testing, but similar studies were performed on rabbits and cats. The canines, premolars, and incisors were moved predominantly. Open orthodontic springs and nickel and titanium wires were usually used for this purpose, with some studies using bone anchoring, i.e., mini implants. Some of the animals had their blood tested, others had their vaginal mucus tested. The tooth movement ranges were measured on gypsum dental casts poured on the examination day or inside the animals' mouths using a caliper. Most animals were euthanized after the research. Only healthy females took part in testing, with their own heat cycle or sterilized and exogenously administered with hormones [15,19–29].

The publications were divided into groups and specific publications were selected based on the scheme shown in Table 1.

Table 1. List of research methods and results obtained in papers analyzed as part of the review.

Authors	Animals and Their Division	Tooth and Type of Tooth Shift	Material Used	Method Used to Test the Hormone Level	Results
Olyaee et al. [15]	Wistar rats Age—3 months N = 48 m = 250 ± 25 g females	Central incisors tipped distally	SS spring with a diameter of 0.35 mm	Administration of ethinyl estradiol/norgestrel	Ethinyl estradiol/norgestrel (oral contraceptives) can decrease the amount of tooth movement
Guo, Zhao, Chen [19]	Wistar rats Age—not given N = 120 m = not given females	Not given	Not given	Estradiol level in serum and periodontium tissue using radioimmune and immune-cytochemical methods	Estrogen affects teeth movement
Zhao, Than, Guo, Chen [20]	Wistar rats Age—not given N = not given m = not given females	Not given	Not given	Estradiol level in serum and periodontium tissue using radioimmune and immune-cytochemical methods	Estrogen affects teeth movement
Guo, Zhao, Chen [21]	Wistar rats Age—3 months N = 80 m = not given females	Left upper incisor and left upper molar Tipped	Not given	Not given	Teeth movement dependent on cycle

Table 1. *Cont.*

Authors	Animals and Their Division	Tooth and Type of Tooth Shift	Material Used	Method Used to Test the Hormone Level	Results
Zhao, Than, Guo, Chen [22]	Wistar rats Age—not given N = not given m = not given females	Not given	Not given	Not given	Estrogen affects teeth movement
Guo, Zhao, Chen [23]	Rats, strain not given Age—not given Not given N = 200 m = not given females	Not given	Not given	Estradiol level in serum	Estrogen affects teeth movement
Haruyama et al. [24]	Wistar rats Age—10 weeks N = 85 m = 136 g females	right and left, first upper molar Tipped	NiTi spring with a diameter of 0.012 inch	Acc. to estrous cycle with vaginal smear	Estrus can increase tooth movement
Tan et al. [25]	Wistar rats Age—3 months N = 200 m = 300 g females	Left distal incisor and first left molar Tipped	NiTi spring with a diameter of 0.012 inch	Monitoring estrus cycle and vaginal smear	Estrus can increase tooth movement
Celebi [26]	Domestic cat Age—2–4 years N = 18 m = not given females	Jaw canine and mini implant Tipped	NiTi spring with a diameter of 0.2 inch	According to estrus cycle	Teeth movement speed was higher in sterilized specimens
Sirisoontorn [27]	Wistar rats Age—10 weeks N = 10 m = 170–190 g females	Left distal incisor and and first left molar Tipped	NiTi spring with a diameter of 0.012 inch	Monitoring estrus cycle and vaginal smear	Teeth movement speed was higher in sterilized specimens
Mackie et al. [28]	Sprague Dawley rats Age—6 weeks N = 55 m = 160 ± 20 g females	Right distal incisor and first right molar Tipped	NiTi spring with a diameter of 0.03 × 0.01 inch	Not given	Estrus can increase tooth movement
Poosti et al. [29]	Rabbit Age—8 weeks N = 24 m = 1850 g females	Central incisors tipped distally	SS spring with a diameter of 0.014 inch	Administration of progesterone	Progesterone affects tooth movement

Legend: N—number of animals used in testing; m—body mass of animal used in testing; SS—stainless steel; NiTi—nitinol.

3. Cross-Species Comparison with Humans

Factors crucial for hormonal and orthodontic experiments, differentiating women from female animals selected for testing, are shown in Table 2.

Table 2. Comparison of laboratory animals in terms of dentition, periodontium, reproductive cycle, body temperature.

Species	Size	Dental Formula	Teeth	Periodontium	Hormones	Body Temperature
Human (*Homo sapiens*)	Target species	$\frac{212}{212}\ \frac{2123}{2123}$	Diphyodont Heterodont Bunodont	Thecodont Brachydont	Cycle 24–33 days Ovulation 24 h	36.6 °C
Laboratory rat (*Rattus*)	Acceptable size	X $\frac{1003}{1003}$	Monophyodont heterodont	Thecodont Monophyodont Incisors—hypsodont, elodont Molars—brachydont	Cycle 4–5 days Spontaneous ovulation Estrus 10–20 h	37.5–39 °C
Rabbit (*Oryctolagus cuniculus*)	Acceptable size	X $\frac{2033}{1032}$	Monophyodont Heterodont	Thecodont Hypsodont Elodont	Cycle 16 days Induced ovulation	38.5–40 °C
Domestic cat (*Felis catus*)	Acceptable size	$\frac{313}{312}\ \frac{3131}{3121}$	Diphiodont Heterodont Secodont	Thecodont Brachydont	14–21 days	38–39 °C

Definition of terms (acc. to Kobryń, H.; Kobryńczuk, F.; Krysiak, K. *Anatomia Zwierząt Tom 1–3 (Animal Anatomy Vol. 1–3)*; PWN: Warsaw, Poland, 2011 [30]): X- does not occur; monophyodont—having one set of teeth; diphyodont—having two set of teeth; heterodont—with tooth shape differences between incisors, canines, premolars, and molars; thecodont—tooth embedded in a socket; brachydont—having short crowns with short growth time; hypsodont—having high crowns with long growth time; secodont—having sharp enameled teeth; bunodont—having rounded enameled teeth; elodont—teeth with an open top. An animal's dentition for either deciduous (first fraction) or permanent (second fraction) teeth expressed as a dental formula, written in the form of a fraction, as $\frac{I.C.P.M}{I.C.P.M}$ maxillary arch (above the line) and mandibular arch (below the line) I—incisors, C—canine, P—premolar, and M—molar, f.ex human 2123—2-I, 1-C, 2-C, 3-M.

4. Rats

Rats (Figure 1a,b) procreate quickly and their genes have also been well-mapped. Their transgenic strains, possible to be created nearly exclusively in small rodents, regardless of certain difficulties in maintaining their permanence, are still an important aspect favoring the selection of rats as experimental animals. This choice is also determined by the size of the animal, which facilitates the conduct of treatments and the preparation of histopathological materials. Rats are also cheap, which helps in testing their large groups. This is likely why rats were experimental animals in the majority of reviewed studies [15,19–25,27,28]. The evaluated studies theoretically demonstrated that the forced tooth movement rates are dependent on menstrual cycle or hormonal contraception. However, studies by Guo et al. [19,20,22] and Zhao et al. [21,23], performed at various intervals on various groups of rats, do not provide data for the measurements of the tooth movement and orthodontic biomechanics in relation to the heat cycle. The full-text articles are available only in Chinese, so we were able to review their abstracts, which merely summarized the research and did not fully report the results. We attempted to contact the researchers, but they did not reply to our emails. In turn, Mackie et al. [28] performed their tests on a strain other than Wistar, namely Sprague Dawley characterized by longer jaws, which naturally increases the distance between the incisors and the molars. Longer wires and springs are more easily deformed during chewing, when the so-called trampoline effect caused by the occlusal forces becomes evident [31]. Furthermore, the study used very young, 6-week-old specimens additionally subjected to stress that might drastically change their hormone levels; rats are less willing to reproduce when experiencing distress. As many as 4 specimens out of 55 died from stress, which undermines the entire test results, as the intensity of the metabolism is significantly increased during stress. In another study, also done on young rats, Olyaee et al. [15] used a spring between two incisors, which also challenges the reliability of the outcomes; due to the high probability of the not-fused palatal suture, the diastema quite likely resulted not from the tooth movement itself, but from separating the maxillary halves. Thus, it may be concluded that choosing a rat as an animal model for hormonal and orthodontic research is also unfounded. Although, similarly to humans, gaps between rat teeth (Figure 1c–e) are sufficiently wide to secure the natural drift of molars, and the teeth may be affected by caries; this is where the similarities end. Aside from the many evident differences (Table 2), the incisors in rats have enamel only on their front surfaces and they grow throughout the animals' lifetimes, which requires continuous sharpening of those teeth; in addition, the enamel is harder than metals such as iron, platinum, and copper. Continuous tooth eruption does not provide anchorage and sufficient control over the direction of force, which may lead to bias while interpreting the published data.

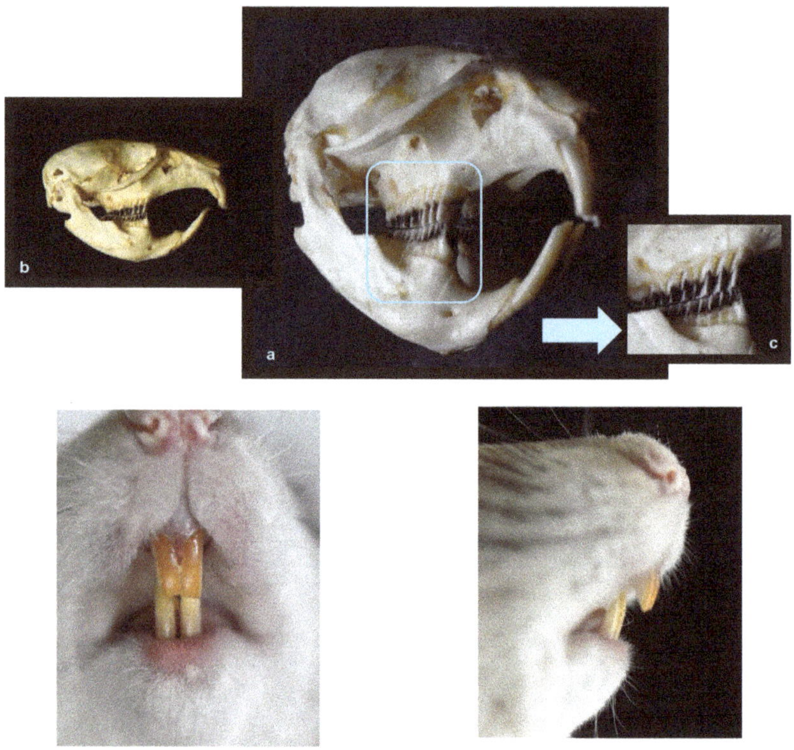

(**d**) (**e**)

Figure 1. Rat teeth in a skull: semi profile view (**a**), lateral view (**b**), premolars and molars view (**c**). Rat teeth: incisors, frontal view—enlarged overbite (**d**) and incisors, lateral view—enlarged overjet (**e**) (by Małgorzata Peruga).

As for the periodontal ligaments (PDL) of rats, they are built of connective tissue collagen fibers, which requires vitamin C (synthesized by rats in their kidneys or liver) to grow. Rats therefore do not have to obtain vitamin C exogenously, in contrast to humans, who do not produce L-gulonolactone oxidase (GULO), an enzyme contributing to vitamin C synthesis. It is worth noting here that although rats without GULO have been bred, they did not correctly reproduce the vitamin deficits seen in humans, which made the extrapolation of the obtained results impossible.

5. Rabbits

Rabbits, and specifically their thigh bones, are used in dentistry as a material for research into the osseointegration of implants. However, rabbit is an animal with a fragile anatomical structure, in particular its limbs, which often fracture under a load. Rabbits not only show little aggression toward humans but are also the smallest and the cheapest animals whose sperm can be harvested and used for artificial insemination. They produce tears and have large eyeballs, which facilitates the testing of chemical substances. However, apart from research into irritating substances, there is scant information available on other experimental studies. Poosti et al. [29] demonstrated that orthodontic movement changed after a female rabbit was provided with hormones from a human female that have a different chemical structure to their own hormones. The teeth of rabbits (Figure 2a–c), namely the incisors and the molars, can grow throughout their lifetimes. The incisors grow even two to three millimeters per week. Rabbits' teeth consist of clinical and anatomic crowns almost entirely covered by a layer of enamel absent at the top of the tooth, in the growth center. The PDL area is very limited (Figure 3a–c), which modifies the tooth behavior under loading with occlusal or orthodontic forces [32–34]. Mastication is also very different compared with humans. After reaching occlusal interdigitation, the rabbits' mandibular teeth rest on pegs (second part of the maxillary incisors) in a reversed overjet. To achieve normal occlusion, a rabbit must unilaterally and partially dislocate the mandibular condylar process from the fossa to close the arcade (Figure 4a,b).

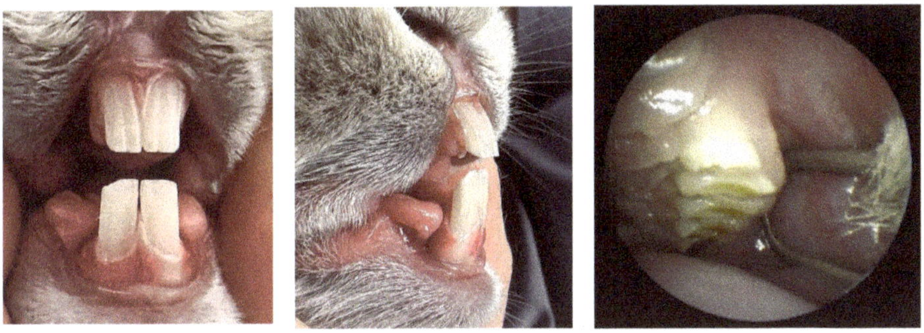

(a) (b) (c)

Figure 2. Rabbit teeth: incisors frontal view—overbite (**a**) incisors lateral view—overjet (**b**) and premolars and molars view (**c**) (by Jakub Kotowicz).

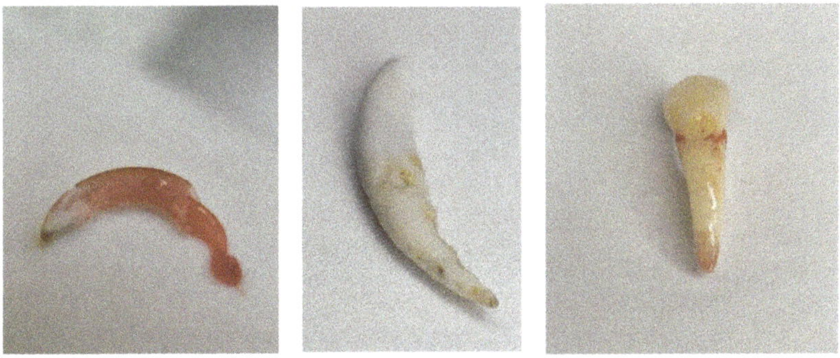

(a) (b) (c)

Figure 3. Rodent incisor with open top and living pulp (**a**). Comparison with thecodont and brachydont dog tooth (**b**) and human tooth (**c**) (by Małgorzata Peruga).

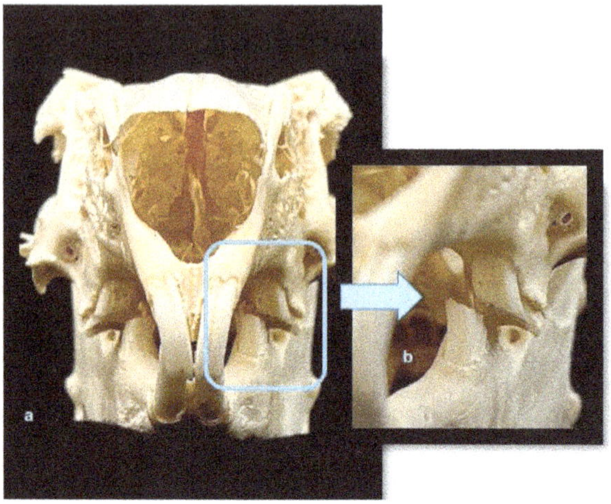

Figure 4. Anisognathism in rabbits: in 1:1 scale (**a**), enlarged (**b**) (by Małgorzata Peruga and Jakub Kotowicz).

Choosing rats and rabbits as test animals to assess the tooth movement during either a menstrual cycle or the administration of hormonal contraceptives is also controversial due to the fact that the estrus of these animals is too short [15,19–25,27–29,32–35] (Table 2) to notice visible changes in the three-stage process caused by orthodontic forces. Moreover, small animals do not match human biologically and do not live for long, not allowing for longitudinal studies.

6. Cats

Celebi et al. [26] used the domestic cat as a research model in hormonal testing. However, cats have no masticating surfaces (Figure 5a–c). The arrangement of their teeth in the dental arches and high degree of nodularity precludes using cats for orthodontic research, unless the bite is raised, which the authors did not mention describing their methodology (Table 2). Although cats are used as laboratory animals, due to the fact that their brain demonstrates the closest similarity to humans, these animals are primarily used in neurological, ophthalmological, and immune deficiency research, and not in hormonal and orthodontic studies.

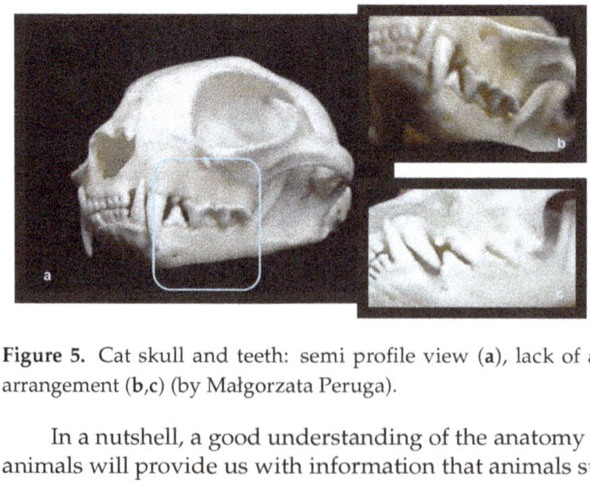

Figure 5. Cat skull and teeth: semi profile view (**a**), lack of adhesion of the side teeth—scissor arrangement (**b**,**c**) (by Małgorzata Peruga).

In a nutshell, a good understanding of the anatomy and physiology of experimental animals will provide us with information that animals such as rats, rabbits, or cats were lacking validity as laboratory animals for orthodontic movement research. The analyzed articles (Table 1) found numerous errors in the research assumptions, due to the fact that the authors did not take into account the important aspects that were pointed out, such as the different anatomical structure of teeth and periodontal and completely different occlusions.

7. Hormone Cycle

Most studies into the relation between the rate of the tooth movement and the action of hormones secreted during the menstrual cycle were done on animals, who were administered human progesterone, estrogens, and relaxin [36–41]. As far as progesterone is concerned, experimental research on rats has demonstrated that the hormone modified the orthodontic movement of their teeth by affecting the periodontium and elasticity of the cortical plate of the alveolar process. On the other hand, the long-term administration of progesterone to rabbits resulted in a reduced rate of the tooth movement; the authors concluded that it was due to the fact that osteoclasts are observed primarily 2 days after an orthodontic force is applied [31]. Administering relaxin to rats resulted in an increase in the rate of orthodontic tooth movement when compared to control groups, as well as the stretching of periodontium made of soft tissue [42–45]. Unfortunately, as the normal hormonal cycle of animals was disrupted in every reviewed study, it cannot be stated with certainty whether the changes in the rate of tooth movements were caused only by the excessive amount of artificially introduced hormones or a disruption in the natural hormonal balance, particularly given the short observation period.

8. Orthodontic Materials

Most of the reviewed studies used materials made from nickel and titanium (NiTi) alloy, which has been popular in orthodontic treatment since the 1970s. Its elasticity modulus approximately equals 20% of the modulus of the stainless steel, which secures a very wide scope of working elasticity. The complex metallurgical nature of nickel and titanium materials and its relation to clinical application have been the subject of many scientific studies. It has been demonstrated that the NiTi alloy has two phases. The first phase, austenitic, has an ordered structure, whereas the second phase, martensitic, is a highly strained body-centered tetragonal form. Shape memory is linked to the reversible transformation of martensite into austenite, which occurs as a result of a crystallographic process. The microstructure of alloys in the temperature found in the human oral cavity (36.1–37 °C) is not fully austenitic. The temperature of the complete transformation to austenite is 40 °C, which is evidently higher than the temperature in the oral cavity. That is why the alloy will behave differently in patients breathing through their mouth (27 °C) or consuming hot food (40 °C). The issue of the natural bodily temperature of animals [46] cannot therefore be ignored, as it changes the behavior of the nickel and titanium spring.

9. Conclusions

Many experimental tests have so far been done on various species of animals to obtain a better understanding of their biological reactions to orthodontic forces. Unfortunately, one of the main problems related to animal experiments is the fact that their results cannot be extrapolated to humans. The two- or even four-year lifespans of rats and rabbits, respectively, together with short estrus, enable us only to observe the immediate effects of the tooth loading, which are impossible to extrapolate on humans due to severe dissimilarities of the tooth anatomy, their PDL, as well as female hormone secretion cycles. Additionally, the lack of long-term results, which are the most reliable scientifically, is a serious limitation of the so-far designed experimental studies. The orthodontic treatment of humans with fixed appliances has been ethically accepted since at least the beginning of the previous century. Thus, evaluating the relationship between the rate of tooth movement upon loading with orthodontic forces, and hormones either secreted during the menstrual cycle of women or released from the contraceptives already present in the market, does not require experimental testing. It is enough to interview the patient and to determine the adequate timing of a force application, with subsequent measuring of the rate of the tooth movement during scheduled appointments. Perhaps a "reductio ad absurdum" argument is irresponsible; however, orthodontic treatment is carried out successfully and, above all, does not involve any risk, because the general principles of mechanics and biomechanics are known. Our results should encourage researchers to analyze the methods and selection of animals for research in more detail. There is no moral justification for performing orthodontic examinations on rats, rabbits, or cats, which we have proved.

Author Contributions: Conceptualization M.P.; methodology, M.P. and J.L.; software, M.P.; validation, M.P.; formal analysis, M.P. and J.L.; investigation, M.P.; resources, M.P.; data curation, M.P. and J.L.; writing—original draft preparation, M.P.; writing—review and editing, J.L., M.S. and B.K.; visualization, M.P. and J.K.; supervision, B.K., M.S. and J.L.; project administration, M.P. and J.L. All authors have read and agreed to the published version of the manuscript.

Funding: This research received no external funding.

Institutional Review Board Statement: Ethical review and approval were waived for this study because it is a retrospective analysis of medical records.

Informed Consent Statement: Not applicable.

Data Availability Statement: Not applicable.

Acknowledgments: Thanks to Jerzy Nadolski, Katarzyna Majecka and Robert Kamiński from Natural History Museum of the University of Lodz https://www.muzeumprzyrodnicze.uni.lodz.pl (accessed on 9 February 2023) for their help.

Conflicts of Interest: The authors declare no conflict of interest.

References

1. Midgley, M. *Animals and Why They Matter*; University of Georgia Press: Georgia, GA, USA, 1998; p. 28.
2. Directive 2010/63/EU of the European Parliament and of the Council of 22 September 2010 on the Protection of Animals Used for Scientific Purposes. Available online: https://eur-lex.europa.eu/legal-content/EN/TXT/PDF/?uri=CELEX:32010L0063&from=pl (accessed on 9 February 2023).
3. Ustawa z Dnia 15 Stycznia 2015 r. o Ochronie Zwierząt Wykorzystywanych do Celów Naukowych lub Edukacyjnych. Available online: https://isap.sejm.gov.pl/isap.nsf/DocDetails.xsp?id=WDU20150000266 (accessed on 9 February 2023).
4. Goñi-Balentziaga, O.; Ortega-Saez, I.; Vila, S.; Azkona, G. A survey on the use of mice, pigs, dogs and monkeys as animal models in biomedical research in Spain. *Lab. Anim. Res.* **2022**, *38*, 14. [CrossRef]
5. Ober, R.A.; Ho, J.W.; Kemp, M.T.; Keeney-Bonthrone, T.P.; Geist, G.E.; Alam, H.B. Culture and collaboration between the clinician-scientist and veterinary specialist: An essential interprofessional partnership in the translational sciences. *Lab. Anim.* **2022**, *51*, 95–97. [CrossRef]
6. Azkona, G.; Sanchez-Pernaute, R. Mice in translational neuroscience: What R we doing? *Prog. Neurobiol.* **2022**, *217*, 102330. [CrossRef] [PubMed]
7. Peruga, M.; Antoszewska-Smith, J. Orthodontic tooth movement and changes in hormone levels during the menstrual cycle. *Clin. Orthod.* **2018**, *4*, 23–27.
8. Magowan, B.A.; Owen, P.; Thomson, A. *Clinical Obstetrics and Gynaecology*; Elsevier: Amsterdam, The Netherlands, 2022.
9. Tal, R.; Taylor, H.S. Endocrinology of Pregnancy. In *Endotext [Internet]*; Feingold, K.R., Anawalt, B., Boyce, A., Chrousos, G., de Herder, W.W., Dhatariya, K., Dungan, K., Hershman, J.M., Hofland, J., Kalra, S., et al., Eds.; MDText.com, Inc.: South Dartmouth, MA, USA, 2000.
10. Speroff, L.; Fritz, M.; Fritz, M.A. *Clinical Gynecologic Endocrinology and Infertility*; Lippincott Williams & Wilkins: Philadelphia, PA, USA, 2005.
11. Sperling, M. *Pediatric Endocrinology*; Elsevier: Amsterdam, The Netherlands, 2014.
12. Calina, D.; Docea, A.O.; Golokhvast, K.S.; Sifakis, S.; Tsatsakis, A.; Makrigiannakis, A. Management of Endocrinopathies in Pregnancy: A Review of Current Evidence. *Int. J. Environ. Res. Public Health* **2019**, *16*, 781. [CrossRef] [PubMed]
13. Hamed, N.A.; Mirza, K.B.; AL-Rubaie, M.S. Effects of Oral Contraceptives Intake on the Gingiva. *Iraqi Postgrad. Med. J.* **2010**, *9*, 335–341.
14. Brunet, L.; Miranda, J.; Farre, M. Gingival Enlargement induced by drugs. *Drug Saf.* **1996**, *15*, 219–231. [CrossRef]

15. Olyaee, P.; Mirzakouchaki, B.; Ghajar, K.; Seyyedi, S.A.; Shalchi, M.; Garjani, A.; Dadgar, E. The effect of oral contraceptives on orthodontic tooth movement in rat. *Med. Oral Patol. Oral Cir. Bucal* **2013**, *18*, e146–e150. [CrossRef]
16. WMA. *The Declaration of Helsinki*; WMA: Helsinki, Finland, 1964.
17. Available online: https://ec.europa.eu/environment/chemicals/lab_animals/pdf/SWD_%20part_A_and_B.pdf (accessed on 9 February 2023).
18. Available online: https://eur-lex.europa.eu/legal-content/EN/TXT/PDF/?uri=CELEX:52020DC0015&from=EN (accessed on 9 February 2023).
19. Guo, J.; Zhao, Q.; Chen, Y.X.; Zeng, X.L. Effects of orthodontic tooth movements on serum and local estrogen expression. *Shanghai Kou Qiang Yi Xue* **2007**, *16*, 618–622.
20. Guo, J.; Zhao, Q.; Chen, Y.X. A bio-mechanism study of differential orthodontic tooth moving speed during the estrous cycle. *Hua Xi Kou Qiang Yi Xue Za Zhi* **2008**, *26*, 327–330.
21. Zhao, Q.; Tan, Z.; Guo, J.; Chen, Y.X. Influences of applying force during the different stages of estrous cycle on orthodontic tooth movement of rats. *Hua Xi Kou Qiang Yi Xue Za Zhi* **2005**, *23*, 480–482. [PubMed]
22. Guo, J.; Che, X.X.; Zhao, Q.; Chen, Y.X. An experimental study on the relationship between the orthodontic tooth movements and menstrual cycle. *Shanghai Kou Qiang Yi Xue* **2007**, *16*, 187–191.
23. Zhao, Q.; Tan, Z.; Guo, J.; Chen, Y.X. Influences of orthodontic tooth movement on estrous cycle and estrogen in rats. *Chung Hua Kou Chiang Hsueh Tsa Chih* **2006**, *41*, 90–91. [PubMed]
24. Haruyama, N.; Igarashi, K.; Saeki, S.; Otsuka-Isoya, M.; Shinoda, H.; Mitani, H. Estrous-cycle-dependent variation in orthodontic tooth movement. *J. Dent. Res.* **2002**, *81*, 406–410. [CrossRef]
25. Tan, Z.; Zhao, Q.; Chen, Y. The mutual effects between orthodontic tooth movement and estrous cycle or estrogen. *Biol. Rhythm. Res.* **2010**, *41*, 75–81. [CrossRef]
26. Celebi, A.A.; Demirer, S.; Catalbas, B.; Arikan, S. Effect of ovarian activity on orthodontic tooth movement and gingival crevicular fluid levels of interleukin-1β and prostaglandin E(2) in cats. *Angle Orthod.* **2013**, *83*, 70–75.
27. Sirisoontorn, I.; Hotokezaka, H.; Hashimoto, M.; Gonzales, C.; Luppanapornlarp, S.; Darendeliler, M.A.; Yoshida, N. Tooth movement and root resorption; The effect of ovariectomy on orthodontic force application in rats. *Angle Orthod.* **2011**, *81*, 570–577. [CrossRef] [PubMed]
28. Mackie, M.A.; Momeni Danaei, S.H.; Habibagahi, S.; Tanide, N.; Montazeri, M. Effect of Ovariectomy on Orthodontic Tooth Movement in Rats. *J. Iran. Dent. Assoc.* **2016**, *28*, 86–91. [CrossRef]
29. Poosti, M.; Basafa, M.; Eslami, N. Progesterone effects on experimental tooth movement in rabbits. *J. Calif. Dent. Assoc.* **2009**, *37*, 483–486.
30. Kobryń, H.; Kobryńczuk, F.; Krysiak, K. *Anatomia Zwierząt Tom 1–3 (Animal Anatomy Vol. 1–3)*; PWN: Warsaw, Poland, 2011.
31. Bennett, J.C.; McLaughlin, R.P. *Fundamentals of Orthodontic Treatment Mechanics*; Legrander Publishing: London, UK, 2014.
32. Mapara, M.; Thomas, B.S.; Bhat, K.M. Rabbit as an animal model for experimental research. *Dent. Res. J.* **2012**, *9*, 111–118.
33. Mizoguchi, Y.; Matsuoka, T.; Asano, Y. Changes in blood parameters in New Zealand White rabbits during pregnancy. *Lab. Anim.* **2010**, *44*, 33–39. [CrossRef] [PubMed]
34. Hanke, H.; Hanke, S.; Finking, G.; Muhic-Lohrer, A.; Muck, A.O. Different Effects of Estrogen and Progesterone on Experimental Atherosclerosis inFemale Versus Male Rabbits. *Circulation* **1996**, *2*, 175–181. [CrossRef]
35. Ren, Y.; Maltha, J.C.; Kuijpers-Jagtman, A.N. The rat as a model for orthodontic tooth movement—A critical review and a proposed solution. *Eur. J. Orthod.* **2004**, *26*, 483–490. [CrossRef] [PubMed]
36. Hellsing, E.; Hammarström, L. The effects of pregnancy and fluoride on orthodontic tooth movements in rats. *Eur. J. Orthod.* **1991**, *13*, 223–230. [CrossRef] [PubMed]
37. He, Z.; Chen, Y.; Luo, S. Effects of pregnancy on orthodontic tooth movements: Effects of progesterone on orthodontic tooth movements in pregnant rats. *Hua Xi Kou Qiang Yi Xue Za Zhi* **1998**, *16*, 124–126.
38. Bartzela, T.; Türp, J.C.; Motschall, E.; Maltha, J.C. Medication effects on the rate of orthodontic tooth movement: A systematic literature review. *Am. J. Orthod. Dentofac. Orthop.* **2009**, *135*, 16–26. [CrossRef]
39. Dschietzig, T.; Bartsch, C.; Baumann, G.; Stangl, K. Relaxin—A pleiotropic hormone and its emerging role for experimental and clinical therapeutics. *Pharmacol. Ther.* **2006**, *112*, 38–56. [CrossRef] [PubMed]
40. Liu, Z.J.; King, G.J.; Gu, G.M.; Shin, J.Y.; Stewart, D.R. Does human relaxin accelerate orthodontic tooth movement in rats? *Ann. N. Y. Acad. Sci.* **2005**, *1041*, 388–394. [CrossRef]
41. Stewart, D.R.; Sherick, P.; Kramer, S.; Breining, P. Use of relaxin in orthodontics. *Ann. N. Y. Acad. Sci.* **2005**, *1041*, 379–387. [CrossRef] [PubMed]
42. Nicozisis, J.L.; Nah-Cederquist, H.D.; Tuncay, O.C. Relaxin affects the dentofacial sutural tissues. *Clin. Orthod. Res.* **2000**, *3*, 192–201. [CrossRef] [PubMed]
43. Tyrovola, J.B.; Spyropoulos, M.N. Effects of drugs and systemic factors on orthodontic treatment. *Quintessence Int.* **2001**, *32*, 365–371. [PubMed]
44. Yamashiro, T.; Takano-Yamamoto, T. Influences of ovariectomy on experimental tooth movement in the rat. *J. Dent. Res.* **2001**, *80*, 1858–1861. [CrossRef] [PubMed]

45. Arslan, S.G.; Arslan, H.; Ketani, A.; Hamamci, O. Effects of estrogen deficiency on tooth movement after force application: An experimental study in ovariectomized rats. *Acta Odontol. Scand.* **2007**, *65*, 319–323. [CrossRef] [PubMed]
46. Brantley, W.; Eliades, T. *Orthodontic Materials: Scientic and Clinical Aspects*; Thieme: Stuttgart, Germany, 2001.

Disclaimer/Publisher's Note: The statements, opinions and data contained in all publications are solely those of the individual author(s) and contributor(s) and not of MDPI and/or the editor(s). MDPI and/or the editor(s) disclaim responsibility for any injury to people or property resulting from any ideas, methods, instructions or products referred to in the content.

animals MDPI

Article

Brain Organoids to Evaluate Cellular Therapies

Ana Belén García-Delgado [1,2], Rafael Campos-Cuerva [1,3], Cristina Rosell-Valle [1], María Martin-López [1], Carlos Casado [1], Daniela Ferrari [4], Javier Márquez-Rivas [3,5], Rosario Sánchez-Pernaute [1,*] and Beatriz Fernández-Muñoz [1,*]

1 Unidad de Producción y Reprogramación Celular (UPRC), Red Andaluza de Diseño y Traslación de Terapias Avanzadas (RAdytTA), Fundación Progreso y Salud, 41092 Sevilla, Spain
2 Departamento de Farmacia y Tecnología Farmacéutica, Facultad de Farmacia, Universidad de Sevilla, 41012 Sevilla, Spain
3 IBiS, Instituto de Biomedicina de Sevilla, 41013 Sevilla, Spain
4 Department of Biotechnology and Biosciences, University Milan-Bicocca, 20126 Milan, Italy
5 Departamento de Neurocirugía, Hospital Virgen del Rocío, 41013 Sevilla, Spain
* Correspondence: rossapernaute@gmail.com (R.S.-P.); beatriz.fernandez.munoz@juntadeandalucia.es (B.F.-M.)

Simple Summary: Animal models are routinely used in pre-clinical studies to evaluate the safety and efficacy of novel therapies, such as cell transplantation, but have limited predictive value. In this study, we set up an experimental model using human stem cells grown in 3D, which form rudimentary brain structures in vitro, called brain organoids. We investigated the possibility of using these brain organoids to evaluate the safety of a cell therapy product, by comparing the results obtained in our model with the standard mouse model. Our results suggest that brain organoids can be informative in the evaluation of cell therapies, helping to reduce the number of animals used in regulatory studies.

Abstract: Animal models currently used to test the efficacy and safety of cell therapies, mainly murine models, have limitations as molecular, cellular, and physiological mechanisms are often inherently different between species, especially in the brain. Therefore, for clinical translation of cell-based medicinal products, the development of alternative models based on human neural cells may be crucial. We have developed an in vitro model of transplantation into human brain organoids to study the potential of neural stem cells as cell therapeutics and compared these data with standard xenograft studies in the brain of immunodeficient NOD.Cg-*Prkdc*^{scid} *Il2rg*^{tm1Wjl}/SzJ (NSG) mice. Neural stem cells showed similar differentiation and proliferation potentials in both human brain organoids and mouse brains. Our results suggest that brain organoids can be informative in the evaluation of cell therapies, helping to reduce the number of animals used for regulatory studies.

Keywords: brain organoids; cell therapy; neural stem cells; neural progenitors; translation; 3 Rs; reduction

Citation: García-Delgado, A.B.; Campos-Cuerva, R.; Rosell-Valle, C.; Martin-López, M.; Casado, C.; Ferrari, D.; Márquez-Rivas, J.; Sánchez-Pernaute, R.; Fernández-Muñoz, B. Brain Organoids to Evaluate Cellular Therapies. *Animals* **2022**, *12*, 3150. https://doi.org/10.3390/ani12223150

Academic Editor: Garikoitz Azkona

Received: 6 October 2022
Accepted: 10 November 2022
Published: 15 November 2022

Copyright: © 2022 by the authors. Licensee MDPI, Basel, Switzerland. This article is an open access article distributed under the terms and conditions of the Creative Commons Attribution (CC BY) license (https://creativecommons.org/licenses/by/4.0/).

1. Introduction

Cell therapies are perceived as medicines of the future and many pharmaceutical companies are now investing in the development of advanced therapies to treat different pathologies that to date have no cure [1]. However, for their translation into clinical practice, these therapies face numerous challenges such as their high cost, regulatory requirements for authorization and lack of predictive in vivo models to achieve effective translation [2,3].

For the development of cell therapeutics, a regulatory requirement is to perform efficacy and safety studies. These experiments are usually performed in animal models, most often in murine models, and may include homologous and heterologous models. Homologous studies in which mouse cells are transplanted in mouse models of the disease generally use mice with a C57BL/6J genetic background [4,5]. Heterologous models are required to test the proliferation and differentiation potential of the cell therapeutic that will be used in humans and are commonly evaluated in nude and NSG immunodeficient mice to avoid rejection of human cells [6–8]. These animal models have many limitations and it has been shown that results from mice experiments are rarely predictive of the human outcome [9]. These problems are even more accentuated in translational neuroscience research, since the brain structure lacks much complexity in rodents and fails to mimic many key features of the human brain [10,11]. Non-human primates can provide more relevant information because of their homology to humans. However, high costs and the need for special animal facilities, among other limitations, hinder generalization of experiments with these animals. In addition to the lack of predictability, animal experiments are expensive and laborious, and there is an increasing demand from society and regulatory bodies to reduce the number of animals used for research [11].

In vitro models based on human cells cultured in 2D have been proposed as alternative systems to test specific safety and efficacy attributes of new therapeutics. However, 2D cultures also have important limitations. Growing cells as monolayers can lead to alterations in the cytoskeleton, nucleus and cell shape, resulting in altered gene expression patterns [12,13]. Furthermore, 2D cultures fail to reproduce the 3D nature of in vivo environments and lack crucial cell–cell as well cell–matrix interactions, limiting their usefulness [14,15]. Therefore, for an effective translation of new cell therapies to the clinic, the development of more complex and predictive human models is crucial.

In recent years, alternative 3D in vitro models, which could replace the animal model in certain studies have been developed. Among these, the most promising are organoids, which are stem cell-derived mini-organs developed in the laboratory [16]. The emergence of organoids has generated tremendous interest in the field of regenerative medicine [17]. Brain organoids are 3D in vitro aggregates generated from human induced pluripotent stem cells (hiPSC), or human embryonic stem cells (hESC) that reproduce specific brain structures simulating different human brain regions in their composition and cytoarchitecture [18]. They produce neural stem cells (NSC) that develop into mature cerebral cell types and recapitulate both the transcriptome and the epigenome of the fetal brain [19,20]. In this way, brain organoids allow obtaining in vitro information about the human brain to study brain development and cellular interconnections, and are emerging as an excellent in vitro model for the investigation of mechanisms involved in neurological diseases and for drug testing [21]. For example, cortical organoids have been successfully used as drug screening tools to test different chemical drugs for Rett Syndrome [22,23]. However, the applicability of organoids for the evaluation of stem cell-based therapeutics has not been explored to date.

NSC are multipotent cells that can potentially differentiate into neurons and glial cells and secrete trophic and immunomodulatory factors. All these properties make NSC attractive for regenerative therapies and several clinical-grade NSC-based medicinal products have been tested in clinical trials [24]. Evidence of their safety and efficacy profiles has been collected in all those studies from experiments performed in immunodeficient animals, mostly rodents.

Here, we have investigated whether human brain organoids can be used as in vitro models for the evaluation of human cell-based therapeutics. We successfully transplanted human NSC into brain organoids and compared these data with standard xenograft studies in the brain of NOD.Cg-*Prkdc*^{scid} *Il2rg*^{tm1Wjl}/SzJ (NSG) mice. Our results show that brain organoids can be informative when exploring human NSC therapeutic potential, helping to facilitate preclinical studies and to reduce the number of animals used for regulatory studies.

2. Materials and Methods

2.1. Organoid Generation

The hiPSC line CBiPSS1sv-4F-5, derived from CD133$^+$ umbilical cord cells (CB-hiPSC), was used to generate cortical organoids. This cell line is available from the Spanish National Repository, and data regarding characterization can be downloaded at https://eng.isciii.es

(accessed on 1 November 2022). The use of CB-hiPSC was approved by the Andalusian Ethical Committee of Research with Biological Samples of an Embryonic Origin and Similar Cells (PR-02-2016). CB-hiPSC were maintained on Matrigel-coated plates, cultured in mTeSR^{TM} Plus medium (Stemcell Technologies, Vancouver, British Columbia, Canada, Cat. #100-0276). Normal karyotype and pluripotency were verified before organoid generation (Figure S1A,B).

Brain organoids were generated and cultured following a protocol slightly modified from that described by Lancaster et al. [19,25] (Figure 1A). Briefly, on day 0, 3000 hiPSC/well were seeded in mTeSR^{TM} plus medium with 10 µM ROCK inhibitor Y-27632 (Tocris Bioscience, Minneapolis, MN, USA, Cat. 1254) in agarose microwells to generate uniform embryoid bodies (EB). Agarose microwells were prepared with a Micro Tissues 3D Petri Dish spheroids micro-mold (size L, 9 × 9 array; Sigma, St. Louis, MO, USA, Cat. #Z764019-6EA) with 2% D1 Low EEO agarose (Pronadisa by Conda, Torrejón de Ardoz, Madrid, Spain, Cat 8010.22) in Dulbecco's PBS without $CaCl_2$ and $MgCl_2$ (Sigma, St. Louis, MO, USA, Cat. D8537). The solidified microwells were equilibrated with 2.5 mL mTeSR^{TM} plus medium for 15 min at 37 °C before use. For 6 days, EB were fed every other day with mTeSR^{TM} plus medium without ROCK inhibitor and then changed to neural induction media (NIM) containing Dulbecco's modified Eagle Medium (DMEM):F12 (Gibco by Thermo Fisher Scientific, Waltham, MA, USA, Cat. 21331020), 1X N2 supplement (Gibco by Thermo Fisher Scientific, Waltham, MA, USA, Cat. 17502-048), Glutamax (Gibco by Thermo Fisher Scientific, Waltham, MA, USA, Cat. 35050-061), minimum essential media–non essential amino acids (MEM-NEAA) (Sigma, St. Louis, MO, USA, Cat. P4999) and 1 µg/mL heparin 1000 UI/mL (Rovi, Madrid, Spain, Cat. 641747). Then, EB were fed every other day with NIM for 4–5 days. On day 10 of the protocol, organoids were covered with Matrigel (Corning, New York, NY, USA, Cat. 354277) and grown in differentiation media containing a 1:1 mixture of DMEM/F12 and Neurobasal (Gibco by Thermo Fisher Scientific, Waltham, MA, USA, Cat. 21103-049) with 1X N2 supplement, 2X B27 supplement without vitamin A (Gibco by Thermo Fisher Scientific, Waltham, MA, USA, Cat. 12587-10), 1X Glutamax, 1X MEM-NEAA, 1X P/S (Sigma, St. Louis, MO, USA, Cat. P4333), 0.09% 2-mercaptoethanol (Gibco by Thermo Fisher Scientific, Waltham, Massachusetts, USA, Cat. 21985-023) and 0.025% insulin (Sigma, St. Louis, MO, USA, Cat. I9278). After 4 days of stationary growth, the organoids were transferred to an orbital shaker (Celltron, Infors HT, Bottmingen, Switzerland) installed in a 37 °C incubator with differentiation media, but adding B27 supplement with vitamin A (Gibco by Thermo Fisher Scientific, Waltham, MA, USA, Cat. 17504-044).

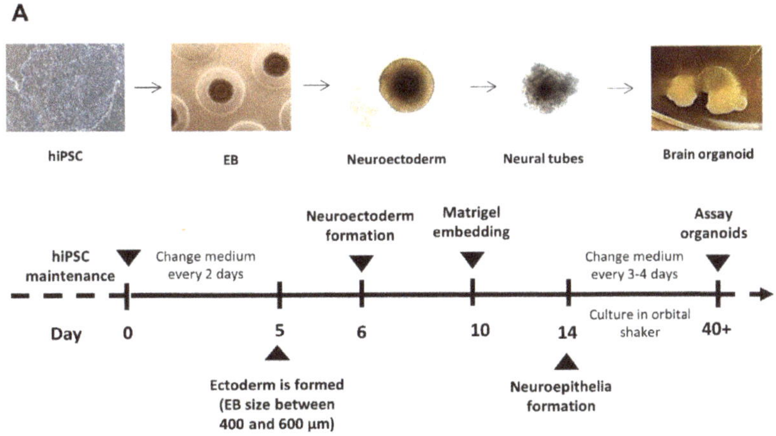

Figure 1. *Cont.*

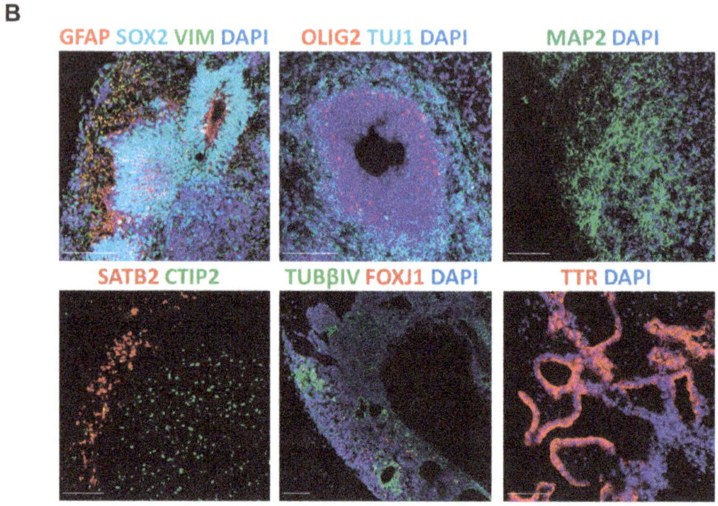

Figure 1. Generation of human brain organoids. (**A**) Protocol used for the generation of brain organoids. EB: embryoid body; hiPSC: human induced pluripotent stem cells. (**B**) Formation of brain structures was assessed by detection of different markers by immunofluorescence: Glial fibrillary acidic protein (GFAP), SOX2, Vimentin (VIM), OLIG2, Beta-III-tubulin (TUJ1), MAP2, SATB2, CTIP2, Tubulin-beta-IV (TUBβIV) and transthyretin (TTR). Scale bar: 100 µm.

2.2. Transplantation into Brain Organoids

For transplantation experiments, we used human NSC isolated from the germinal zone of the ventral forebrain (Gz-NSC) and purified for the expression of the stem cell marker $CD133^{+}$ [26]. For transplantation, two different NSC lines were first expanded and transduced with a lentiviral vector with the green fluorescent protein containing a nuclear localization sequence (EGFP-NLS) under the constitutive promoter of spleen focus-forming virus (SFFV). NLS allows EGFP to enter in the nucleus of NSC facilitating the quantification of transplanted cells. Lentiviral particles were pseudotyped with the VSVG protein (vesicular stomatitis virus G protein). Viral particles were added to the culture medium at a concentration of 25,000 TU/mL (transduction units/mL) for 6 h. Subsequently, the medium was removed and EGFP expression was confirmed under a fluorescence microscope at 24–48 h.

For cell transplantation, brain organoids were selected and individually transferred to a low-adhesion 24-well plate without culture medium. Brain organoids were injected under a stereomicroscope (SMZ1500, Nikon, Tokyo, Japan) located inside a laminar flow cabinet using a Hamilton syringe with a 30GA small Hub RN sterile needle. Then, 1 µL of cell suspension (1×10^5 NSC-EGFP in hypothermosol (HTS, Biolife Solutions, Bothell, WA, USA Cat. 101102) was slowly injected (0.25 µL/30 seg). At least 3 brain organoids were transplanted with each of the NSC-EGFP lines and with HTS alone (1 µL) to control for background fluorescence. After transplantation, medium was slowly added to the wells and organoids were maintained 24 h without agitation. Then, they were cultured again under agitation until the end of the experiment. The presence of fluorescent transplanted cells ($NSC\text{-}EGFP^{+}$) inside the brain organoids was evaluated using a Nikon Eclipse Ti-S inverted fluorescence microscope at different time points after transplantation.

Three weeks after the transplant, brain organoids were fixed with 4% paraformaldehyde (PFA, Santa Cruz Biotechnology, Dallas, TX, USA, Cat. sc-281692) for 20 min at room temperature (RT), equilibrated in 30% sucrose (VWR, Radnor, PL, USA, Cat. M117-1KG) in phosphate buffer saline (PBS, Gibco by Thermo Fisher Scientific, Waltham, MA, USA, Cat. 14190-144), embedded in optimal cutting temperature compound (OCT, Sakura Finetek Inc.,

Torrance, CA, USA, Cat. 4583) and kept at −80 °C. Once frozen, serial sections of 20 µm thickness were cut in the cryostat (CM 3050 S, Leica, Wetzlar, Germany) and processed for immunofluorescence analysis. At least 2 organoids transplanted with each NSC line were analyzed.

2.3. Transplantation into the Brain of Immunodeficient Animals

Animals were anesthetized by inhalation with sevoflurane (5% sevoflurane, 2% oxygen) in an anesthesia induction chamber and were later transferred to a stereotaxic frame where 3 µL of cell suspension (3×10^5 NSC in HTS) were injected into the caudate-putamen (anteroposterior: 0 mm; dorsolateral: −2.5 mm from bregma; dorso-ventral −3 mm from dura matter) [27] close to the lateral ventricle. In total, 8 animals were transplanted with each of the NSC lines. Three weeks after transplantation, animals were anesthetized by injection of a sublethal dose of 30–40 mg/kg of thiobarbital intraperitoneally and transcardially perfused with a saline solution followed by 4% PFA diluted in cold PBS. The brains were removed and post-fixed in 4% PFA overnight, cryoprotected in 30% sucrose in PBS and then, embedded in OCT and kept at −80 °C. Once frozen, 20 µm-thick serial sections were cut in the cryostat and immunofluorescence studies were carried out. Expression of the different markers was investigated in at least 2 animals transplanted with each NSC line. Animal care and experimental procedures were conducted according to the current National and International Animal Ethics Guidelines and approved by the Research Ethics Committee of University Hospital Virgen Macarena and Virgen del Rocío (0287-N-20).

2.4. Immunofluorescence

Immunofluorescence experiments were performed on mouse brains and human brain organoids serial sections, which were permeabilized and blocked in PBS with 10% Donkey Serum (DKS, Sigma, St. Louis, MO, USA, Cat. D9663) and 0.1% Triton X-100 (Sigma, St. Louis, MO, USA, Cat. T8787) for 90 min. An antigen retrieval step with 10mM citrate buffer (Thermo Fisher Scientific, Waltham, Massachusetts, USA, Cat. J63950) was performed previously to overnight incubation with the primary antibodies at 4°C. Samples were subsequently incubated with secondary antibodies and Hoechst for 1 h at RT and mounted with ProLong™ Gold Antifade Mountant (Thermo Fisher Scientific, Waltham, MA, USA, Cat. P36930). Primary and secondary antibodies are listed in Table S1. Acquisition of fluorescence images was performed in a Nikon Eclipse Ti-S or a Leica TCS-SP5 confocal microscope. Images were processed using Image J 1.53t software developed at the National Institutes of Health and the Laboratory for Optical and Computational Instrumentation (LOCI, University of Wisconsin, Madison, WI, USA) [28]. For cell counting, at least 100 transplanted cells were counted for each condition.

2.5. Statistics

Data are represented as mean ± standard deviation (SD). Significance was determined using two-tailed Student's *t* test for comparisons between two groups. A two-way analysis of variance (ANOVA) was used to compare cell proliferation of each line in the two model systems (mouse brain versus human brain organoids). $p < 0.05$ was considered significant. All statistical analyses were performed using the GraphPad Prism 8.01 software (San Diego, CA, USA).

3. Results

3.1. Transplantation into Human Brain Organoids

We generated brain organoids from hiPSC following the original protocol designed by Lancaster and colleagues [19,25] with few modifications, the most important of them being the use of agarose wells to favor uniform EB formation (Figure 1A).

We maintained organoids for 1–2 months in culture. The presence of several human brain tissues and cell types was characterized by immunofluorescence analysis. We identified structures resembling the ventricular zone (VZ) with many NSC positive for the stem cell transcription factor SOX2 and relatively more mature cell types located in the outer part of the VZ, such as neuroblasts positive for doublecortin (DCX) or neuron specific beta III tubulin (TUJ1), precursors of oligodencrocytes positive for OLIG2 and more mature neurons positive for the neuronal marker, microtubule-associated protein 2 (MAP2). We also detected more mature structures such as cerebral cortex, with early-born and late-born cortical neurons identified by the transcriptional regulators SATB2 and CTIP2 forming rudimentary layers, ependyma (positive for TUBβIV and FOXJ1) and non-neuronal cuboid epithelia characteristic of the choroid plexus, marked by the expression of transthyretin (TTR) (Figure 1B). We next analyzed several markers expressed during human brain development at two time points. We found that the expression of the early forebrain marker FOXG1 and the NSC marker SOX2 decreased with time, while the expression of the neuroblast markers DCX and GAD increased with time, indicating progressive maturation of cells in the brain organoids (Figure S1C).

For transplantation experiments, we used human NSC derived from the germinal zones (Gz-NSC) of the developing human brain. These cells have been purified for the expression of the stem cell marker CD133^{+} and have been extensively characterized by us [26]. Gz-NSC lines isolated from different donors showed different levels of expression of the ventral marker NKX2.1, depending on the gestational age of the donor. For this study, we selected a line with high expression of NKX2.1 and another line with low expression of NKX2.1 to investigate whether results were independent of the line differentiation bias. Before transplantation, we transduced the two lines of Gz-NSC with a lentiviral vector containing EGFP with an NLS in order to detect the grafted cells in the organoid. Gz-NSC were effectively transduced, and we verified that after transduction, they did not lose the expression of the NSC markers Nestin and SOX2 and maintained their inherent regional identity determined by the expression of NKX2.1 (Figure 2A).

For cell transplantation into brain organoids, we evaluated two technical options: (1) organoid–cell coculture, where the Gz-NSC forms a neurosphere that fuses with the surface and cells slowly penetrate the organoid (Figure 2B); (2) injection into the organoid parenchyma (Figure 2C).

For this study we decided to perform the injection into the organoid parenchyma because it can help cell integration and it is more similar to the procedure in the animal model. We injected the two Gz-NSC lines transduced with EGFP in 3-month-old organoids. The brain organoids generated with our protocol were around 5 mm in size and injection was easily performed with a Hamilton microsyringe under a stereomicroscope (Figure 2C). The organoids were individually located in wells of a 24-well low binding plate without medium and injections were performed by gently immobilizing the organoid against the wall of the well (Figure 2C). Some organoids were mock transplanted with vehicle to control for background fluorescence. After injection, medium was slowly added, and organoids were maintained for 24 h without agitation to favor recovery after the puncture. In preliminary studies, we slowly injected 3 μL per organoid, but the organoids collapsed and most of the fluid came out. In this study, we injected 1 μL and there was no evident reflux, and the organoid structure was maintained with no visible signs of damage. However, 24 h after injection, we could identify some neurospheres floating around the organoid, indicating that some cells were ejected out of the organoid. Using live cell imaging, we confirmed that NSC-EGFP^{+} injected into brain organoids survived, and were integrated within the human tissue (Figure 2D). We detected cells in all transplanted organoids 24 h after cell injection, but 3 weeks later, NSC-EGFP^{+} cells were visible in only 61% (19/31) of all transplanted organoids.

Analysis by immunofluorescence 3 weeks after transplantation confirmed that cells from both donor cell lines survived and integrated into the organoids (Figure 3A). Injected cells mostly differentiated into DCX^{+} neuroblasts with some OLIG2^{+} oligodendrocyte precursor cells and few GFAP^{+} astrocytes, while other cells maintained an immature, Nestin^{+} NSC phenotype (Figure 3B).

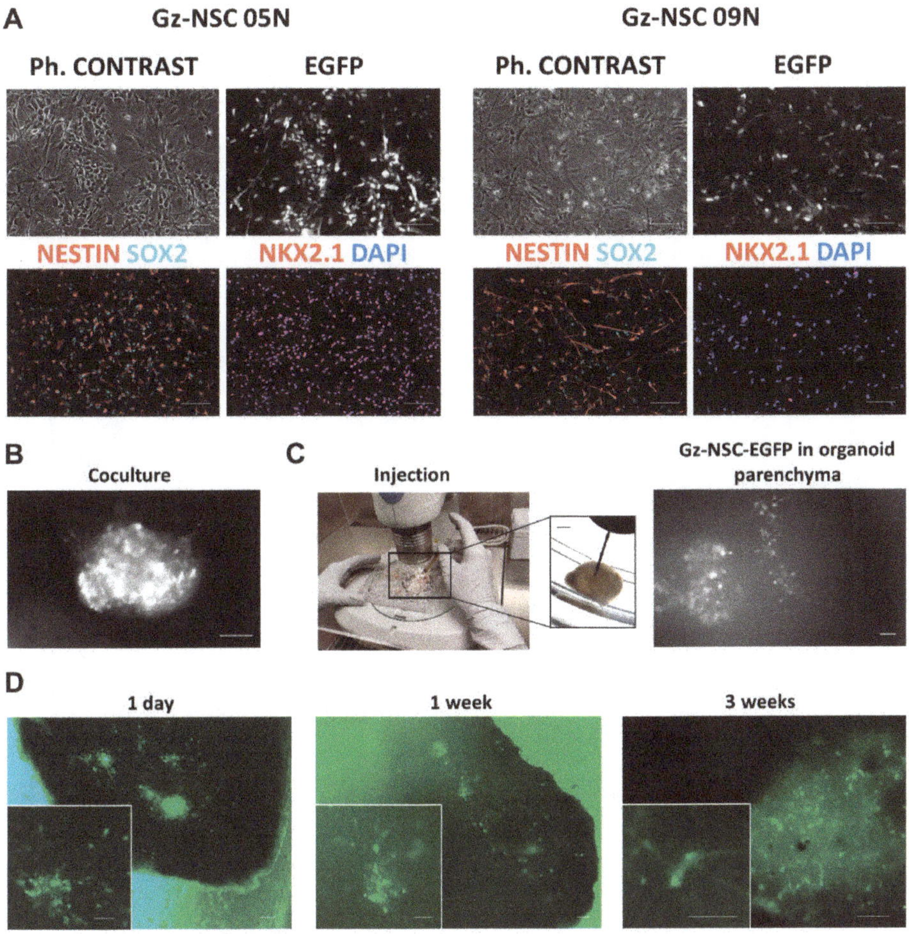

Figure 2. Gz-NSC transplantation in brain organoids. (**A**) Gz-NSC were transduced with a lentiviral vector containing EGFP with a NLS to allow the detection of transplanted cells in the human tissue. Upper panels show phase contrast images and fluorescent images of EGFP-NLS. Lower panels show expression of the NSC markers Nestin and SOX2 and the ventral regional transcription factor NKX2.1 which is expressed at the level of the medial ganglionic eminence. The two Gz-NSC lines used for transplantation experiments expressed different levels of NKX2.1. Scale bar: 100 µm. (**B**) Coculture of Gz-NSC-EGFP and brain organoids. Gz-NSC EGFP form a neurosphere that fuses with the organoid. Scale bar: 100 µm. (**C**) Cell injection under a stereomicroscope inside a laminar flow cabinet. Scale bar: 100 µm and 1 mm in the inset (**D**) Detection of injected EGFP^{+} cells at different time points under an inverted fluorescence microscope. Scale bar: 100 µm and 50 µm in the insets.

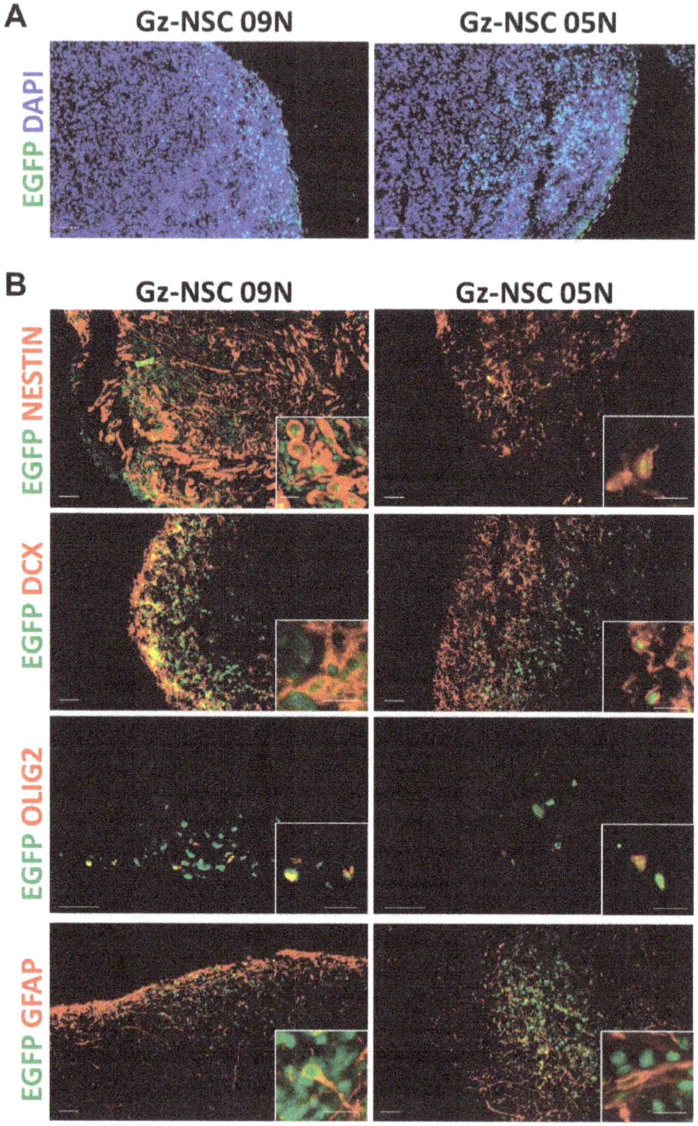

Figure 3. Gz-NSC survive and differentiate in the human brain organoids. (**A**) Survival and localization of two different lines of human Gz-NSC (EGFP^{+}) transplanted into brain organoids. (**B**) Some transplanted cells maintain the stem cell phenotype as shown by the expression of Nestin while other cells differentiate into doublecortin (DCX)$^{+}$ neuroblasts, OLIG2^{+} oligodendrocyte precursors and glial fibrillary acid (GFAP)$^{+}$ astrocytes. Insets show colocalization of EGFP in green and the different markers in red. Scale bar: 50 µm and 20 µm in the insets.

3.2. Transplantation into Immunodeficient Mouse Brains

We compared the data from transplantation experiments into brain organoids with data obtained from standard transplantation studies of the same two Gz-NSC lines into the brains of adult NSG mice.

In mice brains, Gz-NSC mostly remained at the transplantation site, which was surrounded by reactive astrocytes positive for GFAP and activated microglia positive for IBA1 (Figure 4A). Three weeks after transplantation, Gz-NSC transplanted into the mouse brain, mostly differentiated into DCX$^+$ neuroblasts with some OLIG2$^+$ oligodendrocyte progenitor cells and few GFAP$^+$ astrocytes, with some cells still maintaining a more immature Nestin$^+$ phenotype (Figure 4B).

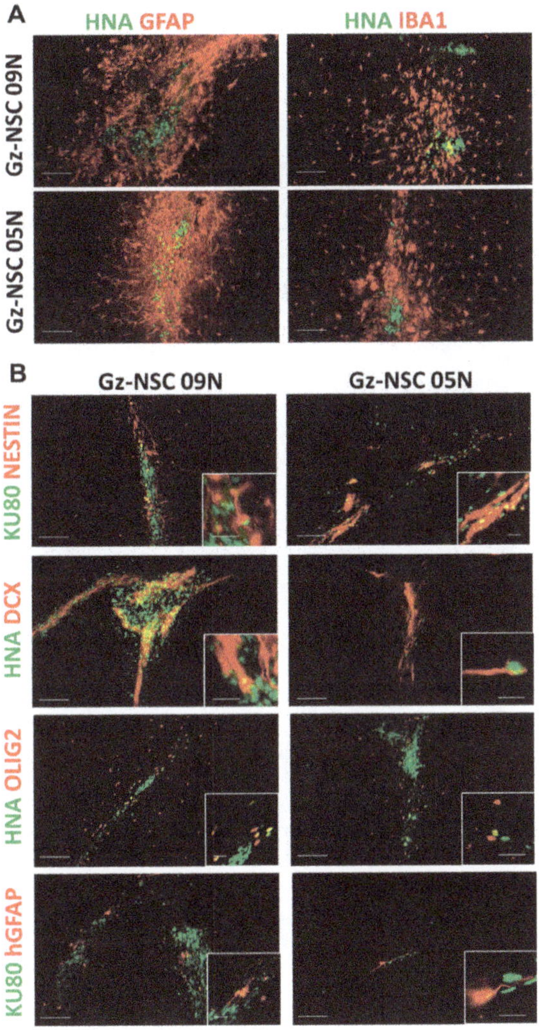

Figure 4. Gz-NSC survive and differentiate in the brain of immunodeficient mice. (**A**) Transplantation site of two different lines of human Gz-NSC identified by the expression of human nuclear antigen (HNA in green), surrounded by mouse reactive GFAP$^+$ astrocytes and IBA1$^+$ microglia (in red). (**B**) Some transplanted cells maintain a stem cell phenotype as shown by the expression of Nestin while other cells differentiate to DCX$^+$ neuroblasts, OLIG2+ oligodendrocyte precursors and human specific (h)GFAP$^+$ astrocytes. Insets show colocalization of nuclear human markers, HNA or KU80 respectively, in green, with the differentiation markers in red. Scale bar: 100 μm and 20 μm in the insets.

Since proliferation capability is an important safety issue for the development of cell therapeutics, we studied the proliferation rate of the Gz-NSC lines when transplanted in the human and mice models. In organoids, quantification of Ki67^{+} transplanted cells showed a remarkably similar proliferation activity for both lines (6.54% of 2179 NSC-EGFP^{+} cells and 6.54% of 1493 NSC-EGFP^{+} cells, respectively) (Figure 5A). Quantification of Ki67 expression in cells transplanted in mice revealed a variable proliferation rate (12.6% and 7.4%), with a similar inter-line (between groups) and inter-subject (within groups) variability (Figure 5B). The proliferation rate of transplanted cells was slightly higher in mice than in organoids for both Gz-NSC lines, although the differences were not statistically significant (Figure 5C).

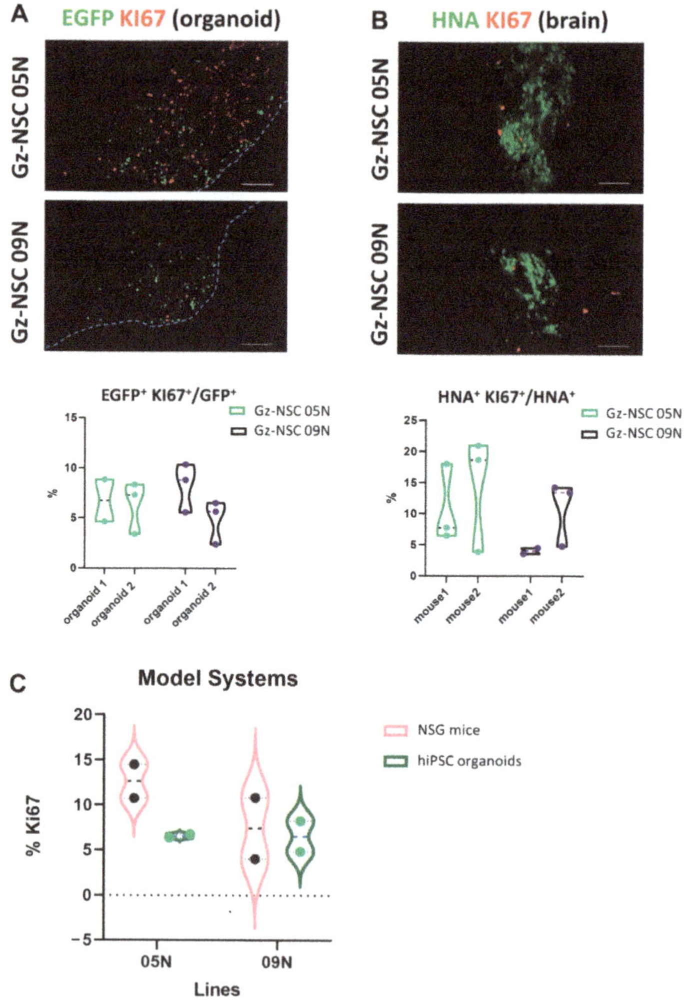

Figure 5. Gz-NSC proliferation rate after transplantation in the human brain organoids and in mice brain. (**A**) Gz-NSC proliferate within the brain organoids as shown by the expression of the proliferation marker KI67 in EGFP^{+} grafted cells 3 weeks after transplantation. Scale bar: 100 μm. (**B**) Gz-NSC proliferate at 3 weeks after transplantation in mice brain as shown by expression of the proliferation marker KI67 in HNA^{+} grafted cells. (**C**) Comparison of two Gz-NSC lines proliferation rate in human organoid versus mouse model.

4. Discussion

In this work, we have explored whether human brain organoids can be used as in vitro models for cell transplantation. We have successfully transplanted NSC lines into brain organoids by injection and compared these data with standard xenograft studies in the brain of NSG mice. NSC proliferation and differentiation potential were similar in both models, suggesting that brain organoids can be informative when studying the proliferative and differentiation potential of cell products, which are important safety and efficacy indicators. Our results suggest that brain organoids can help reduce the number of animals used for regulatory efficacy and safety studies in cell therapy and, importantly, to facilitate preclinical studies.

The proliferation activity of transplanted cells was more uniform in brain organoids than in mouse brains, probably reflecting the higher complexity of the animal as a model system, with many factors affecting the reproducibility of the outcome. In this regard, brain organoids may be advantageous for addressing some specific aspects of NSC biology.

Our brain organoid model for cell transplantation would facilitate regulatory studies, since it is a relatively simple method that can be carried out in a cell culture cabinet without the need for highly regulated and specialized animal facilities. Other in vitro models for cell transplantation, such as brain slices, have been proposed, but that 3D model is usually generated from mice brains. Brain organoids have the advantage of being a human model, and provide more accurate information of human-specific cell–cell and cell–matrix interactions. Furthermore, unlike brain slices, organoids can be kept in culture for a longer time [25].

A limitation of our system is that, in the brain organoids, we cannot assess the host immune reaction to the cell graft, as the organoids lack a vascular and hematopoietic system, although with this protocol we can occasionally observe the presence of microglial cells in some organoids [29]. In animals, we can better evaluate the immunogenicity of our product, but we do not know if it will predict the reaction in immunocompetent hosts, as we usually employ immunodeficient or immunosuppressed animals. In this regard, the use of human brain organoids that consistently include microglial cells generated more recently [30] could be informative in evaluating the immune reaction to cell transplantation, and the interaction of the transplanted cells with the immune cells in the organoid.

Another limitation of our study is that our experiments have been performed with brain organoids generated by a default protocol; therefore, different brain structures and non-neural structures are formed. Furthermore, the generation of brain organoids is a variable process, with significant batch-to-batch and organoid-to-organoid differences [31]. Attempts are being made in the scientific community to decrease this variability and to automate the process. The use of small molecules to generate organoids of a specific brain structure (e.g., cerebral cortex) with no contamination with non-neural tissues can normalize identity and decrease variability [32], probably being a more robust tool to systematically evaluate differentiation and maturation of the NSC progeny. Finally, a technical consideration is that, for the transplantation into organoids, we used EGFP-transduced cells, and it is known that viral transduction can affect cell proliferation [33]. This can lead to an underestimation of the proliferative state of transplanted cells. Nevertheless, our preliminary findings might help guide future studies on the evaluation of safety and efficacy of cell therapeutics using human in vitro models.

Overall, our results suggest that the use of brain organoids as in vitro human models for study the safety and efficacy profile of new cell therapeutics is feasible and may help reduce animal experimentation. This provides unprecedented opportunities for elucidating mechanisms and studying cell integration in neural circuits. However, future studies should determine key factors for performing an effective transplantation, such as the age of the organoid in which the transplant is most effective and the optimal volume and number of transplanted cells. Furthermore, larger experiments will be required to validate the system and discern whether brain organoids increase predictability for future treatments.

5. Conclusions

NSC differentiate into neuroblasts and oligodendrocyte precursors and generate few astrocytes in both human brain organoids and mouse brains. Similar proliferation potential was detected in both models, although it was more variable in animals. Overall, our results suggest that brain organoids can be useful in the evaluation of cell therapy approaches, facilitating preclinical studies and helping to reduce the number of animals used for testing, being aligned with the philosophy of the 3Rs.

Supplementary Materials: The following supporting information can be downloaded at: https://www.mdpi.com/article/10.3390/ani12223150/s1, Figure S1: iPSC and organoid quality controls; Table S1: List of antibodies used.

Author Contributions: B.F.-M., J.M.-R. and R.S.-P. designed the study. A.B.G.-D. and B.F.-M. performed the experiments in Figures 1 and S1. A.B.G.-D., C.C. and B.F.-M. performed the experiments in Figure 2. A.B.G.-D., C.C., R.C.-C. and B.F.-M. performed the experiments in Figure 3. A.B.G.-D., C.C., R.C.-C., C.R.-V., M.M.-L., D.F., R.S.-P. and B.F.-M. performed the experiments in Figures 4 and 5. A.B.G.-D., B.F.-M. and R.S.-P. wrote the manuscript. B.F.-M. assembled the figures of the article. All authors have read and agreed to the published version of the manuscript.

Funding: This work was supported by research funds from the Fundación Alicia Koplowitz and the Instituto de Salud Carlos III- FEDER funds "Una manera de hacer Europa" through the projects "DTS20/00108" and "PT20/00065" (Plataforma de Biobancos y Biomodelos, Plataformas ISCIII de apoyo a la I+D+I en Biomedicina y Ciencias de la Salud, Biobanco del Sistema Sanitario Público de Andalucía).

Institutional Review Board Statement: Animal care and experimental procedures were conducted according to the current National and International Animal Ethics Guidelines and approved by the Research Ethics Committee of University Hospital Virgen Macarena and Virgen del Rocío (0287-N-20).

Informed Consent Statement: Not applicable.

Data Availability Statement: The data that support the findings of this study are available upon reasonable request to the corresponding author.

Acknowledgments: The authors are grateful to Daniel Rodríguez and Paloma Domínguez for technical assistance with the experiments performed in the Histology and Microscopy Core Facilities of the Andalusian Center of Molecular Biology and Regenerative Medicine (CABIMER), to Francisco García Cozar and Lucía Olvera from Cádiz University/INIBICA for the production of lentiviral particles containing EGFP-NLS and to all the members of the Unidad de Producción y Reprogramación Celular (UPRC) for technical help and support. Drawings in Figure 2 were created through BioRender.com, accessed on 26 May 2022.

Conflicts of Interest: J.M.-R., R.S.-P. and B.F.-M are authors of a patent application for the use of Gz-NSC (nº application European Patent Office: 200930943). The other authors indicate no potential conflict of interest.

References

1. Bashor, C.J.; Hilton, I.B.; Bundukwala, H.; Smith, D.M.; Veiseh, O. Engineering the next generation of cell-based therapeutics. *Nat. Rev. Drug Discov.* **2022**, *21*, 655–675. [CrossRef]
2. Mak, I.W.Y.; Evaniew, N.; Ghert, M. Lost in translation: Animal models and clinical trials in cancer treatment. *Am. J. Transl. Res.* **2014**, *6*, 114.
3. Ransohoff, R.M. All (animal) models (of neurodegeneration) are wrong. Are they also useful? *J. Exp. Med.* **2018**, *215*, 2955–2958. [CrossRef]
4. Donega, V.; Nijboer, C.H.; Van Velthoven, C.T.J.; Youssef, S.A.; De Bruin, A.; Van Bel, F.; Kavelaars, A.; Heijnen, C.J. Assessment of long-term safety and efficacy of intranasal mesenchymal stem cell treatment for neonatal brain injury in the mouse. *Pediatr. Res.* **2015**, *78*, 520–526. [CrossRef]
5. Nabar, N.R.; Yuan, F.; Lin, X.; Wang, L.; Bai, G.; Mayl, J.; Li, Y.; Zhou, S.F.; Wang, J.; Cai, J.; et al. Cell Therapy: A Safe and Efficacious Therapeutic Treatment for Alzheimer's Disease in APP+PS1 Mice. *PLoS ONE* **2012**, *15*, e0243343. [CrossRef]
6. Rosati, J.; Ferrari, D.; Altieri, F.; Tardivo, S.; Ricciolini, C.; Fusilli, C.; Zalfa, C.; Profico, D.C.; Pinos, F.; Bernardini, L.; et al. Establishment of stable iPS-derived human neural stem cell lines suitable for cell therapies. *Cell Death Dis.* **2018**, *9*, 937. [CrossRef]

7. Aguilera, Y.; Mellado-Damas, N.; Olmedo-Moreno, L.; López, V.; Panadero-Morón, C.; Benito, M.; Guerrero-Cázares, H.; Márquez-Vega, C.; Martín-Montalvo, A.; Capilla-González, V. Preclinical safety evaluation of intranasally delivered human mesenchymal stem cells in juvenile mice. *Cancers* **2021**, *13*, 1169. [CrossRef]
8. Barrett, D.M.; Zhao, Y.; Liu, X.; Jiang, S.; Carpenito, C.; Kalos, M.; Carroll, R.G.; June, C.H.; Grupp, S.A. Treatment of advanced leukemia in mice with mRNA engineered T cells. *Hum. Gene Ther.* **2011**, *22*, 157–1586. [CrossRef]
9. Van Norman, G.A. Limitations of Animal Studies for Predicting Toxicity in Clinical Trials: Is it Time to Rethink Our Current Approach? *JACC Basic to Transl. Sci.* **2019**, *4*, 845–854. [CrossRef]
10. Dawson, T.M.; Golde, T.E.; Tourenne, C.L.; Programs, S.C.; Sciences, M.; Snyder, S.H.; Malvin, H.; Orleans, N.; Hospital, M.G. Animal Models of Neurodegenerative Diseases. *Nat. Neurosci.* **2018**, *21*, 1370–1379. [CrossRef] [PubMed]
11. Azkona, G.; Sanchez-Pernaute, R. Mice in translational neuroscience: What R we doing? *Prog. Neurobiol.* **2022**, *217*, 102330. [CrossRef]
12. Knight, E.; Przyborski, S. Advances in 3D cell culture technologies enabling tissue-like structures to be created in vitro. *J. Anat.* **2015**, *14*, 384–395. [CrossRef]
13. Vergani, L.; Grattarola, M.; Nicolini, C. Modifications of chromatin structure and gene expression following induced alterations of cellular shape. *Int. J. Biochem. Cell Biol.* **2004**, *36*, 1447–1461. [CrossRef]
14. Doss, M.; Sachinidis, A. Current Challenges of iPSC-Based Disease Modeling and Therapeutic Implications. *Cells* **2019**, *8*, 403. [CrossRef]
15. de Leeuw, S.M.; Davaz, S.; Wanner, D.; Milleret, V.; Ehrbar, M.; Gietl, A.; Tackenberg, C. Increased maturation of iPSC-derived neurons in a hydrogel-based 3D culture. *J. Neurosci. Methods* **2021**, *360*, 109254. [CrossRef]
16. Morata Tarifa, C.; López Navas, L.; Azkona, G.; Sánchez Pernaute, R. Chimeras for the twenty-first century. *Crit. Rev. Biotechnol.* **2020**, *40*, 283–291. [CrossRef]
17. Bai, J.; Wang, C. Organoids and Microphysiological Systems: New Tools for Ophthalmic Drug Discovery. *Front. Pharmacol.* **2020**, *11*, 407. [CrossRef]
18. Rabeling, A.; Goolam, M. Cerebral organoids as an in vitro model to study autism spectrum disorders. *Gene Ther.* **2022**, *2022*, 1–11. [CrossRef]
19. Lancaster, M.A.; Renner, M.; Martin, C.A.; Wenzel, D.; Bicknell, L.S.; Hurles, M.E.; Homfray, T.; Penninger, J.M.; Jackson, A.P.; Knoblich, J.A. Cerebral organoids model human brain development and microcephaly. *Nature* **2013**, *493*, 385–388. [CrossRef]
20. Luo, C.; Lancaster, M.A.; Castanon, R.; Nery, J.R.; Knoblich, J.A.; Ecker, J.R. Cerebral Organoids Recapitulate Epigenomic Signatures of the Human Fetal Brain. *Cell Rep.* **2016**, *17*, 3369–3384. [CrossRef]
21. Shou, Y.; Liang, F.; Xu, S.; Li, X. The Application of Brain Organoids: From Neuronal Development to Neurological Diseases. *Front. Cell Dev. Biol.* **2020**, *8*, 579659. [CrossRef]
22. Trujillo, C.A.; Adams, J.W.; Negraes, P.D.; Carromeu, C.; Tejwani, L.; Acab, A.; Tsuda, B.; Thomas, C.A.; Sodhi, N.; Fichter, K.M.; et al. Pharmacological reversal of synaptic and network pathology in human MECP2-KO neurons and cortical organoids. *EMBO Mol. Med.* **2021**, *13*, e12523. [CrossRef]
23. Samarasinghe, R.A.; Miranda, O.A.; Buth, J.E.; Mitchell, S.; Ferando, I.; Watanabe, M.; Allison, T.F.; Kurdian, A.; Fotion, N.N.; Gandal, M.J.; et al. Identification of neural oscillations and epileptiform changes in human brain organoids. *Nat. Neurosci.* **2021**, *24*, 1488–1500. [CrossRef]
24. Fernandez-Muñoz, B.; Garcia-Delgado, A.B.; Arribas-Arribas, B.; Sanchez-Pernaute, R. Human neural stem cells for cell-based medicinal products. *Cells* **2021**, *10*, 2377. [CrossRef]
25. Lancaster, M.A.; Knoblich, J.A. Generation of cerebral organoids from human pluripotent stem cells. *Nat. Protoc.* **2014**, *9*, 2329–2340. [CrossRef]
26. Fernández-Muñoz, B.; Rosell-Valle, C.; Ferrari, D.; Alba-Amador, J.; Montiel, M.Á.; Campos-Cuerva, R.; Lopez-Navas, L.; Muñoz-Escalona, M.; Martín-López, M.; Profico, D.C.; et al. Retrieval of germinal zone neural stem cells from the cerebrospinal fluid of premature infants with intraventricular hemorrhage. *Stem Cells Transl. Med.* **2020**, *9*, 1085–1101. [CrossRef]
27. Paxinos, G.F.K. *The Mouse Brain in Stereotaxic Coordinates: Compact Second Edition*; Acadademic Press: San Diego, CA, USA, 2004.
28. Schneider, C.A.; Rasband, W.S.; Eliceiri, K.W. NIH Image to ImageJ: 25 years of image analysis. *Nat. Methods* **2012**, *9*, 671–675. [CrossRef]
29. Ormel, P.R.; Vieira de Sá, R.; van Bodegraven, E.J.; Karst, H.; Harschnitz, O.; Sneeboer, M.A.M.; Johansen, L.E.; van Dijk, R.E.; Scheefhals, N.; Berdenis van Berlekom, A.; et al. Microglia innately develop within cerebral organoids. *Nat. Commun.* **2018**, *9*, 4167. [CrossRef]
30. Cakir, B.; Tanaka, Y.; Kiral, F.R.; Xiang, Y.; Dagliyan, O.; Wang, J.; Lee, M.; Greaney, A.M.; Yang, W.S.; duBoulay, C.; et al. Expression of the transcription factor PU.1 induces the generation of microglia-like cells in human cortical organoids. *Nat. Commun.* **2022**, *13*, 430. [CrossRef]
31. Lancaster, M.A.; Corsini, N.S.; Wolfinger, S.; Gustafson, E.H.; Phillips, A.W.; Burkard, T.R.; Otani, T.; Livesey, F.J.; Knoblich, J.A. Guided self-organization and cortical plate formation in human brain organoids. *Nat. Biotechnol.* **2017**, *35*, 659–666. [CrossRef]
32. Kim, J.; Sullivan, G.J.; Park, I.H. How well do brain organoids capture your brain? *iScience* **2021**, *24*, 102063. [CrossRef] [PubMed]
33. Emmett, S.R.; Dove, B.; Mahoney, L.; Wurm, T.; Hiscox, J.A. The cell cycle and virus infection. *Methods Mol. Biol.* **2005**, *296*, 197–218. [CrossRef]

animals

MDPI

Article

Progress towards the Replacement of the Rabbit Blood Sugar Test for the Quantitative Determination of the Biological Activity of Insulins (USP <121>) with an In Vitro Assay

Sabrina Rüggeberg [1,*], Antje Wanglin [1], Özlem Demirel [1], Rüdiger Hack [2], Birgit Niederhaus [3], Bernd Bidlingmaier [4], Matthias Blumrich [5] and Dirk Usener [1,*]

1 CMC-Bioanalytics, R&D Sanofi, 65926 Frankfurt, Germany
2 TIM Global Compliance and Policy, R&D Sanofi, 65926 Frankfurt, Germany
3 CMC-Statistics, R&D Sanofi, 65926 Frankfurt, Germany; birgit.niederhaus@sanofi.com
4 MSAT Analytics , M&S Sanofi, 65926 Frankfurt, Germany
5 MSAT Project Nitrosamines, M&S Sanofi, 65926 Frankfurt, Germany
* Correspondence: sabrina.rueggeberg2@sanofi.com (S.R.); dirk.usener@sanofi.com (D.U.); Tel.: +49-69-305-28336 (S.R.); +49-69-305-25764 (D.U.)

Simple Summary: The recent United States Pharmacopeia general chapter <121> requires a non-quantitative bioidentity test either as a rabbit blood sugar assay or as an in vitro insulin cell-based assay using in-cell Western (ICW) technology for insulin batch release. However, for quantification during stability or comparability studies, the rabbit blood sugar test is still required using a minimum of 24 rabbits to obtain one result. Based on the 3R principle (replace, reduce, and refine), this study sought to qualify the in vitro ICW cell-based bioassay approach for quantifying insulin activity. A bridging study with different insulins and stress samples revealed a clear correlation between the in vitro and in vivo test results. The replacement of the animal-based assay with the quantitative in vitro ICW cell-based bioassay for batch quality control saved cost, reduced cycle times while obtaining more meaningful and reliable data, and, above all else, reduced the suffering of many rabbits.

Citation: Rüggeberg, S.; Wanglin, A.; Demirel, Ö.; Hack, R.; Niederhaus, B.; Bidlingmaier, B.; Blumrich, M.; Usener, D. Progress towards the Replacement of the Rabbit Blood Sugar Test for the Quantitative Determination of the Biological Activity of Insulins (USP <121>) with an In Vitro Assay. *Animals* **2023**, *13*, 2953. https://doi.org/10.3390/ani13182953

Academic Editor: Garikoitz Azkona

Received: 2 August 2023
Revised: 12 September 2023
Accepted: 13 September 2023
Published: 18 September 2023

Copyright: © 2023 by the authors. Licensee MDPI, Basel, Switzerland. This article is an open access article distributed under the terms and conditions of the Creative Commons Attribution (CC BY) license (https://creativecommons.org/licenses/by/4.0/).

Abstract: For the quantification of insulin activity, United States Pharmacopeia (USP) general chapter <121> continues to require the rabbit blood sugar test. For new insulin or insulin analogue compounds, those quantitative data are expected for stability or comparability studies. At Sanofi, many rabbits were used to fulfil the authority's requirements to obtain quantitative insulin bioactivity data until the in vivo test was replaced. In order to demonstrate comparability between the in vivo and in vitro test systems, this study was designed to demonstrate equivalency. The measurement of insulin lispro and insulin glargine drug substance and drug product batches, including stress samples (diluted or after temperature stress of 30 min at 80 °C), revealed a clear correlation between the in vitro and in vivo test results. The recovery of quantitative in vitro in-cell Western (ICW) results compared to the in vivo test results was within the predefined acceptance limits of 80% to 125%. Thus, the in vitro ICW cell-based bioassay leads to results that are equivalent to the rabbit blood sugar test per USP <121>, and it is highly suitable for insulin activity quantification. For future development compounds, the in vitro in-cell Western cell-based assay can replace the rabbit blood sugar test required by USP <121>.

Keywords: USP <121>; insulin cell-based bioassay; in-cell western; ICW; rabbit blood sugar test; 3R; replacement; reduction; refinement; method bridging

1. Introduction

The early concept of replacing, reducing, or refining animal use in research and testing was set in the late 1950s by Russell and Burch with their book *The Principles of Humane*

Experimental Technique [1]. The concept of replacement aims to substitute traditional animal models with non-animal systems such as biochemical or cell-based systems, the concept of reduction aims to decrease the number of animals required for testing, and the concept of refinement attempts to eliminate pain or distress in animals and enhance animal well-being.

This 3R principle comprises the ethical guidelines for animal experiments, which are only allowed if no alternative method is available, as regulated by European legislation. The EU directive 2010/63/EU on the protection of animals used for scientific purpose is to be considered, and it represents "*. . .an important step towards achieving the final goal of full replacement of procedures on live animals for scientific and educational purposes as soon as it is scientifically possible to do so. . .*".

Thus, the three Rs for animal-based experiments are a common goal of health authorities and pharmaceutical industries worldwide [2–6]. The European Union article 13 of the new directive 2010/63/EU further states that "*Member States shall ensure that a procedure is not carried out if another method or testing strategy for obtaining the result sought, not entailing the use of a live animal, is recognized under the legislation of the Union*" [7]. Consequently, the European Medicines Agency has recommended to marketing authorization holders to ensure compliance with the 3R methods described in the European Pharmacopoeia, and it points to the fact that competent authorities for granting the approval of animal testing will request the more animal-friendly European Pharmacopoeia method to be used [8].

Progress has been achieved for refinement [9,10], reduction [11], and replacement in research and as quality controls for vaccines [12] or pharma products [13].

Nevertheless, animal tests still play an essential role in the research (e.g., to assess toxicity [14]), development [15], and quality control of vaccines [16] or pharma products [17]. There are still legislative demands which require animal testing for quality control [10,15].

The animal-based biological assays such as the rabbit blood sugar test have a long history in the testing and development of human insulin and insulin analogues for release [18]. The current USP general chapter <121> requires the use of either the in vivo rabbit blood sugar test or the in vitro in-cell Western cellular assay with a specification of not less than 15 units (U) per mg of insulin as a qualitative bioidentity measurement for the release of insulin or insulin analogue batches manufactured for the United States market [19]. Quantitative measurements are needed to assess the long-term activity of insulin or an insulin analogue during stability or comparability studies. For these quantitative measurements, the in vivo rabbit blood sugar test is still mandatory. Alternatively, the in vitro in-cell Western (ICW) cell-based bioassay could be used for quantitative measurements by applying the same statistical significance as defined by USP general chapter <121>, with a confidence limit of $\pm 10\%$.

Within Sanofi, the in vivo rabbit blood sugar test had still been performed for the development and approval of new insulins or insulin analogues to fulfil the regulatory requirements of authorities in several countries, such as the United States, China, and Japan. Until 2018, many thousands of rabbits were used to support these studies. Since 2018, Sanofi has relied on in vitro ICW cell-based bioassay data.

The in vitro ICW cell-based bioassay procedure was introduced in USP general chapter <121> in 2020 [19]. Its use and the ICW methodology have been described previously [13,20,21] in articles, and they were further characterized by the FDA [22] and alternative cell-based assays have been proposed [23]. The ICW method measures the biological effect of human insulin through the activation of the human insulin receptor (hIR). The binding of human insulin or insulin analogs to the hIR induces a conformational change that stimulates the auto-phosphorylation of the hIR on three tyrosine (Tyr) residues [24]. This auto-phosphorylation leads to the full activation of the hIR and enhances its activity towards the intracellular substrates involved in the downstream signaling cascade, leading to the metabolic end effects and, finally, to a decrease in blood sugar level [25,26]. Measuring the auto-phosphorylation of hIR reflects the activity of the hormone, and it can be used for mimicking the biological activity of human insulin or insulin analogs [24]. The initial step of human insulin action, the activation of hIR auto-phosphorylation, is a highly efficient read-out for such biological activity because it is a direct approach that leads to accurate data with low background noise.

The objective of this study was to demonstrate analytical equivalence between the in vivo quantitative rabbit blood sugar test according to USP <121> (in short, the in vivo test) and the in vitro quantitative bioassay by ICW (in short, the in vitro qICW).

The study was divided into two parts. The first part evaluated regular test items. The second part evaluated decreased potent/stressed test items with the in vitro qICW and the in vivo test.

2. Materials and Methods

2.1. Samples Set

2.1.1. The First Part: The Regular Test Items

Samples from the primary stability studies of Suliqua (a lixisenatide/insulin glargine combination), the insulin lispro Sanofi drug product (DP), and the insulin lispro Sanofi drug substance (DS) were analyzed with the in vitro insulin cell-based bioassay and in vivo rabbit blood sugar test.

2.1.2. The Second Part: The Stressed Test Items

The insulin lispro Sanofi DP (LISDP001) was stressed at 80 °C for 30 min or diluted to 50% of its initial concentration. The in vitro qICW and in vivo tests were performed to compare the loss of potency.

The relevant parameters (appearance of solution (i.e., clarity and color), assay insulin lispro by HPLC, product-related substances by HPLC, product-related impurities, and particulate matter (i.e., visible particles)) of the temperature-stressed insulin lispro were monitored to confirm the stability of the stress-induced changes.

2.2. Statistical Considerations

The objective of the evaluation was to demonstrate that the in vitro test results are statistically comparable to the in vivo test results. To achieve this, an equivalence approach was used as an appropriate statistical test (for the use of equivalence instead of difference testing, also see USP <1033> [27] and ICH-Q2 [28]) as set out below.

In order to compare the in vitro and in vivo test results, recoveries were calculated. The recoveries were calculated by dividing the in vitro qICW results by the in vivo test results (ratio), followed by a multiplication by 100 in order to obtain the percentage values, as follows:

$$\text{recovery} = \frac{\text{in vitro qICW result}}{\text{in vivo test result}} \times 100.$$

A recovery of 100% therefore corresponded to identical results for the in vitro and in vivo tests. To statistically demonstrate equivalence between the two methods (in vitro and in vivo), a predefined acceptance criterion had to be applied to the mean recovery, with a 90% confidence interval, by applying the following decision procedure:

- the in vitro and in vivo test results were considered comparable at the 5% significance level and the in vitro tests were considered as performing acceptably if the 90% two-sided CI was totally within the acceptance interval
- the difference between the in vitro and in vivo test results was considered as significant in cases where the point estimates of the recoveries were outside the acceptance interval, and the performance of the in vitro test was considered as unsatisfactory for using the method
- there was insufficient information to conclude that there were no relevant differences between the in vitro and in vivo tests, beyond a reasonable doubt, in cases where the point estimates of the recoveries lied within the acceptance interval and the CI overlapped the acceptance interval, and the decision had to be based on the magnitude

of the observed recovery and other supporting information (i.e., further experiments may have been needed)

The below acceptance criteria (see Table 1) needed to be met to demonstrate equivalency between the in vitro qICW and the in vivo test based on USP <1090> for demonstrating bioequivalency.

Table 1. Acceptance criteria for showing similarity.

Parameters	Acceptance Criterion
Recovery means of in vitro qICW in a percentage of the in vivo tests for the regular test items	80% ≤ recovery mean ≤ 125%
Recovery of in vitro qICW in a percentage of the in vivo test for the stressed test items *	80% ≤ recovery mean ≤ 125%

*, described as ≤20% of the difference (0.0792 on a log scale).

The number of measurements (i.e., the sample size) for showing equivalence was calculated to obtain accurate results with a power of 80%. A minimum of 12 measurements were needed for a valid result. In this study, 19 samples were tested with each method to demonstrate equivalency between the in vitro qICW and the in vivo test.

The mean recovery was calculated using the internally developed and validated software BioSt@t-Stars version 2.6. The in vitro qICW data were obtained on a logarithmic scale and the in vivo test data were obtained on a linear scale. Since the acceptance criterion was given on the logarithmic scale, further calculations were continued on the logarithmic scale.

2.3. *In Vitro Cell-Based Bioassay using the In-Cell Western Cell-Based Method (USP <121> Method)*

Chinese hamster ovary (CHO) cells expressing human insulin receptor B (hIR, genebank accession number M10051) (ATCC, Manassas, VA, USA, CRL-3307™) were cultivated with 90% Ham's F12 nutrient mixture with glutamax, 10% fetal bovine serum, and 0.6% hygromycin B (Life Technologies, Darmstadt, Germany) at 37 °C in a humidified atmosphere of 5% CO_2.

In order to determine the tyrosin (Tyr) phosphorylation status of the hIR [29], the cells were seeded into 96-well microplates with a density of ~0.5 to 1.5 × 10^5 cells/mL and grown for 2 to 4 days. The cells were serum-starved with serum-free Ham's F12 nutrient mixture with glutamax for 3 to 5 h at 37 °C in a humidified atmosphere of 5% CO_2. The cells were subsequently treated with serial dilutions of human insulin or insulin analog prepared in serum-free Ham's F12 nutrient mixture with glutamax (insulin glargine) or in 0.1% BSA in D-PBS (insulin lispro) for 20 min at 37 °C and 5% CO_2. For the potency determination of the insulin or the insulin analog, two independent measurements of triplicates on different microplates were performed. An insulin reference standard was analyzed in parallel on each plate.

After stimulation, the medium was discarded and the cells were fixed in 3.7% freshly prepared para-formaldehyde (Merck, Darmstadt, Germany) in Dulbecco's phosphate-buffered saline without calcium and magnesium (D-PBS; Life Technologies, Darmstadt, Germany) for 20 min. After permeabilization with 0.1% Triton®-X-100 (Merck, Darmstadt, Germany) in D-PBS for 2 × 10 min, blocking was performed with a blocking solution containing 2% bovine serum albumin (BSA; Sigma Aldrich, Taufkirchen, Germany) in D-PBS overnight at + 2–8 °C. Immersion in an incubation mixture with the anti-p-Tyr 4G10 mouse monoclonal antibody (Millipore, Schwalbach, Germany) [30] prepared in D-PBS and 0.1% (v/v) polysorbate 20 (AppliChem, Darmstadt, Germany) for insulin glargine or in 2% BSA in D-PBS and 0.1% (v/v) polysorbate 20 for insulin lispro for 2 h at room temperature was followed by a washing step with D-PBS containing 0.1% polysorbate 20. Incubation with an IRdye 800 CW goat anti-mouse IgG antibody (Li-Cor, Bad Homburg, Germany) and cell/DNA staining dye prepared in D-PBS and 0.2% (v/v) polysorbate 20 for insulin glargine or in 2% BSA in D-PBS and 0.2% (v/v) polysorbate 20 for insulin lispro was performed for 1 h at room temperature. The application of near-infrared-labeled antibodies had the distinct advantage of a high signal-to-noise ratio due to very little auto-fluorescence from both cellular materials and plastics. Fluorescence was detected by the Odyssey Infrared Imaging System (Li-Cor, Bad Homburg, Germany) using the 800 nm channel for the detection of the tyrosine phosphorylation at the hIR. The results were normalized to the cell number by combined cell- and DNA-staining with Sapphire700™ (Li-Cor, Bad Homburg, Germany) and DraQ5™ (BioStatus, Leicestershire, UK) and detected with the 700 nm channel.

The normalized data of the dilution curves of a reference standard and the test samples were used to perform a 4-parameter logistic (4-PL) regression analysis and calculate the EC_{50} value (the half maximal effective concentration), which represented the potency of the insulin and insulin analog samples. The relative potencies were calculated based on dividing the EC_{50} value of the reference standard by the EC_{50} value of the sample, which was then multiplied by 100%. The obtained relative potency was further calculated to units/mg (or U/mg) with the known activity of the reference standard. The final potency results were reported either as relative potencies in percentages or in U/mg.

The Suliqua and insulin lispro DS and DP were measured with the in vitro qICW method. For the Suliqua, eight replicate measurements for the in vitro qICW and insulin lispro were calculated using the same requirements, which are described in USP <121> for the in vivo test, with 95% CI $\leq$ 0.082 (CL within $\pm$10%).

2.4. HPLC Method for Assay Quantification

The assays for the insulin, product-related substances, and product-related impurities, including the degradation products of the insulin lispro in the drug product, were quantified using validated reversed phase liquid chromatography (HPLC) methods at Sanofi (Frankfurt, Germany). The assay for the insulin was determined by external standard calibration whereas the product-related substances and product-related impurities, including the degradation products of the insulin lispro, were quantified by the 100% peak area method (i.e., the normalization method). Three independent determinations per testing time-point were performed for each sample, and the mean values of those measurements were calculated. As per internal Sanofi rules for analytics procedures, the individual measurements did not differ from each other by more than 2%.

2.5. HPSEC for Determination of High-Molecular-Weight Proteins

The determination of any impurity with a molecular mass greater than that of insulin was performed by size-exclusion chromatography as prescribed in the Ph. Eur. and USP. The use of the methods for the individual insulins was verified or validated, respectively. The quantification was performed by the 100% peak area method (i.e., the normalization method). The sum of the areas of the peaks with retention times less than those of the principal peaks (i.e., the peaks due to the insulin monomer) was regarded to be the sum of the high-molecular-weight proteins. Any peak with a retention time greater than that of the peak due to the insulin monomer was disregarded.

2.6. In Vivo Rabbit Blood Sugar Test

The experimental design and management procedures were approved by the District Government of South-Hesse in Darmstadt under animal use permit nos. HMR-7/Anz. 02 and FH-1009 according to the German Animal Welfare Legislation implementing the European Directive 2010/63/EU. The studies were conducted in an AAALAC International-accredited facility of Sanofi in Frankfurt.

2.6.1. Animals and Husbandry

New Zealand White (NZW) female rabbits aged 10–16 weeks with body weights of at least 1.8 kg were purchased from two rabbit breeders (Bauer, Neuenstein, Germany and Zimmermann, Untergröningen, Germany). After arrival, the animals were kept in groups of a maximum of 32 animals in solid-floor pens with elevated platforms for at least 7 days before the study start and between the study parts. The animals were housed in environmental conditions as follows: a temperature range of 15–21 °C, a relative humidity rate of 40–70%, and an air change rate 18–21 times/hour. Water and food (sniff rabbit maintenance diet, Soest, Germany) were provided ad libitum, and hay and aspen sticks were given as environmental enrichment. All the animals were inspected daily by skilled personnel.

2.6.2. Experimental Design

The quantitative rabbit blood sugar test was performed according to USP general chapter <121>. For the testing of one test article, the rabbits were randomly assigned to 4 groups of at least 6 animals.

Fourteen hours before the dosing food was withdrawn and approximately 1 h before dosing, the animals were transferred into a rabbit restrainer and arterial catheters were implanted in their central ear arteries. Two solutions of standard preparations and two solutions of test articles were prepared and applied in volumes of 0.5 mL subcutaneously to two respective groups of rabbits.

At 1 and 2.5 h, blood samples (1.3 mL) were collected from the rabbits' central ear arteries into tubes coated with fluoride-heparin (microtubes from Sarstedt). After blood sampling, the catheters were removed, and the animals were brought back to their home areas. The whole period in the restrainer normally lasted 3.5 h.

Two to six days later, the second part of the study was performed using a twin cross-over design (see Figure 1).

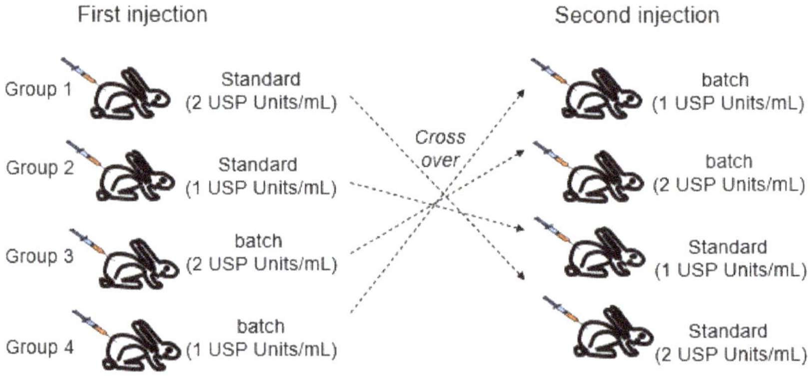

Figure 1. Double cross-over design of the rabbit blood sugar test according to USP <121>.

The human insulin and the other insulin analogues used are xenoproteins for rabbits, and as with all xenoproteins, there is a risk of the production of anti-drug antibodies (ADA) against these xenoproteins, especially after several exposures. These ADAs were considered as a risk for under-estimating the activity of the insulin in the samples, and therefore, they represent a risk for patients who receive too much insulin. Finally, after extensive discussions with immunologists and the animal welfare authority, it was decided to re-use the animals once. After the second test, the possibility of animal re-use for the other rabbit experiments was evaluated. If re-use was not possible, then the animals were handed over to a local zoo in accordance with local veterinarian authorities to be used for the feeding of carnivores.

After blood centrifugation, the plasma samples were analyzed for their blood glucose concentrations with the Hexokinase method using a multianalyser (KonelabPrime 30) (Thermo Fisher, Dreiech, Germany). The basic principle of this method is the phosphorylation of glucose to glucose-6-phosphate in the presence of ATP and hexokinase, followed by the oxidation of glucose-6-phosphate to 6-phosphogluconate by glucose-

6-phosphatedehydrogenase. In this reaction, an equimolar amount of NADP was reduced to NADPH2, with a resulting increase in the absorbance at 340 nm. The increase was measured by the KonelabPrime 30 instrument.

The final calculations were completed according to USP <121>. The 95% confidence interval of the final result needed to be smaller than 0.082, which corresponded to confidence limits of approximately ± 10%. For all regular test items, the in vivo test was performed in the course of the respective stability studies.

3. Results

The regular test items and stressed test items were tested, and the results are summarized below.

3.1. Regular Test Items

Nineteen different insulin/insulin analogue batches were tested during the stability studies in order to cover the different conditions such as the storage temperature (−20 °C, 5 °C, 25 °C, 37 °C, and 40 °C), age (freshly produced (T0), 36 months), formulations (DS and DP), and concentrations (approximately 25 U/mL to 100 U/mL). Two different insulin analogues (insulin lispro and insulin glargine) were used for this study. All the single results for the contents of the bioactive insulins are listed in Table 2. The individual recoveries varied between 80.11% to 118.75%. The mean recovery of the regular test items was 95%, with a 90% confidence interval (CI) ranging from 91% to 99% and a geometric coefficient of variation (CV%) of 12% (95% one-sided upper confidence limit (CL) of 17%).

Table 2. Single results of each measurement and recovery (the in vitro qICW and the in vivo test) for each insulin lispro and insulin glargine batch.

	Batch	Comment		Content–In Vitro qICW (U/mg)	Content–In Vivo Test (U/mg)	Recovery: In Vitro qICW/ In Vivo Test (%)
Insulin lispro Sanofi	LISDP001	5 °C, t_0	DP	101.28 *	103.30	98.04
	LISDP001	37 °C, 1 month	DP	89.48 *	104.85	85.34
	LISDP001	25 °C, 6 months	DP	95.48 *	98.48	96.85
Insulin Suliqua (lixisenatide/insulin glargine combination)	GLADP001	5 °C, 12 months	DP	118.76	113.41	104.72
	GLADP002	5 °C, 12 months	DP	123.63	104.11	118.75
	GLADP003	5 °C, 12 months	DP	125.00	108.18	115.55
	GLADP004	5 °C, t_0	DP	107.04	106.18	100.81
	GLADP005	5 °C, t_0	DP	93.16	111.07	83.88
	GLADP006	5 °C, t_0	DP	93.97	105.48	89.08
	GLADP005	40 °C, 1 month	DP	108.03	98.27	109.93
	GLADP004	5 °C, 12 months	DP	108.03	109.04	99.07
	GLADP005	5 °C, 12 months	DP	109.02	110.85	98.35
	GLADP006	5 °C, 12 months	DP	102.87	122.77	83.79
Insulin lispro Sanofi	LISDS001	−20 °C, 24 months	DS	27.4	28.4	96.64
	LISDS002	−20 °C, 24 months	DS	27.5	30.0	91.72
	LISDS003	−20 °C, 24 months	DS	24.9	29.0	85.79
	LISDS001	−20 °C, 36 months	DS	26.2	28.2	92.81
	LISDS002	−20 °C, 36 months	DS	26.1	30.0	86.88
	LISDS003	−20 °C, 36 months	DS	23.2	29.0	80.11

* the in vitro qICW data were obtained with four replicates based on USP <1032> requirements [27].

The data were statistically analyzed, as shown in Figure 2. The 19 final recovery results with the geometric mean in the interquartile range are given in a box blot for the data.

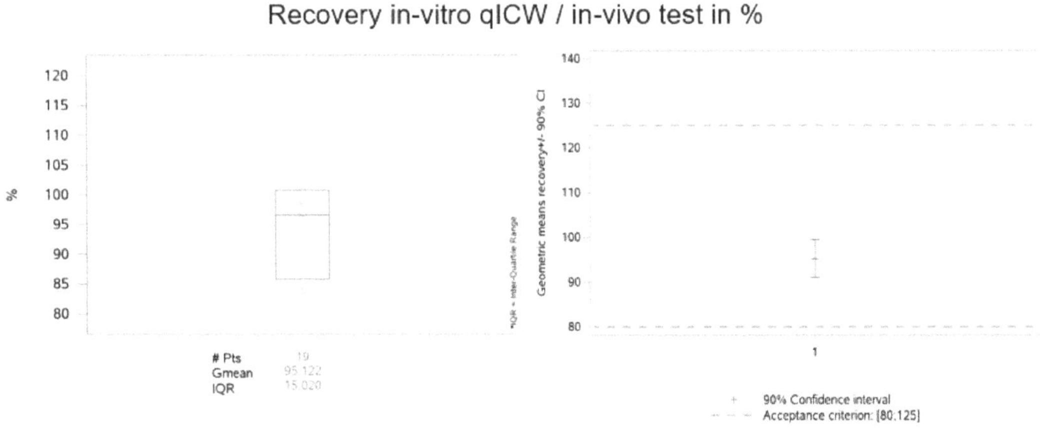

Parameters	Acceptance Criterion	Result
Recovery means of the in vitro qICW in a percentage of the in vivo test for the regular test items	80% ≤ recovery mean ≤ 125%	95% [91%; 99%]

Figure 2. The results for the mean recovery and the 90% CIs for the regular test items are displayed. A total number of 19 drug substance and drug product batches of either insulin lispro or insulin glargine were compared by calculating the recovery of the in vitro qICW concentration (U/mg) in the in vivo test concentration (U/mg), as shown in Table 2.

3.2. Stressed Test Items

Different reduced biological potencies were determined using either the stressed or the diluted samples. As for the regular test items, the recoveries were calculated to show similarity. Furthermore, a set of physicochemical tests were performed to demonstrate the sample suitability for the study and to monitor the heat-stressed test items.

As these measurements were dedicated only to this study, the number of experiments was kept as low as possible to comply with the 3R principle for the in vivo tests.

The results of the physicochemical characterization are listed in Table 3, and they show highly stressed insulin lispro. The amount of the product-related substance 3^B-Asp insulin increased to 4.28% from 0.79%, and other impurities increased to 1.78% from 0.35% while the aggregates (i.e., the high-molecular-weight proteins) increased to 32.48%.

All the single results obtained for the contents by HPLC and the bioactivity by the in vitro qICW or the in vivo test are listed in Table 4. While the unstressed starting material revealed potencies of 101.28 U/mL and 103.3 U/mL as determined by the in vitro qICW and the in vivo rabbit blood sugar test, respectively, the recovery was calculated to be 98.04%. Lower potency values, either by dilution (47.45 vs. 51.87 U/mL) or stress (63.59 vs. 64.77 U/mL), showed very good recovery rates of 91.48% and 98.18%, respectively.

The clear correlation between the in vitro qICW and the in vivo rabbit blood is visualized in Figure 3. The contents in U/mL are given for the unstressed items on the right, for 50% potency in the middle, and for the stressed insulin lispro Sanofi on the left. The in vitro qICW test system is shown in black, the in vivo rabbit blood sugar test is shown in white, and the RP-HPLC contents are shown in grey.

Thus, the stressed samples method confirmed the similarity between the in vitro qICW and the in vivo rabbit blood sugar test.

Table 3. HPLC results for the heat-stressed LISDP001.

Test	Acceptance Criterion *	Results from the Certificate of Analysis (CoA)		Results from the Heat-Stressed Test Items	
Assay insulin lispro (HPLC)	3.30 to 3.64 mg/mL (95.0% to 105.0% of the label claim)	3.53 mg/mL (101.7% of the label claim) 101.7 USP insulin lispro units/mL		1.97 mg/mL (56.7% of the label claim)	
Product-related substances (HPLC)					
Total of 3^B-Asp insulin lispro and 3^B-iso-Asp insulin lispro	≤1.5%	0.79%		4.28%	
27^B-allo-Thr insulin lispro	≤0.5%	0.35%		0.24%	
Product-related impurities and degradation products (HPLC)					
21^A-Asp insulin lispro	≤1.00%	0.21%		0.27%	
Any other unspecified unidentified impurity	≤0.5%	RRT ** 0.90 0.95 1.10 1.17 1.15	% 0.07% 0.06% 0.07% 0.05% 0.10%	RRT ** 1.25	% 1.78%
Total of other impurities	≤2.00%	0.34%		7.40% ***	
High molecular weights	≤0.50%	0.10%		32.48% ***	

* only for scientific information; ** RRT (relative retention time); *** the observed level was found to be out of the validated range of the analytical method; Asp and Thr refer to aspartic acid and threonine, respectively, at amino acid positions 3, 21, or 27.

Table 4. Results and calculated ratios of the stressed test items.

Substance	Batch	Comment		Content–In Vitro qICW (U/mL)	Content–In Vivo Test (U/mL)	Recovery: In Vitro qICW/In Vivo Test (%)	Content–HPLC Assay (U/mL)
Insulin lispro	LISDP001	No stress	DP	101.28 [a]	103.30	98.04	101.7 [b]
Insulin lispro	LISDP001	50% diluted sample	DP	47.45	51.87	91.48	50.9 [c]
Insulin lispro	LISDP001	80 °C, 30 min (heat stress)	DP	63.59	64.77	98.18	56.7

[a], data taken from the stability start; [b], data taken from the release certificate; [c], initial value of 101.7 U/mg divided by 2 to obtain the 50% content.

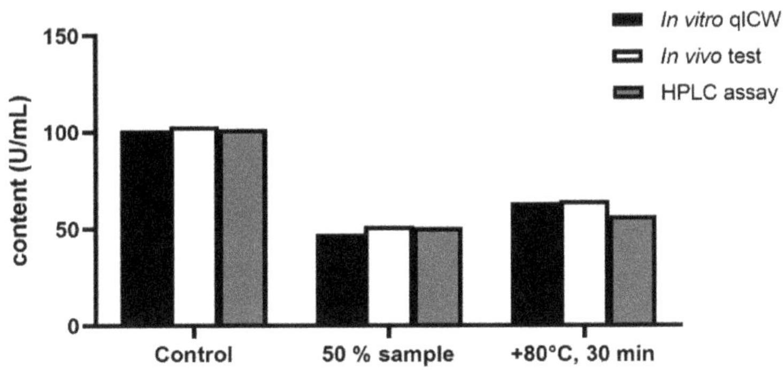

Parameters	Acceptance Criterion	Result
Recovery of the in vitro qICW as a percentage of the in vivo test for the stressed test items	80% ≤ recovery ≤ 125%	91% and 98%

Figure 3. The results for the stressed and unstressed insulin lispro Sanofi LISDP001 and the recoveries.

4. Discussion

All acceptance criteria were met for the regular test items, with a mean recovery of 95% and a 90% confidence interval of 91% to 99%. Furthermore, each single recovery met the predefined acceptance criteria. It could be concluded that the in vitro qICW and the in vivo test deliver similar results.

Seven different DP batches of either insulin glargine or insulin lispro and three different DS batches of insulin lispro were used for this study. The in vivo tests were performed on a regular basis for the primary stabilities during development. No additional in vivo tests could be performed for the regular test items due to animal welfare regulations and ethical reasons. Hence, the in vivo data from existing studies were used. The in vitro qICW data for Suliqua consisted of the initial Bio-ID (two plates) combined with additional experiments (six plates) to reach USP <121> precision, and all plates of the in vitro qICW for the insulin lispro Sanofi DS and DP were performed at the same time. The insulin lispro batch LISDP001 was tested with fewer replicates, which led to higher variability in the test results and represented a worst-case scenario for this bridging study.

Many of the in vitro qICW results showed lower potencies than those of the in vivo tests. It appeared that the in vitro qICW was more susceptible to stability-induced changes in the insulin lispro.

The acceptance criterion of a ≤20% difference (0.0792 on a log scale) was equivalent to the acceptance criterion of an 80% ≤ recovery ≤ 125%. For the purpose of harmonization, the acceptance criterion expressed as 80% ≤ recovery ≤ 125% is used.

Both stressed test items met the acceptance criterion of 80% ≤ recovery ≤ 125%, with recovery rates of 91% (the 50% test item) and 98% (the heat-stressed test items).

The 50% test item was used as a positive control for the potency reduction. Both the in vitro qICW and the in vivo test showed potency results very close to 50%.

The heat-stressed test items showed slightly increased potencies (from 7% to 8%) with the in vitro qICW and the in vivo tests compared to the HPLC data. This may have been caused by degraded or aggregated insulin lispro molecules which were still functional.

Potency measurements are needed to reflect the integrity of a complex three-dimensional structure in a solution, and this cannot be measured using chromatographic techniques [31]. Thus, for biotherapeutics, a potency assay for measuring biological activity is required for all release, stability, or study measurements [32]. As insulin/insulin analogues are manufactured with very well-established and robust processes and are characterized by well -defined and stable conformational structures, HPLC data could be representative of the biological activity. Therefore, many authorities rely on physicochemical determination for the biological activity of an insulin [18]. Nevertheless, insulins are a class of biotherapeutics where the integrity determines the biological activity, clinical efficacy, and safety.

The biological activities of human insulin and insulin analogues are currently assessed by the in vivo rabbit blood sugar test for quantitative determinations [19]. The in vitro cell-based ICW bioassay based on detecting human insulin receptor auto-phosphorylation in intact cells as the first step of the insulin signaling pathway is quite representative of the mode of action already used for batch release bioidentity measurements [13,19,21,22].

This ICW as a quantitative version (qICW) was investigated as an equivalent test system to the in vivo rabbit blood sugar bioassay test for human insulin and insulin analogues. At Sanofi, several thousand rabbits per year (ranging from 1000 to 4000) were used for quantitative insulin bioassay determinations depending on the number of studies performed in research and development until the rabbit blood sugar test was replaced with the proposed in vitro insulin cell-based bioassay (qICW). Thus, this alternative should also be applied by other companies developing or producing insulin for marketed use.

Furthermore, this ICW is more precise, at least as accurate, and as robust as the in vivo rabbit blood sugar bioidentity test, and it allows for quantitation of the results with high reproducibility, which is beneficial for patients.

The assay is favorable from an ethical point of view because it can replace animal testing for the quality control of new insulin batches. This is especially relevant since many initiatives are ongoing and great efforts are being made to replace, reduce, and refine animal-based testing in line with the 3R principle.

Nevertheless, this replacement does not replace all animal experiments as, for example, in vivo toxicity tests (e.g., embryotoxicity, reproductive toxicity, carcinogenicity, etc.) are still needed to assess the safety profiles of new insulin analogues, for example, those that have been performed for insulin glargine [33] and insulin lispro [34].

In summary, the in vitro qICW can be regarded as a superior alternative to the currently prescribed rabbit blood sugar test to quantify the biological activities of insulins and insulin analogues.

5. Conclusions

All batches tested to demonstrate method similarity were within the predefined acceptance criteria, as shown in Figures 2 and 3. Thus, the in vitro test system is similar and comparable to the in vivo test system for testing insulin bioassay activity according to USP general chapter <121>.

The data in this study clearly indicated that the rabbit blood sugar test can be replaced with the in vitro insulin cell-based bioassay (ICW) for the quantitative measurement of new insulins or insulin analogues during development or after post-approval changes. This will reduce the suffering of many rabbits and provide more meaningful and precise data for the determination of insulin potency for patient safety and efficacy.

Author Contributions: Conceptualization, S.R., R.H. and D.U.; ICW method, S.R., A.W. and Ö.D.; HPLC method, B.B.; rabbit blood sugar method, M.B.; statistics, B.N.; writing, S.R. and D.U.; review, A.W., Ö.D., B.B., R.H., M.B. and B.N.; final editing, D.U.; project administration, D.U. and R.H. All authors have read and agreed to the published version of the manuscript.

Funding: This research received no external funding.

Institutional Review Board Statement: This study received approval from the relevant ethics committee. The animal study protocols were reviewed and approved by the District Government of South-Hesse in Darmstadt under the animal use permit nos. HMR-7/Anz. 02 and FH-1009, approved on 29 May 2006 and 18 March 2016 respectively.

Informed Consent Statement: Not applicable.

Data Availability Statement: Not applicable.

Acknowledgments: The authors thank Christine Graf and Ana Serchinger for their scientific bioassay support; Noreen Stenzel, Constanze Pfaff, and Franziska Löbel for their excellent technical assistance; and Jochen Maas for his continuous support of the project over many years.

Conflicts of Interest: All authors involved in providing, analyzing and interpreting data as well as writing this manuscript are employees of Sanofi. All authors declare no conflict of interest.

References

1. Russel, W.; Burch, R. *The Principles of Humane Experimental Technique*; Methuen Publishing: London, UK, 1959.
2. Bruckner, L. European pharmacopoeia achievements and perspectives. In Proceedings of the European Directorate for the Quality of Medicines & Health Care International Conference on Quality of Medicines in a Globalised World: Dreams and Reality; Workshop 4; Prague, Czech Republic, 14–15 October 2010; p. 46.
3. Chumakov, K. Food and Drug Administration (FDA) perspective. In Proceedings of the European Directorate for the Quality of Medicines & Health Care International Conference on Quality of Medicines in a Globalised World: Dreams and Reality; Workshop 4; Prague, Czech Republic, 14–15 October 2010; p. 47.
4. Grein, K. Achieving the 3Rs in a globalised word: Feedback from VICH. In Proceedings of the European Directorate for the Quality of Medicines & Health Care International Conference on Quality of Medicines in a Globalised World: Dreams and Reality; Workshop 4; Prague, Czech Republic, 14–15 October 2010; p. 48.
5. Joung, J. Association of SouthEast Asian Nations (ASEAN) perspective. In Proceedings of the European Directorate for the Quality of Medicines & Health Care International Conference on Quality of Medicines in a Globalised World: Dreams and Reality; Workshop 4; Prague, Czech Republic, 14–15 October 2010; p. 49.

6. Ponsar, C. European vaccine manufacturer's perspectives. In Proceedings of the European Directorate for the Quality of Medicines & Health Care International Conference on Quality of Medicines in a Globalised World: Dreams and Reality; Workshop 4; Prague, Czech Republic, 14–15 October 2010; p. 49.
7. European Parliament and the Council of the European Union. *Directive 2010/63/EU of the European Parliament and of the Council of 22 September 2010 on the Protection of Animals Used for Scientific Purposes*; Official Journal of the European Union L 276; European Parliament and the Council of the European Union: Brussels, Belgium; Strasbourg, France, 2010; pp. 33–79.
8. European Medicines Agency. *Recommendation to Marketing Authorisation Holders, Highlighting the Need to Ensure Compliance with 3Rs Methods Described in the European Pharmacopoeia*; European Medicines Agency: London, UK, 2012; Available online: https://www.ema.europa.eu/en/documents/scientific-guideline/recommendation-marketing-authorisation-holders-highlighting-need-ensure-compliance-3rs-methods_en.pdf (accessed on 17 February 2015).
9. Diez-Solinska, A.; Vegas, O.; Azkona, G. Refinement in the European Union: A Systematic Review. *Animals* **2022**, *12*, 3263. [CrossRef] [PubMed]
10. Herrmann, K. Refinement on the way towards replacement: Are we doing what we can? In *Animal Experimentation: Working towards a Paradigm Change*; Brill: Leiden, The Netherlands, 2019; pp. 1–64. [CrossRef]
11. Verderio, P.; Lecchi, M.; Ciniselli, C.M.; Shishmani, B.; Apolone, G.; Manenti, G. 3Rs Principle and Legislative Decrees to Achieve High Standard of Animal Research. *Animals* **2023**, *13*, 277. [CrossRef] [PubMed]
12. Metz, B.; Hendriksen, C.F.; Jiskoot, W.; Kersten, G.F. Reduction of animal use in human vaccine quality control: Opportunities and problems. *Vaccine* **2002**, *20*, 2411–2430. [CrossRef] [PubMed]
13. Hack, R.; Rueggeberg, S.; Schneider, L.; Mayert, D.; Tennagels, N.; Welte, S.; Niederhaus, B.; Arz, W.; Usener, D.; Troschau, G.; et al. Progress towards the replacement of the rabbit blood sugar bioidentity assay by an in vitro test for batch release of insulins in quality control. *ALTEX* **2017**, *34*, 565–566. [CrossRef]
14. Khabib, M.N.H.; Sivasanku, Y.; Lee, H.B.; Kumar, S.; Kue, C.S. Alternative animal models in predictive toxicology. *Toxicology* **2022**, *465*, 153053. [CrossRef]
15. Swaters, D.; van Veen, A.; van Meurs, W.; Turner, J.E.; Ritskes-Hoitinga, M. A History of Regulatory Animal Testing: What Can We Learn? *Altern. Lab. Anim.* **2022**, *50*, 322–329. [CrossRef]
16. Akkermans, A.; Chapsal, J.M.; Coccia, E.M.; Depraetere, H.; Dierick, J.F.; Duangkhae, P.; Goel, S.; Halder, M.; Hendriksen, C.; Levis, R.; et al. Animal testing for vaccines. Implementing replacement, reduction and refinement: Challenges and priorities. *Biologicals* **2020**, *68*, 92–107. [CrossRef]
17. Da Silva, C.C.; Presgrave, O.A.; Hartung, T.; de Moraes, A.M.; Delgado, I.F. Applicability of the Monocyte Activation Test (MAT) for hyperimmune sera in the routine of the quality control laboratory: Comparison with the Rabbit Pyrogen Test (RPT). *Toxicol. In Vitro* **2016**, *32*, 70–75. [CrossRef]
18. Hamza, A.A. Biological Assay of Insulin: An Old Problem ReDiscovered. *J. Adv. Technol. Endocrinol. Res.* **2019**, *1*, 180002.
19. U.S. Pharmacopeial Convention. *Insulin Assays <121>*; USP 43-NF38; US Pharmacopeial Convention: Rockville, MD, USA, 2019.
20. Sommerfeld, M.R.; Müller, G.; Tschank, G.; Seipke, G.; Habermann, P.; Kurrle, R.; Tennagels, N. In vitro metabolic and mitogenic signaling of insulin glargine and its metabolites. *PLoS ONE* **2010**, *5*, e9540. [CrossRef]
21. Tennagels, R.; Rueggeberg, S.; Schneider, L.; Mayert, D.; Welte, S.; Hack, R.; Arz, W.; Usener, D. Stimuli to the Revision Process: A Proposed In Vitro Insulin Cell-Based Bioassay to Be Included in USP General Chapter Insulin Assays ⟨121⟩. *Pharmacopeial Forum* **2017**, *43*, 507–520. Available online: https://www.uspnf.com/sites/default/files/usp_pdf/EN/USPNF/pf-legacy-pdf/pf-2017_vol-43.pdf (accessed on 17 July 2018).
22. Garige, M.; Ghosh, S.; Roelofs, B.; Rao, V.A.; Sourbier, C. Protocol to Assess the Biological Activity of Insulin Glargine, Insulin Lispro, and Insulin Aspart In Vitro. *Methods Protoc.* **2023**, *6*, 33. [CrossRef]
23. Yie, J.; Dey, M.; Su, J.; Sergi, J.; Zhang, Y.; Le, T.H.; Kashi, S.; Gurney, K. Development of a robust functional cell-based assay for replacing the rabbit blood sugar bioidentity test of insulin glargine drug substance. *J. Pharm. Biomed. Anal.* **2020**, *186*, 113328. [CrossRef] [PubMed]
24. Rask-Madsen, C.; Kahn, C.R. Tissue-specific insulin signaling, metabolic syndrome, and cardiovascular disease. *Arterioscler. Thromb. Vasc. Biol.* **2012**, *32*, 2052–2059. [CrossRef] [PubMed]
25. Chen, H.; Kovar, J.; Sissons, S.; Cox, K.; Matter, W.; Chadwell, F.; Luan, P.; Vlahos, C.J.; Schutz-Geshwender, A.; Olive, D.M. A cell-based immunocytochemical assay for monitoring kinase signaling pathways and drug efficacy. *Anal. Biochem.* **2005**, *338*, 136–142. [CrossRef]
26. Müller, G.; Herling, A.W.; Sandow, J. Antidiabetic activity. In *Drug Discovery and Evaluation: Pharmacological Assays*, 3rd ed.; Vogel, H.G., Ed.; Springer: Berlin/Heidelberg, Germany; New York, NY, USA, 2007; pp. 1323–1607.
27. U.S. Pharmacopeial Convention. *Design and Development of Biological Assays ⟨1032⟩, Biological Assay Validation ⟨1033⟩, Analysis of Biological Assays ⟨1034⟩*; USP 40–NF 35; US Pharmacopeial Convention: Rockville, MD, USA, 2017; pp. 917–963.
28. ICH. *International Congress on Harmonization Q2(R2) on Validation of Analytical Procedures–Step2b*; ICH: Geneva, Switzerland, 2022.
29. DeMeyts, P.; Whittaker, J. Structural biology of insulin and IGF1 receptors: Implications for drug design. *Nat. Rev. Drug Discov.* **2002**, *1*, 769–783. [CrossRef]
30. Issad, T.; Boute, N.; Pernet, K. The activity of the insulin receptor assessed by bioluminescence resonance energy transfer. *Ann. N. Y. Acad. Sci.* **2002**, *973*, 120–123. [CrossRef] [PubMed]

31. Manta, B.; Obal, G.; Ricciardi, A.; Pritsch, O.; Denicola, A. Tools to evaluate the conformation of protein products. *Biotechnol. J.* **2011**, *6*, 731–741. [CrossRef] [PubMed]
32. White, J.R.; Abodeely, M.; Ahmed, S.; Debauve, G.; Johnson, E.; Meyer, D.M.; Mozier, N.M.; Naumer, M.; Pepe, A.; Qahwash, I.; et al. Best practices in bioassay development to support registration of biopharmaceuticals. *Biotechniques* **2019**, *67*, 126–137. [CrossRef]
33. Hofmann, T.; Horstmann, G.; Stammberger, I. Evaluation of the reproductive toxicity and embryotoxicity of insulin glargine (LANTUS) in rats and rabbits. *Int. J. Toxicol.* **2002**, *21*, 181–189. [CrossRef] [PubMed]
34. Byrd, R.A.; Blackbourne, J.L.; Knadler, M.P.; Schultze, A.E.; Vahle, J.L. Chronic Toxicology Studies of Basal Insulin Peglispro in Rats and Dogs: A Novel, PEGylated Insulin Lispro Analog with a Prolonged Duration of Action. *Toxicol. Pathol.* **2017**, *45*, 402–415. [CrossRef] [PubMed]

Disclaimer/Publisher's Note: The statements, opinions and data contained in all publications are solely those of the individual author(s) and contributor(s) and not of MDPI and/or the editor(s). MDPI and/or the editor(s) disclaim responsibility for any injury to people or property resulting from any ideas, methods, instructions or products referred to in the content.

animals

MDPI

Systematic Review

Refinement in the European Union: A Systematic Review

Alina Díez-Solinska, Oscar Vegas and Garikoitz Azkona *

Department of Basic Psychological Processes and Their Development, Euskal Herriko Unibertsitatea (UPV/EHU), Tolosa Hiribidea, 20018 Donostia, Spain
* Correspondence: garikoitz.azkona@ehu.eus

Simple Summary: More than 10 years have passed since the publication of Directive 2010/63/EU on the protection of animals used for scientific purposes based on replacement, reduction, and refinement (3Rs). These principles state that if animals have to be used in experiments, researchers should make every effort to replace them with non-sentient alternatives, reduce them to a minimum, and refine experiments and housing conditions so as to cause the minimum possible pain and distress. In this systematic review, we aimed to identify and summarize published advances in the refinement protocols made by European Union-based research groups from 2011 to 2021, and to determine whether or not said research was financially supported. Our results indicated that the majority of advances were related to improvements in experimental procedures for mice, and the research groups were mostly from universities and the United Kingdom. More than two thirds of the studies received financial support, mostly national. There is a clear willingness in the scientific community to improve the welfare of laboratory animals. However, we believe that more progress in refinement would have been made during these years if there had been more specific financial support available at both the national and European Union levels.

Abstract: Refining experiments and housing conditions so as to cause the minimum possible pain and distress is one of the three principles (3Rs) on which Directive 2010/63/EU is based. In this systematic review, we aimed to identify and summarize published advances in the refinement protocols made by European Union-based research groups from 2011 to 2021, and to determine whether or not said research was supported by European or national grants. We included 48 articles, the majority of which were related to improvements in experimental procedures (37/77.1%) for mice (26/54.2%) and were written by research groups belonging to universities (36/57.1%) and from the United Kingdom (21/33.9%). More than two thirds (35/72.9%) of the studies received financial support, 26 (mostly British) at a national level and 8 at a European level. Our results indicated a clear willingness among the scientific community to improve the welfare of laboratory animals, as although funding was not always available or was not specifically granted for this purpose, studies were published nonetheless. However, in addition to institutional support based on legislation, more financial support is needed. We believe that more progress would have been made in refinement during these years if there had been more specific financial support available at both the national and European Union levels since our data suggest that countries investing in refinement have the greatest productivity in successfully publishing refinements.

Keywords: 3Rs; refinement; financial support; European Union

Citation: Díez-Solinska, A.; Vegas, O.; Azkona, G. Refinement in the European Union: A Systematic Review. *Animals* **2022**, *12*, 3263. https://doi.org/10.3390/ani12233263

Academic Editor: Vera Baumans

Received: 3 November 2022
Accepted: 22 November 2022
Published: 23 November 2022

Copyright: © 2022 by the authors. Licensee MDPI, Basel, Switzerland. This article is an open access article distributed under the terms and conditions of the Creative Commons Attribution (CC BY) license (https://creativecommons.org/licenses/by/4.0/).

1. Introduction

In the European Union (EU), the protection of animals used for scientific purposes is regulated by Directive 2010/63/EU [1]. In order to harmonize standards across the EU, member states were required to transpose the Directive into national legislation and most of them did so during 2013 [2]. In 2019, the Directive was amended to increase transparency [3]. This Directive is considered an essential piece of legislation with which anybody who carries out fundamental biological research and preclinical development potentially involving live cyclostomes, cephalopods, and/or vertebrate animals must be familiar [4].

Overall, the Directive promotes both animal welfare and high-quality scientific research and establishes one of the most progressive and stringent mandatory lab animal protection frameworks in the world [4]. It was drafted with four very clear fundamental principles in mind. First is the recognition that the ultimate goal is to replace the use of animals. Second is the acknowledgment that animals, including non-human primates, are still needed today. Third is the acceptance that animals have intrinsic value in themselves and must be respected. Fourth is the agreement that the principle of the Three Rs (3Rs) is the key to ensuring more humane and better science [5].

The principle of the 3Rs—replacement, reduction, and refinement [6]—is the cornerstone of the Directive. These principles state that if animals have to be used in experiments, researchers should make every effort to replace them with non-sentient alternatives, reduce them to a minimum, and refine experiments and housing conditions so as to cause the minimum possible pain and distress. Thus, the 3Rs concept is both a framework designed to minimize the use and suffering of animals (harm to the animal) and a means to support high-quality science and translation (benefit to society). The conflict between these two aims is usually resolved on a case-by-case basis by weighing up the harm to the animals involved and the benefits of the research, or by prioritizing the experience of the animals (i.e., refinement) over reduction [7].

Whereas there is a greater consensus regarding the replacement and reduction principles, the implementation of the refinement principle has caused the greatest controversy. Refinement is an ongoing process that requires input from all those involved in the use of experimental animals [8] and covers all animal and human interactions throughout the entire life of the animal. It is not limited solely to experimental procedures, but rather also encompasses the transport, husbandry, and euthanasia of animals [9]. Recently, we observed that people who work with laboratory animals are clearly aware of this and show great sensitivity to their well-being [10,11]. Moreover, perceived animal stress/pain negatively affects the professional quality of life of people working with laboratory animals [12].

In this systematic review, we aimed to identify and summarize published advances in refinement protocols made by EU-based research groups from 2011 to 2021 and to determine whether or not the research was supported by European or national grants.

2. Materials and Methods

This systematic review is reported in accordance with the Preferred Reporting Items for Systematic Reviews and Meta-Analysis (PRISMA statement) flowsheet.

2.1. Search Strategy

Web of Science and PubMed were chosen for the search. The search was carried out on 13 May 2022. As the main aim of the study was to examine the number of original publications developing and/or improving a refinement technique (considering Russell and Burch's 3Rs) in laboratory animal research, the following search terms were used in both databases: (3 Rs OR 3 R OR 3R OR 3Rs) AND (refinement) AND (animal*) AND (techniqu* OR strateg*). The filters included were publication year (from 2011 to 2021), document type (article), and language (English). In the Web of Science database, the countries/regions filter was also applied (countries from the European Union), whereas PubMed articles from outside the European Union were excluded manually. Since our goal was to select articles refining animal techniques, we also used Web of Science filters to exclude human research. We comprehensively searched for published full-text studies. The study selection was performed by A.D.-S., and G.A., who independently examined the full texts of potentially relevant studies and applied the eligibility criteria in order to select, by consensus, those studies to be included. The information extracted from the articles included the title, authors, year of publication, DOI, animals in which the refinement technique was implemented, country in which the study was performed, the kind of procedure that was refined and a brief description of it, the institution to which the authors were affiliated, whether or not the article had received any kind of financial support, and which type of institution funded the studies. These data are provided in Supplementary Table S1.

2.2. Eligibility Criteria

Original articles were deemed eligible if they met the following criteria:

- The articles were published from 2011 to 2021 (both years included).
- At least one of the authors of the article belonged to an institution/research center/university (public or private) from the European Union.
- The studies were not classified as reviews, meta-analyses, or reports, nor were they guidelines, protocols, surveys, or ethical discussions for research reflection.
- The main topic of the studies was the improvement or development of a refinement technique, not a replacement or reduction method, even when this was linked to the 3Rs.
- The studies were published in English.

2.3. Categorization of Refinement Procedures

We categorized studies based on whether they proposed improvements in experimental procedures, husbandry, transport, or euthanasia. Within these categories, we defined sub-categories with the aim of grouping studies in accordance with the common characteristics of the procedures.

2.4. Statistical Analysis

Frequency (%) statistics were used to describe the sample.

3. Results

3.1. Study Selection

We systematically searched for references related to refinement procedures in laboratory animals. A total of 384 references were identified by electronic search; 338 full-text studies were evaluated in accordance with the eligibility criteria and 290 were excluded. Finally, 48 studies [13–60] complied with all the established eligibility criteria and were included in the study (Figure 1).

3.2. Refinement Procedures by Category

First, we categorized the different refinement procedures into previously defined categories (Table 1). Experimental procedure was the category into which most studies fell (37/77.1%), followed by husbandry (10/20.8%), with only one study being categorized as refinement in transport (2.1%). We did not find any studies refining euthanasia protocols.

Of the experimental procedure sub-categories, sampling encompassed the most studies (10/28.6%), followed by analgesia (4/10.8%) and animal training (4/10.8%). Regarding husbandry, welfare assessment in the animals' natural environment and social housing were the sub-categories containing the highest number of published studies (3/30% each).

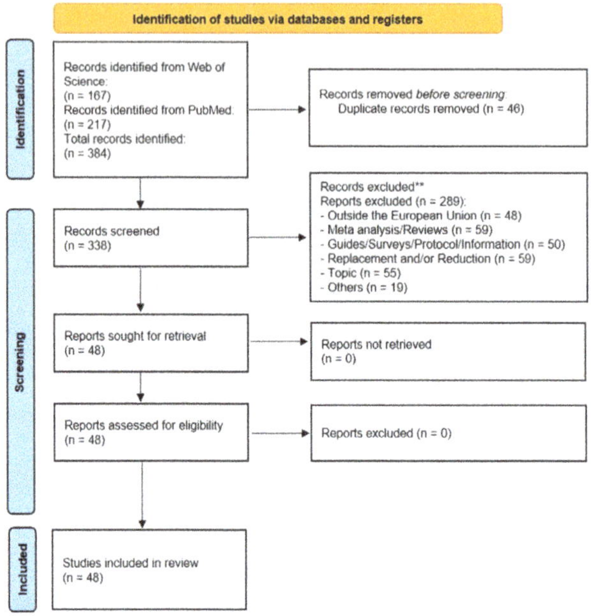

Figure 1. PRISMA flowchart of study selection.

Table 1. Classification of the studies by category and sub-category.

Category	Sub-Category	Brief Description	Reference
Experimental procedure	Analgesia	Buprenorphine is mixed with a desirable nut paste for mice to ensure its voluntary ingestion.	[13]
		Evaluation of different analgesic methods in order to improve their quality in surgical interventions.	[14]
		Refinement approach in the spinal cord injury model.	[15]
		Description of an intrathecal technique to administer morphine in order to minimize the interaction between this substance and the research itself.	[16]
	Behavioral Testing	Refinement of the Radial Arm Test to avoid animal handling and water and food deprivation.	[17]
		Reliable alternatives to the Water Maze Test in order to avoid the stress associated with deep water swimming.	[18]
	Diagnosis	Categorization of animals suffering from either hyperacusis or tinnitus through a model based on operant conditioning with positive reinforcement.	[19]
	Dosing	A new application of a device enabling the long-term enteral administration of drugs or nutritional supplementation.	[20]
		Syringe feeding as an alternative to the commonly used water bottles and oral gavage dosing in the administration of probiotics.	[21]
	Identification	A new tagging protocol to preserve animal welfare.	[22]
		A non-invasive technique for sexing in fish.	[23]
	Imaging	Thermographic imaging as an alternative method of examining tumor growth in a non-invasive way.	[24]
		Magnetic resonance imaging as a non-invasive imaging technique to detect early cancer onset and its development.	[25]
		The early detection of animal disease through imaging techniques.	[26]

Table 1. *Cont.*

Category	Sub-Category	Brief Description	Reference
		No need to use anesthesia for the procedure.	[27]
	Handling	A method to inject substances using a non-restrained technique.	[28]
	Others	Development of an implantable device for long-term Deep Brain Stimulation which provides mice with the freedom of movement.	[29]
		Reduction in exposure time to smoke and analyzing lung lobes separately in tobacco product testing.	[30]
		Refinement of a respiratory monitoring setup for an in vivo bioluminescence and fluorescence imaging device.	[31]
	Sampling	A non-invasive method for determining liver weight gain that reduces the number of animals needed and avoids the post-mortem technique.	[32]
		Blood micro-sampling from the ear capillary.	[33]
		Cutting feathers instead of plucking them from the animal's skin in order to avoid this painful method.	[34]
		Non-invasive collection of murine hair follicle cells to avoid biopsies.	[35]
		Monitoring immunoglobulins via fecal pellets as a new non-invasive way to assess the immune response.	[36]
		Comparison of blood sampling techniques.	[37]
		Non-invasive ocular (tear) sampling for genetic ascertainment of transgenic mice.	[38]
		Refinement of sperm sampling techniques.	[39]
		Blood volume reduction in each sample.	[40]
		Swabbing the skin of non-anesthetized fish for DNA extraction.	[41]
	Surgery	Development of a surgical procedure to induce myocardial infarction in order to reduce mice mortality and distress.	[42]
		Refinement of a heterotopic cardiac transplantation model.	[43]
		Use of non-operated rather than sham-operated controls.	[44]
	Training	Cognitive enrichment allows mice to better lead with their home environment.	[45]
		Training dogs for gavage administration.	[46]
		An approach to alternative rewards (social stimuli instead of fluid droplets) in animal training.	[47]
		Refinement in training pigs prior to the oral glucose tolerance test.	[48]
	Welfare Assessment	Identification of pain and distress signals among macaques during experimental procedures in order to promote their wellbeing.	[49]
Husbandry	Breeding	Three techniques to refine the activation of CreERT2 in maternal mice.	[50]
	Environmental enrichment	An environmental enrichment tool to enhance animal welfare in large-scale mouse husbandry.	[51]
		A comparison of different environmental enrichment techniques and their advantageous effect on mice pups' wellbeing and survival rate.	[52]
	Health	Optimization of a device to prevent mites from penetrating the skin of experimental hens.	[53]
	Social housing	A perforated transparent wall that divides the cage into two compartments and allows olfactory, acoustic, and visual communication between two mice but prevents fighting and injuries.	[54]
		Improvement in social housing in toxicology studies.	[55]
		An automated homecage system to register social alterations after pharmacological exposure.	[56]

Table 1. *Cont.*

Category	Sub-Category	Brief Description	Reference
	Welfare Assessment	A new tool to assess animal wellbeing, drawing on their natural behavior.	[57]
	Welfare Assessment	Use of home cage running wheels to obtain daily measurements of motor function in SOD1G93A mice, with minimal intervention.	[58]
	Welfare Assessment	Voluntary wheel running as an indication of distress suffered by mice in an experimental procedure.	[59]
Transport	Training	Reinforcement (positive and sometimes negative) to help monkeys acclimate more quickly to transport devices.	[60]

3.3. *Refinement Procedures by Animal Species, Institution, and Country*

More than half of the published studies described refinement procedures for mice and just over twenty percent did so for rats, meaning that most of the protocols described were for rodents. The remaining procedures were described as refinements for other mammals such as macaques, dogs, or pigs, as well as for fish and birds (Figure 2a).

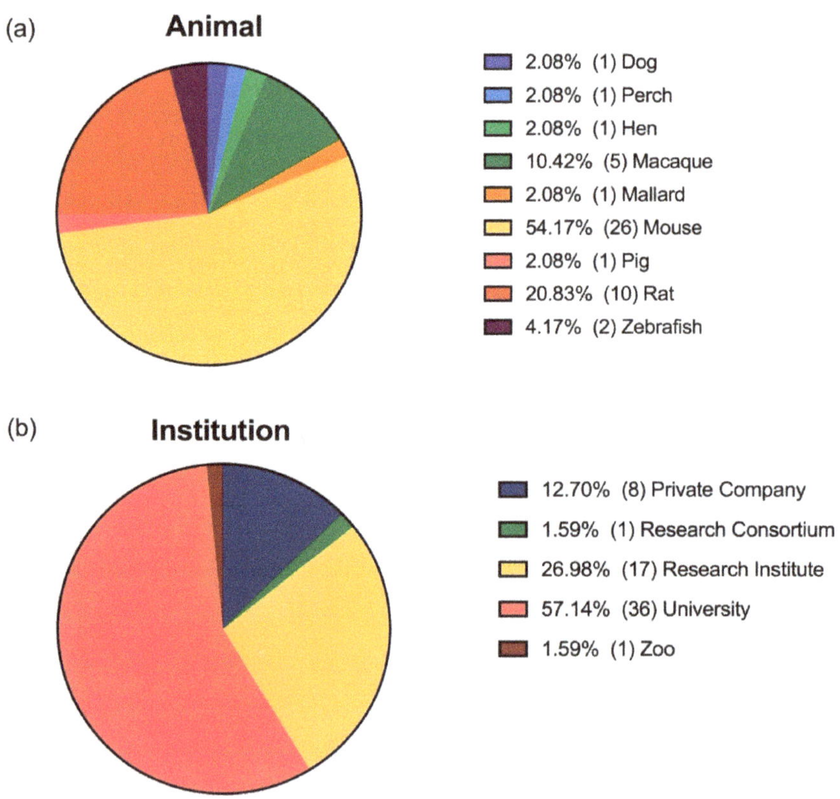

Figure 2. *Cont.*

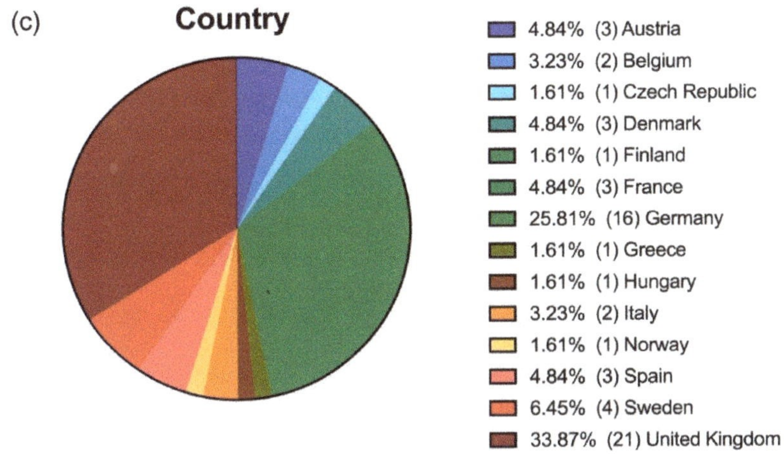

Figure 2. (**a**) Animal species for which the refinement protocols were described; (**b**) country; and (**c**) institution in which the research groups were working. Data are presented in percentages (total number).

More than half of the research groups belonged to universities, just over a quarter belonged to research institutes, and just under ten percent were from private companies (Figure 2b). Moreover, 14 (29.2%) of the studies were collaborations between different institutions: private company and research institute (1); university and private company (2); university, a private company, and research institute (1); university and research institute (9); and university and zoo (1).

In terms of country, most research groups were based in the United Kingdom (UK), followed by Germany (Figure 2c). Of the 48 articles selected, 13 (27.1%) were collaborations between groups from several countries: Austria and Germany (3); Belgium and the UK (2); Czech Republic, Denmark, and Sweden (1); Denmark and Sweden (1); France and Norway (1); Germany and Spain (1); Germany, the UK, and Spain (1); Hungary, Finland, and the UK (1); the UK, Australia, and New Zealand (1); and the UK, Australia, and South Africa (1).

3.4. Financial Support for the Studies

More than two-thirds (35/72.9%) of the studies received financial support, 26 in the form of national funding, 7 from their own institution, and 2 from private foundations. Moreover, 8 received European funding: (1) COST Action (the Behavioral Management and Training of Laboratory non-human Primates and Large Laboratory Animals—PRIMTRAIN) [61], (1) the European Regional Development Fund, (2) the European Research Council (ERC), (1) the EU Integrated Project (Xenome), (1) the Innovative Medicine Initiative (IMI), (1) the Sex'NPerch program by the European Maritime and Fisheries Fund, and (1) the Seven Framework Program (FP7-HEALTH-MITOTARGET). Of the remaining studies, 8 (16.7%) received no financial support and 6 (12.5%) did not specify (Figure 3). Supplementary Table S2 shows the number of un-funded and funded articles per country.

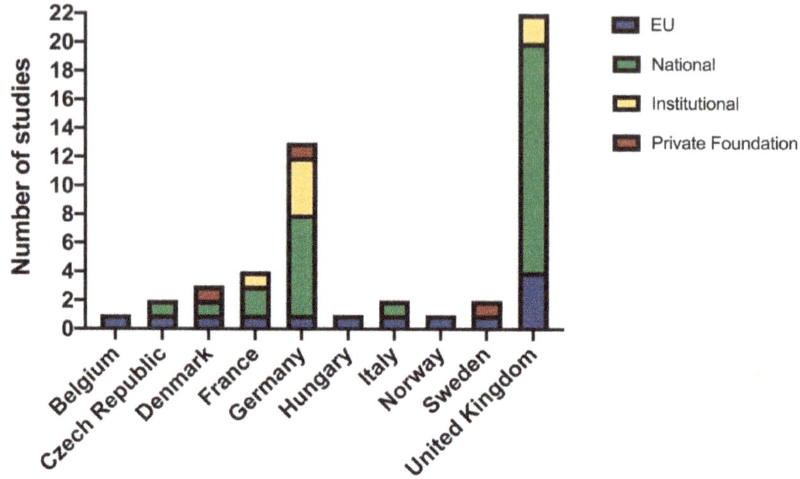

Figure 3. Financial support by country and organization. Data are presented in total numbers.

Studies carried out by the UK research groups obtained the most funding at both the EU and the national level. Of the 21 UK articles published, 18 (85.7%) were funded, 16 (76.2%) of which were funded nationally. Of these 16, half (8/50%) were partially or fully funded by the National Center for the Replacement, Refinement, and Reduction of Animals in Research (NC3Rs), and a quarter (4/25%) were also funded at a European level. The remaining two studies were conducted by groups working in private companies and were funded by their own institution. Of the 16 studies published by German groups, 11 (43.75%) received funding, 7 (63.6%) at a national level. Of these, one also received European funding and another one received private funding. The rest (4/36.4%) received institutional funding. One of these institutions is the Charité 3^{R} of the Universitätsmedizin Berlin, which actively promotes the 3Rs principle in biomedical research and education. One Swedish study received funding from a private foundation that promotes scientific research against painful animal experiments (Torvald and Britta Gahlin's foundation).

4. Discussion

An important component of good scientific practice is to reduce the suffering of laboratory animals through refinement techniques. In our systematic review, we identified 48 studies conducted by EU-based research groups between 2011 and 2021 that aimed to improve the welfare of animals used in research. We chose these 10 years because they correspond to the decade following the publication of Directive 2010/63/EU, which clearly promotes refinement [1].

During these ten years, tissue sampling improvements have been described to minimize the stress and pain associated with this procedure. Many of these methods are non-invasive and do not require great technical skill, thus reducing both the stress on the experimenter during handling and the harm to the animal. Similarly, other studies have sought to improve analgesia protocols by refining drug combinations or administration routes. The use of training both to condition and to habituate animals to a procedure is also worth noting. Although time consuming, training is a very good strategy for reducing animal stress and discomfort. Improvements have also been described in husbandry, with one area of focus being social housing. Many of the animals used in biomedical research belong to social species, and Directive 2010/63/EU recommends their group housing [1]. Technological advances are increasingly enabling animals to be monitored in their home cage, thereby reducing the stress associated with interaction with humans and improving their welfare. A COST (European Cooperation in Science and Technology) action is currently underway for this purpose [62].

Recent statistics on animal use in the EU indicate that mice are the most commonly used animals [63–65], and we observed the same trend in our review; more than half of the selected articles described refinements in mouse protocols. Regarding the origin of the research groups, in terms of institution and country, our results follow the same trend observed in the biomedical area [66], with university-based research groups from the UK being the ones with the most publications to their name.

Scientists are currently working to produce valid data on measuring and improving all laboratory animals' welfare. In addition to the COST actions described above [61,62], the Eurogroup for Animal Welfare, a lobbying organization, is working to implement refinement methods in research [67]. However, we observed that there were only a few studies funded by European entities, and only one was specific to laboratory animals [61]. The rest of the European grants were oriented toward biomedical research. By country, the UK is the leading national funder of projects. It should be noted that the NC3Rs funded half of all the UK projects during this period. Other European countries also have centers dedicated to the achievement of the 3Rs, and a list of these can be found on the Norecopa webpage [68]. Our results show that specific funding for the achievement of the third R during this decade was far from substantial. We should not forget that scientists have been constantly asking for more public resources and interdisciplinary teams to solve the quandary of how to strike a balance between animal welfare and science [69].

The present study has certain limitations. First, since the word refinement can be applied to many scientific fields, the search was restricted to articles that also mentioned either animals or the 3Rs. In this sense, we were unable to identify some types of articles, such as, for example, strategies to reduce singly housed male mice [70], as well as others in which the authors did not identify the term refinement as a keyword [71]. Our strategy also has the possibility of underestimating the scope of refinements because some refinements are often published as part of the scientific work that animals are used in. Furthermore, our search would not have identified the refinements developed by research groups or animal facilities and implemented in their centers that were not published. In this sense, the use of platforms dedicated to the 3Rs may be a good tool to collect and disseminate these protocols, as it is sometimes difficult to publish them in indexed journals. The Animal Welfare Institute's website contains protocols and scientific papers describing methods for reducing or eliminating pain, stress, and discomfort for animals, not only during experimental procedures but also in relation to their daily social and physical environment [72]. Our search did not include patented protocols such as, for example, the project "HaPILLness-Voluntary oral dosing in rodents", which replaces oral gavage with voluntary dosing [73]. Finally, the results of the funded projects completed in 2021 may still be under preparation or under consideration.

Overall, our findings show that, in recent years, advances have been made in the refinement of procedures using laboratory animals. Currently, there are refined protocols that are used on a daily basis in many animal facilities, such as administering substances with sweetened condensed milk [74] or transporting mice through a plastic tube instead of holding them by the tail [75]. Moreover, statistics indicate that protocols classified as severe have been decreasing slightly in recent years, by 1% per year [65]. However, we cannot forget that in order to be able to carry out experiments to refine a technique, in addition to a multidisciplinary team in which veterinarians must play an essential role [76], financial support is still necessary.

5. Conclusions

Our results indicate a clear willingness among the scientific community to improve the welfare of laboratory animals, as although funding was not always available, or was not specifically granted for this purpose, studies were published nonetheless. However, in addition to institutional support based on legislation, more financial support is needed. We believe that more progress would have been made in refinement during these years if there had been more specific financial support available at both the national and EU levels since our data suggest that countries investing in refinement have the greatest productivity in successfully publishing refinements.

Supplementary Materials: The following supporting information can be downloaded at: https://www.mdpi.com/article/10.3390/ani12233263/s1: Table S1: Information extracted from the included articles; Table S2: The number of unfunded and funded articles per country

Author Contributions: Conceptualization, A.D.-S., O.V., and G.A.; methodology and formal analysis, A.D.-S. and G.A.; writing—original draft preparation, review and editing, A.D.-S., O.V., and G.A. All authors have read and agreed to the published version of the manuscript.

Funding: This research received no external funding.

Institutional Review Board Statement: Not applicable.

Informed Consent Statement: Not applicable.

Data Availability Statement: All the data pertaining to the study will be made available upon request.

Conflicts of Interest: The authors declare no conflict of interest.

References

1. EU. Directive 2010/63/EU of the European Parliament and of the Council of 22 September 2010 on the Protection of Animals Used for Scientific Purposes. 2010. Available online: https://eur-lex.europa.eu/LexUriServ/LexUriServ.do?uri=OJ:L:2010:276:0033:0079:en:PDF (accessed on 2 November 2022).
2. Olsson, I.A.S.; Silva, S.P.D.; Townend, D.; Sandøe, P. Protecting Animals and Enabling Research in the European Union: An Overview of Development and Implementation of Directive 2010/63/EU. *ILAR J.* **2016**, *57*, 347–357. [CrossRef] [PubMed]
3. EU. Regulation (EU) 2019/1010 of the European Parliament and of the Council of 5 June 2019 on the Alignment of Reporting Obligations in the Field of Legislation Related to the Environment, and Amending REGULATIONS (EC) No 166/2006 and (EU) No 995/2010 of the European Parliament and of the Council, Directives 2002/49/EC, 2004/35/EC, 2007/2/EC, 2009/147/EC and 2010/63/EU of the European Parliament and of the Council, Council Regulations (EC) No 338/97 and (EC) No 2173/2005, and Council Directive 86/278/EEC. 2019. Available online: https://eur-lex.europa.eu/legal-content/en/TXT/?uri=CELEX%3A32019R1010 (accessed on 2 November 2022).
4. Chlebus, M.; Guillen, J.; Prins, J.B. Directive 2010/63/EU: Facilitating full and correct implementation. *Lab. Anim.* **2016**, *50*, 151. [CrossRef] [PubMed]
5. Louhimies, S. (Ed.) EU Directive 2010/63/EU: "Implementing the Three Rs Through Policy" the Three Rs Together It's Possible. In Proceedings of the 8th World Congress on Alternatives, Montreal, QC, Canada, 21–25 August 2011.
6. Russel, W.; Burch, R. *The Principles of Humane Experimental Technique*; Methuen Publishing: London, UK, 1959.
7. Graham, M.L.; Prescott, M.J. The multifactorial role of the 3Rs in shifting the harm-benefit analysis in animal models of disease. *Eur. J. Pharmacol.* **2015**, *759*, 19–29. [CrossRef]
8. Flecknell, P. Refinement of animal use—Assessment and alleviation of pain and distress. *Lab. Anim.* **1994**, *28*, 222–231. [CrossRef] [PubMed]
9. Tannenbaum, J.; Bennett, B.T. Russell and Burch's 3Rs then and now: The need for clarity in definition and purpose. *J. Am. Assoc. Lab. Anim. Sci.* **2015**, *54*, 120–132. [PubMed]
10. Goñi-Balentziaga, O.; Ortega-Saez, I.; Vila, S.; Azkona, G. Working with laboratory rodents in Spain: A survey on welfare and wellbeing. *Lab. Anim. Res.* **2021**, *37*, 18. [CrossRef]
11. Goñi-Balentziaga, O.; Ortega-Saez, I.; Vila, S.; Azkona, G. A survey on the use of mice, pigs, dogs and monkeys as animal models in biomedical research in Spain. *Lab. Anim. Res.* **2022**, *38*, 14. [CrossRef]
12. Goñi-Balentziaga, O.; Vila, S.; Ortega-Saez, I.; Vegas, O.; Azkona, G. Professional Quality of Life in Research Involving Laboratory Animals. *Animals* **2021**, *11*, 2639. [CrossRef] [PubMed]
13. Abelson, K.; Jacobsen, K.R.; Sundbom, R.; Kalliokoski, O.; Hau, J. Voluntary ingestion of nut paste for administration of buprenorphine in rats and mice. *Lab. Anim.* **2012**, *46*, 349–351. [CrossRef]
14. Bauer, C.; Schillinger, U.; Brandl, J.; Meyer-Lindenberg, A.; Ott, A.; Baumgartner, C. Comparison of pre-emptive butorphanol or metamizole with ketamine + medetomidine and s-ketamine + medetomidine anaesthesia in improving intraoperative analgesia in mice. *Lab. Anim.* **2019**, *53*, 459–469. [CrossRef] [PubMed]
15. Redaelli, V.; Papa, S.; Marsella, G.; Grignaschi, G.; Bosi, A.; Ludwig, N.; Luzi, F.; Vismara, I.; Rimondo, S.; Veglianese, P.; et al. A refinement approach in a mouse model of rehabilitation research. Analgesia strategy, reduction approach and infrared thermography in spinal cord injury. *PLoS ONE* **2019**, *14*, e0224337. [CrossRef] [PubMed]

16. Thomas, A.A.; Detilleux, J.; Sandersen, C.F.; Flecknell, P.A. Minimally invasive technique for intrathecal administration of morphine in rats: Practicality and antinociceptive properties. *Lab. Anim.* **2017**, *51*, 479–489. [CrossRef] [PubMed]
17. Mei, J.; Kohler, J.; Winter, Y.; Spies, C.; Endres, M.; Banneke, S.; Emmrich, J.V. Automated radial 8-arm maze: A voluntary and stress-free behavior test to assess spatial learning and memory in mice. *Behav. Brain Res.* **2020**, *381*, 112352. [CrossRef] [PubMed]
18. Deacon, R.M. Shallow water (paddling) variants of water maze tests in mice. *J. Vis. Exp.* **2013**, *76*, e2608. [CrossRef]
19. Möhrle, D.; Hofmeier, B.; Amend, M.; Wolpert, S.; Ni, K.; Bing, D.; Klose, U.; Pichler, B.; Knipper, M.; Rüttiger, L. Enhanced Central Neural Gain Compensates Acoustic Trauma-induced Cochlear Impairment, but Unlikely Correlates with Tinnitus and Hyperacusis. *Neuroscience* **2019**, *407*, 146–169. [CrossRef]
20. Fante, F.; Baldan, N.; De Benedictis, G.M.; Boldrin, M.; Furian, L.; Sgarabotto, D.; Ravarotto, L.; Besenzon, F.; Ramon, D.; Cozzi, E. Refinement of a macaque transplantation model: Application of a subcutaneous port as a means for long-term enteral drug administration and nutritional supplementation. *Lab. Anim.* **2012**, *46*, 114–121. [CrossRef]
21. Tillmann, S.; Wegener, G. Syringe-feeding as a novel delivery method for accurate individual dosing of probiotics in rats. *Benef. Microbes* **2018**, *9*, 311–315. [CrossRef] [PubMed]
22. Rácz, A.; Allan, B.; Dwyer, T.; Thambithurai, D.; Crespel, A.; Killen, S. Identification of Individual Zebrafish. *Animals* **2021**, *11*, 616. [CrossRef]
23. Ledoré, Y.; Bestin, A.; Haffray, P.; Morvezen, R.; Alix, M.; Schaerlinger, B.; Fontaine, P.; Chardard, D. Sex identification in immature Eurasian perch (*Perca fluviatilis*) using ultrasonography. *Aquac. Res.* **2021**, *52*, 6046–6051. [CrossRef]
24. Hussain, N.; Connah, D.; Ugail, H.; Cooper, P.A.; Falconer, R.A.; Patterson, L.H.; Shnyder, S.D. The use of thermographic imaging to evaluate therapeutic response in human tumour xenograft models. *Sci. Rep.* **2016**, *6*, 31136. [CrossRef]
25. Schmid, A.; Rignall, B.; Pichler, B.J.; Schwarz, M. Quantitative analysis of the growth kinetics of chemically induced mouse liver tumors by magnetic resonance imaging. *Toxicol. Sci.* **2012**, *126*, 52–59. [CrossRef] [PubMed]
26. Akladios, C.; Ignat, M.; Mutter, D.; Aprahamian, M. Survival variability of controls and definition of imaging endpoints for longitudinal follow-up of pancreatic ductal adenocarcinoma in rats. *J. Cancer Res. Clin. Oncol.* **2016**, *143*, 29–34. [CrossRef] [PubMed]
27. Karda, R.; Perocheau, D.P.; Suff, N.; Ng, J.; Delhove, J.M.K.M.; Buckley, S.M.K.; Richards, S.; Counsell, J.R.; Hagberg, H.; Johnson, M.R.; et al. Continual conscious bioluminescent imaging in freely moving somatotransgenic mice. *Sci. Rep.* **2017**, *7*, 6374. [CrossRef]
28. Stuart, S.A.; Robinson, E. Reducing the stress of drug administration: Implications for the 3Rs. *Sci. Rep.* **2015**, *5*, 14288. [CrossRef]
29. Plocksties, F.; Kober, M.; Niemann, C.; Heller, J.; Fauser, M.; Nüssel, M.; Uster, F.; Franz, D.; Zwar, M.; Lüttig, A.; et al. The software defined implantable modular platform (STELLA) for preclinical deep brain stimulation research in rodents. *J. Neural Eng.* **2021**, *18*, 056032. [CrossRef]
30. Dalrymple, A.; Ordoñez, P.; Thorne, D.; Walker, D.; Camacho, O.M.; Büttner, A.; Dillon, D.; Meredith, C. Cigarette smoke induced genotoxicity and respiratory tract pathology: Evidence to support reduced exposure time and animal numbers in tobacco product testing. *Inhal. Toxicol.* **2016**, *28*, 324–338. [CrossRef] [PubMed]
31. Gomes, A.L.; Gilchrist, S.; Kersemans, V.; Westcott, M.; Smart, S. Refinement of in vivo optical imaging: Development of a real-time respiration monitoring system. *Lab. Anim.* **2018**, *52*, 531–535. [CrossRef] [PubMed]
32. Will, O.M.; Damm, T.; Campbell, G.M.; von Schönfells, W.; Açil, Y.; Will, M.; Chalaris-Rissmann, A.; Ayna, M.; Drucker, C.; Glüer, C.-C. Longitudinal micro-computed tomography monitoring of progressive liver regeneration in a mouse model of partial hepatectomy. *Lab. Anim.* **2017**, *51*, 422–426. [CrossRef] [PubMed]
33. Lefevre, A.; Ballesta, S.; Pozzobon, M.; Charieau, J.-L.; Duperrier, S.; Sirigu, A.; Duhamel, J.-R. Blood microsampling from the ear capillary in non-human primates. *Lab. Anim.* **2015**, *49*, 349–352. [CrossRef]
34. Voit, M.; Baumgartner, K.; von Fersen, L.; Merle, R.; Reese, L.; Wiegard, M.; Will, H.; Tallo-Parra, O.; Carbajal, A.; Lopez-Bejar, M.; et al. Comparison of Two Different Feather Sampling Methods to Measure Corticosterone in Wild Greater Flamingos. *Animals* **2021**, *11*, 2796. [CrossRef] [PubMed]
35. Kudlova, N.; Slavik, H.; Duskova, P.; Furst, T.; Srovnal, J.; Bartek, J.; Mistrik, M.; Hajduch, M. An efficient, non-invasive approach for. *Aging* **2021**, *13*, 25004–25024. [CrossRef]
36. Carvalho, J.J.; Walter, M.A.; Baermann-Stapel, Y.; Weller, M.G.; Panne, U.; Schenk, J.; Schneider, R.J. Non-invasive monitoring of immunization progress in mice via IgG from feces. *Vivo* **2012**, *26*, 63–69. [PubMed]
37. Meyer, N.; Kröger, M.; Thümmler, J.; Tietze, L.; Palme, R.; Touma, C. Impact of three commonly used blood sampling techniques on the welfare of laboratory mice: Taking the animal's perspective. *PLoS ONE* **2020**, *15*, e0238895. [CrossRef]
38. Balafas, E.; Katsila, T.; Melissa, P.; Doulou, A.; Moltsanidou, E.; Agapaki, A.; Patrinos, G.P.; Kostomitsopoulos, N. A Noninvasive Ocular (Tear) Sampling Method for Genetic Ascertainment of Transgenic Mice and Research Ethics Innovation. *OMICS A J. Integr. Biol.* **2019**, *23*, 312–317. [CrossRef] [PubMed]
39. Del Val, G.M.; Robledano, P.M. In vivo serial sampling of epididymal sperm in mice. *Lab. Anim.* **2013**, *47*, 168–174. [CrossRef] [PubMed]
40. Dainty, T.C.; Richmond, E.S.; Davies, I.; Blackwell, M.P. Dried blood spot bioanalysis: An evaluation of techniques and opportunities for reduction and refinement in mouse and juvenile rat toxicokinetic studies. *Int. J. Toxicol.* **2012**, *31*, 4–13. [CrossRef] [PubMed]

41. Breacker, C.; Barber, I.; Norton, W.H.; McDearmid, J.R.; Tilley, C.A. A Low-Cost Method of Skin Swabbing for the Collection of DNA Samples from Small Laboratory Fish. *Zebrafish* **2017**, *14*, 35–41. [CrossRef] [PubMed]
42. Curaj, A.; Simsekyilmaz, S.; Staudt, M.; Liehn, E. Minimal invasive surgical procedure of inducing myocardial infarction in mice. *J. Vis. Exp.* **2015**, *99*, e52197. [CrossRef] [PubMed]
43. Nowocin, A.K.; Brown, K.; Edwards, L.A.; Meader, L.; Hill, J.I.; Wong, W. An Extraperitoneal Technique for Murine Heterotopic Cardiac Transplantation. *Am. J. Transplant.* **2015**, *15*, 2491–2494. [CrossRef]
44. Atkinson, J.; Boden, T.; Mocho, J.-P.; Johnson, T. Refining the unilateral ureteral obstruction mouse model: No sham, no shame. *Lab. Anim.* **2021**, *55*, 21–29. [CrossRef]
45. Leidinger, C.; Herrmann, F.; Thöne-Reineke, C.; Baumgart, N. Introducing Clicker Training as a Cognitive Enrichment for Laboratory Mice. *J. Vis. Exp.* **2017**, *121*, e55415. [CrossRef] [PubMed]
46. Hall, L.E.; Robinson, S.; Buchanan-Smith, H.M. Refining dosing by oral gavage in the dog: A protocol to harmonise welfare. *J. Pharmacol. Toxicol. Methods* **2015**, *72*, 35–46. [CrossRef] [PubMed]
47. Gray, H.; Pearce, B.; Thiele, A.; Rowe, C. The use of preferred social stimuli as rewards for rhesus macaques in behavioural neuroscience. *PLoS ONE* **2017**, *12*, e0178048. [CrossRef] [PubMed]
48. Manell, E.; Hedenqvist, P.; Jensen-Waern, M. Training Pigs for Oral Glucose Tolerance Test-Six Years' Experience of a Refined Model. *Animals* **2021**, *11*, 1677. [CrossRef]
49. Descovich, K.A.; Richmond, S.E.; Leach, M.C.; Buchanan-Smith, H.M.; Flecknell, P.; Farningham, D.A.H.; Witham, C.; Gates, M.C.; Vick, S.-J. Opportunities for refinement in neuroscience: Indicators of wellness and post-operative pain in laboratory macaques. *ALTEX* **2019**, *36*, 535–554. [CrossRef]
50. Savery, D.; Maniou, E.; Culshaw, L.H.; Greene, N.D.E.; Copp, A.J.; Galea, G.L. Refinement of inducible gene deletion in embryos of pregnant mice. *Birth Defects Res.* **2020**, *112*, 196–204. [CrossRef]
51. Vogt, M.A.; Mertens, S.; Serba, S.; Palme, R.; Chourbaji, S. The 'Cage Climber'—A new enrichment for use in large-dimensioned mouse facilities. *Appl. Anim. Behav. Sci.* **2020**, *230*, 105078. [CrossRef]
52. Leidinger, C.S.; Thöne-Reineke, C.; Baumgart, N.; Baumgart, J. Environmental enrichment prevents pup mortality in laboratory mice. *Lab. Anim.* **2019**, *53*, 53–62. [CrossRef]
53. Nunn, F.; Bartley, K.; Palarea-Albaladejo, J.; Innocent, G.T.; Turnbull, F.; Wright, H.W.; Nisbet, A.J. A novel, high-welfare methodology for evaluating poultry red mite interventions in vivo. *Veter. Parasitol.* **2019**, *267*, 42–46. [CrossRef]
54. Hohlbaum, K.; Frahm, S.; Rex, A.; Palme, R.; Thöne-Reineke, C.; Ullmann, K. Social enrichment by separated pair housing of male C57BL/6JRj mice. *Sci. Rep.* **2020**, *10*, 11165. [CrossRef] [PubMed]
55. Annas, A.; Bengtsson, C.; Törnqvist, E. Group housing of male CD1 mice: Reflections from toxicity studies. *Lab. Anim.* **2013**, *47*, 127–129. [CrossRef]
56. Mitchell, E.J.; Brett, R.R.; Armstrong, J.D.; Sillito, R.R.; A Pratt, J. Temporal dissociation of phencyclidine: Induced locomotor and social alterations in rats using an automated homecage monitoring system-implications for the 3Rs and preclinical drug discovery. *J. Psychopharmacol.* **2020**, *34*, 709–715. [CrossRef] [PubMed]
57. Jones, S.; Neville, V.; Higgs, L.; Paul, E.S.; Dayan, P.; Robinson, E.S.J.; Mendl, M. Assessing animal affect: An automated and self-initiated judgement bias task based on natural investigative behaviour. *Sci. Rep.* **2018**, *8*, 12400. [CrossRef] [PubMed]
58. Bennett, E.J.; Mead, R.J.; Azzouz, M.; Shaw, P.; Grierson, A.J. Early detection of motor dysfunction in the SOD1G93A mouse model of Amyotrophic Lateral Sclerosis (ALS) using home cage running wheels. *PLoS ONE* **2014**, *9*, e107918. [CrossRef]
59. Häger, C.; Keubler, L.M.; Talbot, S.R.; Biernot, S.; Weegh, N.; Buchheister, S.; Buettner, M.; Glage, S.; Bleich, A. Running in the wheel: Defining individual severity levels in mice. *PLoS Biol.* **2018**, *16*, e2006159. [CrossRef]
60. Mason, S.; Premereur, E.; Pelekanos, V.; Emberton, A.; Honess, P.; Mitchell, A.S. Effective chair training methods for neuroscience research involving rhesus macaques (*Macaca mulatta*). *J. Neurosci. Methods* **2019**, *317*, 82–93. [CrossRef]
61. COST. PRIMTRAIN (Behavioural Management and Training of Laboratory Non-Human Primates and Large Laboratory Animals). 2016. Available online: https://www.cost.eu/actions/CA15131/ (accessed on 2 November 2022).
62. COST. Improving Biomedical Research by Automated Behaviour Monitoring in the Animal Home-Cage (TEATIME). 2021. Available online: https://www.cost.eu/actions/CA20135/ (accessed on 2 November 2022).
63. EC. Report on the Statistics on the Use of Animals for Scientific Purposes in the Member States of the European Union in 2015–2017. 2019. Available online: https://eur-lex.europa.eu/legal-content/EN/TXT/?qid=1581689520921&uri=CELEX:52020DC0016 (accessed on 2 November 2022).
64. EC. Summary Report on the Statistics on the Use of Animals for Scientific Purposes in the Member States of the European Union and Norway in 2018. 2021. Available online: https://ec.europa.eu/environment/chemicals/lab_animals/pdf/SWD_%20part_A_and_B.pdf (accessed on 2 November 2022).
65. EC. Summary Report on the Statistics on the Use of Animals for Scientific Purposes in the Member States of the European Union and Norway in 2019. 2022. Available online: https://ec.europa.eu/environment/chemicals/lab_animals/pdf/SWD2019_Part_A_and_B.pdf (accessed on 2 November 2022).
66. EC. Scientific Output and Collaboration of European Universities 2014. Available online: https://op.europa.eu/en/publication-detail/-/publication/ea818496-b5f1-4ca7-9354-79ba99acd244 (accessed on 2 November 2022).
67. Kolar, R. ECVAM: Desperately needed or superfluous? An animal welfare perspective. *Altern. Lab. Anim.* **2002**, *30*, 169–174. [CrossRef] [PubMed]

68. Norecopa. 2022. Available online: https://norecopa.no/search?fq=cat:%223R%20Center%22&fq=db:%223r%22&sort=name_s%20asc&q=*&facet.limit=125 (accessed on 2 November 2022).
69. Bert, B.; Chmielewska, J.; Hensel, A.; Grune, B.; Schönfelder, G. The animal experimentation quandary: Stuck between legislation and scientific freedom: More research and engagement by scientists is needed to help to improve animal welfare without hampering biomedical research. *EMBO Rep.* **2016**, *17*, 790–792. [CrossRef]
70. Azkona, G.; Caballero, J.M. Implementing strategies to reduce singly housed male mice. *Lab. Anim.* **2019**, *53*, 508–510. [CrossRef]
71. Grífols, R.; Zamora, C.; Ortega-Saez, I.; Azkona, G. Postweaning Grouping as a Strategy to Reduce Singly Housed Male Mice. *Animals* **2020**, *10*, 2135. [CrossRef] [PubMed]
72. Animal Welfare Institute. 2022. Available online: https://awionline.org/ (accessed on 2 November 2022).
73. FMUC. HaPILLness-Voluntary Oral Dosing in Rodents. 2021. Available online: https://www.uc.pt/ucbusiness/hapillness-semi-solid-formulation-for-voluntary-oral-administration-of-bioactive-compounds-to-rodents/ (accessed on 2 November 2022).
74. Scarborough, J.; Mueller, F.; Arban, R.; Dorner-Ciossek, C.; Weber-Stadlbauer, U.; Rosenbrock, H.; Meyer, U.; Richetto, J. Preclinical validation of the micropipette-guided drug administration (MDA) method in the maternal immune activation model of neurodevelopmental disorders. *Brain Behav. Immun.* **2020**, *88*, 461–470. [CrossRef]
75. Hurst, J.L.; West, R.S. Taming anxiety in laboratory mice. *Nat. Methods* **2010**, *7*, 825–826. [CrossRef] [PubMed]
76. Azkona, G.; Sanchez-Pernaute, R. Mice in translational neuroscience: What R we doing? *Prog. Neurobiol.* **2022**, *217*, 102330. [CrossRef] [PubMed]

animals

MDPI

Communication

Stress Evaluation of Mouse Husbandry Environments for Improving Laboratory Animal Welfare

Gwang-Hoon Lee [1], KilSoo Kim [1,2,*] and Woori Jo [1,*]

[1] Preclinical Research Center, Daegu-Gyeongbuk Medical Innovation Foundation, Daegu 41061, Republic of Korea
[2] Department of Veterinary Toxicology, College of Veterinary Medicine, Kyungpook National University, Daegu 41566, Republic of Korea
* Correspondence: kskim728@knu.ac.kr (K.K.); c2dar@kmedihub.re.kr (W.J.)

Simple Summary: It is well recognized that companionship is important to animals and that they need to be provided with an environment accompanied by materials for enrichment, such as toys. However, few studies have evaluated whether specific environments actually benefit animals. Therefore, we designed various environments for laboratory animals and scientifically evaluated which environments reduced these animals' stress. We found that an environment with freer air circulation and the provision of enrichment materials reduced animal stress, and no risk or benefit could be determined for the presence or absence of a companion. We do not consider that our results necessarily indicate the lack of a need for a companion, but, rather, the importance of having a good companion. Our results can serve as a meaningful guideline for the creation of suitable environments for laboratory animals.

Abstract: Animal welfare is recognized as essential for the coexistence of humans and animals. Considering the increased demand and interest in animal welfare, many methods for improving animal welfare are being devised, but which method reduces animal stress has not been scientifically verified. Therefore, reducing animal stress by providing a proper breeding environment and environmental enrichment can be the basis for animal study. In this study, stress levels were assessed based on the mouse-breeding environment. We considered that the higher the body weight and the lower the corticosterone concentration, the lower the stress. According to the results, animals in the individual ventilation cages were determined to have lower serum cortisol concentrations, while the body weight of the animals was increased when in individual ventilation cages compared with individual isolated cages and when providing environmental enrichment compared with group breeding or not providing environmental enrichment. The results provide appropriate guidelines for improving laboratory animal welfare.

Keywords: animal welfare; environmental enrichment; housing; laboratory animals; stress evaluation

Citation: Lee, G.-H.; Kim, K.; Jo, W. Stress Evaluation of Mouse Husbandry Environments for Improving Laboratory Animal Welfare. *Animals* **2023**, *13*, 249. https://doi.org/10.3390/ani13020249

Academic Editor: Garikoitz Azkona

Received: 4 December 2022
Revised: 5 January 2023
Accepted: 6 January 2023
Published: 10 January 2023

Copyright: © 2023 by the authors. Licensee MDPI, Basel, Switzerland. This article is an open access article distributed under the terms and conditions of the Creative Commons Attribution (CC BY) license (https://creativecommons.org/licenses/by/4.0/).

1. Introduction

With the recent increasing interest in animal ethics, there has been a growing focus within the international laboratory animal research community on improving the housing environment and welfare of laboratory animals [1]. Although related laws and systems have been continuously strengthened in the field of laboratory animal studies, issues regarding ethical relations between humans and animals in bioscience research laboratories remain. Since the United States of America first enacted the laboratory animal welfare act in 1966, laws related to animal welfare have continued to be strengthened [2]. The Animal Welfare Act: From Enactment to Enforcement of the Europe Union (E.U) also extensively revised laws regarding animal welfare, including for laboratory animals, since 1974 [3]. The Ministry of Agriculture, Food, and Rural Affairs (MAFRA, Republic of Korea) announced a five-year comprehensive plan for animal welfare to raise public awareness of and sympathy for animal protection and welfare, and to enhance animal laboratory animal ethics by reinforcing the function of the Institutional Animal Care and Use Committee (IACUC). Furthermore, recent debates on laboratory animal welfare have been conducted at the National Assembly.

In addition, unlike domestic companion animals, laboratory animals are limited to a fixed space for the entire experimental period, so laboratory animals are more exposed to stress from the housing environment than companion animals. This constrained environmental condition can cause stress in animals, which affects their physiological indicators and can change the results of experiments. Studies associated with laboratory animal behavioral analysis have also reported that housing conditions, such as a socially isolated environment, group size, or cage size, are an important factor influencing animal behavior [4–7]. Therefore, to obtain reliable experimental results in laboratory animal research, considering animal welfare is important, as it can alleviate the stress experienced by laboratory animals, affecting the experimental results [8]. Furthermore, policies and practices with respect to laboratory animal housing, husbandry, and quality care can enhance animal welfare [9–11]. A previous study demonstrated that the welfare of laboratory animals contributes to improving the quality of life for the people involved in animal research facilities. Animal welfare staff showed a more positive correlation with a professional quality of life than researchers who were reported to perceive animal stress/pain [10].

Laboratory animals are used in a variety of research fields, and stress induced during experiments can change the background data. Stress is defined as the body's nonspecific response to external stimuli, environmental demands, or stimuli beyond the body's ability to cope [12]. Vladimir K et al. reported that stress causes changes in the limbic–hypothalamic–pituitary–adrenal (LHPA) neuroendocrine axis, De Kloet ER et al. described that stress induced structural and functional change in the limbic brain, and B Olivier demonstrated that stress increases the body temperature of laboratory animals [13–15].

Most studies on stress involve experiments with artificially applied stress, including, among others, repeated social defeat stress (RSDS), electric shock, wire netting, and repeated stress [16–20]. In addition, previous studies on the efficacy of newly developed drugs have been based on either the use of an artificially induced stress model [16,20] or measuring the hormonal changes caused by different levels of stress exposure [21]. The interest and demand regarding the welfare of zoo animals and laboratory animals is also elevated because of their stereotypic behavior affected by limited breeding space. Many methods for improving their behavior are devised with environmental enrichment to respect their natural habits [22]. However, studies on stress levels after exposure to various housing environments with regard to the improvement of laboratory animal welfare for animals that live in limited space are lacking. Furthermore, it is necessary to evaluate environmental effects on the stress of laboratory animals depending on the type of cage and/or social isolation.

Stress–response hormones include cortisol and corticosterone. Cortisol is widely considered in studies on large laboratory animals such as beagles [23,24], whereas for smaller animals, such as rodents, corticosterone is the more important glucocorticoid in responding to stress exposure [19].

Therefore, in this study, we aimed to evaluate the changes in the serum corticosterone concentration and body weight as stress indicators according to the presence or absence of environmental enrichment (E.E), different types of cages (individually ventilated cages (IVC) or individually isolated cages (ISO)), and social isolation stress (single breeding or group breeding). It is hoped that these results will be useful for enhancing the environmental conditions for laboratory animals and improving animal welfare by reducing stress on laboratory animals throughout the entire experimental period.

2. Materials and Methods

2.1. Animals and Husbandry

Four-week-old CrljOri:CD1 (ICR) male mice were purchased from Orient Bio (Seongnam, KyungKi, Korea). The animal experiments were reviewed and approved by the Institutional Animal Care and Use Committee at the Daegu Gyeongbuk Medical Innovation Foundation (K-MEDI hub) (approved IACUC Number: DGMIF-20032407, approved date: 23 March 2020), and the animals were maintained in a facility accredited by the Ministry of Food and Drug Safety in Korea as a Korean Excellent Laboratory Animal Facility (KELAF), and by the Association for Assessment and Accreditation of Laboratory Animal Care International (#001796). The animals were fed an autoclaved pellet diet (SAFE+40RMM; SAFE Diets, Augy, France) and provided with drinking water ad libitum. The animals were housed in environmental conditions with a temperature of 22 ± 1 °C, 50 ± 10% humidity, illumination at 150–300 Lux, and a breeding room ventilation cycle of 10–20 times/h. All of the animals were monitored every day and there were no mortality injuries or clinical signs. Cages, shavings, and fresh enrichment materials were exchanged once a week.

2.2. Experimental Design

The experimental design is shown in Figure 1a. The mice were divided into six groups (nine mice/group) after an initial week of acclimatization, as follows: (A) IVC/Single; (B) IVC/Single + (Environmental Enrichment) E.E; (C) IVC/Group (Group: three mice in one cage); (D) IVC/Group + E.E; (E) ISO/Single; and (F) ISO/Group. Both cage systems (Tecniplast, Buguggiate, Varese, Italy), IVC (Cat No: GM500) and ISO (Cat No. ISO cage-N), had an air-circulation system in each cage, with the main difference being whether room air outside the cage was allowed to enter into the cage with mice inside. In the IVC cage, the air in the cage is not only circulated by the automatic blower system, but room air outside the cage also enter into the cage. Room air outside the cage can enter through the HEPA filter (size: 141 mm × 170 mm, efficiency: 0.3 micrometer of the particle at 99.5%) on the IVC lid, while in the ISO cage (size: 73 mm × 73 mm × 24 mm, efficiency: 0.3 micrometer of the particle at 99.97%), room air outside the cage is completely blocked and air in the cage is forcibly circulated only by the automatic blower system. Therefore, the air in the IVC cage is circulates better than the air in the ISO cage.

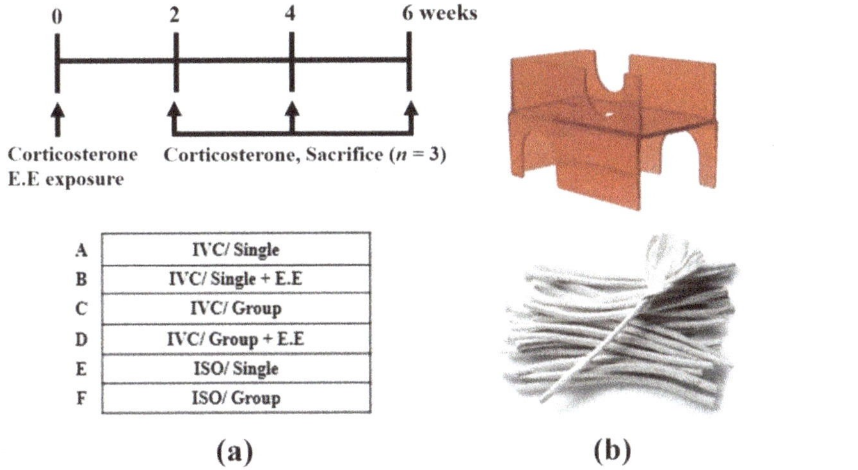

Figure 1. (**a**) Experimental design; (**b**) representative photographs of environmental enrichment (harbor mouse retreat, top; diamond twist, bottom).

Two environmental enrichment (E.E) materials were used simultaneously, the first being a harbor mouse retreat (Figure 1b top, Cat No. K3583, Bio-serv, Frenchtown, NJ, USA) and the second being a diamond twist (Figure 1b bottom, Envigo, Madison, WI, USA), which were provided to groups B and D until the end of the experiment. The harbor mouse retreat and the diamond twist were selected as the E.E materials as the standard environmental enrichment required and an additional enrichment, respectively, with reference to the IACUC policy of the University of California, Irvine (Figure 1).

2.3. *Preparation of the Blood Serum and Corticosterone Assay*

In the present study, blood collection was performed once every 2 weeks, with three mice in each group. The mice were anesthetized using isoflurane at 17:00–18:00 p.m., and blood was rapidly collected to minimize the stress caused by the anesthesia process, and euthanasia was performed by exsanguination under anesthesia. Because of the slight possibility of survival, cervical dislocation was performed under anesthesia. As it is painful to live after bloodletting, cervical dislocation was performed to block the weak possibility of living under anesthesia. Blood was collected from the abdominal vein (approximately 600 μL) in serum-separating tubes (SST tube, Becton, Dickinson and Company, Franklin Lakes, NJ, USA) and was centrifuged (3000 rpm, 10 min, 4 °C) to separate the serum.

Triplicate serum corticosterone assays were conducted using an ELISA kit (Cat No. K014, Arbor assays, Ann Arbor, MI, USA), and the optical density (OD) was read using a synergy H4 microplate reader (BioTek Instruments, Inc., Winooski, VT, USA).

2.4. *Statistical Analysis*

Statistical significance was determined using GraphPad Prism 8 (GraphPad Software Inc., San Diego, CA, USA). All the data are presented as mean ± standard deviation (SD) and passed tests for normality. Two-way ANOVA with Bonferroni's multiple comparisons test and unpaired t-test were used.

3. Results

3.1. *Body Weight*

After an initial week of acclimatization, the body weight was measured weekly until euthanasia. During the experiment period, group D had the highest average body weight, followed by groups A, C, B, F, and E. There was a significant difference ($p < 0.05$) between groups D and F, and E was significantly different from A, B, C, and D (Figure 2).

3.2. *Concentrations of Serum Corticosterone*

From 2 to 4 weeks, the serum corticosterone concentrations decreased in all groups; from 4 to 6 weeks, the concentrations decreased in groups C and D, which had been group-reared, and increased slightly again in the remaining groups. The serum corticosterone concentration showed the order of F, E, A, B, C, and D at week 6, with a significant difference ($p < 0.05$) between groups C and E, C and F, D and E, and D and F (Figure 3).

3.3. *Concentrations of Serum Corticosterone/Body Weight*

The relative corticosterone concentration (corticosterone/body weight) had an overall similar tendency to the serum corticosterone level. The only observed difference was that the order of A and B was reversed at 2 and 4 weeks. In addition, there was a significant ($p < 0.05$) difference between groups C and F, D and E, and D and F (Figure 4).

3.4. *Overall Analysis*

The IVC cage/Group + E.E group mice had the highest body weight and the lowest corticosterone concentration on average, the ISO cage/Single group mice had the lowest body weight, and the ISO cage/Group mice had the lowest corticosterone concentration on average. All IVC cage mice weighed more and had lower corticosterone concentrations than all ISO cage mice on average. The mice from groups with E.E had a higher body weight and lower corticosterone concentration than those of groups without E.E on average (IVC cage/Group + E.E vs. IVC cage/Group, IVC cage/Single + E.E vs. IVC cage/Single) (Table 1).

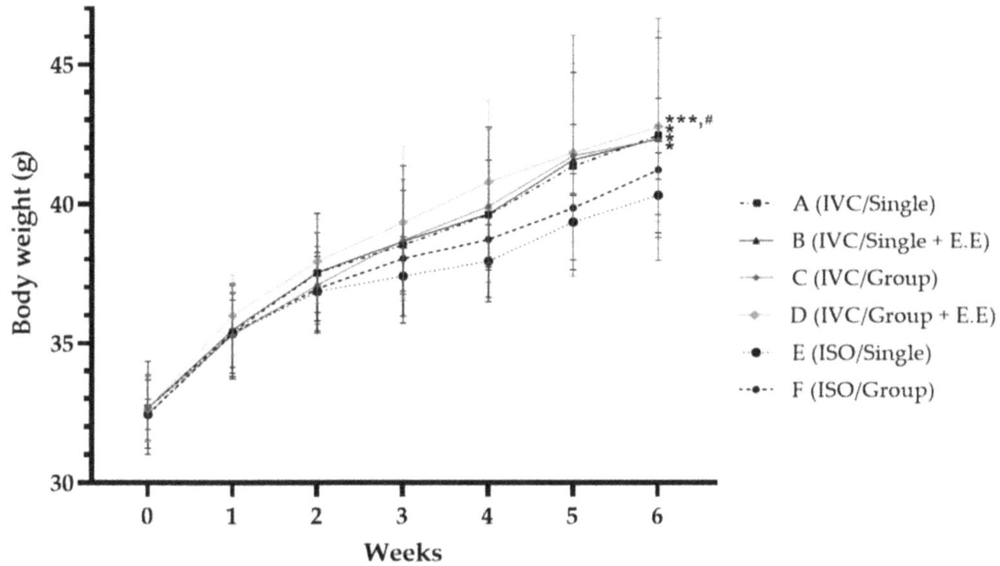

(**a**)

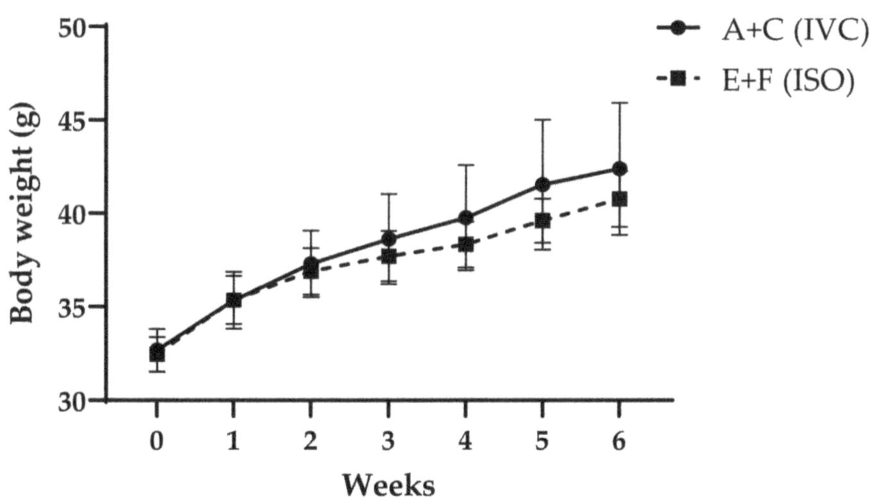

(**b**)

Figure 2. Body weight changes in (**a**) each group and (**b**) groups A + C and E + F (* $p < 0.05$ and *** $p < 0.001$ compared to E, # $p < 0.05$ compared to F), (0~2 weeks: $n = 9$/group, 2~4 weeks: $n = 6$/group, 4~6 weeks: $n = 3$/group).

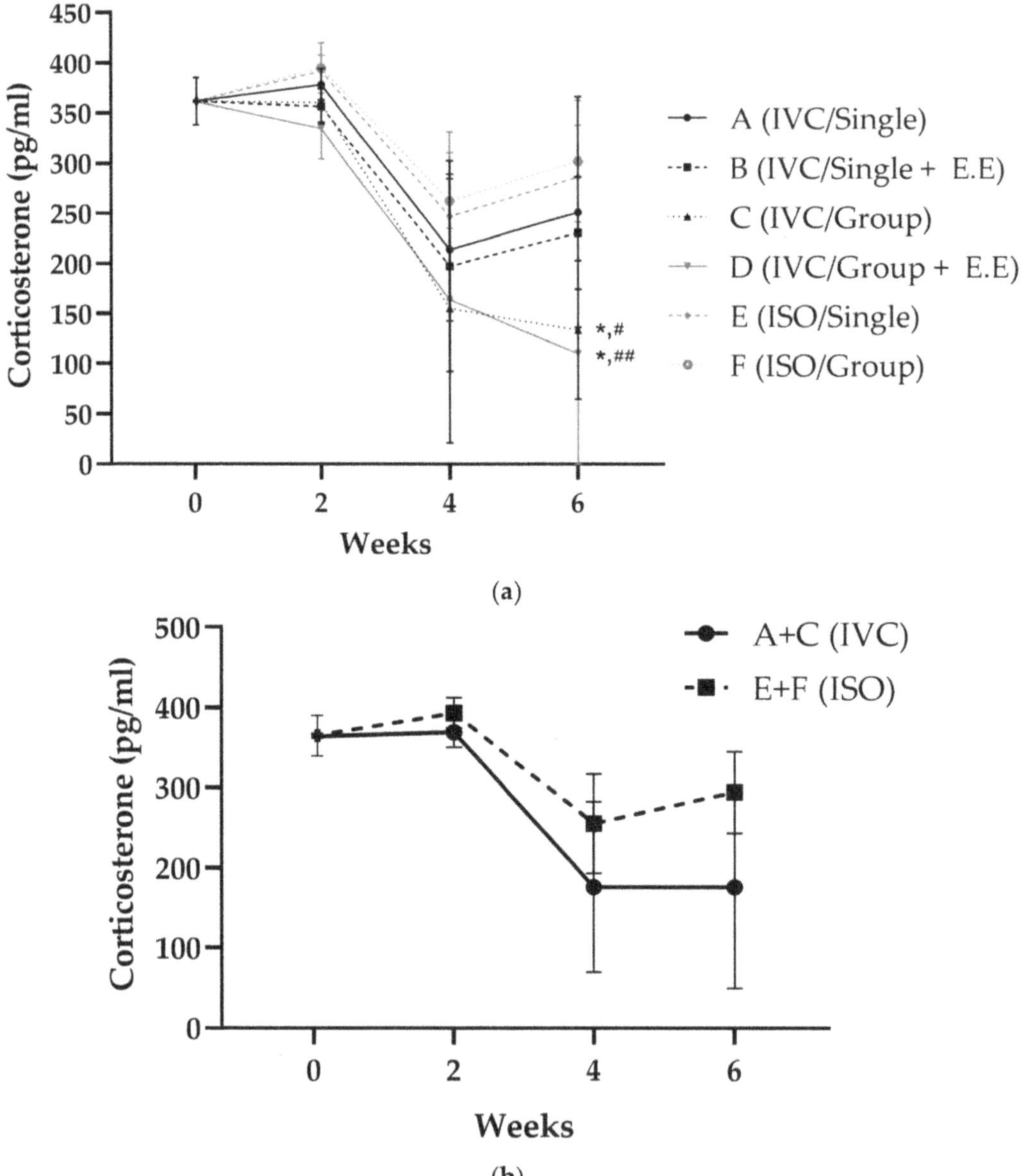

Figure 3. (**a**) Corticosterone concentration in the serum change for each group; (**b**) corticosterone concentration in the serum change for groups A + C and E + F (* $p < 0.05$ compared with E, # $p < 0.05$ compared with F, ## $p < 0.01$ compared with F).

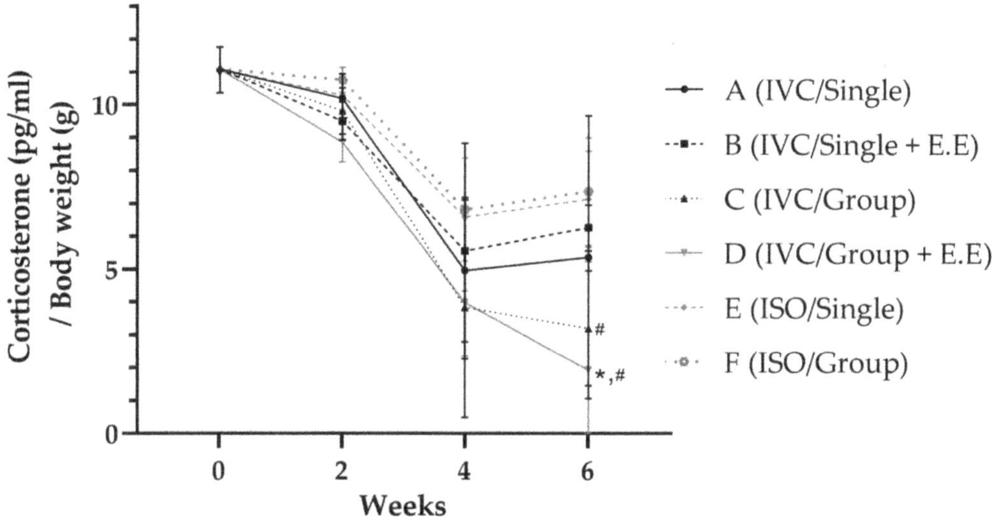

(**a**)

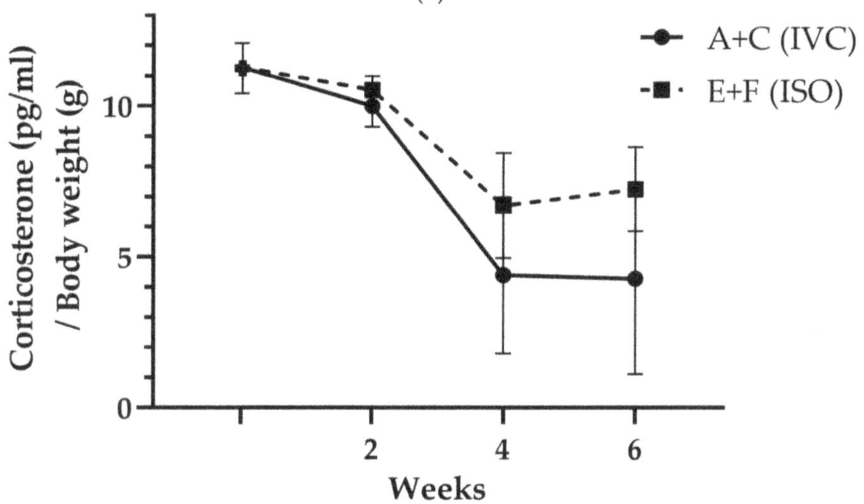

(**b**)

Figure 4. Corticosterone concentration in the serum/body weight change for (**a**) each group and (**b**) groups A + C and E + F (* $p < 0.05$ compared with E, # $p < 0.05$ compared with F).

Table 1. Overall analysis of body weight, serum corticosterone concentration, and corticosterone concentration/body weight.

Stress Level	Less Stressful	←			→	More Stressful
	1st	2nd	3rd	4th	5th	6th
Body weight (g)	IVC cage Group E.E *	IVC cage Single E.E *	IVC cage Single	IVC cage Group	ISO cage Group	ISO cage Single

Table 1. *Cont.*

Stress Level	Less Stressful	←		→		More Stressful
Corticosterone concentration in serum (pg/mL)	6th	5th	4th	3rd	2nd	1st
	IVC cage Group E.E *	IVC cage Group	IVC cage Single E.E *	IVC cage Single	ISO cage Single	ISO cage Group
Corticosterone concentration/body weight	6th	5th	4th	3rd	2nd	1st
	IVC cage Group E.E *	IVC cage Group	ISO cage Single	IVC cage Single E.E *	ISO cage Single	ISO cage Group

* E.E: environmental enrichment.

4. Discussion

In this study, we created six different environmental conditions for mice at the Preclinical Research Center (PRC), K-MEDI hub to evaluate the levels of stress.

Studies have been carried out on the effectiveness of E.E [24–27] at preventing oxidative injury and restoring cholinergic neurotransmission in cognitively impaired aged rats [25], as well as in alleviating the behavioral changes in a mouse model of post-traumatic stress disorder [28]. In addition, studies regarding the relationship between housing conditions, such as single housing or grouped housing, and behavioral phenotypes have been conducted [29]. Both isolation and environmental enrichment have fundamental effects on mouse behavior and should be considered in the course of experimental design with stress-related animal models and in animal welfare assessment.

In this study, body weight can be an indicator of stress level, as it is generally known that stress can cause the amount of feed intake and body weight to decrease [16–18]. This study reports that animals exposed to stress might have a decrease in the body weight gain rate. Based on the results of body weight for the groups, group D (IVC cage/Group + E.E), in which body weight increased the most, was less stressed, and group E (ISO cage/Single), with the least body weight gain, represented the more stressful environment (Table 1). Corticosterone in the blood is a factor widely known to increase when a mouse is exposed to stress [19]. Therefore, based on the results of the serum corticosterone concentration, the mice of group D (IVC cage/Group + E.E), with the lowest concentration, were the least stressed, and those of group F (ISO cage/Group), with the highest concentration, were the most stressed. Overall, the concentration of serum corticosterone was the highest in the second week and then decreased. The results of the relative serum corticosterone (corticosterone/body weight) were generally similar to the serum corticosterone results, with the only change being in the order of group A (IVC/Single) and group B (IVC/Single + E.E). This slight change means that stress had no biased effect on either body weight or corticosterone blood levels. The serum corticosterone concentrations were similar between weeks 0 and 2, but the relative corticosterone levels decreased at week 2. We judged that this was due to the maintained corticosterone concentration contrary to the increased body weight.

Based on these results, group D (IVC/Group + E.E) was the group with the highest body weight and lowest corticosterone concentration, on average, and it was thus the group that was the least stressed.

Furthermore, the groups with E.E showed a less stressful environment than the groups without E.E. However, there was no difference in body weight or corticosterone concentration between group breeding and single breeding. Moreover, it was difficult to judge the results of the relationship between the group housing and single housing conditions, and we did not consider a sufficient number of individuals in our experiment. In addition, opinion is divided regarding the factors attributed to social ranking, which can also be a hindrance when judging the results. Benaroya-Milshtein and Hollander et al. reported that E.E reduces anxiety and weakens stress reactions in mice, while Chapillon et al. reported that environmental energy in mice reduces anxiety [27,30]. Although the mouse is a social animal, it is judged that there can be differences among animals exposed to the same stress due to differences in hierarchy that are determined by the group. Both group and single breeding are known to have pros and cons, and the various experiments show conflicting results [9]. Liu and Wang et al. reported that single mice had a reduced concentration of corticosterone, while Kamakura and Kovalainen et al. reported that group-housed mice had higher levels of corticosterone than single-house mice, indicating that those mice living in groups had higher levels of stress [31,32]. On the other hand, Norman et al. reported that single mice had a significantly reduced memory, and according to a study by Kamal et al., the level of LTP injury was increased in the hippocampus of C57BL/6J mice in the single-breeding group, which also had higher blood corticosterone concentrations [33,34].

In addition, in this study, the average weight of the ISO cage mice was lower than that of the IVC cage mice, while the blood corticosterone concentration was higher on average, so it was judged that the stress was higher because the ISO cage has limited air circulation. The graph pattern for the change in corticosterone value and corticosterone/body weight value was not significantly different, even if the ranking of the average value for each factor was slightly different.

Furthermore, stress effects may differ depending on the mouse strain, so further studies are needed on stress using diverse strains and species of animals with disease as models [15]. In addition, with the growing interest in animal welfare, stress assessments could be expanded to abandoned animals or industrial animals [26].

Full accreditation for The Association for Assessment and Accreditation of the Laboratory Animal Care International (AAALAC-i) is a recognized certification awarded by a nonclinical study institution that encourages the humane treatment of animals in the field of science by maintaining a high level for laboratory animal care and use [35]. In addition to this full AAALAC-i accreditation, the PRC at the K-MEDI hub is devising and continuing to develop ways to promote the welfare of laboratory animals.

These results suggest that IVC cages are preferable to ISO cages and that providing E.E would be more effective for relieving stress. However, the appropriate composition of different animals raised in the same cage is needed because it may influence whether stress is more active. The results of this study are helpful by serving as appropriate guidelines for the management and operation of laboratory animal breeding and the welfare of laboratory animals.

Animals are originally organisms that live in nature, but laboratory animals inevitably live in a limited space. It is difficult to derive reproducible and reliable experimental results from psychologically fatigued animals [36]. We judged that the environmental conditions that reduced stress in this experiment were conditions close to the animal's natural habit. Therefore, our study suggests that environmental enrichment that respects their natural habits should be provided for reliable animal test results. Ultimately, our society already knows that animals are not voluntarily subjected to experiments. We believe that any facility that raises animals should be induced to change to a facility that values the autonomy of animals, and this study can serve as a basis for inducing that change.

Future studies regarding the effects of stress on animals should include behavioral assessments. Luo et al. demonstrated that displacement movement can be predicted by evaluating random behavior permutation in ungulates, and they reported that the differences could be found when comparing the isolated environment considered as stressful and an enriched environment considering their natural habits [37]. Considering their report, we judge that meaningful behavioral data can be achieved if we conduct a study by applying our experimental environmental conditions on mice as a laboratory animal.

5. Conclusions

We have scientifically confirmed a reduction in stress with environmental enrichment, or for animals raised together with other animals, or in a well-ventilated environment in mice.

With the results of this study, we determine that animals raised in limited space such as an animal research facility should be provided with environmental enrichment, a well-ventilated environment. Because the environments that reduce stress are thought as a result of the natural habits of animals, we should devise various methods to respect their natural habits as much as possible, and evaluate whether these methods reduce their stress.

This data will be guideline to all researchers who work in a laboratory animal facility. Laboratory animals with improved environmental enrichment can achieve reliable scientific results.

Author Contributions: Conceptualization, G.-H.L. and K.K.; methodology, G.-H.L.; validation, G.-H.L. and W.J.; resources, K.K.; data curation, W.J.; writing—original draft preparation, G.-H.L.; writing—review and editing, W.J.; supervision, K.K. and W.J.; project administration, K.K. and W.J.; funding acquisition, K.K. All authors have read and agreed to the published version of the manuscript.

Funding: This research was funded by the Daegu Gyeongbuk Medical Innovation Foundation in 2020 (Project Number: E20015).

Institutional Review Board Statement: The animal study protocol was approved by the Institutional Review Board of Daegu Gyeongbuk Medical Innovation Foundation (protocol code DGMIF-20032407 and approved date: 23 March 2020).

Informed Consent Statement: Not applicable.

Data Availability Statement: The data presented in this study are available on request from the corresponding author.

Conflicts of Interest: The authors declare no conflict of interest.

References

1. Hagelin, J.; Hau, J.; Carlsson, H.-E. The refining influence of ethics committees on animal experimentation in Sweden. *Lab. Anim.* **2003**, *37*, 10–18. [CrossRef] [PubMed]
2. Cardon, A.D.; Bailey, M.; Bennett, B. The Animal Welfare Act: From enactment to enforcement. *J. Am. Assoc. Lab. Anim. Sci.* **2012**, *51*, 301–305. [PubMed]
3. Simonin, D.; Gavinelli, A. The European Union legislation on animal welfare: State of play, enforcement and future activities. *Anim. Welf. Sci. Law* **2019**, *1*, 59–70.
4. Peters, S.; Pothuizen, H.; Spruijt, B. Ethological concepts enhance the translational value of animal models. *Eur. J. Pharmacol.* **2015**, *759*, 42–50. [CrossRef] [PubMed]
5. Van Loo, P.L.; Mol, J.A.; Koolhaas, J.M.; Van Zutphen, B.F.; Baumans, V. Modulation of aggression in male mice: Influence of group size and cage size. *Physiol. Behav.* **2001**, *72*, 675–683. [CrossRef]
6. Van Loo, P.L.P.; Van Zutphen, L.; Baumans, V. Male management: Coping with aggression problems in male laboratory mice. *Lab. Anim.* **2003**, *37*, 300–313. [CrossRef]
7. Meer, V.; Raber, J. Mouse behavioural analysis in systems biology. *Biochem. J.* **2005**, *389*, 593–610.
8. Davies, G.; Greenhoughet, B.; Kirk, R.G.W. *Science, Culture, and Care in Laboratory Animal Research: Interdisciplinary Perspectives on the History and Future of the 3Rs*; SAGE Publications Sage CA: Los Angeles, CA, USA, 2018.
9. Kappel, S.; Hawkins, P.; Mendl, M.T. To group or not to group? Good practice for housing male laboratory mice. *Animals* **2017**, *7*, 88. [CrossRef]
10. Goñi-Balentziaga, O.; Vila, S.; Ortega-Saez, I.; Vegas, O.; Azkona, G. Professional quality of life in research involving laboratory animals. *Animals* **2021**, *11*, 2639. [CrossRef]
11. LaFollette, M.R.; Riley, M.C.; Cloutier, S.; Brady, C.M.; O'Haire, M.E.; Gaskill, B.N. Laboratory animal welfare meets human welfare: A cross-sectional study of professional quality of life, including compassion fatigue in laboratory animal personnel. *Front. Vet. Sci.* **2020**, *7*, 114. [CrossRef]
12. Fink, G. Stress, Definitions, Mechanisms, and Effects Outlined: Lessons from Anxiety. In *Stress: Concepts, Cognition, Emotion, and Behavior*; Elsevier: Amsterdam, The Netherlands, 2016; pp. 3–11.
13. Patchev, V.K.; Patchev, A.V. Experimental models of stress. *Dialogues Clin. Neurosci.* **2006**, *8*, 417. [CrossRef]
14. Chrousos, G.P.; Gold, P.W. The concepts of stress and stress system disorders. *JAMA I* **1992**, *993*, 1244–1252. [CrossRef]
15. Olivier, B.; Zethof, T.; Pattij, T.; van Boogaert, M.; van Oorschot, R.; Leahy, C.; Oosting, R.; Bouwknecht, A.; Veening, J.; van der Gugten, J.; et al. Stress-induced hyperthermia and anxiety: Pharmacological validation. *Eur. J. Pharmacol.* **2003**, *463*, 117–132. [CrossRef]
16. Takahashi, A.; Chung, J.-R.; Zhang, S.; Zhang, H.; Grossman, Y.; Aleyasin, H.; Flanigan, M.E.; Pfau, M.L.; Menard, C.; Dumitriu, D.; et al. Establishment of a repeated social defeat stress model in female mice. *Sci. Rep.* **2017**, *7*, 1–12. [CrossRef]

17. Rybkin, I.I.; Zhou, Y.; Volaufova, J.; Smagin, G.N.; Ryan, D.H.; Harris, R.B.S. Effect of restraint stress on food intake and body weight is determined by time of day. *Am. J. Physiol. Regul. Integr. Comp. Physiol.* **1997**, *273*, R1612–R1622. [CrossRef]
18. Harris, R.B.S.; Zhou, J.; Youngblood, B.D.; Rybkin, I.I.; Smagin, G.N.; Ryan, D. Effect of repeated stress on body weight and body composition of rats fed low-and high-fat diets. *Am. J. Physiol. Regul. Integr. Comp. Physiol.* **1998**, *275*, R1928–R1938. [CrossRef]
19. Gong, S.; Miao, Y.-L.; Jiao, G.-Z.; Sun, M.-J.; Li, H.; Lin, J.; Luo, M.-J.; Tan, J.-H. Dynamics and correlation of serum cortisol and corticosterone under different physiological or stressful conditions in mice. *PloS ONE* **2015**, *10*, e0117503. [CrossRef]
20. Kim, D.-H.; Jung, J.-S.; Song, D.-K.; Suh, H.-W.; Huh, S.O.; Kim, Y.-H. Intracerebroventricular injection-induced increase in plasma corticosterone levels in the mouse: A stress model. *J. Pharmacol. Toxicol. Methods* **1998**, *39*, 71–73. [CrossRef]
21. Alfarez, D.; Wiegert, O.; Joëls, M.; Krugers, H. Corticosterone and stress reduce synaptic potentiation in mouse hippocampal slices with mild stimulation. *Neuroscience* **2002**, *115*, 1119–1126. [CrossRef]
22. Swaisgood, R.R.; Shepherdson, D.J. Scientific approaches to enrichment and stereotypies in zoo animals: What's been done and where should we go next? *Zoo Biol. Publ. Affil. Am. Zoo Aquar. Assoc.* **2005**, *24*, 499–518. [CrossRef]
23. Coppola, C.L.; Grandin, T.; Enns, R.M. Human interaction and cortisol: Can human contact reduce stress for shelter dogs? *Physiol. Behav.* **2006**, *87*, 537–541. [CrossRef] [PubMed]
24. Uccheddu, S.; Mariti, C.; Sannen, A.; Vervaecke, H.; Arnout, H.; Gutienrrez Rufo, J.; Haverbeke, A. Behavioural and cortisol responses of shelter dogs to a cognitive bias test after olfactory enrichment with essential oils. *Dog Behav.* **2018**, *2*, 1–14.
25. Fernandez, C.I.; Collazo, J.; Bauza, Y.; Castellanos, M.R.; Lopez, O. Environmental Enrichment-Behavior-Oxidative Stress Interactions in the Aged Rat: Issues for a Therapeutic Approach in Human Aging. *Ann. N. Y. Acad. Sci.* **2004**, *1019*, 53–57. [CrossRef] [PubMed]
26. Rehnberg, L.K.; Robert, K.A.; Watson, S.J.; Peters, R.A. The effects of social interaction and environmental enrichment on the space use, behaviour and stress of owned housecats facing a novel environment. *Appl. Anim. Behav. Sci.* **2015**, *169*, 51–61. [CrossRef]
27. Benaroya-Milshtein, N.; Hollander, N.; Apter, A.; Kukulansky, T.; Raz, N.; Wilf, A.; Yaniv, I.; Pick, C. Environmental enrichment in mice decreases anxiety, attenuates stress responses and enhances natural killer cell activity. *Eur. J. Neurosci.* **2004**, *20*, 1341–1347. [CrossRef]
28. Golub, Y.; Kaltwasser, S.F.; Mauch, C.P.; Herrmann, L.; Schmidt, U.; Holsboer, F.; Czisch, M.; Wotjak, C.T. Reduced hippocampus volume in the mouse model of Posttraumatic Stress Disorder. *J. Psychiatr. Res.* **2011**, *45*, 650–659. [CrossRef]
29. Kulesskaya, N.; Rauvala, H.; Võikar, V. Evaluation of social and physical enrichment in modulation of behavioural phenotype in C57BL/6J female mice. *PLOS ONE* **2011**, *6*, e24755. [CrossRef]
30. Chapillon, P.; Manneché, C.; Belzung, C.; Caston, J. Rearing environmental enrichment in two inbred strains of mice: 1. Effects on emotional reactivity. *Behav. Genet.* **1999**, *29*, 41–46. [CrossRef]
31. Kamakura, R.; Kovalainen, M.; Leppäluoto, J.; Herzig, K.-H.; Mäkelä, K.A. The effects of group and single housing and automated animal monitoring on urinary corticosterone levels in male C57 BL/6 mice. *Physiol. Rep.* **2016**, *4*, e12703. [CrossRef]
32. Liu, N.; Wang, Y.; An, A.Y.; Banker, C.; Qian, Y.; O'Donnell, J.M. Single housing-induced effects on cognitive impairment and depression-like behavior in male and female mice involve neuroplasticity-related signaling. *Eur. J. Neurosci.* **2019**, *52*, 2694–2704. [CrossRef]
33. Kamal, A.; Ramakers, G.; Altinbilek, B.; Kas, M. Social isolation stress reduces hippocampal long-term potentiation: Effect of Animal Strain and Involvement of Glucocorticoid Receptors. *Neuroscience* **2014**, *256*, 262–270. [CrossRef] [PubMed]
34. Kamal, A.; Ramakers, G.; Altinbilek, B.; Kas, M. Social stress and escalated drug self-administration in mice I. Alcohol and corticosterone. *Psychopharmacology* **2015**, *232*, 991–1001.
35. Gettayacamin, M.; Retnam, L. AAALAC international standards and accreditation process. *Toxicol. Res.* **2017**, *33*, 183–189. [CrossRef] [PubMed]
36. Poole, T. Happy animals make good science. *Lab. Anim.* **1997**, *31*, 116–124. [CrossRef]
37. Luo, Y.; Wang, L.; Yang, L.; Li, X.F.; Anselme, P.; Wang, X.; Li, Z. Using a behaviour random permutation model to identify displacement grooming in ungulates. *Current Zoology* **2022**, zoac035. [CrossRef]

Disclaimer/Publisher's Note: The statements, opinions and data contained in all publications are solely those of the individual author(s) and contributor(s) and not of MDPI and/or the editor(s). MDPI and/or the editor(s) disclaim responsibility for any injury to people or property resulting from any ideas, methods, instructions or products referred to in the content.

animals

MDPI

Article

Individualized Housing Modifies the Immune–Endocrine System in CD1 Adult Male Mice

Iván Ortega-Saez [1,†], Alina Díez-Solinska [2,†], Roger Grífols [3], Cristina Martí [3], Carolina Zamora [3], Maider Muñoz-Culla [2,4], Oscar Vegas [2,4] and Garikoitz Azkona [2,*]

1 Parc de Recerca Biomèdica de Barcelona (PRBB) Animal Facility, 08003 Barcelona, Spain
2 Department of Basic Psychological Processes and Their Development, Euskal Herriko Unibertsitatea (UPV/EHU), 20018 Donostia, Spain
3 Charles River Laboratories, Parc de Recerca Biomèdica de Barcelona (PRBB), 08003 Barcelona, Spain
4 Biodonostia Health Research Institute, 20014 Donostia, Spain
* Correspondence: garikoitz.azkona@ehu.eus
† These authors contributed equally to this work.

Simple Summary: In recent years, awareness of laboratory animals' wellbeing and the refinement of their house conditions have increased considerably. Mice (*Mus musculus*) are the most widely used animal species in research in the European Union and are sociable and hierarchical creatures. It is important to determine whether experimental conditions may affect research results and whether housing conditions (isolated or grouped) may be one such condition. The aim of this study was, therefore, to determine whether 4 weeks of social isolation (usual practice in our animal facility and some laboratory procedures) could induce changes in different physiological parameters (body weight, number of blood cells, and stress hormones) in adult mice. Although we did not observe changes in body weight, red blood cells, and platelets, mice that were socially isolated for 4 weeks did have a decreased count of some white blood cells. Moreover, levels of the main stress hormone were higher in single-housed mice after 1 week, although they decreased after 4 weeks to the same levels as those recorded for grouped mice. We can, therefore, conclude that social isolation affects some physiological parameters, and that this should be taken into account in the interpretation of research data.

Abstract: In the last years, different research groups have made considerable efforts to improve the care and use of animals in research. Mice (*Mus musculus*) are the most widely used animal species in research in the European Union and are sociable and hierarchical creatures. During experiments, researchers tend to individualize males, but no consideration is given to whether this social isolation causes them stress. The aim of this study was, therefore, to explore whether 4 weeks of social isolation could induce changes in different physiological parameters in adult Crl:CD1(ICR) (CD1) males, which may interfere with experimental results. Body weight, blood cells, and fecal corticosterone metabolites levels were the analyzed parameters. Blood and fecal samples were collected at weeks 1 and 4 of the experimental procedure. Four weeks of single housing produced a significant time-dependent decrease in monocytes and granulocytes. Fecal corticosterone metabolite levels were higher in single-housed mice after 1 week and then normalized after 4 weeks of isolation. Body weight, red blood cells, and platelets remained unchanged in both groups during this period. We can, therefore, conclude that social isolation affects some immune and endocrine parameters, and that this should be taken into account in the interpretation of research data.

Keywords: CD1 male; single-housed; stress; white blood cells; fecal corticosterone metabolites

Citation: Ortega-Saez, I.; Díez-Solinska, A.; Grífols, R.; Martí, C.; Zamora, C.; Muñoz-Culla, M.; Vegas, O.; Azkona, G. Individualized Housing Modifies the Immune–Endocrine System in CD1 Adult Male Mice. *Animals* **2023**, *13*, 1026. https://doi.org/10.3390/ani13061026

Academic Editor: Vera Baumans

Received: 21 February 2023
Revised: 8 March 2023
Accepted: 9 March 2023
Published: 10 March 2023

Copyright: © 2023 by the authors. Licensee MDPI, Basel, Switzerland. This article is an open access article distributed under the terms and conditions of the Creative Commons Attribution (CC BY) license (https://creativecommons.org/licenses/by/4.0/).

1. Introduction

People working with laboratory animals display a high level of awareness of and sensitivity to their wellbeing [1]. Indeed, perceived animal stress/pain has been found to negatively affect their professional quality of life [2]. In the last few years, different research groups have made considerable efforts to improve the care and use of animals in research, regardless of receiving specific funding for that purpose [3]. In the near future, this new scientific knowledge will provide new evidence to improve the welfare and housing conditions of animals used in scientific procedures. Current European legislation on the protection of animals used for scientific purposes (Directive 210/63/EU) establishes suitable environmental conditions and minimum enclosure measures by age and animal species. It likewise indicates that social laboratory animals must be socially housed in stable groups of compatible individuals. Moreover, procedures in which social animals (e.g., dogs and monkeys) are completely isolated for prolonged periods are classified as "severe" [4]. However, the legislation does not specify what exactly is considered to be a "prolonged period", and it does not mention other social species.

Despite the current debate about their predictive value in basic and regulatory studies [5–10], mice (*Mus musculus*) continue to be the most widely used animal species in research in the European Union [11]. Mice are sociable and hierarchical animals that, in nature, live in small groups. These groups are usually composed of a dominant male, along with various females with their offspring, both young and juvenile. The size of the territory occupied by a mouse family varies according to different factors. These include the availability of different resources such as water and food, as well as the density of the group. Occasionally, depending on the aggressiveness of the dominant male and the density of the group, young males are found in the aforementioned family groups. Generally, however, males are usually rejected from the group when they reach sexual maturity and can be found in the wild alone or in groups of young males. As for the females, they usually become part of the family group once they reach sexual maturity [12].

Unfortunately, in animal facilities, mice are not housed as in their natural environment, thus interfering with their natural ethogram. Standard laboratory protocols stipulate that mice's weaning and maternal separation should occur 21 days after birth. Thereafter, it is recommended that animals should be housed separately by sex and strain in stable groups of 2–5 members, a step that fosters the formation of affiliate relationships between individuals in the same group [13] and reduces aggression between males [14]. The main reason for housing male mice individually is aggression between cage mates [15,16]. Recently, a series of recommendations were published to minimize aggression between males [17].

Keeping newly weaned animals in the company of other animals is important for the correct development of their brains. It has been shown that post-weaning social deprivation by isolating mice induces neurochemical and morphological alterations, which have a behavioral impact in adulthood [13,18–23]. Indeed, the lack of social experiences before adulthood has been used in mice as a model to study some impaired behavioral phenotypes, such as depression and anxiety-like behavior types [21–23], as well as social and cognitive deficits [19,22]. In light of the above, in our animal facility, we implemented two different strategies in order to minimize the number of single-housed newly weaned male mice [24,25].

There is still an ongoing debate about whether adult male mice should be housed individually [15,26]. Years ago, "isolation syndrome" was described, with authors arguing that the inability to interact socially is likely to have a harmful effect on the animal's emotional state [27]. Indeed, it has been proven that adult male mice prefer the proximity of another male over individual housing [28], which is considered a stressor. The gold standard to measure the immediate physiological responses to stress is the activation of the hypothalamic–pituitary–adrenal (HPA) axis, which induces the secretion of corticosterone from the adrenal gland [29]. The effect of solitary versus social housing on corticosterone levels has been explored with varying results. Some studies observed that single-housed male mice had increased corticosterone levels after 14 days [30] and 15 months [31], whereas others found that corticosterone levels remained stable up to 42 days of individual housing [32–36], and two studies reported that single housing caused less stress for mice than group housing [37,38]. Other indications of stress include changes in body weight and a decrease in circulating leukocytes. A meta-analysis of the effects of individual housing on body weight found considerable heterogeneity in different mice strains, with higher, unchanged, or lower body weights being reported after social isolation [39]. Although it is well documented that chronic stress results in immunosuppression [40], differences in the total number of white blood cells have also been observed [36,41]. Among other factors, these discrepancies may be due to differing isolation periods.

In our animal facility, researchers tend to individualize males during experiments for a maximum period of 4 weeks, mainly for reasons of convenience and habit. However, no consideration is given to whether individually housing animals may cause them stress. The aim of the present study was, therefore, to determine if 4 weeks of social isolation could induce changes in body weight, blood cells, or fecal corticosterone metabolite levels in adult Crl:CD1(ICR) (CD1) males, which may interfere with experimental results.

2. Materials and Methods

2.1. Animals

Mice born in our specific pathogen-free (SPF) breeding zone were housed in pressurized and individually ventilated 1145T (403 × 165 × 174 mm; 435 cm^2 floor area; Tecniplast) (PIV) cages (70 air changes/h). We used black poplar/aspen shavings (Lignocel Selectfine; Rettenmaier Ibérica S.L.) as litter bedding, two sheets of tissue (Tork®; Essity Spain S.L) irradiated by Ionisos Iberica as nesting material, and an in-house autoclaved cardboard cylinder (12.5 × 9 × 0.5 cm; Sodispan Research S.L.) as enrichment. Once a week, socially housed mice (four mice per cage), together with their nesting material, were transferred to clean cages by picking them up at the base of their tails. This same procedure was carried out with individually housed mice every other week. New irradiated tissue was added if the nest was dirty or did not have enough material. Similarly, if the cardboard was broken, a new cylinder was provided. Mice had ad libitum access to water and diet (irradiated Special Diet Services RM1). Rooms were maintained under standard environmental conditions (humidity: 55 ± 10%; temperature: 20–24 °C) with a 12 h light/dark cycle (lights on at 8:00 a.m.). Animals were monitored every day. The animal care and use program was accredited by AAALAC International. The Catalan Government and the PRBB Ethics Committees approved the experimental protocol (DAAM 10576).

2.2. General Procedure

Eight-week-old CD1 mice were randomly assigned to two groups (grouped or single; n = 8 per group, 16 in total) and housed in the same room in which they were born. We selected CD1 adult male mice because they are outbred, are the most commonly used strain in toxicology studies [42], and have a high propensity to fight, resulting in suggestions that they may benefit from individual housing [15]. This does not apply to females, since chronic social isolation is used to model separation-induced depression [43].

Animals were weighed on the same day of the week for 5 weeks (weeks 0–4; 9:00–11:00 a.m.). Sampling was carried out in a laboratory adjacent to the room where they were housed, and the animals were transferred there 1 h before sampling, around 8:00 a.m., because the technician started their working day at this time. Sampling was carried out at two different time points to minimize the influence of handling as much as possible. Thus, on weeks 1 and 4 (9:00–11:00 a.m.), whole blood and fecal samples were obtained from each animal (Figure 1). No signs of fighting were observed during the experimental period. None of the animals had adverse events, and all completed the procedure. Animals became part of our colony once the experiment was completed.

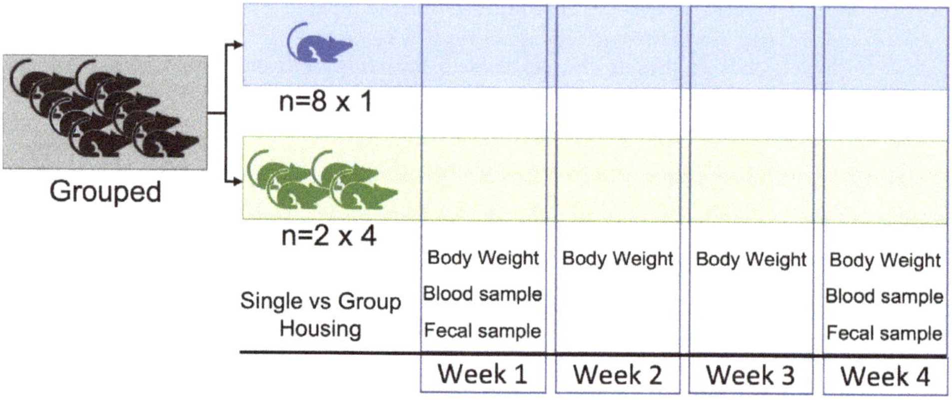

Figure 1. Experimental procedure.

2.3. Hematological Parameters

Blood samples were obtained by facial vein puncture with a 21 G sterile hypodermic needle. We collected blood from the facial vein because this procedure has been found to have the least adverse effects on welfare parameters in mice [44,45]. Samples (15 μL) were collected using a Microvette® 200K3E with potassium salt of ethylenediaminetetraacetic acid (EDTA) as an anticoagulant. After sampling, mice were returned to their home cage. No residual bleeding was noted in any of the animals. The blood was immediately analyzed for complete blood count: white blood cells (WBC), lymphocytes, monocytes, granulocytes, red blood cells (RBC), hemoglobin (HGB), hematocrit (HCT), mean corpuscular volume (MCV), hemoglobin (MCH) and hemoglobin concentration (MCHC), red cell distribution width (RDW), platelets (PLT), mean platelet volume (MPV), platelet distribution width (PDW), and platelet crit (PCT), using the fully automated CVM-Procell analyzer (CVM Diagnóstico Veterinario SL). Since the provider could not give us information about the exact mouse strain, age, or sex where the values were obtained, we first determined if the blood value range of male and female adult mice of different commonly used strains were within the normal range indicated by the analyzer. Our results indicated that the normal range provided for mice by the CVM-Procell analyzer can be used for adult male and female inbred C57BL/6J, outbred CD1, and immunodeficient CB17.Cg-$Prkdc^{scid}Lyst^{bg\text{-}J}$/Crl (SCID Beige) mice (see Supplementary Materials).

2.4. Fecal Corticosterone Metabolites

Fecal samples were obtained by placing each animal on a grid. The fecal boluses were obtained directly, without possible contamination, placed in an Eppendorf, and stored at −80 °C to determine corticosterone metabolite levels. After sampling, mice were returned to their home cage. This sampling method may allow a more accurate interpretation of chronic stress [46]. Moreover, since there is no need to restrain the animals when collecting the samples, this is a good method for enabling repeated sampling without affecting the animal, meaning that fecal samples are less affected by hormone secretion fluctuation or pulsatility. Each fecal sample was homogenized, and an aliquot of 0.05 g was shaken with 1 mL of 80% methanol in Tris/HCl 20 mM, pH 7.5, for 30 min on a multi-vortex. After centrifugation, each aliquot was frozen at −80 °C until analysis. Fecal corticosterone metabolite levels were quantified in duplicate using an enzyme immunoassay (Corticosterone Elisa Kit, Enzo Life Sciences; ADI-900-097), in accordance with the manufacturer's recommendations, and a Synergy HT microplate reader (BioTek Instruments, Inc., Winooski, VT, USA). Data were analyzed by means of a four-parameter logistic curve fit using MyAssays (Data Analysis Tools and Services for Bioassays; available at https://www.myassays.com/ accessed on 10

March 2023). The sensitivity of the assay was 27.0 pg/mL, and the intra- and inter-assay variation coefficients were between 7% and 8%.

2.5. Statistical Analyses

Experimental data were analyzed using GraphPad Prism software (6.01, GraphPad Software, Inc, San Diego, CA, USA). Group comparisons were performed using a two-way repeated-measures ANOVA, followed by Bonferroni's post hoc test. Values of $p < 0.05$ were considered statistically significant (95% confidence). Data are expressed as the mean ± standard deviation (SD). The results are described in accordance with the ARRIVE guidelines [47].

3. Results

3.1. Body Weight

Both groups of animals gained weight over the duration of the experiment ($F_{(4,56)}$ = 34,78, $p < 0.0001$). Grouped mice weighed 36.27 ± 2.46 g at week 0 and 39.46 ± 2.99 g at week 4. Single-housed mice weighed 38.20 ± 3.55 g at week 0 and 41.19 ± 4.29 g at week 4 (Figure 2). No significant differences were observed between grouped or single-housed mice.

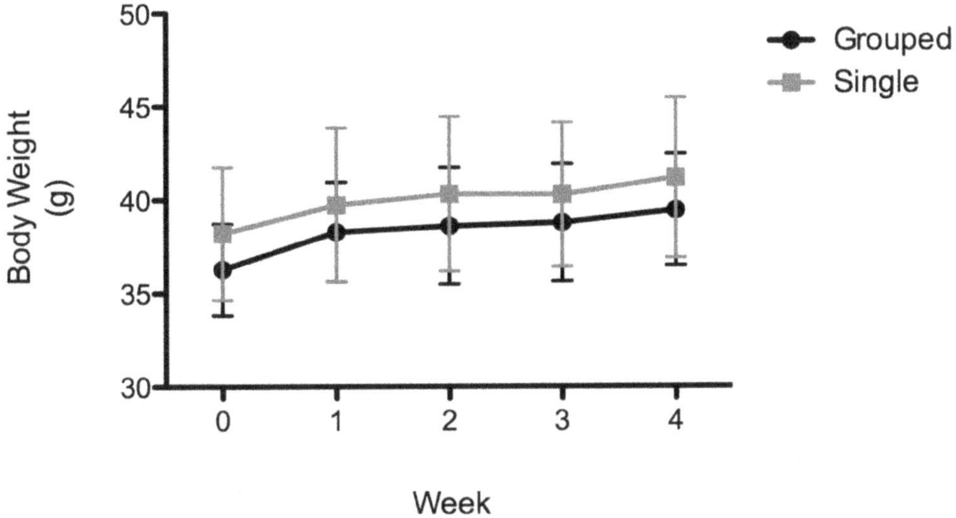

Figure 2. Body weight (g). Data are expressed as the mean ± SD; n = 8 per group.

3.2. Hematological Parameters

The results indicated no significant differences between grouped and single mice in the number of cells in the white series at either week 1 or week 4. However, significant differences were observed as a function of time ($F_{(1,14)}$ = 5.52; $p < 0.05$; Table 1). The post hoc analysis indicated a significant decrease in WBC after 4 weeks of single housing (t = 2.21; $p < 0.05$). When white cell type was analyzed in more detail, significant time-dependent differences were observed in monocytes ($F_{(1,14)}$ = 10.45; $p < 0.01$), and the post hoc analysis indicated a significant drop in monocytes in single-housed mice after 4 weeks (t = 2.714 $p < 0.05$). Similarly, significant time-dependent differences were observed in granulocytes ($F_{(1,14)}$ = 7.63; $p < 0.05$), which dropped in single-housed mice after 4 weeks (t = 2.46, $p < 0.05$).

Table 1. White blood cell population values. Data are expressed as the mean ± SD; n = 8 per group; * $p < 0.05$ (week 1 single vs. week 4 single).

	Week 1		Week 4		Normal Range	Unit
	Grouped	Single	Grouped	Single		
WBC	8.45 ± 3.33	8.15 ± 4.59	7.18 ± 2.59	4.50 ± 1.83 *	0.8–6.8	10^9/L
Lymph	5.81 ± 2.42	5.73 ± 3.08	4.76 ± 0.81	4.13 ± 0.52	0.7–5.7	10^9/L
Mon	0.33 ± 0.14	0.34 ± 0.31	0.26 ± 0.11	0.10 ± 0.08 *	0.0–0.3	10^9/L
Gran	2.33 ± 0.93	1.96 ± 1.50	1.70 ± 0.48	1.07 ± 0.56 *	0.1–1.8	10^9/L

* White blood cell (WBC), lymphocyte (Lymph), monocyte (Mon), and granulocyte (Gran).

The results indicated no significant differences between groups or timepoints in terms of the number of red blood cells and platelets (Table 2).

Table 2. Red blood cell and platelet values.

	Week 1		Week 4		Normal Range	Unit
	Grouped	Single	Grouped	Single		
RBC	8.79 ± 1.18	8.32 ± 1.42	8.76 ± 0.95	8.53 ± 0.93	6.36–9.42	10^{12}/L
HGB	14.59 ± 1.85	13.81 ± 2.79	14.73 ± 1.67	13.71 ± 1.42	11–14.3	g/dL
HCT	43.20 ± 5.24	42.15 ± 6.79	44.08 ± 4.64	42.09 ± 4.65	34.6–44.6	%
MCV	49.30 ± 1.09	50.83 ± 0.98	50.20 ± 1.69	49.41 ± 1.88	48.2–58.3	fL
MCH	16.58 ± 1.42	16.38 ± 0.42	16.46 ± 0.36	14.79 ± 0.78	15.8–19	pg
MCHC	337.13 ± 4.78	325.63 ± 5.06	328.86 ± 18.8	325.63 ± 11.4	302–353	g/L
RDW	13.30 ± 0.77	14.74 ± 0.33	13.55 ± 1.13	12.86 ± 1.27	13–17	%
PLT	1021.8 ± 582.3	762.4 ± 498.0	1140.0 ± 226.6	892.4 ± 273.7	450–1590	10^9/L
MPV	4.91 ± 0.35	5.39 ± 0.15	5.31 ± 0.430	5.13 ± 0.53	3.8–6	fL
PDW	16.88 ± 1.10	17.31 ± 1.47	16.67 ± 1.14	17.21 ± 1.28	-	-
PCT	0.41 ± 0.20	0.30 ± 0.27	0.44 ± 0.25	0.31 ± 0.23	-	%

Hemoglobin (HGB), hematocrit (HCT), mean corpuscular volume (MCV), hemoglobin (MCH) and hemoglobin concentration (MCHC), red cell distribution width (RDW), platelets (PLT), mean platelet volume (MPV), platelet distribution width (PDW), and platelet crit (PCT).

3.3. *Fecal Corticosterone Metabolites*

The statistical study of fecal corticosterone metabolite levels revealed a significant interaction between variables ($F_{(1,14)}$ = 11,40, $p < 0.01$). The post hoc analysis indicated significantly higher corticosterone metabolite levels in single-housed (0.225 ± 0.05 ng/mg) than in grouped animals (0.132 ± 0.02 ng/mg) after 1 week (t = 4.523; $p < 0.001$). At 4 weeks, no differences were observed between groups (grouped: 0.165 ± 0.06 ng/mg vs. single: 0.168 ± 0.04 ng/mg; t = 0.488, $p > 0.05$), and single-housed corticosterone metabolite levels were normalized (Figure 3).

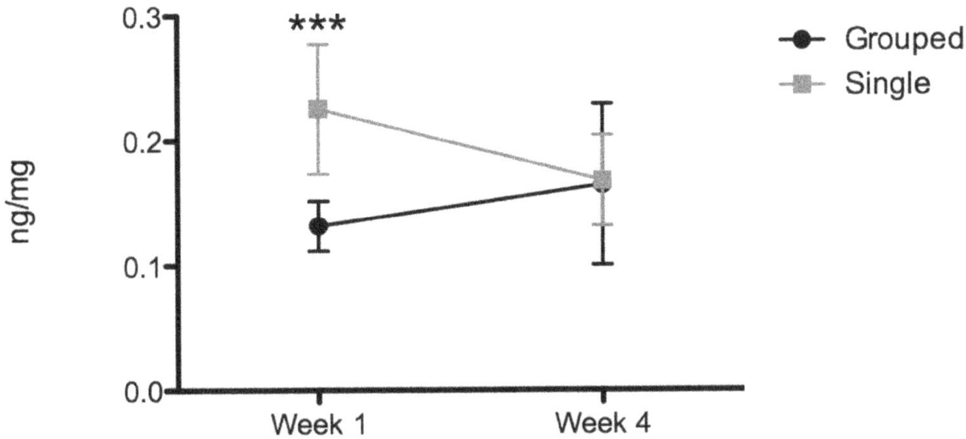

Figure 3. Fecal corticosterone metabolites (ng/mg feces). Data are expressed as the mean ± SD; n = 8 per group; *** $p < 0.001$.

4. Discussion

It is well known that animal welfare has an effect on the outcome of experiments. We must, therefore, always consider this factor when designing and carrying out experimental procedures. However, many researchers systematically tend to individualize animals in their experiments. Thus, the question we aimed to answer in this study was whether a lack of social interaction may modify physiological parameters, which may in turn interfere with experimental results. Our findings indicate that social isolation modifies some physiological parameters.

As previously reported for CD1 male mice [48–50], social isolation for 4 weeks did not affect body weight gain. Similarly, our results revealed that social isolation did not modify RBC parameters. As far as we are aware, this is the first study in mice to analyze RBC parameters; thus, we cannot compare our results with previous findings.

Mice that were changed from sharing a cage with littermates to living alone showed higher fecal corticosterone metabolites than those maintained in the group after the first week, although levels normalized after 1 month. These same results were recently observed in adult CD1 mice housed in the same conditions as our animals, in a ventilated rack with environmental enrichment [50], which may indicate habituation to the new situation. Due to the nature of our experimental design, we were unable to determine when exactly corticosterone metabolite levels normalized, and this is one of our study's limitations. However, data from a previous study [33] indicated that fecal corticosterone metabolite levels start to decrease and remain stable from the second week onward. These data are consistent with those described previously in relation to the return of plasma glucocorticoids to baseline values during the first week after transport or translocation [51–54]. Among the grouped animals, no significant changes were observed across individuals, and the standard deviation within groups was very small. Our data, therefore, seem to suggest that, in contrast to observations by some authors [37,38], remaining grouped together does not appear to cause the animals any stress. We believe the main reason for this is that, as has indeed been pointed out previously [32], our mice were littermates and were grouped together from weaning.

It is well known that increased glucocorticoid levels suppress cellular immunity [55]. No changes in monocytes and granulocytes were observed in single-housed animals after 7 days, although changes were found after 4 weeks. A previous study found no significant differences in the overall number of blood-circulating leukocytes between CD1 male mice that were socially isolated for 2 weeks and their socially housed counterparts [36]. However,

C57BL6/J adult mice separated into individual cages for 2 h every day for 25 days were found to have a decrease in T cells, B cells, monocytes, and neutrophils [41]. Unfortunately, our system is not able to distinguish between the different types of lymphocytes and granulocytes; however, overall, our results are consistent with these findings and highlight the fact that isolation time is a factor to be considered. Another limitation of the study is that we did not study humoral immunity; previous studies found that fecal immunoglobulin A (IgA) excretion (a marker of long-term stress) takes at least 4 weeks to normalize [53]. It is important to note that CD1 adult males isolated for 21 days and subjected to mild psychological stress had lower splenocyte proliferation and lower IL-2 and IL-4 cytokine plasma levels than their grouped counterparts [32]. The same results were reported using shock as a stressor [55].

In addition to the limitations outlined above, our study had some further limitations. When designing the experiment, we wanted it to be as realistic as possible in terms of the day-to-day management of our animal facility technicians and researchers. Therefore, the animals were moved from dirty to clean cages by picking them up by the tail. In recent years, less aversive handling methods (e.g., tunnel or cup handling) have been shown to mitigate anxiety and depressive-like behaviors [56–58]. However, a recent study showed that picking mice up by their tail may not be a significant source of chronic husbandry stress [59]. In view of the results of this study and our daily practice, we decided to change the location of animals in this experiment by picking them up by their tail. We are all aware that efforts have to be made to implement less aversive methods of handling in daily practice in animal facilities. Nevertheless, it should also be kept in mind that this procedure takes more time; hence, the amount of work assigned to each technician when changing cages should also be reviewed. In our work, we did not study whether social isolation induced behavioral changes in our animals, because we were more interested in peripheral biomarkers than behavioral parameters. In a recent study performed on C57BL/6JRj mice housed singly for 10 weeks, no behavioral changes were observed in exploratory activity, anxiety, working memory, and fear memory [60]. However, a previous study using C57BL/6J and DBA/2 kept in individual housing for 7 weeks revealed that individual housing has strong strain- and test-specific effects on emotional behavior and impaired memory in certain tasks. Single-housed mice were hyperactive and displayed reduced habituation to novel environments. Reduced anxiety was established in the elevated plus-maze, but not in the dark/light test. Immobility in the forced swimming test was reduced by social isolation. Novel object recognition and fear conditioning were impaired in the single-housed mice, whereas water-maze learning was not affected [61]. In the same way, 2 weeks of single housing plus acute injection stress induced anxiety-like behavior in C57BL6/J mice [30]. Mouse strain and social environment also influence depression-like behavior caused by an immune challenge. In this sense, group-housed CD1 mice exhibited depression-like behavior 1 day after bacterial lipopolysaccharide (LPS) injection, while the behavior of single-housed CD1 mice was little affected during the 4 weeks of the experiment. In contrast, both grouped and single-housed C57BL/6 mice responded to LPS with an increase in depression-like behavior [62]. It would be interesting to conduct future behavioral studies to determine if, under our conditions, single-housed CD1 male mice show any behavioral changes. Another parameter we did not measure was body temperature. In recent years, it has been observed that laboratory mice suffer from thermal stress, and that this affects their immune system, among other physiological parameters [63,64]. In this sense, huddling, a form of social thermoregulation, is a major contributor to mice's thermal physiology. Thus, single-housed mice are usually more affected by cold temperatures than grouped mice [65]. In order to mitigate this effect, two sheets of tissue were added to their home cage, and we ensured that they made a proper nest.

In light of all these data, we recommend keeping males in stable groups from weaning onward. Researchers should be aware that the change from grouping to living alone induces stress and mild immunosuppression in CD1 male mice; hence, if the mice need to be separated for experimental reasons, these factors should be taken into consideration.

5. Conclusions

We conclude that social isolation has an effect on the immune–endocrine system. Consequently, the stress associated with the new social situation should be taken into consideration in the interpretation of research data.

Supplementary Materials: The following supporting information can be downloaded at https://www.mdpi.com/article/10.3390/ani13061026/s1: Determination of CVM-Procell analyzer reference values for each strain. References [66–71] are cited in the supplementary materials

Author Contributions: Conceptualization, G.A.; methodology, I.O.-S., A.D.-S., M.M.-C. and G.A.; acquisition and analysis of the data, I.O.-S., A.D.-S., R.G., C.M. and C.Z.; analysis and interpretation of the data, I.O.-S., A.D.-S., O.V. and G.A.; interpretation of the data and writing—original draft preparation, I.O.-S., A.D.-S., M.M.-C., O.V. and G.A.; writing—review and editing, G.A. All authors have read and agreed to the published version of the manuscript.

Funding: This study was supported by the University of the Basque Country (UPV/EHU) GIU18/103 grant.

Institutional Review Board Statement: The study was conducted in accordance with the Declaration of Helsinki and approved by the Catalan Government and the PRBB Ethics Committees (DAAM 10576).

Informed Consent Statement: Not applicable.

Data Availability Statement: All the data of the study can be made available upon request.

Conflicts of Interest: The authors declare no conflict of interest.

References

1. Goñi-Balentziaga, O.; Ortega-Saez, I.; Vila, S.; Azkona, G. Working with laboratory rodents in Spain: A survey on welfare and wellbeing. *Lab. Anim. Res.* **2021**, *37*, 18. [CrossRef] [PubMed]
2. Goñi-Balentziaga, O.; Vila, S.; Ortega-Saez, I.; Vegas, O.; Azkona, G. Professional Quality of Life in Research Involving Laboratory Animals. *Animals* **2021**, *11*, 2639. [CrossRef] [PubMed]
3. Díez-Solinska, A.; Vegas, O.; Azkona, G. Refinement in the European Union: A Systematic Review. *Animals* **2022**, *12*, 3263. [CrossRef]
4. EU. Directive 2010/63/EU of the European Parliament and of the Council of 22 September 2010 on the Protection of Animals Used for Scientific Purposes. 2010. Available online: https://eur-lex.europa.eu/legal-content/EN/TXT/?uri=celex%3A32010L0063 (accessed on 9 February 2023).
5. Kafkafi, N.; Agassi, J.; Chesler, E.J.; Crabbe, J.C.; Crusio, W.E.; Eilam, D.; Gerlai, R.; Golani, I.; Gomez-Marin, A.; Heller, R.; et al. Reproducibility and replicability of rodent phenotyping in preclinical studies. *Neurosci. Biobehav. Rev.* **2018**, *87*, 218–232. [CrossRef]
6. Perlman, R.L. Mouse Models of Human Disease: An Evolutionary Perspective. *Evol. Med. Public Health* **2016**, *2016*, 170–176. [CrossRef] [PubMed]
7. Perrin, S. Preclinical research: Make mouse studies work. *Nature* **2014**, *507*, 423–425. [CrossRef] [PubMed]
8. Garner, J.P. The Significance of Meaning: Why Do Over 90% of Behavioral Neuroscience Results Fail to Translate to Humans, and What Can We Do to Fix It? *ILAR J.* **2014**, *55*, 438–456. [CrossRef]
9. Gururajan, A.; Reif, A.; Cryan, J.F.; Slattery, D.A. The future of rodent models in depression research. *Nat. Rev. Neurosci.* **2019**, *20*, 686–701. [CrossRef]
10. Azkona, G.; Sanchez-Pernaute, R. Mice in translational neuroscience: What R we doing? *Prog. Neurobiol.* **2022**, *217*, 102330. [CrossRef]
11. EC. Summary Report on the Statistics on the Use of Animals for Scientific Purposes in the Member States of the European Union and Norway in 2019. Available online: https://ec.europa.eu/environment/chemicals/lab_animals/pdf/SWD2019_Part_A_and_B.pdf (accessed on 9 February 2023).
12. Van Zegeren, K. Variation in Aggressiveness and the Regulation of Numbers in House Mouse Populations. *Neth. J. Zool.* **1980**, *30*, 635–770. [CrossRef]
13. Arakawa, H. Ethological approach to social isolation effects in behavioral studies of laboratory rodents. *Behav. Brain Res.* **2018**, *341*, 98–108. [CrossRef] [PubMed]
14. Van Loo, P.L.; Mol, J.A.; Koolhaas, J.M.; Van Zutphen, B.F.; Baumans, V. Modulation of aggression in male mice: Influence of group size and cage size. *Physiol. Behav.* **2001**, *72*, 675–683. [CrossRef] [PubMed]

15. Kappel, S.; Hawkins, P.; Mendl, M.T. To Group or Not to Group? Good Practice for Housing Male Laboratory Mice. *Animals* **2017**, *7*, 88. [CrossRef] [PubMed]
16. Zidar, J.; Weber, E.M.; Ewaldsson, B.; Tjäder, S.; Lilja, J.; Mount, J.; Svensson, C.I.; Svensk, E.; Udén, E.; Törnqvist, E. Group and Single Housing of Male Mice: Collected Experiences from Research Facilities in Sweden. *Animals* **2019**, *9*, 1010. [CrossRef]
17. Weber, E.M.; Zidar, J.; Ewaldsson, B.; Askevik, K.; Udén, E.; Svensk, E.; Törnqvist, E. Aggression in Group-Housed Male Mice: A Systematic Review. *Animals* **2022**, *13*, 143. [CrossRef]
18. Haj-Mirzaian, A.; Amiri, S.; Kordjazy, N.; Rahimi-Balaei, M.; Haj-Mirzaian, A.; Marzban, H.; Aminzadeh, A.; Dehpour, A.R.; Mehr, S.E. Blockade of NMDA receptors reverses the depressant, but not anxiogenic effect of adolescence social isolation in mice. *Eur. J. Pharmacol.* **2015**, *750*, 160–166. [CrossRef]
19. Okada, R.; Fujiwara, H.; Mizuki, D.; Araki, R.; Yabe, T.; Matsumoto, K. Involvement of dopaminergic and cholinergic systems in social isolation-induced deficits in social affiliation and conditional fear memory in mice. *Neuroscience* **2015**, *299*, 134–145. [CrossRef]
20. Medendorp, W.P.; Petersen, E.D.; Pal, A.; Wagner, L.-M.; Myers, A.R.; Hochgeschwender, U.; Jenrow, K.A. Altered Behavior in Mice Socially Isolated During Adolescence Corresponds with Immature Dendritic Spine Morphology and Impaired Plasticity in the Prefrontal Cortex. *Front. Behav. Neurosci.* **2018**, *12*, 87. [CrossRef]
21. Ago, Y.; Takahashi, K.; Nakamura, S.; Hashimoto, H.; Baba, A.; Matsuda, T. Anxiety-Like and Exploratory Behaviors of Isolation-Reared Mice in the Staircase Test. *J. Pharmacol. Sci.* **2007**, *104*, 153–158. [CrossRef]
22. Koike, H.; Ibi, D.; Mizoguchi, H.; Nagai, T.; Nitta, A.; Takuma, K.; Nabeshima, T.; Yoneda, Y.; Yamada, K. Behavioral abnormality and pharmacologic response in social isolation-reared mice. *Behav. Brain Res.* **2009**, *202*, 114–121. [CrossRef]
23. Amiri, S.; Haj-Mirzaian, A.; Rahimi-Balaei, M.; Razmi, A.; Kordjazy, N.; Shirzadian, A.; Mehr, S.E.; Sianati, H.; Dehpour, A.R. Co-occurrence of anxiety and depressive-like behaviors following adolescent social isolation in male mice; possible role of nitrergic system. *Physiol. Behav.* **2015**, *145*, 38–44. [CrossRef] [PubMed]
24. Grífols, R.; Zamora, C.; Ortega-Saez, I.; Azkona, G. Postweaning Grouping as a Strategy to Reduce Singly Housed Male Mice. *Animals* **2020**, *10*, 2135. [CrossRef] [PubMed]
25. Azkona, G.; Caballero, J.M. Implementing strategies to reduce singly housed male mice. *Lab. Anim.* **2019**, *53*, 508–510. [CrossRef] [PubMed]
26. Brain, P. What does individual housing mean to a mouse? *Life Sci.* **1975**, *16*, 187–200. [CrossRef]
27. Valzelli, L. The "isolation syndrome" in mice. *Psychopharmacology* **1973**, *31*, 305–320. [CrossRef]
28. Van Loo, P.L.P.; de Groot, A.C.; van Zutphen, B.F.M.; Baumans, V. Do Male Mice Prefer or Avoid Each Other's Company? Influence of Hierarchy, Kinship, and Familiarity. *J. Appl. Anim. Welf. Sci.* **2001**, *4*, 91–103. [CrossRef]
29. Beery, A.K.; Kaufer, D. Stress, social behavior, and resilience: Insights from rodents. *Neurobiol. Stress* **2015**, *1*, 116–127. [CrossRef]
30. Hebda-Bauer, E.K.; Dokas, L.A.; Watson, S.J.; Akil, H. Adaptation to single housing is dynamic: Changes in hormone levels, gene expression, signaling in the brain, and anxiety-like behavior in adult male C57Bl/6J mice. *Horm. Behav.* **2019**, *114*, 104541. [CrossRef]
31. Diwakarla, S.; Finkelstein, D.I.; Constable, R.; Artaiz, O.; Di Natale, M.; McQuade, R.M.; Lei, E.; Chai, X.; Ringuet, M.T.; Fothergill, L.J.; et al. Chronic isolation stress is associated with increased colonic and motor symptoms in the A53T mouse model of Parkinson's disease. *Neurogastroenterol. Motil.* **2020**, *32*, e13755. [CrossRef]
32. Bartolomucci, A.; Palanza, P.; Sacerdote, P.; Ceresini, G.; Chirieleison, A.; Panerai, A.; Parmigiani, S. Individual housing induces altered immuno-endocrine responses to psychological stress in male mice. *Psychoneuroendocrinology* **2002**, *28*, 540–558. [CrossRef]
33. Hunt, C.; Hambly, C. Faecal corticosterone concentrations indicate that separately housed male mice are not more stressed than group housed males. *Physiol. Behav.* **2006**, *87*, 519–526. [CrossRef]
34. Arndt, S.S.; Laarakker, M.C.; van Lith, H.A.; van der Staay, F.J.; Gieling, E.; Salomons, A.R.; Klooster, J.V.; Ohl, F. Individual housing of mice—Impact on behaviour and stress responses. *Physiol. Behav.* **2009**, *97*, 385–393. [CrossRef] [PubMed]
35. McQuaid, R.J.; Audet, M.-C.; Jacobson-Pick, S.; Anisman, H. The differential impact of social defeat on mice living in isolation or groups in an enriched environment: Plasma corticosterone and monoamine variations. *Int. J. Neuropsychopharmacol.* **2012**, *16*, 351–363. [CrossRef] [PubMed]
36. Hamilton, A.; Rizzo, R.; Brod, S.; Ono, M.; Perretti, M.; Cooper, D.; D'Acquisto, F. The immunomodulatory effects of social isolation in mice are linked to temperature control. *Brain. Behav. Immun.* **2022**, *102*, 179–194. [CrossRef] [PubMed]
37. Kamakura, R.; Kovalainen, M.; Leppäluoto, J.; Herzig, K.-H.; Mäkelä, K.A. The effects of group and single housing and automated animal monitoring on urinary corticosterone levels in male C57BL/6 mice. *Physiol. Rep.* **2016**, *4*, e12703. [CrossRef] [PubMed]
38. Mertens, S.; Vogt, M.A.; Gass, P.; Palme, R.; Hiebl, B.; Chourbaji, S. Effect of three different forms of handling on the variation of aggression-associated parameters in individually and group-housed male C57BL/6NCrl mice. *PLoS ONE* **2019**, *14*, e0215367. [CrossRef]
39. Schipper, L.; Harvey, L.; van der Beek, E.M.; van Dijk, G. Home alone: A systematic review and meta-analysis on the effects of individual housing on body weight, food intake and visceral fat mass in rodents. *Obes. Rev.* **2018**, *19*, 614–637. [CrossRef] [PubMed]
40. Cain, D.W.; Cidlowski, J.A. Immune regulation by glucocorticoids. *Nat. Rev. Immunol.* **2017**, *17*, 233–247. [CrossRef] [PubMed]
41. Bowers, S.L.; Bilbo, S.D.; Dhabhar, F.S.; Nelson, R.J. Stressor-specific alterations in corticosterone and immune responses in mice. *Brain Behav. Immun.* **2008**, *22*, 105–113. [CrossRef]

42. Annas, A.; Bengtsson, C.; Törnqvist, E. Group housing of male CD1 mice: Reflections from toxicity studies. *Lab. Anim.* **2013**, *47*, 127–129. [CrossRef]
43. Martin, A.L.; Brown, R.E. The lonely mouse: Verification of a separation-induced model of depression in female mice. *Behav. Brain Res.* **2010**, *207*, 196–207. [CrossRef] [PubMed]
44. Harikrishnan, V.S.; Hansen, A.K.; Abelson, K.S.; Sørensen, D.B. A comparison of various methods of blood sampling in mice and rats: Effects on animal welfare. *Lab. Anim.* **2017**, *52*, 253–264. [CrossRef]
45. Jo, E.J.; Bae, E.; Yoon, J.-H.; Kim, J.Y.; Han, J.S. Comparison of murine retroorbital plexus and facial vein blood collection to mitigate animal ethics issues. *Lab. Anim. Res.* **2021**, *37*, 12. [CrossRef] [PubMed]
46. Rowland, N.E.; Toth, L.A. Analytic and Interpretational Pitfalls to Measuring Fecal Corticosterone Metabolites in Laboratory Rats and Mice. *Comp. Med.* **2019**, *69*, 337–349. [CrossRef] [PubMed]
47. Du Sert, N.P.; Hurst, V.; Ahluwalia, A.; Alam, S.; Avey, M.T.; Baker, M.; Browne, W.J.; Clark, A.; Cuthill, I.C.; Dirnagl, U.; et al. The ARRIVE guidelines 2.0: Updated guidelines for reporting animal research. *PLoS Biol.* **2020**, *18*, e3000410. [CrossRef]
48. Barnhart, J.E.; Pizzi, W.J. The Monosodium L-glutamate (MSG) syndrome in mice develops independently of housing condition. *Neurobehav. Toxicol. Teratol.* **1982**, *4*, 549–556.
49. Bartolomucci, A.; Cabassi, A.; Govoni, P.; Ceresini, G.; Cero, C.; Berra, D.; Dadomo, H.; Franceschini, P.; Dell'Omo, G.; Parmigiani, S.; et al. Metabolic Consequences and Vulnerability to Diet-Induced Obesity in Male Mice under Chronic Social Stress. *PLoS ONE* **2009**, *4*, e4331. [CrossRef]
50. Lee, G.-H.; Kim, K.; Jo, W. Stress Evaluation of Mouse Husbandry Environments for Improving Laboratory Animal Welfare. *Animals* **2023**, *13*, 249. [CrossRef]
51. Peng, X.; Lang, C.M.; Drozdowicz, C.K.; Ohlsson-Wilhelm, B.M. Effect of cage population density on plasma corticosterone and peripheral lymphocyte populations of laboratory mice. *Lab. Anim.* **1989**, *23*, 302–306. [CrossRef]
52. Tuli, J.S.; Smith, J.A.; Morton, D.B. Stress measurements in mice after transportation. *Lab. Anim.* **1995**, *29*, 132–138. [CrossRef]
53. Bundgaard, C.J.; Kalliokoski, O.; Abelson, K.S.P.; Hau, J. Acclimatization of mice to different cage types and social groupings with respect to fecal secretion of IgA and corticosterone metabolites. *In Vivo* **2012**, *26*, 883–888. [PubMed]
54. Obernier, J.A.; Baldwin, R.L. Establishing an Appropriate Period of Acclimatization Following Transportation of Laboratory Animals. *ILAR J.* **2006**, *47*, 364–369. [CrossRef]
55. Viveros-Paredes, J.M.; Puebla-Pérez, A.M.; Gutiérrez-Coronado, O.; Sandoval-Ramírez, L.; Villaseñor-García, M.M. Dysregulation of the Th1/Th2 cytokine profile is associated with immunosuppression induced by hypothalamic-pituitary-adrenal axis activation in mice. *Int. Immunopharmacol.* **2006**, *6*, 774–781. [CrossRef]
56. Gouveia, K.; Hurst, J.L. Improving the practicality of using non-aversive handling methods to reduce background stress and anxiety in laboratory mice. *Sci. Rep.* **2019**, *9*, 20305. [CrossRef] [PubMed]
57. Clarkson, J.M.; Dwyer, D.M.; Flecknell, P.A.; Leach, M.C.; Rowe, C. Handling method alters the hedonic value of reward in laboratory mice. *Sci. Rep.* **2018**, *8*, 2448. [CrossRef]
58. Hurst, J.L.; West, R.S. Taming anxiety in laboratory mice. *Nat. Methods* **2010**, *7*, 825–826. [CrossRef] [PubMed]
59. Novak, J.; Jaric, I.; Rosso, M.; Rufener, R.; Touma, C.; Würbel, H. Handling method affects measures of anxiety, but not chronic stress in mice. *Sci. Rep.* **2022**, *12*, 20938. [CrossRef]
60. Buckinx, A.; Van Schuerbeek, A.; Bossuyt, J.; Allaoui, W.; Herrewegen, Y.V.D.; Smolders, I.; De Bundel, D. Exploring Refinement Strategies for Single Housing of Male C57BL/6JRj Mice: Effect of Cage Divider on Stress-Related Behavior and Hypothalamic-Pituitary-Adrenal-Axis Activity. *Front. Behav. Neurosci.* **2021**, *15*, 743959. [CrossRef]
61. Võikar, V.; Polus, A.; Vasar, E.; Rauvala, H. Long-term individual housing in C57BL/6J and DBA/2 mice: Assessment of behavioral consequences. *Genes Brain Behav.* **2004**, *4*, 240–252. [CrossRef]
62. Painsipp, E.; Köfer, M.J.; Sinner, F.; Holzer, P. Prolonged Depression-Like Behavior Caused by Immune Challenge: Influence of Mouse Strain and Social Environment. *PLoS ONE* **2011**, *6*, e20719. [CrossRef]
63. Gordon, C.J. The mouse thermoregulatory system: Its impact on translating biomedical data to humans. *Physiol. Behav.* **2017**, *179*, 55–66. [CrossRef]
64. Seeley, R.J.; MacDougald, O.A. Mice as experimental models for human physiology: When several degrees in housing temperature matter. *Nat. Metab.* **2021**, *3*, 443–445. [CrossRef]
65. Škop, V.; Xiao, C.; Liu, N.; Gavrilova, O.; Reitman, M.L. The effects of housing density on mouse thermal physiology depend on sex and ambient temperature. *Mol. Metab.* **2021**, *53*, 101332. [CrossRef]
66. Croy, B.A.; Chapeau, C. Evaluation of the pregnancy immunotrophism hypothesis by assessment of the reproductive performance of young adult mice of genotype scid/scid.bg/bg. *J. Reprod Fertil.* **1990**, *88*, 231–239. [CrossRef]
67. Santos, E.W.; de Oliveira, D.C.; Hastreiter, A.; da Silva, G.B.; Beltran, J.S.; Tsujita, M.; Crisma, A.R.; Neves, S.M.P.; Fock, R.A.; Borelli, P. Valores de referencia hematologicos e bioquimicos para camundongos das linhagens C57BL/6, Swiss Webster e BALB/c. *Braz. J. Vet. Res. Anim. Sci.* **2016**, *53*, 138–145. [CrossRef]
68. Serfilippi, L.M.; Pallman, D.R.; Russell, B. Serum clinical chemistry and hematology reference values in outbred stocks of albino mice from three commonly used vendors and two inbred strains of albino mice. *Contemp. Top. Lab. Anim. Sci.* **2003**, *42*, 46–52. [PubMed]

69. Mazzaccara, C.; Labruna, G.; Cito, G.; Scarfò, M.; De Felice, M.; Pastore, L.; Sacchetti, L. Age-Related Reference Intervals of the Main Biochemical and Hematological Parameters in C57BL/6J, 129SV/EV and C3H/HeJ Mouse Strains. *PLoS ONE* **2008**, *3*, e3772. [CrossRef] [PubMed]
70. Balcombe, J.P.; Barnard, N.D.; Sandusky, C. Laboratory routines cause animal stress. *Contemp. Top. Lab. Anim. Sci.* **2004**, *43*, 42–51.
71. White, J.R.; Gong, H.; Colaizy, T.T.; Moreland, J.G.; Flaherty, H.; McElroy, S.J. Evaluation of hematologic variables in newborn C57/BL6 mice up to day 35. *Vet. Clin. Pathol.* **2016**, *45*, 87–95. [CrossRef]

Disclaimer/Publisher's Note: The statements, opinions and data contained in all publications are solely those of the individual author(s) and contributor(s) and not of MDPI and/or the editor(s). MDPI and/or the editor(s) disclaim responsibility for any injury to people or property resulting from any ideas, methods, instructions or products referred to in the content.

animals

MDPI

Systematic Review

Aggregation in Group-Housed Male Mice: A Systematic Review

Elin M. Weber [1,†], Josefina Zidar [2,†], Birgit Ewaldsson [3], Kaisa Askevik [4], Eva Udén [4], Emma Svensk [4,*] and Elin Törnqvist [5,6,7]

[1] Department of Animal Environment and Health, Swedish University of Agricultural Sciences, 532 23 Skara, Sweden
[2] Department of Animal Environment and Health, Swedish University of Agricultural Sciences, 750 07 Uppsala, Sweden
[3] Department of Animal Science and Technology, AstraZeneca, 431 83 Mölndal, Sweden
[4] Swedish 3Rs Center, Swedish Board of Agriculture, 553 29 Jönköping, Sweden
[5] Swedish National Committee for the Protection of Animals Used for Scientific Purposes, Swedish Board of Agriculture, 553 29 Jönköping, Sweden
[6] Department of Animal Health and Antimicrobial Strategies, National Veterinary Institute (SVA), 751 89 Uppsala, Sweden
[7] Institute of Environmental Medicine, Karolinska Institutet, 171 77 Stockholm, Sweden
* Correspondence: 3rcenter@jordbruksverket.se
† These authors contributed equally to this work.

Simple Summary: When male mice are kept in groups at animal facilities, aggressive interactions between cage mates are not uncommon. Systematically reviewing previous studies that explored the cause of male mice aggression, we found that studies were disparate, using several different strains, a diverse set of environmental enrichments and different ways of grouping and housing mice, as well as different ways to observe aggression. Understanding the cause of male mice aggression is difficult when researchers use different methods and study designs. Nevertheless, our results suggest that home cage aggression is best studied in home cage environments and not by introducing unfamiliar mice to each other in a novel environment. In addition, while we were able to provide recommendations on how to minimize aggression, our assessment was that there is no universal solution that could be used by all animal facilities. Instead, it is important to realize that aggression is complex and that animal facilities might have to try different possible solutions to find what works best under their specific conditions.

Abstract: Aggression among group-housed male mice is a major animal welfare concern often observed at animal facilities. Studies designed to understand the causes of male mice aggression have used different methodological approaches and have been heterogeneous, using different strains, environmental enrichments, housing conditions, group formations and durations. By conducting a systematic literature review based on 198 observed conclusions from 90 articles, we showed that the methodological approach used to study aggression was relevant for the outcome and suggested that home cage observations were better when studying home cage aggression than tests provoking aggression outside the home cage. The study further revealed that aggression is a complex problem; one solution will not be appropriate for all animal facilities and all research projects. Recommendations were provided on promising tools to minimize aggression, based on the results, which included what type of environmental enrichments could be appropriate and which strains of male mice were less likely to be aggressive.

Keywords: male mice; group housing; aggression; animal welfare; environmental enrichment; group formation; housing conditions; resident-intruder; social dominance; wound scoring

Citation: Weber, E.M.; Zidar, J.; Ewaldsson, B.; Askevik, K.; Udén, E.; Svensk, E.; Törnqvist, E. Aggression in Group-Housed Male Mice: A Systematic Review. *Animals* **2023**, *13*, 143. https://doi.org/10.3390/ani13010143

Academic Editor: Garikoitz Azkona

Received: 14 November 2022
Revised: 23 December 2022
Accepted: 24 December 2022
Published: 30 December 2022

Copyright: © 2022 by the authors. Licensee MDPI, Basel, Switzerland. This article is an open access article distributed under the terms and conditions of the Creative Commons Attribution (CC BY) license (https://creativecommons.org/licenses/by/4.0/).

1. Introduction

In recent years, several publications have addressed the challenges of keeping male mice in groups (e.g., [1–5]). Mice are social animals and, according to current legislation and guidelines [6–8], mice should be group housed when used in research. However, aggression between male cage mates is one of the main problems in laboratory mouse husbandry, affecting both animal welfare and scientific quality [2].

Some general recommendations on how to prevent and minimize aggression have been listed [3,4,9,10]. These include keeping siblings together or grouping familiar mice before sexual maturity, housing male mice in small groups, transferring nesting material at cage cleaning, avoiding disturbances, handling mice with care and using strains with low level of aggression. However, despite being housed according to the general recommendations, mice sometimes fight when kept in groups, illustrating the complexity of the problem.

Aggression in group-housed male mice has been studied using several different methodological approaches [2]. One such approach is to observe undisturbed groups of mice in their home cages; these are usually referred to as home cage observations. There are also a number of different test protocols used to measure aggression and social dominance, such as the resident intruder test [11], the social defeat test [12], the novel arena social encounter test [13] and the tube test [14]. These types of tests mainly focus on territorial aggression and are designed to induce stress, emotional conflict and frustration, with aggression as a readout [2]. They are commonly used in studies where mice are used as animal models to study aggression, social defeat, deficits in social interaction or mood-related disorders (e.g., [12,15]). In these test situations, mice are removed from the social context in their home cage, placed in a novel cage or test arena, and often exposed to unfamiliar conspecifics in different staged encounters. Another way of measuring aggression is to score wounds on the body of the mice, continuously throughout the study and/or after the mice have been euthanized [16]. Sometimes a combination of home cage observations, different test protocols and wound scoring is used.

Aside from differences in methodology, there is a wide variation in different treatments used to study the effects on aggression, such as group composition, differences between strains, handling and cleaning routines, housing conditions and access to environmental enrichment. Environmental enrichment is used to enhance animal welfare by enabling animals to perform positive natural behaviors and increase their ability to gain certain control over their environment [17]. Transferring nesting material at cage cleaning has been shown to decrease aggression [9]. However, the effects of different structural enrichment items on aggression in male mice remain unclear [17]. For example, different studies using shelter as enrichment report both increased aggression [18,19] and no effect on aggression [20–22], while combinations of enrichment items with shelter included have been reported to decrease aggression [23,24].

In general, few experimental studies have addressed the problem of home cage aggression under normal husbandry conditions [5]. This might be another reason why results from experimental studies can be difficult to interpret and implement to daily practice in the lab.

To our knowledge, this is the first systematic review on aggression in group-housed male laboratory mice. In this dataset, data were systematically collected from articles investigating this subject. We also included articles that did not have aggression as the primary outcome, but rather as an additional finding. These articles were selected because they could contain valuable information and might not have been included in previous reviews with a focus on animal welfare. Our aim was to map how the literature in the field support, or do not support, available recommendations on how to prevent aggression in group-housed male mice, and to detect knowledge gaps that ought to be filled. We also wanted to address and describe how aggression has been measured in the literature, since this could influence the possibility of translating outcomes to normal husbandry conditions and contribute to useful recommendations.

2. Materials and Methods

2.1. Literature Search

We conducted a systematic literature search on the 12th of December 2018 using the online databases Medline, Embase and Web of Science. The search included words in titles, abstracts, author keywords and keywords of scientific publications (henceforth referred to as records) and was performed via the University Library at Karolinska Institutet, Sweden. We used the Preferred Reporting Items for Systematic reviews and Meta-Analyses, PRISMA 2020, statement [25]. Details on the search strings are available in the Supplementary Information. Details of the systematic review protocol have also been registered in the International Platform of Registered Systematic Review and Meta-analysis Protocols (INPLASY) with registration number INPLASY2022120078.

Additional articles from reference lists of three relevant and recent literature reviews [1,2,9] were also included. Duplicate records were removed so that each reference was only represented once.

2.2. Inclusion and Exclusion Criteria

The records were screened in a two-step procedure (Figures 1 and S1). In the first step, titles and abstracts were screened to identify empirical studies on group- or single-housed male mice, investigating aggression, social dominance, wounds, stereotypies, stress, physiological parameters or details on husbandry, such as group size, cage cleaning procedures or enrichment. All records that matched these criteria were preliminarily included and the reports were assessed for eligibility in full text following set inclusion and exclusion criteria (Supplementary Figure S1). To be included the article had to: (i) be available as full text and written in English; (ii) investigate the effects of group housing of male mice (or male mice and castrated male/female mice); (iii) investigate aggression or social dominance. Studies that did not have a specific aim to measure aggression but still drew conclusions about aggression were included. Studies of wild mice or other rodents than mice were excluded.

2.3. Data Extraction

The methodology, treatment, outcome of experiments and additional relevant information were extracted from the included articles. Each experimental outcome will hereafter be referred to as an observation.

Sometimes, more than one observation was extracted from the same article, e.g., if the same study explored how enrichment affected aggression in two different strains, or if one study investigated the effect of both group size and kinship.

The 90 included articles were divided between two authors, who extracted information independently. The data were then verified by the author who had not performed the data extraction. The complete dataset, with all extracted information, is present in Supplementary Table S1.

2.4. Data Analysis

The author's original conclusions were extracted from the articles and summarized and have in no way been reanalyzed.

The methods used to assess aggression were divided into four categories: home cage observations, test for aggression, wounding and general observation. General observation refers to studies where no specific method could be identified, but the author had made a comment about aggression in the specific experiment.

The treatments used to investigate the effects on aggression were categorized as follows: strain (different strains, substrains or genetically modified strains), enrichment (enriched/non-enriched conditions or different enrichment items used, as well as cage complexity and resource distribution), time spent in group (amount of time spent in a group), group formation (group size, sibling/non-sibling, communal nesting, weaning age, age at group formation, group composition and applying changes to the group composition), housing condition (cage size, density, cage setup, cleaning procedure, female mice close to cage, lightning paradigm, location, scent and identification method) and other (castration, castrated male or female partner, brain weight and general observations where no specific treatment had been applied).

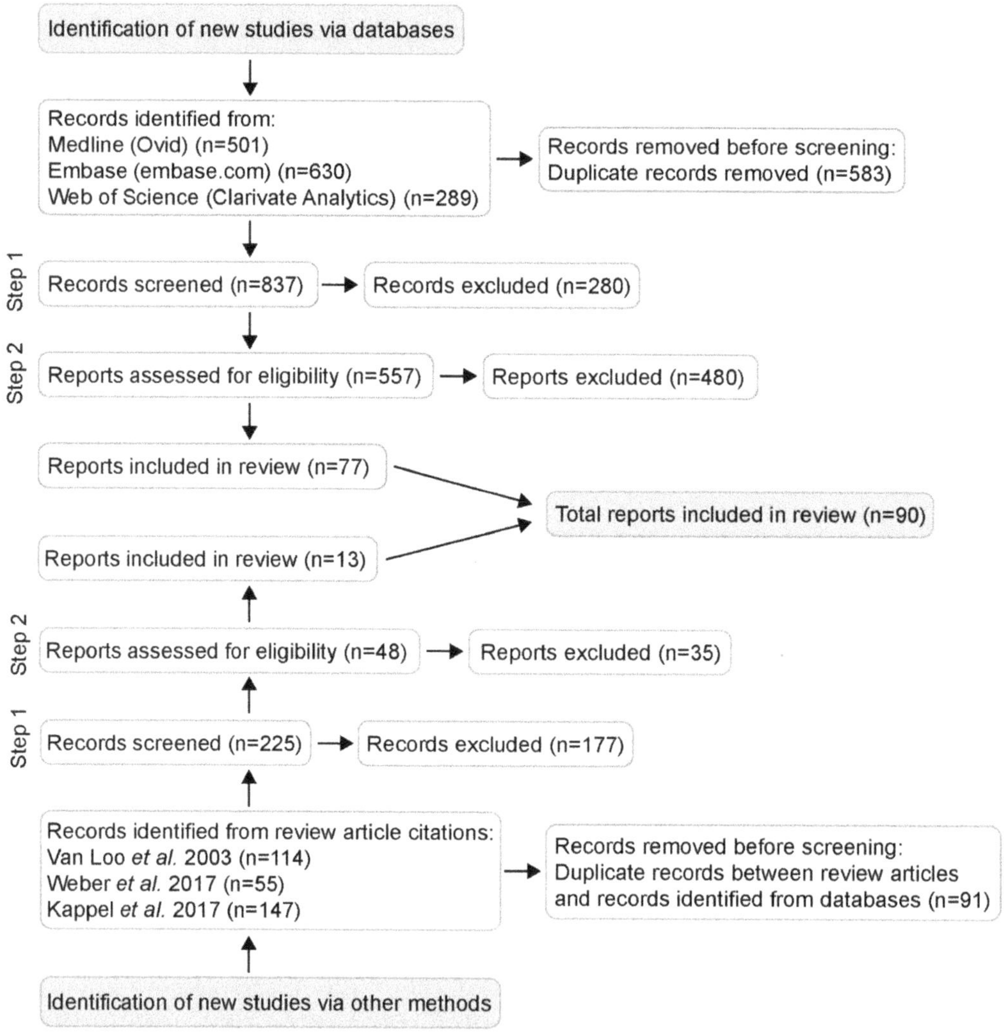

Figure 1. PRISMA flow diagram. Potentially relevant records collected from a systematic literature search using three databases (Medline, Embase and Web of Science) and from scanning the reference lists of three recent literature reviews. Selection performed in two steps; 90 articles suitable for inclusion. Records refer to title, abstract, author keywords and keywords of scientific publications, while reports refer to the full text article.

The enrichment items used were first categorized into five major types, related to which behavioral needs they fulfilled: social (contact, non-contact), occupational (psychological, exercise), physical (cage, accessories), sensory (visual, auditory, other stimuli) and nutritional (delivery, type) [26]. To handle the large variation in enrichment items used and facilitate comparisons within these groups, the enrichments used were then further categorized into the following eight subgroups, according to the supposed purpose of the item: shelter, feed enrichment, nesting material, climbing structure, hiding device, locomotor enrichment, gnawing device and other (Supplementary Tables S3 for categorization and S4 for references). Shelter included items specifically designed to function as a shelter (e.g., nest box), whereas hiding device were structures that mice could hide behind, in or under, but not specifically designed for use as a shelter (e.g., tube).

3. Results

3.1. Search and Study Selection

The database literature search revealed 1420 potentially relevant articles. The reference lists of three recent literature reviews [1,2,9] added another 316, for a total of 1736 articles. After removal of duplicates, 1062 abstracts were screened and 605 articles were preliminarily included. In the full text screening, 90 articles were identified as suitable for inclusion in the systematic review (Figure 1).

3.2. Description of Data Set

In total, 198 observations were extracted from the 90 articles (Supplementary Table S1). Each observation corresponded to the outcome of an experiment with respect to the effect on aggression in group-housed male mice. In most cases, the observations referred to a specific strain and a specific treatment used to affect aggression. Thus, one article, and even one experiment, could include several observations (Table 1). Therefore, comparisons of the outcome on aggression had to be performed among the observations and not between articles.

Table 1. Treatments used to study the effect on aggression. The different treatments are described in detail in the Methods section. One article can contain observations from more than one treatment category.

Treatments	Number of Articles	Number of Observations
Enrichment	33	64
Time in the group	27	38
Strain	26	35
Group formation	23	30
Housing condition	16	23
Other	7	8

Data extracted from the articles were divided into different categories based on different treatments used to study the effect on aggression (Tables 1 and S2). The most common treatment used to study effects on aggression was enrichment, followed by time spent in group, strain and factors related to group formation such as group size and kinship of groups. Details of the housing conditions, such as cage size and cleaning procedure, were less commonly studied. Thirty-one articles contained observations from more than one treatment category.

3.3. Different Methodological Approaches Used to Study Aggression

The methodological approaches used to study aggression can be divided into three categories: home cage observations, test protocols to measure aggression and scoring of wounds. Home cage observations were most common, alone or in combination with other methodological approaches (Figure 2a, Supplementary Table S2). Nineteen articles used methodological approaches from more than one category e.g., home cage observations and wound scoring.

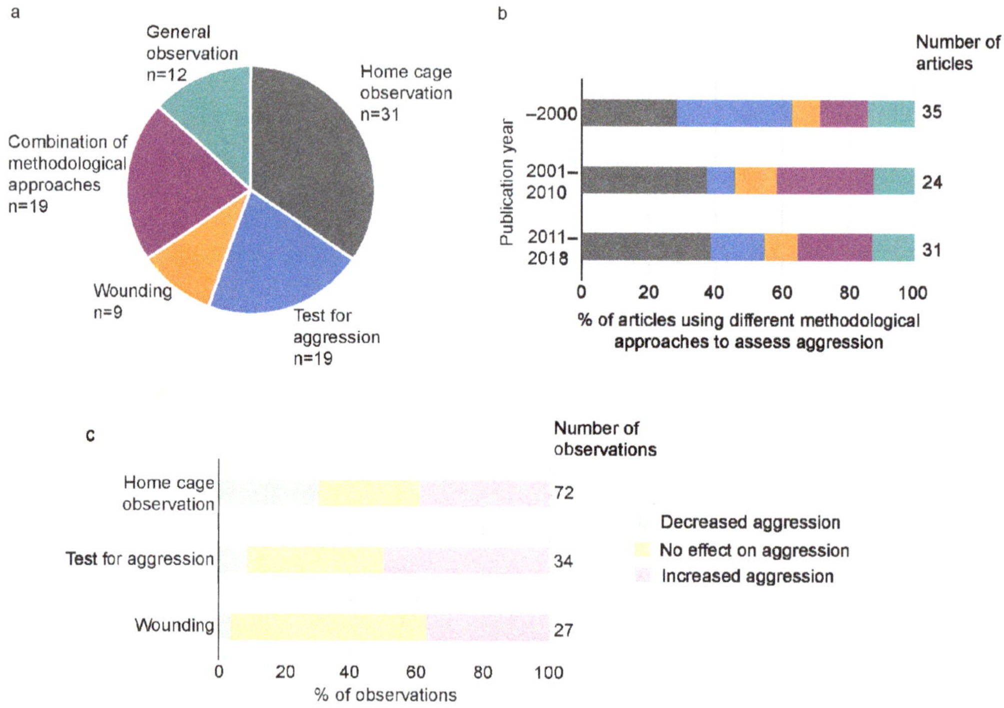

Figure 2. Different methodological approaches to study aggression in the data set. Number of articles (**a**) using the different methodological approaches to assess aggression, and proportion of articles (**b**) using the different methodological approaches before 2000, between 2001 and 2010 and after 2011. The color scheme for the different methodological approaches used in (**a**) also applies to (**b**). The proportion of increased, decreased and no effect on aggression varied between the use of home cage observation, test for aggression and wounding for assessment (**c**). In panel (**c**), observations using more than one methodological approach have been excluded.

The use of different methodological approaches has changed over time (Figure 2b). Home cage observations, alone or in combination with other approaches, have increased in publications after 2001 as well as the practice of using more than one type of method to study aggression. Contrarily, the use of tests to measure aggression as a single methodological approach has decreased.

When using home cage observations as a single approach to study aggression, observations showed decreased aggression, increased aggression and no effect on aggression (Figure 2c). On the other hand, when using tests to measure aggression outside the home cage and wound scoring, as single approaches to study aggression, decreased aggression was seldom observed, while observations of increased aggression or no effect were more commonly reported (Figure 2c). In the sections describing the effects of different treatments below, the methodological approaches used are not described for each observation. Details can be found in Supplementary Tables S1 and S2.

3.4. Details and Description of Housing and Husbandry

Details on cage interiors were included in 70% of the articles published before 2011 and in 80% of the articles published after 2011 (Figure 3). During the period 2001–2018, two-thirds of the articles included information on cleaning routines, compared with only one third of the articles published before 2001. On the contrary, information about kinship—

whether groups were put together with male mice from different litters or with siblings—were included in a larger proportion of the publications before 2011, compared with articles published between 2011–2018 (from 75–78% before 2011 to 68% during 2011–2018; Figure 3).

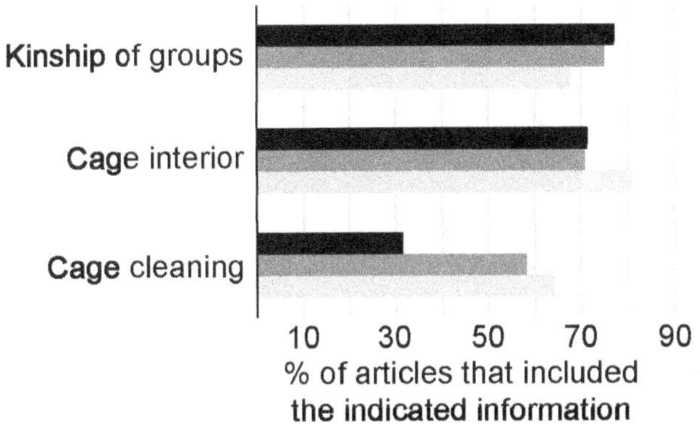

Figure 3. The availability of details on housing and husbandry has changed over time.

3.5. The Effect of Enrichment on Aggression

In many of the articles included in this systematic review, mice were housed in cages with enrichment (Supplementary Table S1). However, the results presented in this section only include papers that specifically studied the effects of enrichment on aggression. In these studies, the standard cages may have had a baseline level of enrichment (not only bedding). Thus, the treatment was defined as additional items added to the enriched cages.

Enrichment was the most commonly used treatment to study effects on aggression (Table 1). More than 40 different enrichment items were used in the different articles (Supplementary Table S3) and different types of enrichment items were often studied together (Supplementary Table S4). The majority of all enrichment items used fell under the major type physical enrichment (38 items in total), including the subgroups nesting material (5 items, 18 observations), shelter (6, 18), hiding device (10, 23), climbing structure (9, 13), locomotor enrichment (1, 10) and other (7, 11). The remaining belonged to nutritional enrichments (4 items in total), including the subgroups feed enrichment (2, 9) and gnawing device (2, 2). Of the eight subgroups, hiding device was the enrichment category that most frequently resulted in decreased aggression (Figure 4). This category also had the highest number of observations, some with only one enrichment item and others where the hiding device was tested in combination with other items.

From all articles that studied enrichment, there were 13 observations of decreased aggression [21,23,24,27–31], 32 where the enrichment did not have any effect [20–23,29,32–41] and 15 observations of increased aggression [18,19,22,32,34,37,42–47]. In addition, there were 4 observations with different outcomes depending on the method used to measure aggression [48–51] (Supplementary Table S4). More than half of the observations of decreased aggression were based on groups formed from non-sibling mice, or substrains of BALB/c mice. Non-sibling groups and groups with BALB/c mice were commonly used also in the observations reporting no effect on aggression. Conversely, sibling groups were more common among the observations of increased aggression and all observations except two were based on other strains than BALB/c. When looking at experiments that investigated the effect of enrichment in C57BL/6, no effect on aggression was most often observed. In one case there was a decrease in aggression, but increased aggression was never observed (Supplementary Table S4).

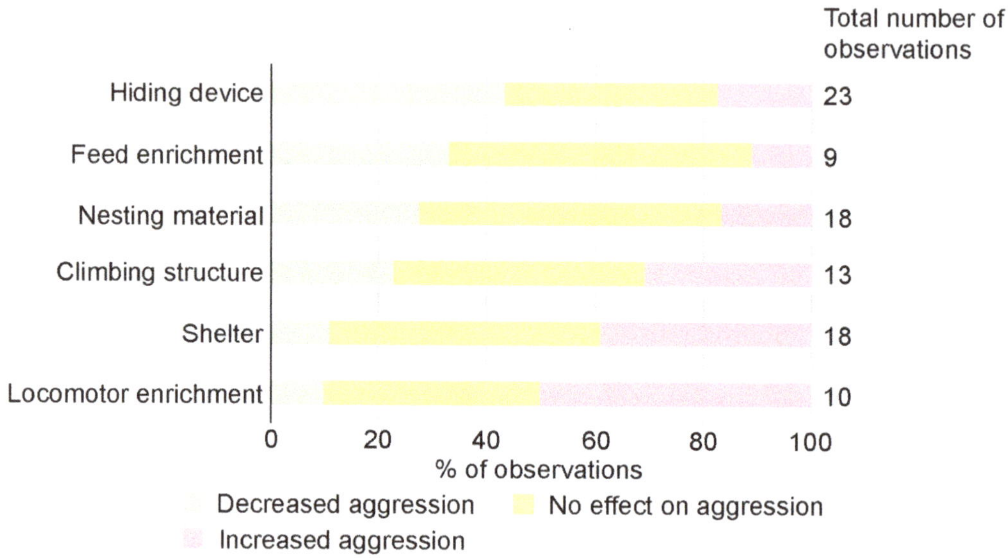

Figure 4. Effect on aggression using different types of enrichment. Note, this includes experiments where either one specific enrichment is used, or when several enrichments are used in combination. In the figure, the category other (11 observations) is excluded due to the variation of enrichments included in that category. Gnawing device is excluded due to a low number of observations (2). Four observations with different outcomes, dependent on the method used to measure aggression, are also excluded.

Cage complexity and resource distribution have also been studied [29,52]. Nadiah et al. (2014) used two connected cages and found that distributing a combination of enrichment items over both cages reduced aggression compared to putting all enrichment items in one cage [29]. Bergmann et al. (1995) studied an enriched setting with one or nine passages to the fodder rack. The enriched condition resulted in an increased proportion of wounded mice and the setting with only one passage to the fodder rack had the highest proportion of injuries [52].

3.6. Aggression Observed over Time

Changes in aggression over time have been studied in multiple articles (Table 1). However, the amount of time the mice were observed varied from a couple of days to weeks or months, and were therefore seldom comparable. With no exceptions, a within approach was used where aggression was measured multiple times and each mouse or group was compared with itself. In addition, in these studies, there were often interacting factors, such as enrichment items in the home cage, which sometimes had a direct influence on the results (Supplementary Table S1). For example, Haemisch et al. in two publications from 1994, reported increased aggression with time in enriched cages but did not see any effect in non-enriched cages [32,48]. Ambrose and Morton (2000), on the other hand, reported increased aggression with time in both enriched and non-enriched cages [24].

From all articles that studied the impact of time spent in a group, there were nine observations of decreased aggression over time [36,53–61], ten with no effect [32,48,59,61–65] and 19 observations of increased aggression [13,18,20,24,32,33,48,57,66–72] (Supplementary Table S5). Aggression as an effect of time has been studied repeatedly in BALB/c [18,20,24,33,36,65,69,70], C57BL/6 [33,61,65] and CD-1 [13,20,67,71] mice. BALB/c and CD-1 were more often observed to become increasingly aggressive with time [13,20,24,33,67,69–71].

3.7. Strain Differences in Aggression

The most commonly used strains in this dataset of 90 articles were BALB/c (24 articles), C57BL/6 (17 articles) and CD-1 (14 articles) (Supplementary Table S2). Other strains were used in ≤7 articles.

Differences in aggression between strains, substrains or genetically modified lines were reported in 34 observations from 26 different articles (Supplementary Table S2). This included 22 observations of comparisons between two strains (Table 2). Similar to the complete dataset, BALB/c and C57BL/6 were the most commonly used strains. C57BL/6 was observed to be less aggressive than another strain on six occasions [23,73–75] and more aggressive only once [33]. BALB/c was observed to be less aggressive than another strain on five occasions [20,33,73,74,76] and more aggressive than another strain on three occasions [73–75]. Six direct comparisons between C57BL/6 and BALB/c were identified; three identified BALB/c as more aggressive [73–75], two did not see any difference [30,65] and one identified C57BL/6 as more aggressive [33]. C57BL/6 and BTBR have also been repeatedly compared but no differences in overall aggression were observed [61,77].

Table 2. Aggressive behavior, comparing two strains. BALB/c, C57BL/6, and CD-1 were most commonly used in the complete dataset and are therefore color-coded in this table; BALB/c in purple, C57BL/6 in pink and CD-1 in grey.

		Methodological Approach Used to Study Aggression				
Less Aggressive	**More Aggressive**	**Home Cage Behavior**	**Test For Aggression**	**Wounding**	**General Observation**	**Reference**
C57BL/6	BALB/c			x		[75]
C57BL/6	BALB/c	x	x			[73]
C57BL/6	BALB/c	x				[74]
C57BL/6	129S [a]		x			[23]
C57BL/6	Swiss	x	x			[73]
C57BL/6	NIH	x				[74]
BALB/c	CD-1 [b]	x		x		[20]
BALB/c	FVB	x			x	[76]
BALB/c	Swiss	x	x			[73]
BALB/c	NIH	x				[74]
BALB/c	C57BL/6	x	x	x		[33]
CBA/J	DBA2		x			[32]
CD-1	BKW	x				[78]
NOD	FVB	x			x	[76]
No Difference		**Home Cage Behavior**	**Test for Aggression**	**Wounding**	**General Observation**	**Reference**
C57BL/6	129S [c]		x			[23]
C57BL/6	BALB/c	x				[30]
C57BL/6	BALB/c	x				[65]
C57BL/6	BTBR	x				[77]
C57BL/6	BTBR	x				[61]
C57BL/6 [d]	BTBR	x				[77]
C57BL/6	DBA/2		x			[79]
BALB/c	NOD	x			x	[76]

[a] In non-enriched cages only. [b] In severity of wounds. BALB/c had higher number of wounds. [c] In cages enriched with a combination of several types of enriched items. [d] More chasing observed in C57BL/6.

Differences in aggression between different substrains of C57BL/6 or BALB/c were studied on two occasions [21,41]. Gaskill et al. (2017) compared six different substrains of C57BL/6 and observed no difference in aggression [41] while Giles et al. (2018) saw more aggression after cage change in BALB/cJ as compared to BALB/cByJ [21].

Differences in aggression between genetically modified strains, compared to the corresponding wildtype, have also been observed [72,80–87]. Out of ten observations, six reported increased aggression in the genetically modified strain [72,83–87]. Martinez-Cue et al. (2005) observed a decrease of aggression in a trisomic strain compared to its disomic control [80] while Trainor et al. (2007) saw no effect on aggression comparing a knockout strain with its wildtype [81]. Lewejohann et al. (2010) observed a decrease in aggression in a homozygous knockout strain but no difference in the heterozygous knockout strain [82]. In addition, Sorensen et al. (2005) saw decreased aggression in a transgenic C57BL/6 strain when housed in mixed groups with wildtype animals [72].

3.8. Size and Group Composition

An effect of group size on aggression was identified in twelve observations from ten articles (Table 3). However, the definition of small or large groups tended to vary between articles and, while small groups usually refer to three or sometimes four or five mice, large groups can refer to five mice all the way up to 10–12 or even 20 mice. Five observations saw the least aggression in small groups [54,59,70,88,89] while six observed the least aggression in large groups [42,90–92].

Table 3. Articles that studied the effects of group size on aggression. For each study, the group size with least aggression is marked in green. Yellow represents no difference between groups.

	Group Size										Methodological Approach Used to Study Aggression				
Strain	**2**	**3**	**4**	**5**	**6**	**8**	**9**	**10**	**12**	**20**	**Home Cage Behavior**	**Test for Aggression**	**Wounding**	**General Observation**	**Reference**
BALB/c		x		x		x					x		x		[70]
BALB/c		x		x		x								x	[88]
Varying strains	x	x	x	x							x		x		[89]
Randomly bred		x	x	x			x		x		x				[54]
DD/S strain		x		x							x				[59]
MF1		x			x						x				[38]
BALB/c		x					x				x				[92]
CFLP		x			x			x			x				[42]
CFLP		x			x			x			x				[91]
DUB/ICR Swiss	x			x				x			x	x			[90]
DUB/ICR Swiss	x		x			x				x	x	x			[90]
DUB/ICR Swiss	x			x				x			x	x			[90]

In this dataset, 65 articles presented information on kinship, whether groups were formed from sibling or mice from different litters (Figure 3, Supplementary Table S2). It was more than twice as common to form groups from non-siblings as compared to siblings: 41 and 16 articles respectively. Only six articles studied the effects of group formation on aggression [58,59,63,64,69,93]. In four articles, no effect on aggression was observed when comparing groups of siblings with groups of mice from different litters [59,63,64,69]. In the remaining two articles, one using wild type CD-1 and the other a transgenic line of CD-1, there was increased aggression in the non-sibling groups [58,93].

Bartolomucci et al. (2004) studied group formation from non-sibling CD-1 mice at weaning or adolescence and observed increased aggression in groups formed from juvenile CD-1 mice [93]. The effects of other early life events, such as weaning age [41,94] and nesting condition (standard or communal) [95], were also studied. The two articles studying weaning age showed contrary results; one observed increased aggression after early weaning [94] while the other observed increased aggression after late weaning [41]. When comparing standard or communal nesting, increased aggression was reported in communal nesting [95].

The effects of regrouping mice were studied in seven articles [41,89,94,96–99]. The group formation in the experiments varied widely and included single housing at delivery to be grouped later [96], rearranging group-housed mice into new groups [41,94,97,98] and removal of alpha males [99]. Only one of these treatments, breaking up groups of fighting mice into smaller groups with wounded mice only, resulted in decreased aggression [89].

3.9. Housing Conditions and Male Mouse Aggression

In this dataset, procedures and factors that related to housing conditions were the least studied category of treatments that could affect aggression (Table 1).

Aggression in relation to different cleaning procedures was studied in four articles [24,69,72,100]. Transferring the mice to a clean cage and moving nesting material to the new cage reduced aggression [69,100]. Placing clean sawdust in a soiled home cage and moving soiled sawdust or enrichment items to the clean cage increased aggression [24,69,100].

Another aspect that relates to housing conditions is the cage size, or density in the cage. Eleven observations from seven articles were related to this aspect: three defined as density [38,75] and eight as cage size [43,55,70,88,101]. All three observations of density, as well as three observations of cage size, showed no effect on aggression [38,43,75,101]. These six observations were made in six different strains. Increased aggression with increasing cage size was observed three times [70,76,88]. Two of these observations were based on BALB/c [70,88]. Contrarily, Poole and Morgan (1976) observed decreased aggression with increasing cage size in LACA/CFW mice, both comparing different groups and from within analysis of the same mice in different-sized cages [55].

Other practices relating to housing conditions have also been studied [36,41,102]. Disturbed lighting and having female mice close to the male mice cage increased aggression [36,102]. Gaskill et al. (2017) studied several parameters of the housing condition and observed increased aggression in mice identified by ear notch as compared to tail tattoo and increased aggression in lavender-scented cages, as well as a difference in aggression between racks. No differences were observed within the same rack [41].

4. Discussion

In the articles included in this systematic review, researchers investigated how aggression among group-housed male mice was affected by environmental enrichment, strain, age, housing conditions, group formation and time spent in groups. In all treatment categories, there were observations of decreased aggression, increased aggression, as well as no effect on aggression. When compiling the results, it was, in many cases, not possible to single out a clear effect of a specific treatment because of potential bias presented from other parameters. For example, effects of enrichment in the home cage could be affected by characteristics of the strain tested, the amount of time the mice were grouped together before the study or the length of the study itself. Altogether, this confirmed what has been pointed out in previous review articles on aggression in male mice [1,2]: that the problem is complex and that it is unlikely that one solution will fit every situation. We also found indications that the method used to assess aggression could influence the outcome, further complicating interpretation of results. Still, aggression was decreased when using certain environmental enrichment items and certain strains.

4.1. Methods Used for Assessment of Aggressive Behaviour

In the included articles, assessments of aggression among group-housed male mice were performed by observations in the home cage, using tests for aggression outside the home cage, by registration of wounds, or by a combination of these methods. Home cage observations of aggression were used in more than half of the articles and this approach has been more frequently used in recent years (Figure 2b). Studying aggressive behavior in the home cage is likely more relevant for understanding aggression under normal husbandry than using aggression tests outside the home cage, as the fighting observed in the home cage will be linked to social behavior in the group, whereas fighting in a novel environment or with an intruder captures aggressive behavior in relation to those factors.

Many conclusions of aggression in male mice have historically been drawn from assessment of aggressive behavior outside the home cage, using tests such as the resident intruder test and social defeat test (Figure 2b). This approach was used, alone or in combination, in about one third of the articles in the present systematic review (Supplementary

Table S2). Not surprisingly, decreased aggression was rarely seen as an outcome in articles where the aggression was measured outside the home cage (Figure 2c). The large proportion of increased aggression seen when studying aggression outside the home cage could be a result of the method itself rather than a treatment effect, since the purpose of these tests is to provoke aggression. Aggression assessed in the home cage, on the other hand, resulted in decreased, increased or no difference in aggression, evenly distributed between the included articles. The importance of using reliable methods relevant for assessing aggression in the home cage when developing/evaluating preventative measures for group housing male mice was therefore elucidated in this systematic review. The observed effect of the treatments used to study the effects on aggression varied between articles regarding strain, environmental enrichment and group formation. In some cases, the results were contradictory even when using the same comparison/hypothesis. It is thus possible that the increased aggressive behavior was a result of the aggression test methods rather than true effects of the specific treatment.

4.2. Environmental Enrichment

Enrichment is commonly used as a means to improve animal welfare and for refinement of animal use in research [103,104]. However, general recommendations on how enrichment should be used to prevent aggression are not straightforward, as enrichment has been reported to have both positive results, negative results or no impact on aggression [2,17]. This was consistent with the results in this review; in almost half of the observations, enrichment had neither positive nor negative observed effects on aggression. Interestingly, enrichment increased aggression in only about 20% of the observations. It is worth mentioning that some research groups are reluctant to use enrichments because of a strong belief that enrichments will increase aggression. The presented results indicate that the type of enrichment used was important, e.g., hiding devices, feed enrichment and nesting materials, used alone or in combination with other enrichments, were associated with decreased aggression in male mouse groups (Figure 4). These enrichments resulted in increased aggression in only seven out of 50 observations (Supplementary Table S4). Enrichment items that mice can interact and work with, such as nesting material, often decrease aggression and are generally included in advice for reducing aggression among male mice [3,4,9,10]. In a study based on data from group-housed male mice in 40 animal facilities by Lidster et al. (2019), transfer of clean and dry nesting material at cage cleaning had a clear positive effect, with reduced aggression [3]. Transfer of nesting material was also perceived as the most effective approach to prevent fighting and aggression in male mouse groups when asking animal technicians and researchers at Swedish universities at workshops and in a survey [4]. The workshop and survey respondents also mentioned adding extra nesting material as efficient. In other species, such as non-human primates, visual barriers were recommended to minimize aggression when housing animals in social groups [105]. It could be that providing hiding devices to allow mice to remove themselves from sight of one another mimics the natural behavior response of fleeing when threatened by a conspecific [2,41], thus decreasing aggression in mice.

In the present review, locomotor enrichment (wheel) and shelter were associated with increased aggression in more than half of the observations. Interestingly, a combination of shelter and wheel were used in five of these nine observations. Mice can defend and monopolize valuable physical enrichment items, and only a limited number of items can fit in a conventional mouse cage. Lower levels of aggression when enrichment items were dispersed was found in both male [29] and female [106] mice. It might be that the shelter and wheel were only accessible for one or a few mice at a time, especially when the wheel was connected to the house, resulting in mice displaying resource aggression. If there were only one entrance, it might also be difficult to escape an aggressive encounter, with the risk of escalating aggression as a consequence. This could explain why increased aggressive behavior was found when these types of enrichments were used.

Interestingly, the results presented herein could indicate that environmental enrichment had a more pronounced positive effect when used in non-sibling male groups. There were no observations of decreased aggression in sibling groups when assessing effects of environmental enrichment. This could be due to the overall lower levels of aggression in groups of siblings. On the contrary, increased aggression in enriched environments was observed for both sibling and non-sibling groups (Supplementary Table S4).

Recommendations to prevent aggression that considered housing conditions for mice include transfer of nesting material at cage cleaning [3,4,9,10], avoiding disturbances and handling mice with care [9]. However, there are no clear recommendations for enrichment. This could be due to the wide variety of enrichment items and combinations of different items studied. In the reviewed studies, 42 different enrichment items were used in combination with different strains and group constellations, making it very difficult to interpret results and give clear recommendations. Many parameters of normal husbandry can affect aggression in group-housed laboratory mice. Some, such as enrichment, are often represented in empirical studies. Despite this, systematic evaluation of many enrichment items is lacking. Experienced personnel working with mice on a daily basis learn, through trial and error, what type of enrichments reduce or trigger aggression—but this type of information is rarely collected systematically by researchers. Information on basic husbandry and routines were missing in many articles in the present review. A complete methods section is a prerequisite to be able to repeat experiments, or to implement suggested practices in normal husbandry. The ARRIVE guidelines [107], originally published in 2010 [108], should ensure that studies are reported in enough detail. However, several important parameters were often missing from the reviewed studies.

4.3. Strain

A previously published recommendation advises the use of strains with low prevalence of aggression whenever possible [3]. In a study from 2019, Lidster et al. collected significant data on incidents of aggressive behavior in group-housed mice at animal facilities and determined strain to be one of the key factors influencing prevalence of aggression. They concluded that C57BL/6 and BALB/c were the two strains that showed lower prevalence of aggression [3]. Among the herein included articles, similar tendencies were found. For example, C57BL/6 mice were less aggressive than their counterparts in all but one study investigating the difference in aggression between strains. BALB/c mice also showed low levels of aggression when compared to other strains. In fact, BALB/c mice were more aggressive only when compared with C57BL/6, indicating that both these strains showed low levels of aggression. Contrary to these results, C57BL/6 has sometimes been reported as an aggressive strain when doing surveys and workshops at animal facilities [1,4]. Because C57BL/6 is a commonly used strain, it is possible that this conception among survey respondents reflected the large number of C57BL/6 housed at facilities, rather than it being a more aggressive strain. That being said, if animal technicians experience aggression among these strains, this should be taken seriously. It is possible that these strains are less aggressive overall and, under certain conditions, display low levels of aggression, but that aggression can be triggered in other circumstances. When choosing what strain to use for animal experiments, we still know very little about what triggers aggression in the various strains. Empirical studies focused on comparing aggression among strains have shown somewhat inconclusive results (Tables Table 2 and S2). Considering that studies vary in a large number of factors other than strain—differences in number of animals per cage, age, cage size, enrichment items, etc.—it was difficult to interpret results and impossible to draw any major conclusions based on this dataset alone. Researchers must explore how aggressive behavior in various strains is influenced by environmental factors. For example, among the included articles herein, we noted a repeated negative relationship between cage size and aggression in BALB/c, something that could be further explored. Preferably, we would like to see systematic studies comparing multiple strains under varying environmental conditions to help us better understand variations in aggressive behavior and how it is influenced by other factors in group-housed male mice.

4.4. Group Formation

General recommendations exist, aimed at preventing and minimizing aggression in group-housed male mice. These are related to group formation, such as housing in small groups formed from siblings kept together or familiar mice grouped before sexual maturity [3,9,10]. Recent publications also suggested that male mice can be grouped with unfamiliar mice at weaning [4], shortly after weaning with one-week age difference between males [109], or before sexual maturity [110]. In the present review, there was only one article evaluating age-related aggression or grouping before and after sexual maturity. This article reported increased aggression in groups formed from unfamiliar mice at weaning, as compared to groups formed from littermates or groups formed from somewhat older unfamiliar mice [93]. However, time spent in groups was also studied in some articles, with a time frame varying from a few days up to 40 weeks (Supplementary Table S5). Scattered results were observed, including increased aggression, decreased aggression, and no observed change in aggression, indicating that the incidence of aggressive behavior was influenced by factors other than time spent together.

In the present review, there were inconclusive results on preferable group size for male mice. Decreased aggression was observed in smaller groups with 2–5 mice per group as well as in larger groups of 10–20 mice per group (Table 3). The literature on group size was divided; there were recommendations for 3 mice per group in some cases (e.g., [70]), while the incidence of aggression was significantly higher in groups of three, compared to groups of 4 and 5, in others [3].

4.5. Recommendations Based on Results from the Systematic Review

Based on this literature review, a few recommendations stand out as promising tools to minimize aggression among group-housed male mice and should be considered.

- Enrichment: Use nesting material and other enrichment items that the mice can manipulate and work with. Use hiding devices where the mice can hide. Be careful when using enrichments that can be monopolized by one or a few individuals Note that environmental enrichment seem to be even more important in non-sibling groups.
- Strain: Use less aggressive strains whenever possible. This literature review indicates that C57BL/6 and BALB/c show low prevalence of aggression, in line with recently published advice [3].
- Solutions may vary: Diverging and varying opinions exist between research facilities concerning what works to prevent aggression. Try different solutions and adjust to specific needs in your lab. Be aware that the same solution might not work for all strains and research projects.
- Methods used to study aggression: Be aware that the method used to study aggression could influence the relevance of the results for home cage aggression. It is therefore important to evaluate the method used to study aggression before drawing conclusions on whether or not the reported results should be implemented in daily practice. Recent studies using home cage observations could provide more reliable information.

4.6. Important Notes for Future Research

To facilitate the understanding of home cage aggression, the most relevant approach is to study aggression in groups of male mice using home cage observations, where the social group is kept intact (and not by aggression tests outside the home cage with unfamiliar mice).

It is important to include all data regarding housing and husbandry in publications so that results can be interpreted correctly, and comparisons can be made between studies. Follow the ARRIVE guidelines.

More systematic studies are needed, using various strains for the same treatments and housing conditions and the same methods to study aggression.

5. Conclusions

Aggression among group-housed male mice is a major welfare concern. Identifying factors that prevent aggression and thus enable mice to be held in stable social groups will not only minimize pain and suffering but also contribute to a reduction in the number of animals used in research.

The keys to success for male mouse group-housing are multifactorial. There is no one solution that fits all; solutions can and will vary and must be adjusted depending on animal facilities and research areas. Finding the factors that work at a given facility may take time and several factors should be evaluated before resorting to single housing. In this literature review, a number of recommendations have been identified as promising tools to minimize aggression among group-housed male mice. Taken together with other current information from the literature, this study could be an important complement when developing guidelines on how to prevent aggression when male mice are housed in groups.

Supplementary Materials: The following supporting information can be downloaded at: https://www.mdpi.com/article/10.3390/ani13010143/s1, Supplementary information: Search strings; Figure S1: Outline of the screening strategy with inclusion and exclusion criteria; Table S1: Raw data - information extracted from all included articles; Table S2: Strains, treatments and methods used to study aggression in all included articles; Table S3: Enrichment items used in the articles that specifically studied enrichment; Table S4 Articles that study the effect of enrichment items on aggression; Table S5: Articles that study the effect of time spent in group.

Author Contributions: The study was designed by E.T. and J.Z. who planned the systematic search with support from the Karolinska Institutet. J.Z. and E.M.W. determined the inclusion and exclusion criteria with input from E.T. and B.E. Screening of records was conducted by E.S. and E.U. and screening of reports was conducted by K.A. and E.S., in both cases with support from E.M.W. and J.Z., K.A. and E.S. extracted data from the selected articles. The paper was written by E.M.W., J.Z., K.A., E.S. and E.T. and commented on by E.U. and B.E. All authors have read and agreed to the published version of the manuscript.

Funding: This research received no external funding.

Institutional Review Board Statement: Not applicable.

Informed Consent Statement: Not applicable.

Data Availability Statement: All data generated or analyzed during this study are included in this published article (and its supplementary information files).

Acknowledgments: We are grateful to librarians at the Karolinska Institutet University Library for professional help with the systematic search. Additionally, sincere thanks to Erika Roman for valuable comments on the manuscript.

Conflicts of Interest: The authors declare no conflict of interest.

References

1. Kappel, S.; Hawkins, P.; Mendl, M.T. To Group or Not to Group? Good Practice for Housing Male Laboratory Mice. *Anim. Open Access J. MDPI* **2017**, *7*, 88. [CrossRef] [PubMed]
2. Weber, E.M.; Dallaire, J.A.; Gaskill, B.N.; Pritchett-Corning, K.R.; Garner, J.P. Aggression in group-housed laboratory mice: Why can't we solve the problem? *Lab Anim.* **2017**, *46*, 157–161. [CrossRef] [PubMed]
3. Lidster, K.; Owen, K.; Browne, W.J.; Prescott, M.J. Cage aggression in group-housed laboratory male mice: An international data crowdsourcing project. *Sci. Rep.* **2019**, *9*, 15211. [CrossRef] [PubMed]
4. Zidar, J.; Weber, E.M.; Ewaldsson, B.; Tjäder, S.; Lilja, J.; Mount, J.; Svensson, C.; Svensk, E.; Udén, E.; Törnqvist, A.E. Group and Single Housing of Male Mice: Collected Experiences from Research Facilities in Sweden. *Animals* **2019**, *9*, 1010. [CrossRef]
5. Theil, J.H.; Ahloy-Dallaire, J.; Weber, E.M.; Gaskill, B.N.; Pritchett-Corning, K.R.; Felt, S.A.; Garner, J.P. The epidemiology of fighting in group-housed laboratory mice. *Sci. Rep.* **2020**, *10*, 16649. [CrossRef]

6. EN. Directive 2010/63/EU of the European Parliament and of the Council of 22 September 2010 on the Protection of Animals Used for Scientific Purposes. *Off. J. Eur. Union* **2010**, *276*, 33–76.
7. National Research Council (US) Committee for the Update of the Guide for the Care and Use of Laboratory Animals. *Guide for the Care and Use of Laboratory Animals*, 8th ed.; National Academies Press: Washington, DC, USA, 2011. [CrossRef]
8. UK Home Office Code of Practice for the Housing and Care of Animals Bred, Supplied or Used for Scientific Purposes Rodents and Rabbits. Available online: https://assets.publishing.service.gov.uk/government/uploads/system/uploads/attachment_data/file/389807/44389_unact_animals_Rodents_and_Rabbits.pdf (accessed on 23 December 2022).
9. Van Loo, P.L.P.; Van Zutphen, L.F.M.; Baumans, V. Male management: Coping with aggression problems in male laboratory mice. *Lab. Anim.* **2003**, *37*, 300–313. [CrossRef]
10. Gaskill, B. Aggression in Laboratory Mice: Potential Influences and How to Manage It. *Enrich. Rec.* **2014**, *18*, 22–25.
11. Koolhaas, J.M.; Coppens, C.M.; de Boer, S.F.; Buwalda, B.; Meerlo, P.; Timmermans, P.J. The resident-intruder paradigm: A standardized test for aggression, violence and social stress. *J. Vis. Exp. JoVE* **2013**, e4367. [CrossRef]
12. Golden, S.A.; Covington, H.E.; Berton, O.; Russo, S.J. A standardized protocol for repeated social defeat stress in mice. *Nat. Protoc.* **2011**, *6*, 1183–1191. [CrossRef]
13. Terranova, M.L.; Laviola, G.; de Acetis, L.; Alleva, E. A description of the ontogeny of mouse agonistic behavior. *J. Comp. Psychol.* **1998**, *112*, 3–12. [CrossRef] [PubMed]
14. Lindzey, G.; Winston, H.; Manosevitz, M. Social dominance in inbred mouse strains. *Nature* **1961**, *191*, 474–476. [CrossRef] [PubMed]
15. Kwiatkowski, C.C.; Akaeze, H.; Ndlebe, I.; Goodwin, N.; Eagle, A.L.; Moon, K.; Bender, A.R.; Golden, S.A.; Robison, A.J. Quantitative standardization of resident mouse behavior for studies of aggression and social defeat. *Neuropsychopharmacology* **2021**, *46*, 1584–1593. [CrossRef] [PubMed]
16. Gaskill, B.N.; Stottler, A.; Pritchett-Corning, K.R.; Wong, L.K.; Geronimo, J.; Garner, J.P. He's getting under my skin! Comparing the sensitivity and specificity of dermal vs subcuticular lesions as a measure of aggression in mice. *Appl. Anim. Behav. Sci.* **2016**, *183*, 77–85. [CrossRef]
17. Olsson, I.A.; Dahlborn, K. Improving housing conditions for laboratory mice: A review of "environmental enrichment". *Lab. Anim.* **2002**, *36*, 243–270. [CrossRef]
18. Van Loo, P.L.P.; Kruitwagen, C.; Koolhaas, J.M.; Van de Weerd, H.A.; Van Zutphen, L.F.M.; Baumans, V. Influence of cage enrichment on aggressive behaviour and physiological parameters in male mice. *Appl. Anim. Behav. Sci.* **2002**, *76*, 65–81. [CrossRef]
19. Howerton, C.L.; Garner, J.P.; Mench, J.A. Effects of a running wheel-igloo enrichment on aggression, hierarchy linearity, and stereotypy in group-housed male CD-1 (ICR) mice. *Appl. Anim. Behav. Sci.* **2008**, *115*, 90–103. [CrossRef]
20. Van Loo, P.L.P.; Van der Meer, E.; Kruitwagen, C.; Koolhaas, J.M.; Van Zutphen, L.F.M.; Baumans, V. Strain-specific aggressive behavior of male mice submitted to different husbandry procedures. *Aggress. Behav.* **2003**, *29*, 69–80. [CrossRef]
21. Giles, J.M.; Whitaker, J.W.; Moy, S.S.; Fletcher, C.A. Effect of environmental enrichment on aggression in BALB/cJ and BALB/cByJ mice monitored by using an automated system. *J. Am. Assoc. Lab. Anim. Sci.* **2018**, *57*, 236–243. [CrossRef]
22. Swetter, B.J.; Karpiak, C.P.; Cannon, J.T. Separating the effects of shelter from additional cage enhancements for group-housed BALB/cJ mice. *Neurosci. Lett.* **2011**, *495*, 205–209. [CrossRef]
23. Abramov, U.; Puussaar, T.; Raud, S.; Kurrikoff, K.; Vasar, E. Behavioural differences between C57BL/6 and 129S6/SvEv strains are reinforced by environmental enrichment. *Neurosci. Lett.* **2008**, *443*, 223–227. [CrossRef] [PubMed]
24. Ambrose, N.; Morton, D.B. The Use of Cage Enrichment to Reduce Male Mouse Aggression. *J. Appl. Anim. Welf. Sci.* **2000**, *3*, 117–125. [CrossRef]
25. Page, M.J.; McKenzie, J.E.; Bossuyt, P.M.; Boutron, I.; Hoffmann, T.C.; Mulrow, C.D.; Shamseer, L.; Tetzlaff, J.M.; Akl, E.A.; Brennan, S.E.; et al. The PRISMA 2020 statement: An updated guideline for reporting systematic reviews. *BMJ* **2021**, *372*, n71. [CrossRef]
26. Bloomsmith, M.A.; Brent, L.Y.; Schapiro, S.J. Guidelines for developing and managing an environmental enrichment program for nonhuman primates. *Lab. Anim. Sci.* **1991**, *41*, 372–377. [PubMed]
27. Armstrong, K.R.; Clark, T.R.; Peterson, M.R. Use of Corn-Husk Nesting Material to Reduce Aggression in Caged Mice. *Contemp. Top Lab. Anim. Sci.* **1998**, *37*, 64–66.
28. Lima, F.B.; Spinelli de Oliveira, E. What is the impact of low testosterone levels on the anatomical and behavioral repertoire of long-term enriched housing of male mice? *Behav. Process.* **2014**, *108*, 57–64. [CrossRef]
29. Nadiah, M.Y.N.; Saadiah, M.B.H.; Nurdiana, S. Effects of Different Resource Distribution on Behaviour and Corticosterone Level of Male Mice (*Mus musculus*). *APCBEE Procedia* **2014**, *8*, 189–193. [CrossRef]
30. Clipperton-Allen, A.E.; Ingrao, J.C.; Ruggiero, L.; Batista, L.; Ovari, J.; Hammermueller, J.; Armstrong, J.N.; Bienzle, D.; Choleris, E.; Turner, P.V. Long-Term Provision of Environmental Resources Alters Behavior but not Physiology or Neuroanatomy of Male and Female BALB/c and C57BL/6 Mice. *J. Am. Assoc. Lab. Anim. Sci. JAALAS* **2015**, *54*, 718–730.
31. Nakamura, Y.; Ueno, A.; Nunomura, Y.; Nakagaki, K.; Takeda, S.; Suzuki, K. Effects of inducing exercise on growing mice by means of three-dimensional structure in rearing environment. *Exp. Anim.* **2016**, *65*, 403–411. [CrossRef]

32. Haemisch, A.; Gartner, K. The cage design affects intermale aggression in small groups of male laboratory mice: Strain specific consequences on social organization, and endocrine activations in two inbred strains (DBA/2J and CBA/J). *J. Exp. Anim. Sci.* **1994**, *36*, 101–116.
33. Eskola, S.; Kaliste-Korhonen, E. Aspen wood-wool is preferred as a resting place, but does not affect intracage fighting of male BALB/c and C57BL/6J mice. *Lab. Anim.* **1999**, *33*, 108–121. [CrossRef] [PubMed]
34. Marashi, V.; Barnekow, A.; Ossendorf, E.; Sachser, N. Effects of different forms of environmental enrichment on behavioral, endocrinological, and immunological parameters in male mice. *Horm. Behav.* **2003**, *43*, 281–292. [CrossRef] [PubMed]
35. Marashi, V.; Barnekow, A.; Sachser, N. Effects of environmental enrichment on males of a docile inbred strain of mice. *Physiol. Behav.* **2004**, *82*, 765–776. [CrossRef] [PubMed]
36. Van der Meer, E.; Van Loo, P.L.P.; Baumans, V. Short-term effects of a disturbed light-dark cycle and environmental enrichment on aggression and stress-related parameters in male mice. *Lab. Anim.* **2004**, *38*, 376–383. [CrossRef]
37. Kaliste, E.K.; Mering, S.M.; Huuskonen, H.K. Environmental modification and agonistic behavior in NIH/S male mice: Nesting material enhances fighting but shelters prevent it. *Comp. Med.* **2006**, *56*, 202–208.
38. Hunt, C.; Hambly, C. Faecal corticosterone concentrations indicate that separately housed male mice are not more stressed than group housed males. *Physiol. Behav.* **2006**, *87*, 519–526. [CrossRef]
39. Workman, J.L.; Fonken, L.K.; Gusfa, J.; Kassouf, K.M.; Nelson, R.J. Post-weaning environmental enrichment alters affective responses and interacts with behavioral testing to alter nNOS immunoreactivity. *Pharmacol. Biochem. Behav.* **2011**, *100*, 25–32. [CrossRef]
40. Lockworth, C.R.; Kim, S.-J.; Liu, J.; Palla, S.L.; Craig, S.L. Effect of Enrichment Devices on Aggression in Manipulated Nude Mice. *J. Am. Assoc. Lab. Anim. Sci. JAALAS* **2015**, *54*, 731–736.
41. Gaskill, B.N.; Stottler, A.M.; Garner, J.P.; Winnicker, C.W.; Mulder, G.B.; Pritchett-Corning, K.R. The effect of early life experience, environment, and genetic factors on spontaneous home-cage aggression-related wounding in male C57BL/6 mice. *Lab Anim.* **2017**, *46*, 176–184. [CrossRef]
42. Barnard, C.J.; Behnke, J.M.; Sewell, J. Environmental enrichment, immunocompetence, and resistance to Babesia microti in male mice. *Physiol. Behav.* **1996**, *60*, 1223–1231. [CrossRef]
43. McGregor, P.K.; Ayling, S.J. Varied cages result in more aggression in male CFLP mice. *Appl. Anim. Behav. Sci.* **1990**, *26*, 277–281. [CrossRef]
44. Haemisch, A.; Gartner, K. Effects of cage enrichment on territorial aggression and stress physiology in male laboratory mice. *Acta Physiol. Scand. Suppl.* **1997**, *640*, 73–76.
45. McQuaid, R.J.; Audet, M.C.; Anisman, H. Environmental enrichment in male CD-1 mice promotes aggressive behaviors and elevated corticosterone and brain norepinephrine activity in response to a mild stressor. *Stress-Int. J. Biol. Stress* **2012**, *15*, 354–360. [CrossRef] [PubMed]
46. Mesa-Gresa, P.; Perez-Martinez, A.; Redolat, R. Environmental enrichment improves novel object recognition and enhances agonistic behavior in male mice. *Aggress. Behav.* **2013**, *39*, 269–279. [CrossRef] [PubMed]
47. Leduc, R.Y.M.; Rauw, G.; Baker, G.B.; McDermid, H.E. What goes around can come around: An unexpected deleterious effect of using mouse running wheels for environmental enrichment. *J. Am. Assoc. Lab. Anim. Sci.* **2017**, *56*, 194–201.
48. Haemisch, A.; Voss, T.; Gartner, K. Effects of environmental enrichment on aggressive behavior, dominance hierarchies, and endocrine states in male DBA/2J mice. *Physiol. Behav.* **1994**, *56*, 1041–1048. [CrossRef]
49. McQuaid, R.J.; Audet, M.-C.; Jacobson-Pick, S.; Anisman, H. Environmental enrichment influences brain cytokine variations elicited by social defeat in mice. *Psychoneuroendocrinology* **2013**, *38*, 987–996. [CrossRef]
50. McQuaid, R.J.; Audet, M.-C.; Jacobson-Pick, S.; Anisman, H. The differential impact of social defeat on mice living in isolation or groups in an enriched environment: Plasma corticosterone and monoamine variations. *Int. J. Neuropsychopharmacol.* **2013**, *16*, 351–363. [CrossRef]
51. McQuaid, R.J.; Dunn, R.; Jacobson-Pick, S.; Anisman, H.; Audet, M.-C. Post-weaning Environmental Enrichment in Male CD-1 Mice: Impact on Social Behaviors, Corticosterone Levels and Prefrontal Cytokine Expression in Adulthood. *Front. Behav. Neurosci.* **2018**, *12*, 145. [CrossRef]
52. Bergmann, P.; Militzer, K.; Buttner, D. Environmental enrichment and aggressive behaviour: Influence on body weight and body fat in male inbred HLG mice. *J. Exp. Anim. Sci.* **1995**, *37*, 69–78.
53. Page, D.L.; Glenner, G.G. Social interaction and wounding in the genesis of "spontaneous" murine amyloidosis. *Am. J. Pathol.* **1972**, *67*, 555–567.
54. Poole, T.B.; Morgan, H.D.R. Differences in aggressive behaviour between male mice (*Mus musculus* L.) in colonies of different sizes. *Anim. Behav.* **1973**, *21*, 788–795. [CrossRef] [PubMed]
55. Poole, T.B.; Morgan, H.D.R. Social and territorial behaviour of laboratory mice (*Mus musculus* L.) in small complex areas. *Anim. Behav.* **1976**, *24*, 476–480. [CrossRef]
56. Bishop, M.J.; Chevins, P.F.D. Urine odours and marking patterns in territorial laboratory mice (*Mus musculus*). *Behav. Process.* **1987**, *15*, 233–248. [CrossRef] [PubMed]
57. Burright, R.G.; Freeman, M.J.; Donovick, P.J. Repeated tests of intermale aggression in mice (*Mus musculus*) are influenced by housing and test conditions. *J. Comp. Psychol.* **1988**, *102*, 303–305. [CrossRef]

58. Hilakivi-Clarke, L.; Dickson, R.B. Stress influence on development of hepatocellular tumors in transgenic mice overexpressing TGF alpha. *Acta Oncol.* **1995**, *34*, 907–912. [CrossRef]
59. Grimm, M.S.; Emerman, J.T.; Weinberg, J. Effects of social housing condition and behavior on growth of the Shionogi mouse mammary carcinoma. *Physiol. Behav.* **1996**, *59*, 633–642. [CrossRef]
60. Vaughan, L.M.; Dawson, J.S.; Porter, P.R.; Whittaker, A.L. Castration promotes welfare in group-housed male Swiss outbred mice maintained in educational institutions. *J. Am. Assoc. Lab. Anim. Sci. JAALAS* **2014**, *53*, 38–43.
61. Bove, M.; Ike, K.; Eldering, A.; Buwalda, B.; de Boer, S.F.; Morgese, M.G.; Schiavone, S.; Cuomo, V.; Trabace, L.; Kas, M.J.H. The Visible Burrow System: A behavioral paradigm to assess sociability and social withdrawal in BTBR and C57BL/6J mice strains. *Behav. Brain Res.* **2018**, *344*, 9–19. [CrossRef]
62. Svare, B.B.; Leshner, A.I. Behavioral correlates of intermale aggression and grouping in mice. *J. Comp. Physiol. Psychol.* **1973**, *85*, 203–210. [CrossRef]
63. O'Donnell, V.; Blanchard, R.J.; Blanchard, D.C. Mouse aggression increases after 24 hours of isolation or housing with females. *Behav. Neural Biol.* **1981**, *32*, 89–103. [CrossRef] [PubMed]
64. Kareem, A.M.; Barnard, C.J. The importance of kinship and familiarity in social interactions between mice. *Anim. Behav.* **1982**, *30*, 594–601. [CrossRef]
65. Fairless, A.H.; Katz, J.M.; Vijayvargiya, N.; Dow, H.C.; Kreibich, A.S.; Berrettini, W.H.; Abel, T.; Brodkin, E.S. Development of home cage social behaviors in BALB/cJ vs. C57BL/6J mice. *Behav. Brain Res.* **2013**, *237*, 338–347. [CrossRef] [PubMed]
66. Zwirner, P.P.; Porsolt, R.D.; Loew, D.M. Inter-group aggression in mice: A new method for testing the effects of centrally active drugs. *Psychopharmacologia* **1975**, *45*, 133–138. [CrossRef] [PubMed]
67. Terranova, M.L.; Laviola, G.; Alleva, E. Ontogeny of amicable social behavior in the mouse: Gender differences and ongoing isolation outcomes. *Dev. Psychobiol.* **1993**, *26*, 467–481. [CrossRef] [PubMed]
68. Benus, R.F.; Henkelmann, C. Litter composition influences the development of aggression and behavioural strategy in male Mus domesticus. *Behaviour* **1998**, *135*, 1229–1249.
69. Van Loo, P.L.P.; Kruitwagen, C.; Van Zutphen, L.F.M.; Koolhaas, J.M.; Baumans, V. Modulation of aggression in male mice: Influence of cage cleaning regime and scent marks. *Anim. Welf.* **2000**, *9*, 281–295.
70. Van Loo, P.L.; Mol, J.A.; Koolhaas, J.M.; Van Zutphen, B.F.; Baumans, V. Modulation of aggression in male mice: Influence of group size and cage size. *Physiol. Behav.* **2001**, *72*, 675–683. [CrossRef]
71. Garris, D.R. Aggression-associated changes in murine olfactory tubercle bioamines. *Brain Res.* **2003**, *963*, 150–155. [CrossRef]
72. Sorensen, D.B.; Johnsen, P.F.; Bibby, B.M.; Bottner, A.; Bornstein, S.R.; Eisenhofer, G.; Pacak, K.; Hansen, A.K. PNMT transgenic mice have an aggressive phenotype. *Horm. Metab. Res. = Horm.-Und Stoffwechs. = Horm. Et Metab.* **2005**, *37*, 159–163. [CrossRef]
73. Bisazza, A. Social organization and territorial behaviour in three strains of mice. *Boll. Zool.* **1981**, *48*, 157–167. [CrossRef]
74. Mondragón, R.; Mayagoitia, L.; López-Luján, A.; Díaz, J.L. Social structure features in three inbred strains of mice, C57Bl/6J, Balb/cj, and NIH: A comparative study. *Behav. Neural Biol.* **1987**, *47*, 384–391. [CrossRef]
75. Nicholson, A.; Malcolm, R.D.; Russ, P.L.; Cough, K.; Touma, C.; Palme, R.; Wiles, M.V. The response of C57BL/6J and BALB/cJ mice to increased housing density. *J. Am. Assoc. Lab. Anim. Sci.* **2009**, *48*, 740–753. [PubMed]
76. Smith, A.L.; Mabus, S.L.; Muir, C.; Woo, Y. Effects of housing density and cage floor space on three strains of young adult inbred mice. *Comp. Med.* **2005**, *55*, 368–376. [PubMed]
77. Babineau, B.A.; Yang, M.; Berman, R.F.; Crawley, J.N. Low home cage social behaviors in BTBR T+tf/J mice during juvenile development. *Physiol. Behav.* **2013**, *114–115*, 49–54. [CrossRef]
78. Fitchett, A.E.; Collins, S.A.; Mason, H.; Barnard, C.J.; Cassaday, H.J. Urinary corticosterone measures: Effects of strain and social rank in BKW and CD-1 mice. *Behav. Process.* **2005**, *70*, 168–176. [CrossRef] [PubMed]
79. Kempf, E.; Puglisi-Allegra, S.; Cabib, S.; Schleef, C.; Mandel, P. Serotonin levels and turnover in different brain areas of isolated aggressive or non-aggressive strains of mice. *Prog. Neuro-Psychopharmacol. Biol. Psychiatry* **1984**, *8*, 365–371. [CrossRef]
80. Martinez-Cue, C.; Rueda, N.; Garcia, E.; Davisson, M.T.; Schmidt, C.; Florez, J. Behavioral, cognitive and biochemical responses to different environmental conditions in male Ts65Dn mice, a model of Down syndrome. *Behav. Brain Res.* **2005**, *163*, 174–185. [CrossRef]
81. Trainor, B.C.; Workman, J.L.; Jessen, R.; Nelson, R.J. Impaired Nitric Oxide Synthase Signaling Dissociates Social Investigation and Aggression. *Behav. Neurosci.* **2007**, *121*, 362–369. [CrossRef]
82. Lewejohann, L.; Kloke, V.; Heiming, R.S.; Jansen, F.; Kaiser, S.; Schmitt, A.; Lesch, K.P.; Sachser, N. Social status and day-to-day behaviour of male serotonin transporter knockout mice. *Behav. Brain Res.* **2010**, *211*, 220–228. [CrossRef]
83. Rodriguez-Arias, M.; Navarrete, F.; Daza-Losada, M.; Navarro, D.; Aguilar, M.A.; Berbel, P.; Minarro, J.; Manzanares, J. CB1 cannabinoid receptor-mediated aggressive behavior. *Neuropharmacology* **2013**, *75*, 172–180. [CrossRef] [PubMed]
84. Rodriguez-Arias, M.; Navarrete, F.; Blanco-Gandia, M.C.; Arenas, M.C.; Aguilar, M.A.; Bartoll-Andres, A.; Valverde, O.; Minarro, J.; Manzanares, J. Role of CB2 receptors in social and aggressive behavior in male mice. *Psychopharmacology* **2015**, *232*, 3019–3031. [CrossRef] [PubMed]
85. McNamara, G.I.; John, R.M.; Isles, A.R. Territorial Behavior and Social Stability in the Mouse Require Correct Expression of Imprinted Cdkn1c. *Front. Behav. Neurosci.* **2018**, *12*, 28. [CrossRef] [PubMed]
86. Tan, S.; Xiao, Y.; Yin, H.H.; Chen, A.I.; Soong, T.W.; Je, H.S. Postnatal TrkB ablation in corticolimbic interneurons induces social dominance in male mice. *Proc. Natl. Acad. Sci. USA* **2018**, *115*, E9909–E9915. [CrossRef] [PubMed]

87. Rodriguiz, R.M.; Chu, R.; Caron, M.G.; Wetsel, W.C. Aberrant responses in social interaction of dopamine transporter knockout mice. *Behav. Brain Res.* **2004**, *148*, 185–198. [CrossRef] [PubMed]
88. Bailoo, J.D.; Murphy, E.; Varholick, J.A.; Novak, J.; Palme, R.; Wurbel, H. Evaluation of the effects of space allowance on measures of animal welfare in laboratory mice. *Sci. Rep.* **2018**, *8*, 713. [CrossRef]
89. Blankenberger, W.B.; Weber, E.M.; Chu, D.K.; Geronimo, J.T.; Theil, J.; Gaskill, B.N.; Pritchett-Corning, K.; Albertelli, M.A.; Garner, J.P.; Ahloy-Dallaire, J. Breaking up is hard to do: Does splitting cages of mice reduce aggression? *Appl. Anim. Behav. Sci.* **2018**, *206*, 94–101. [CrossRef]
90. Welch, B.L.; Welch, A.S. Graded effect of social stimulation upon d-amphetamine toxicity, aggressiveness and heart and adrenal weight. *J. Pharmacol. Exp. Ther.* **1966**, *151*, 331–338.
91. Barnard, C.J.; Behnke, J.M.; Sewell, J. Social behaviour and susceptibility to infection in house mice (*Mus musculus*): Effects of group size, aggressive behaviour and status-related hormonal responses prior to infection on resistance to *Babesia microti*. *Parasitology* **1994**, *108 Pt 5*, 487–496. [CrossRef]
92. Saldívar-González, J.A.; Posadas-Andrews, A.; Rojas, J.A.; Yoldi-Negrete, M.; Álvarez-Sekely, A.; Flores-Hernández, V.; Ortiz-León, S.; Foilloux-Morales, C.; Mayagoitia, L.; Mondragón-Ceballos, R. Effect of imipramine and electro convulsive stimulation in mice under social stress conditions. *Curr. Top. Pharmacol.* **2007**, *11*, 57–70.
93. Bartolomucci, A.; Chirieleison, A.; Gioiosa, L.; Ceresini, G.; Parmigiani, S.; Palanza, P. Age at group formation alters behavior and physiology in male but not female CD-1 mice. *Physiol. Behav.* **2004**, *82*, 425–434. [CrossRef] [PubMed]
94. Kikusui, T.; Takeuchi, Y.; Mori, Y. Early weaning induces anxiety and aggression in adult mice. *Physiol. Behav.* **2004**, *81*, 37–42. [CrossRef] [PubMed]
95. D'Andrea, I.; Alleva, E.; Branchi, I. Communal nesting, an early social enrichment, affects social competences but not learning and memory abilities at adulthood. *Behav. Brain Res.* **2007**, *183*, 60–66. [CrossRef] [PubMed]
96. Hilakivi-Clarke, L.A.; Lister, R.G. Are there preexisting behavioral characteristics that predict the dominant status of male NIH Swiss mice (*Mus musculus*)? *J. Comp. Psychol.* **1992**, *106*, 184–189. [CrossRef]
97. Kaliste-Korhonen, E.; Eskola, S. Fighting in NIH/S male mice: Consequences for behaviour in resident-intruder tests and physiological parameters. *Lab. Anim.* **2000**, *34*, 189–198. [CrossRef]
98. Lee, W.; Khan, A.; Curley, J.P. Major urinary protein levels are associated with social status and context in mouse social hierarchies. *Proceedings. Biol. Sci.* **2017**, *284*, 20171570. [CrossRef]
99. Williamson, C.M.; Romeo, R.D.; Curley, J.P. Dynamic changes in social dominance and mPOA GnRH expression in male mice following social opportunity. *Horm. Behav.* **2017**, *87*, 80–88. [CrossRef]
100. Gray, S.; Hurst, J.L. The effects of cage cleaning on aggression within groups of male laboratory mice. *Anim. Behav.* **1995**, *49*, 821–826. [CrossRef]
101. Smith, A.L.; Corrow, D.J. Modifications to husbandry and housing conditions of laboratory rodents for improved well-being. *ILAR J.* **2005**, *46*, 140–147. [CrossRef]
102. Koyama, S.; Kamimura, S. Influence of social dominance and female odor on the sperm activity of male mice. *Physiol. Behav.* **2000**, *71*, 415–422. [CrossRef]
103. Wurbel, H.; Garner, J.P. Refinement of rodent research through environmental enrichment and systematic randomization. *Natl. Cent. Replace. Refinement Reduct. Anim. Res.* **2007**, *9*, 1–9.
104. Ratuski, A.S.; Weary, D.M. Environmental Enrichment for Rats and Mice Housed in Laboratories: A Metareview. *Animals* **2022**, *12*, 414. [CrossRef] [PubMed]
105. Olsson, I.A.S.; Westlund, K. More than numbers matter: The effect of social factors on behaviour and welfare of laboratory rodents and non-human primates. *Appl. Anim. Behav. Sci.* **2007**, *103*, 229–254. [CrossRef]
106. Akre, A.K.; Bakken, M.; Hovland, A.L.; Palme, R.; Mason, G. Clustered environmental enrichments induce more aggression and stereotypic behaviour than do dispersed enrichments in female mice. *Appl. Anim. Behav. Sci.* **2011**, *131*, 145–152. [CrossRef]
107. Percie du Sert, N.; Hurst, V.; Ahluwalia, A.; Alam, S.; Avey, M.T.; Baker, M.; Browne, W.J.; Clark, A.; Cuthill, I.C.; Dirnagl, U.; et al. The ARRIVE guidelines 2.0: Updated guidelines for reporting animal research. *PLoS Biol.* **2020**, *18*, e3000410. [CrossRef]
108. Kilkenny, C.; Browne, W.J.; Cuthill, I.C.; Emerson, M.; Altman, D.G. Improving Bioscience Research Reporting: The ARRIVE Guidelines for Reporting Animal Research. *PLoS Biol.* **2010**, *8*, e1000412. [CrossRef] [PubMed]
109. Azkona, G.; Caballero, J.M. Implementing strategies to reduce singly housed male mice. *Lab Anim* **2019**, *53*, 508–510. [CrossRef]
110. Grífols, R.; Zamora, C.; Ortega-Saez, I.; Azkona, G. Postweaning Grouping as a Strategy to Reduce Singly Housed Male Mice. *Animals* **2020**, *10*, 2135. [CrossRef]

Disclaimer/Publisher's Note: The statements, opinions and data contained in all publications are solely those of the individual author(s) and contributor(s) and not of MDPI and/or the editor(s). MDPI and/or the editor(s) disclaim responsibility for any injury to people or property resulting from any ideas, methods, instructions or products referred to in the content.

animals MDPI

Article

Clicker Training Mice for Improved Compliance in the Catwalk Test

Jana Dickmann, Fernando Gonzalez-Uarquin, Sandra Reichel, Dorothea Pichl, Konstantin Radyushkin, Jan Baumgart and Nadine Baumgart *

Translational Animal Research Center, University Medical Center of the Johannes Gutenberg University, 55131 Mainz, Germany
* Correspondence: nadine.baumgart@uni-mainz.de

Simple Summary: Refinement-oriented research remains essential for animal welfare and data reproducibility. When evaluating mouse locomotion, the implementation of the CatWalk XT is helpful for gait assessment, but its application requires eliciting movement of the animals across the corridor, usually by forcing them with unpleasant stimuli. In this study, we tested the efficacy of clicker training to increase performance with the CatWalk test while assessing behavioral changes in the Open Field and Elevated Plus Maze to address the well-being of trained and untrained mice. Clicker training improved running speed on the CatWalk for both sexes. Interestingly, clicker training appeared to reduce anxiety and improve general well-being parameters in the Open Field and the Elevated Plus Maze tests to a greater extent in females. We conclude that clicker training enhances the performance of mice on the CatWalk and is a promising alternative for welfare improvement.

Abstract: The CatWalk test relies on the run of mice across the platform to measure a constant speed with low variation. Mice usually require a stimulus to walk to the end of the catwalk. However, such stimuli are usually aversive and can impair welfare. Positive reinforcement training of laboratory animals is a thriving tool for refinement and contributes to meeting the demands instituted by Directive 2010/63/EU. We have already demonstrated the positive effects of clicker training. In this study, we trained male and female mice to complete the CatWalk protocol while assessing the effects of training on their well-being (Open Filed and Elevated Plus Maze). In the CatWalk test, we observed that clicker training improved the running speed of the mice. In addition, clicker training reduced the number of runs required by mice, which was more pronounced in males. Clicker training lowered anxiety-like behaviors in our mice, especially in females, where a significant difference was observed between trained and untrained ones. Based on our findings, we hypothesize that clicker training is an effective tool to motivate mice and increase performance on the CatWalk test without potentially impairing their welfare (e.g., by puffing them).

Keywords: clicker training; 3Rs; CatWalk test; refinement; welfare

Citation: Dickmann, J.; Gonzalez-Uarquin, F.; Reichel, S.; Pichl, D.; Radyushkin, K.; Baumgart, J.; Baumgart, N. Clicker Training Mice for Improved Compliance in the Catwalk Test. *Animals* **2022**, *12*, 3545. https://doi.org/10.3390/ani12243545

Academic Editor: Garikoitz Azkona

Received: 21 November 2022
Accepted: 12 December 2022
Published: 15 December 2022

Copyright: © 2022 by the authors. Licensee MDPI, Basel, Switzerland. This article is an open access article distributed under the terms and conditions of the Creative Commons Attribution (CC BY) license (https://creativecommons.org/licenses/by/4.0/).

1. Introduction

Brain damage of different natures—trauma, stroke, degenerative diseases, and genetic manipulations—has often caused an impairment of motor functions. Quantitative measurement of locomotion is an essential method to better understand the mechanisms of this impairment and, most importantly, to evaluate functional recovery after treatment. The Noldus CatWalk XT is a computer-assisted gait analysis setup that allows rapid and objective quantification of many gait parameters in laboratory rodents [1–4]. The CatWalk XT system detects actual footprints by video recording the animal from below while it traverses a glass plate. Animal compliance is a critical factor for the success of gait analysis. Individual animals must run on a glass plate at the highest possible speed. Only in this case could moderate and subtle deficiencies be appropriately quantified. Moreover, for successful analysis, an animal must maintain its speed constant, that is, with low variation. In this way, a standard practice to promote the movement of mice along the CatWalk includes using air puffs [5], which makes the animals anxious and thus affects their welfare. In addition, it takes a considerable amount of time for the experimenter to make each individual comply to complete the test. If compliance cannot be achieved for all animals, this increases the number of experimental animals required.

Recently, Noldus company has offered an alternative approach: the home cage of the individual mouse could be placed at one end of the glass plate in the hope that the mouse will be motivated to run on the plate in order to reach its home cage; however, it may not be sufficient to motivate the animal to run with the highest velocity. This approach is based on anxiety, thus, making mice escape to a safe place (home cage). The question we are concerned with is how to increase mice's compliance while preserving maximum animal performance in the most welfare-friendly way. Clicker training showed promising approaches to improving animal performance, as is reported in rats [6–10]; however, researchers reporting successful protocols did not describe them in detail [3–5,11–14]. From our experience, animal-friendly handling improves mild and moderate procedures in mice. For instance, we implemented positive reinforcement training to decrease anxiety-like behaviors, suggesting a less stressful experience for our mice and the experimenter [15].

Positive reinforcement training to improve the performance of mice in the CatWalk test may involve the implementation of the 3Rs principle (replace, reduce, and refine) while promoting animal welfare and scientific reproducibility, as required by Directive 2010/63/EU [16]. Refining with positive reinforcement requires, among others, the adoption of strategies to habituate and train animals to perceive fewer threats by gaining partial control over a situation with a reduction in stress [17,18]. Clicker training is a form of positive reinforcement in which expected behavior is compensated with a reward [15]. The researchers reported its successful application in companion, zoo, and laboratory animals [15,19,20]. As mice are notoriously fearful, clicker training of mice appears to be a genuine and accurate alternative to our laboratory routine [15]. Furthermore, factors that affect normal animal behavior, such as the experimenter and the environment, can subtly confound the experimental results [21], so training animals to achieve greater interaction with their experimenter and environment can improve welfare (by reducing stress) and science (by improving reproducibility) [22].

Currently, we apply clicker training protocols in mice to the CatWalk test with promising results. In this study, we evaluated the implementation of our clicker training protocol to improve participation in the CatWalk test in male and female mice. Subsequently, we evaluated the individual performance of mice on the Open Field (OF) and the Elevated Plus Maze (EPM) tests as indicators of stress and anxiety behaviors. We hypothesized that our clicker training protocol improved CatWalk performance and could potentially decrease the distress or anxiety caused in mice by the experimental setup and the experimenter.

2. Materials and Methods

This study was carried out following the ARRIVE guidelines [23]. The experimental design and management procedures were approved by the Rhineland-Palatinate State Authority (permit numbers: G-18-1-065) following the European Directive 2010/63/EU for the protection of animals used for scientific purposes.

2.1. Mice

Forty-eight C57BL/6JRj mice (24 eight-week-old males and 24 eight-week-old females) were purchased from a verified international breeder (Janvier Labs, Le Genest-Saint-Isle, France). All mice were raised following the recommendations of the Federation of Laboratory Animal Science Associations (FELASA). The mice were randomly housed in groups of four in type II long filter-top cages (Tecniplast, Buguggiate, Italy; SealSafe Plus, polyphenylsulfone, 36.5 cm × 21 cm × 14 cm), equipped with red transparent shelters, cocoons, and opaque 10 cm PVC tubes (tunnels) to transport the mice in an animal-friendly way. Housing followed a 12/12 light–dark cycle (200 lux from 7:00 to 19:00) in a temperature and humidity-controlled animal room (22–24 °C and 50–55%, respectively). Water and food (ssniff M-Z Extrudat, ssniff, Soest, Germany) were provided ad libitum. The mice were kept in same-sex groups of four. All animals were allowed a habituation period of one week before the experimental phase began.

2.2. Clicker Training Protocol

The present protocol was carried out over 11 days; 12 male and 12 female mice were clicker trained; the other 24 mice only received control handling. The clicker training protocol was conducted as previously reported [15]. Before the training period started, the reward (white chocolate cream) was placed into the home cages for 2–3 days. The same person carried out this protocol in a quiet room to reduce the stress of a new environment. The training lasted 5 min per mouse, divided into a series of 45 sec of training followed by a 15-second break during which all training equipment was removed from the cage. Each mouse was trained individually in the home cage, while the remaining group animals were transferred to a separate cage with their familiar enrichment. One training cycle was conducted per day. The reward was presented only for as long as it took the mouse to take one bite, except for the first display of learning a new task, which was rewarded with a "jackpot" reward (reward for three seconds).

Clicker training was established in sequential steps. A step was considered successfully learned after the animal showed 10 repetitions of the trained behavior within two minutes. Only after successfully training each mouse we moved to the next step. The clicker training protocol for the current experiment is depicted in Figure 1. In brief: (1) we established a conditioned connection between the "click" sound and the food reward by continuously clicking at the exact moment of white chocolate intake (4–5 s). This was repeated 3–4 times on day one; (2) we placed the familiar tunnel in the empty home cage and took the natural thigmotaxis of mice into account by placing it along a wall. As soon as the animal entered the tunnel, we clicked and presented the reward; (3) we placed a target stick (clicker device with an extendable arm and a small plastic ball attached) on one opening of the tunnel; when the mouse entered the other side, crossed the tunnel and touched the plastic ball with its nose, we administered the reward; (4) we placed the target stick close to a wall in the cage, and as soon as the animal touched it, we clicked, removed the stick, and rewarded. We proceeded by varying the position of the target stick in the cage (no tunnel was needed for this task); (5) As soon as each mouse touched the target stick, we started slowly moving the stick away from the mouse within the cage, leading the mouse to follow it. The first reward was given when following 1 cm; later, we extended the distance; (6) we transported the animals individually to the CatWalk and left them roaming freely for 1 min; (7) we repeated the target stick without tunnel on the CatWalk using the full length of the pathway without stopping in-between. Rewards were given at the two outer ends of the walkway immediately after the mouse walked the corridor.

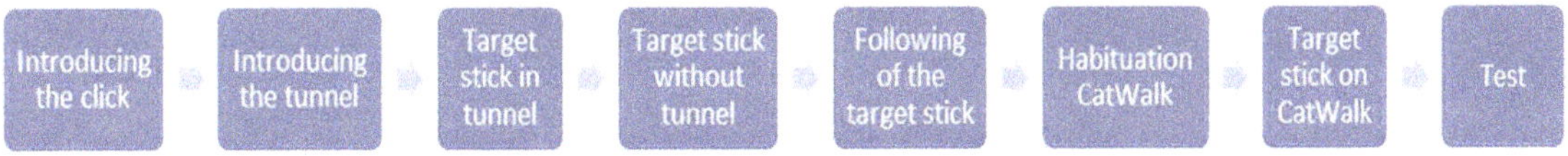

Figure 1. Graphical illustration of the clicker training protocol.

2.3. CatWalk Test

The CatWalk XT (produced by Noldus Information Technology BV, Wageningen, The Netherlands) is a computerized gait analysis system for assessing forelimb–hindlimb coordination. The CatWalk structure is comprised of a 130 × 20 × 0.5 cm glass plate, a 120 × 5 cm plastic corridor (with no floor and ceiling) to narrow the running area on the glass plate, the moveable cover with inbuilt red light providing a background illumination for video acquisition, and a high-speed video camera mounted below the glass plate

(Figures 2 and 3a). A source of green LED light is mounted alongside the long edge of the glass plate in such a way that a green light enters the glass from the edge [24]. Thus, a green light is internally reflected inside the glass plate (the same principle is used in fiber optics technology). However, in those areas where the animal paws make contact with the glass plate, the green light is reflected at about 90° down and thus detected by the camera. The high-speed (100 frames per second) color camera captures these areas and sends the data to the CatWalk XT software. The red lamp, mounted on top of the mouse, provides good contrast between the paw prints and the rest of the body.

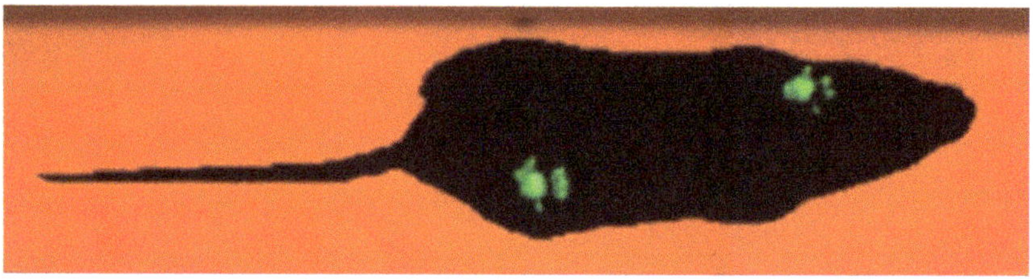

Figure 2. Mouse performing the Catwalk test. The footprint of each paw is visualized in green color.

All 48 mice were moved to the neighboring test room in their home cages and habituated there for 1 h. In our experiment, each mouse underwent only one testing trial in the Catwalk. The light in the room was turned off, and after recording the background, one mouse was randomly selected among not yet tested individuals in its home cage, placed on one side of the corridor, and left undisturbed for the whole duration of the trial. The cover was closed, and the software recording started. During the recording period, the mouse was voluntarily moving back and forth in the corridor. Each time mouse moved through the recording area, a "run" was registered. The recording was stopped automatically after three compliant runs were reached. The trial was finished, and the mouse was placed back in its home cage. The software considers the runs to be compliant when the mouse moves from one side of the corridor to the other without hesitation, showing no rearing against the bounding walls or the side walls of the corridor, change in direction, straightening up on the bounding walls, or other substitute behaviors. A maximum speed variation of 60% was allowed. The number of runs to reach three compliant runs was recorded. Once each mouse accumulated three runs, data acquisition stopped automatically. Further, the running speed data was calculated. CatWalk XT software (Noldus Inc., Wageningen, The Netherlands) was used to analyze gait patterns. Before starting the next animal, the running path was cleaned with water.

2.4. Open Field (OF) Test

Running speed, wall latency, and time spent in the center and periphery of the OF test were evaluated in a white square plexiglass arena (dimensions: 40 × 40 × 40 cm) (Figure 4a). The mice were placed in the OF center (20 × 20 cm) and allowed to explore it for five minutes. Target behaviors were analyzed using the Ethovision XT software version 8.5.614 (Noldus Inc., Wageningen, The Netherlands). After each animal, the boxes were cleaned with water. No experimentalists were present in the room during the test.

2.5. Elevated Plus Maze Test (EPM)

The day after the OF test, the EPM was performed. The mice were placed in the central intersection (5 × 5 cm) from which the animal had free access to four arms (30 × 5 cm each) (Figure 5a). Two opposing arms were surrounded by opaque walls (15 cm), while the other two had no walls. The targeted behaviors were recorded for five minutes using an overhead video camera (ICD-49, Ikegami Electronics (Europe), Neuss, Germany) and analyzed by the Ethovision XT software version 8.5.614 (Noldus Inc., Wageningen, The Netherlands). The EPM was cleaned with water between animal tests. No mice were excluded from the test.

2.6. Statistical Analysis

Prior to statistical analysis, data were tested for normal distribution by D'Agostino and Pearson test. All the data shown in this manuscript were normally distributed. Clicker training data between males and females (Section 3.1) was evaluated by the student's *t*-test. For all the other data (Sections 3.2–3.4), we used two-way ANOVA (main factors: training and sex) followed by Tukey's Honest Significant Difference (Tukey HSD) test. Pearson's correlation analyses were performed to establish relationships between clicker training and catwalk running speed. Statistical analyses were performed with GraphPad Prism, version 9.0, for Windows (GraphPad Software, San Diego, CA, USA). The F values indicated the variance ratio between and within groups. Degrees of freedom are shown as subscripted F values. Exact *p*-values were reported in the results and the figures. In all the figures, the values were expressed as means ± SD. We excluded one untrained female mouse from the OFT and one trained male mouse from the EPM test (open arms duration) based on the outlier Grubb's test (using the log-normal correction). All staff involved in collecting data in the main study protocol were blinded whenever possible (e.g., video files and behavioral tests were analyzed by people not directly involved in the housing and training of mice).

3. Results

3.1. Clicker Training

Male and female mice participated well in training. Student *t*-test indicated that male mice showed slightly higher average repetitions per day of the desired behavior compared to female mice (Males: 5.7 ± 3.08; females: 7.8 ± 1.48 (mean ± SD); $n = 12$; $t_{22} = 2.1$; $p = 0.02$).

3.2. Performance on the CatWalk

We represent the scheme of the CatWalk test in Figure 3a. Both sexes showed higher values for running speed when trained compared to untrained (Figure 3b). The two-way ANOVA results indicated a significant effect of training and sex factors (Training $F_{1,44} = 20$, $p \leq 0.001$; Sex $F_{1,44} = 8.7$, $p = 0.005$; Interaction $F_{1,42} = 0.4$, $p = 0.52$). Tukey HSD test revealed significant differences between trained and untrained animals regardless of sex ($p = 0.04$ and $p = 0.004$ for females and males, respectively). Furthermore, there was a significant relationship between the number of training repetitions per day (from Section 3.1.) and the running speed on the CatWalk (Pearson $r = 0.48$; $p = 0.01$; Supplementary Figure S1).

Clicker training decreased the number of runs necessary to complete a CatWalk trial in both sexes (Training $F_{1,44} = 7.3$, $p = 0.009$; Sex $F_{1,44} = 0.50$, $p = 0.48$; Interaction $F_{1,44} = 3.9$, $p = 0.05$). Tukey HSD analyses showed a significant decrease in males but not in females compared to the untrained group ($p < 0.01$; Figure 3c).

3.3. Open Field (OF) Test

We represent the scheme of the OF test in Figure 4a. Clicker training significantly increased the time spent in the center (Training $F_{1,43} = 5.70$, $p = 0.02$; Sex $F_{1,43} = 0.71$, $p = 0.40$; Interaction $F_{1,43} = 1.6$, $p = 0.21$). However, when comparing the groups by Tukey HSD test, we only observed a trend between trained and untrained females ($p = 0.06$; Figure 4b).

Clicker training significantly increased the distance traveled in the center (Training $F_{1,43} = 15$, $p = 0.0004$; Sex $F_{1,43} = 0.08$, $p = 0.77$; Interaction $F_{1,43} = 1.1$, $p = 0.29$). Tukey HSD test revealed specific differences between trained and untrained females ($p = 0.01$; Figure 4c), although such difference was not observed in males. Moreover, a subsequent evaluation of the ratio between distance traveled in the center and total distance traveled indicated a training factor effect (Training $F_{1,43} = 8.6$, $p = 0.005$; Sex $F_{1,43} = 0.11$, $p = 0.74$; Interaction $F_{1,43} = 2.2$, $p = 0.14$), but after a multiple comparison test, we found only a statistical trend between trained and untrained females ($p = 0.06$; Figure 4d).

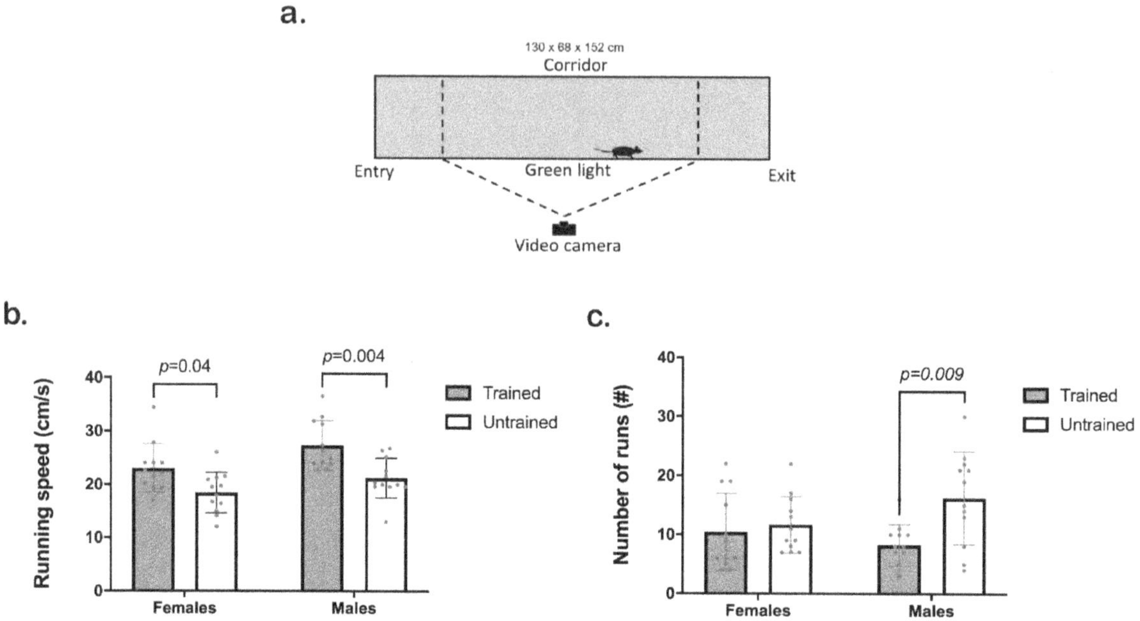

Figure 3. CatWalk XT performance of trained (gray) and untrained (white) C57BL/6J male and female mice. (**a**) Scheme of the CatWalk XT platform. (**b**) Running speed (cm/s). (**c**) Number of runs per group (#). We used two-way ANOVA followed by the Tukey HSD test for statistical analysis and multiple comparisons, $n = 12$. Each point represents an individual. Bars indicate the means ± SD. The exact p-value is provided when significant differences are given.

3.4. Elevated Plus Maze (EPM)

We assessed specific parameters in the EPM test (running speed, distance traveled, and duration in open arms. Figure 5a). Clicker training increased the time mice spent in the open arms (Training $F_{1,43} = 4.8$, $p = 0.03$; Sex $F_{1,43} = 0.74$, $p = 0.39$; Interaction $F_{1,43} = 0.23$; $p = 0.63$), although Tukey HSD analyses showed no statistical differences (Figure 5b). The two-way ANOVA test from a distance traveled in the EPM revealed training and interaction effects (Training $F_{1,44} = 17$, $p = 0.0002$; Sex $F_{1,44} = 0.41$, $p = 0.52$; Interaction $F_{1,44} = 7.4$; $p = 0.009$). Tukey HSD analyses revealed a significant increase in the traveled distance of females (not males) when they were trained ($p < 0.001$; Figure 5c). Finally, trained mice had higher running speeds than untrained ones (Training $F_{1,44} = 18$; $p = 0.0001$; Sex $F_{1,44} = 0.35$; $p = 0.55$; Interaction $F_{1,44} = 7.8$; $p = 0.007$. Figure 5d). When assessing the number of entries in open arms (Supplementary Figure S2), we found no statistical differences.

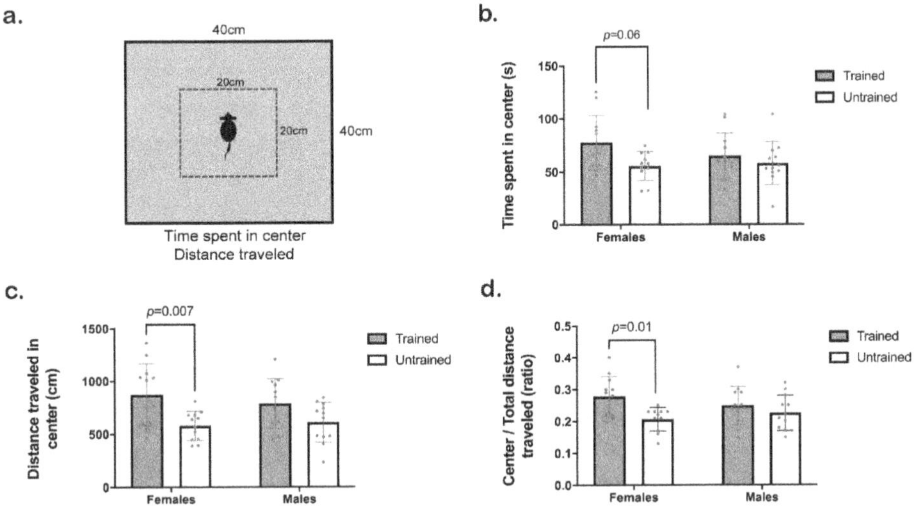

Figure 4. Behavioral activity of trained (gray) and untrained (white) C57BL/6J male and female mice in the open field (OF) test. (**a**) Schematic representation of the OF experiment. (**b**) Time the mice spent in the center (s). (**c**) Distance traveled in the center (cm). (**d**) Distance traveled in the center/distance traveled in total. Two-way ANOVA followed by the Tukey-HSD test was used for statistical analysis, $n = 12$ (untrained females $n = 11$). Each point represents an individual. Bars indicate the means ± SD. The exact p-value is provided when significant differences or trends are given.

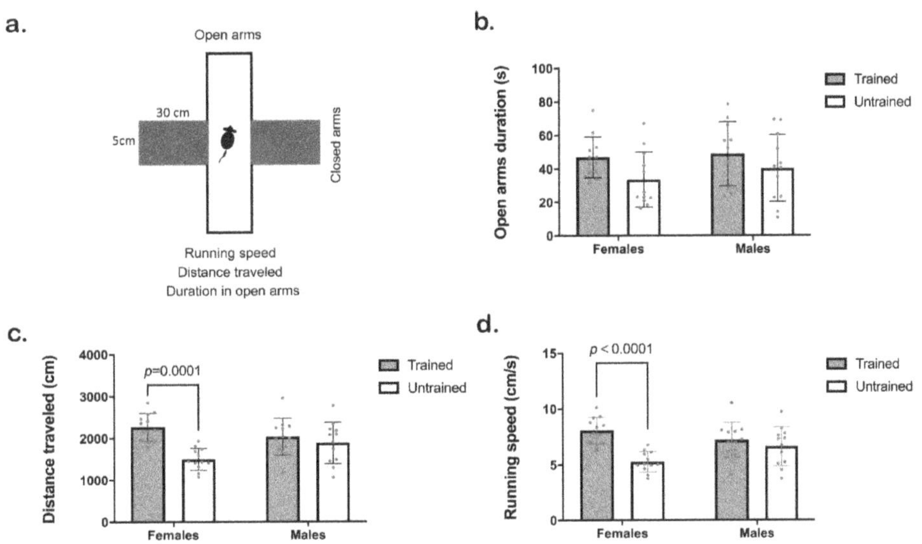

Figure 5. Behavioral activity of trained (gray) and untrained (white) C57BL/6J male and female mice in the Elevated Plus Maze (EPM) test. (**a**) Schematic representation of the EPM experiment. (**b**) Time spent in open arms (s). (**c**) Distance traveled in the EPM (cm). (**d**) Running speed (cm/s). We used Two-way ANOVA followed by the Tukey HSD test for statistical analysis, $n = 12$ (trained males in 5b $n = 11$). Each point represents an individual. Bars indicate the means ± SD. The exact p-value is provided when significant differences are given. Two-way ANOVA analysis for open arm duration (**b**) indicated an effect of training, although the Tukey HSD test did not show differences.

4. Discussion

We demonstrated for the first time that clicker training improved the performance of mice in the CatWalk XT test.

Clicker training started in the home cage before transferring to the CatWalk, simplifying the learning process while maintaining a familiar environment and a parallel reduction in stress factors. In other words, we simultaneously trained the mice in the CatWalk as habituation. We found that clicker-trained mice increased their running speed and decreased the number of runs. The average speed of mice crossing the corridor increased, suggesting that they increased their motivation to obtain the reward and thus ran faster to the exit at the end of the corridor. The improvement in motivation is a powerful indicator that our clicker-trained mice crossed the corridor without hesitation.

Our findings indicated that clicker training lowered the number of runs to complete the CatWalk. It means that clicker training may offer promising alternatives for both mice and experimenters. On one side, animals are not forced to cross the corridor (e.g., by puffing them), improving well-being along the test. On the other side, even if training requires a time investment, it reduces the number of runs per mouse, optimizing the time required by the experimenter to assess each mouse in the CatWalk test. We can further hypothesize that our protocol has the potential to decrease animal numbers by reducing non-compliant individuals or animals failing to meet run criteria. The fact that we identified clicker training was adequate for training mice for CatWalk challenges led to the assumption that, unlike rats, mice cannot be trained to make uninterrupted runs [25]. In this study, we did not evaluate specific gait parameters, but we hypothesize that clicker training would improve mice performance in the CatWalk test. We are addressing such a topic in our current research.

As mentioned, clicker training increased running speed in the CatWalk. Previous experiments reported changes in the running speed of rats when the reward changed either in quality or quantity, indicating a negative contrast effect [26,27]. Furthermore, studies that incorporated self-selection of music (a potential reward) during exercise increased running speed and general performance in humans [28,29]. In this regard, we must be careful not to humanize mouse data, but our findings may suggest that positive reinforcement promoted increases in running speed, suggesting positive welfare statuses.

We observed a significant effect of training in running speed and distance traveled on the CatWalk, OF, and EMP. Differences were significant in female (trained vs. untrained) but not in male mice. A recent study demonstrated that the gait performance of young mice assessed by CatWalk depended on age and sex, suggesting that sex hormones and genes on the X and Y chromosomes may impact behavioral outcomes [30]. Konhilas et al. (2004) [31] argued that females had higher aerobic capacity than males, most likely due to intrinsic differences in heart and skeletal muscle. Furthermore, differences in behavior depend on the hormonal status of young adult mice [32]. In addition, ovariectomized mice and rats significantly reduced wheel running compared to their non-ovariectomized and ovariectomized estrogen-receiving counterparts [31,32]. Although we found no substantial differences between females and males, we may speculate that positive reinforcement may modulate sex-intrinsic traits.

Our assumption that training may be beneficial for the well-being relies on the fact that when trained (compared to those not trained), mice increased the distance traveled on the EPM and the OF tests while showing a strong tendency to stay in the center of the OF. When comparing within sex, we found that female mice were significantly more influenced by training than male mice.

We can infer that clicker training exerted potential reductions in female mice's stress and anxiety-like behaviors due to their tendency to explore and interact with the stimuli [33]. Evidently, the physiological and hormonal status may play a sex-specific role in these behaviors, making female mice more susceptible to clicker training. However, it is a matter of further research.

According to our observations, an unexpected finding was the lower behavioral response between trained and untrained males (compared with trained and untrained females). We were surprised because the males performed better in home cage training and required fewer runs to comply with the CatWalk test. Based on our qualitative observations, an explanation of these findings was that separating the males from the group and putting the animals back together (after training or conducting a test) resulted in fights, which could have caused stress to the male animals.

Our present results indicate that clicker training can improve performance in the CatWalk test and may positively influence mouse welfare. However, we acknowledge the limitation of the absence of stimuli on untrained animals so that we can directly compare puffing vs. clicker training, which is a matter of our current experiments.

5. Conclusions

Male and female mice benefited from clicker training. From the point of view of welfare, the cooperation of animals with the clicker training protocol emerges as a promising way to reduce stress and anxiety-like behaviors. However, we recommend more research on the potential influence of clicker training (such as environmental enrichment) on specific experimental arrangements. We also recommend additional studies modulating the frequency, quality, and quantity of rewards in male and female mice to establish the benchmark by which we improve welfare without threatening the reproducibility of the results.

Supplementary Materials: The following supporting information can be downloaded at: https://www.mdpi.com/article/10.3390/ani12243545/s1; Figure S1: Pearson's correlation test between clicker training and running speed; Figure S2: Number of entries in the open arms of the Elevated Plus Maze.

Author Contributions: Conceptualization, J.D, N.B. and K.R.; methodology, J.D. and K.R.; software, K.R.; validation, N.B., K.R. and J.B.; formal analysis, J.D.; investigation, J.D., D.P. and S.R.; resources, J.B.; data curation, N.B.; writing—original draft preparation, F.G.-U. and N.B.; writing—review and editing, F.G.-U., K.R. and N.B.; visualization, F.G.-U.; supervision, N.B. and J.B.; project administration, N.B. All authors have read and agreed to the published version of the manuscript.

Funding: This research received no external funding.

Institutional Review Board Statement: The experimental design and management procedures were approved by the Rhineland-Palatinate State Authority (permit numbers: G-18-1-065).

Informed Consent Statement: Not applicable, as this research did not involve humans.

Data Availability Statement: The data presented in this study are openly available in FigShare at doi 10.6084/m9.figshare.21719957.

Conflicts of Interest: The authors declare no conflict of interest.

References

1. Hamers, F.P.; Koopmans, G.C.; Joosten, E.A. CatWalk-assisted gait analysis in the assessment of spinal cord injury. *J. Neurotrauma* **2006**, *23*, 537–554. [CrossRef] [PubMed]
2. Herold, S.; Kumar, P.; Jung, K.; Graf, I.; Menkhoff, H.; Schulz, X.; Bahr, M.; Hein, K. CatWalk gait analysis in a rat model of multiple sclerosis. *BMC Neurosci.* **2016**, *17*, 78. [CrossRef] [PubMed]
3. Gabriel, A.F.; Marcus, M.A.; Honig, W.M.; Walenkamp, G.H.; Joosten, E.A. The CatWalk method: A detailed analysis of behavioral changes after acute inflammatory pain in the rat. *J. Neurosci. Methods* **2007**, *163*, 9–16. [CrossRef] [PubMed]
4. Miyagi, M.; Ishikawa, T.; Kamoda, H.; Suzuki, M.; Sakuma, Y.; Orita, S.; Oikawa, Y.; Aoki, Y.; Toyone, T.; Takahashi, K.; et al. Assessment of pain behavior in a rat model of intervertebral disc injury using the CatWalk gait analysis system. *Spine* **2013**, *38*, 1459–1465. [CrossRef] [PubMed]
5. Batka, R.J.; Brown, T.J.; Mcmillan, K.P.; Meadows, R.M.; Jones, K.J.; Haulcomb, M.M. The need for speed in rodent locomotion analyses. *Anat. Rec.* **2014**, *297*, 1839–1864. [CrossRef] [PubMed]
6. Tschon, M.; Salamanna, F.; Martini, L.; Giavaresi, G.; Lorenzini, L.; Calzà, L.; Fini, M. Boosting the Intra-Articular Efficacy of Low Dose Corticosteroid through a Biopolymeric Matrix: An In Vivo Model of Osteoarthritis. *Cells* **2020**, *9*, 1571. [CrossRef] [PubMed]
7. Moritz, M.S.; Tepp, W.H.; Inzalaco, H.N.; Johnson, E.A.; Pellett, S. Comparative functional analysis of mice after local injection with botulinum neurotoxin A1, A2, A6, and B1 by catwalk analysis. *Toxicon* **2019**, *167*, 20–28. [CrossRef]

8. Guo, Y.; Hu, H.; Wang, J.; Zhang, M.; Chen, K. Walking Function after Cervical Contusion and Distraction Spinal Cord Injuries in Rats. *J. Exp. Neurosci.* **2019**, *13*, 117906951986961. [CrossRef]
9. Heinzel, J.C.; Oberhauser, V.; Keibl, C.; Swiadek, N.; Längle, G.; Frick, H.; Kolbenschlag, J.; Prahm, C.; Grillari, J.; Hercher, D. Evaluation of Functional Recovery in Rats After Median Nerve Resection and Autograft Repair Using Computerized Gait Analysis. *Front. Neurosci.* **2021**, *14*, 593545. [CrossRef]
10. Boix, J.; von Hieber, D.; Connor, B. Gait Analysis for Early Detection of Motor Symptoms in the 6-OHDA Rat Model of Parkinson's Disease. *Front. Behav. Neurosci.* **2018**, *12*, 39. [CrossRef]
11. Bozkurt, A.; Deumens, R.; Scheffel, J.; O'Dey, D.M.; Weis, J.; Joosten, E.A.; Fuhrmann, T.; Brook, G.A.; Pallua, N. CatWalk gait analysis in assessment of functional recovery after sciatic nerve injury. *J. Neurosci. Methods* **2008**, *173*, 91–98. [CrossRef] [PubMed]
12. Garcia-Ovejero, D.; González, S.; Paniagua-Torija, B.; Lima, A.; Molina-Holgado, E.; de Nicola, A.F.; Labombarda, F. Progesterone Reduces Secondary Damage, Preserves White Matter, and Improves Locomotor Outcome after Spinal Cord Contusion. *J. Neurotrauma* **2014**, *31*, 857–871. [CrossRef] [PubMed]
13. Hestehave, S.; Abelson, K.S.P.; Bronnum Pedersen, T.; Finn, D.P.; Andersson, D.R.; Munro, G. The influence of rat strain on the development of neuropathic pain and comorbid anxio-depressive behaviour after nerve injury. *Sci. Rep.* **2020**, *10*, 20981. [CrossRef]
14. Timotius, I.K.; Bieler, L.; Couillard-Despres, S.; Sandner, B.; Garcia-Ovejero, D.; Labombarda, F.; Estrada, V.; Müller, H.W.; Winkler, J.; Klucken, J.; et al. Combination of Defined CatWalk Gait Parameters for Predictive Locomotion Recovery in Experimental Spinal Cord Injury Rat Models. *Eneuro* **2021**, *8*, ENEURO.0497-20.2021. [CrossRef] [PubMed]
15. Leidinger, C.; Herrmann, F.; Thone-Reineke, C.; Baumgart, N.; Baumgart, J. Introducing Clicker Training as a Cognitive Enrichment for Laboratory Mice. *J. Vis. Exp.* **2017**, *121*, e55415. [CrossRef]
16. European Commission. Directive 2010/63/EU of the European Parliament and of the Council of 22 September 2010 on the protection of animals used for scientific purposes. *Off. J. Eur. Union* **2010**, *50*, 33–79.
17. Dalm, S.; de Kloet, E.R.; Oitzl, M.S. Post-training reward partially restores chronic stress induced effects in mice. *PLoS ONE* **2012**, *7*, e39033. [CrossRef]
18. Laule, G. Positive Reinforcement Training for Laboratory Animals. In *The UFAW Handbook on the Care and Management of Laboratory and Other Research Animals*, 8th ed.; Hubrecht, R., Kirkwood, J., Eds.; Wiley-Blackwell: Oxford, UK, 2010; Volume 36, pp. 206–218. [CrossRef]
19. Feng, L.C.; Howell, T.J.; Bennett, P.C. How clicker training works: Comparing Reinforcing, Marking, and Bridging Hypotheses. *Appl. Anim. Behav. Sci.* **2016**, *181*, 34–40. [CrossRef]
20. Schapiro, S.J.; Perlman, J.E.; Thiele, E.; Lambeth, S. Training nonhuman primates to perform behaviors useful in biomedical research. *Lab. Anim.* **2005**, *34*, 37–42. [CrossRef]
21. Nigri, M.; Åhlgren, J.; Wolfer, D.P.; Voikar, V. Role of Environment and Experimenter in Reproducibility of Behavioral Studies with Laboratory Mice. *Front. Behav. Neurosci.* **2022**, *16*, 835444. [CrossRef]
22. Wiltgen, B.J.; Sinclair, C.; Lane, C.; Barrows, F.; Molina, M.; Chabanon-Hicks, C. The Effect of Ratio and Interval Training on Pavlovian-Instrumental Transfer in Mice. *PLoS ONE* **2012**, *7*, e48227. [CrossRef] [PubMed]
23. Du Sert, N.P.; Hurst, V.; Ahluwalia, A.; Alam, S.; Avey, M.T.; Baker, M.; Browne, W.J.; Clark, A.; Cuthill, I.C.; Dirnagl, U.; et al. The Arrive Guidelines 2.0: Updated Guidelines for Reporting Animal Research. *PLoS Biol.* **2020**, *18*, e3000410. [CrossRef]
24. Clarke, K.A. A technique for the study of spatiotemporal aspects of paw contact patterns, applied to rats treated with a TRH analogue. *Behav. Res. Methods Instrum. Comput.* **1992**, *24*, 407–411. [CrossRef]
25. Deumens, R.; Marinangeli, C.; Bozkurt, A.; Brook, G.A. Assessing motor outcome and functional recovery following nerve injury. *Methods Mol. Biol.* **2014**, *1162*, 179–188. [CrossRef]
26. Binkley, K.A.; Webber, E.S.; Powers, D.D.; Cromwell, H.C. Emotion and relative reward processing: An investigation on instrumental successive negative contrast and ultrasonic vocalizations in the rat. *Behav. Processes* **2014**, *107*, 167–174. [CrossRef]
27. Capaldi, E.D.; Singh, R. Percentage body weight and the successive negative contrast effect in rats. *Learn Motiv.* **1973**, *4*, 405–416. [CrossRef]
28. Bonnette, R.; Smith, M.C., III; Spaniol, F.; Ocker, L.; Melrose, D. The effect of music on running performance and rating of perceived exertion of college students. *Sport J.* **2012**, *24*, 1–7. [CrossRef]
29. Lee, S.; Kimmerly, D.S. Influence of music on maximal self-paced running performance and passive post-exercise recovery rate. *J. Sports Med. Phys. Fit.* **2014**, *56*, 39–48.
30. Pitzer, C.; Kurpiers, B.; Eltokhi, A. Gait performance of adolescent mice assessed by the CatWalk XT depends on age, strain and sex and correlates with speed and body weight. *Sci. Rep.* **2021**, *11*, 21372. [CrossRef]
31. Konhilas, J.P.; Maass, A.H.; Luckey, S.W.; Stauffer, B.L.; Olson, E.N.; Leinwand, L.A. Sex modifies exercise and cardiac adaptation in mice. Am. J. Physiol. *Heart Circ. Physiol.* **2004**, *287*, H2768–H2776. [CrossRef]
32. Kadi, F.; Karlsson, C.; Larsson, B.; Eriksson, J.; Larval, M.; Billig, H.; Jonsdottir, I.H. The effects of physical activity and estrogen treatment on rat fast and slow skeletal muscles following ovariectomy. *J. Muscle Res. Cell Motil.* **2002**, *23*, 335–339. [CrossRef] [PubMed]
33. Gouveia, K.; Hurst, J.L. Reducing Mouse Anxiety during Handling: Effect of Experience with Handling Tunnels. *PLoS ONE* **2013**, *8*, e66401. [CrossRef] [PubMed]

animals

MDPI

Communication

Acclimation and Blood Sampling: Effects on Stress Markers in C57Bl/6J Mice

Nerea Marin [1], Amparo Moragon [2], Domingo Gil [2], Francisco Garcia-Garcia [3] and Viviana Bisbal [1,*]

[1] Animal Facility IISLAFE, Hospital La Fe Research Institute, 46026 Valencia, Spain; nerea_marin@iislafe.es
[2] Animal Facility CIPF, Prince Felipe Research Center, 46012 Valencia, Spain; amoragon@cipf.es (A.M.); dgil@cipf.es (D.G.)
[3] Bioinformatics & Biostatistics Unit CIPF, Prince Felipe Research Center, 46012 Valencia, Spain; fgarcia@cipf.es
* Correspondence: viviana_bisbal@iislafe.es; Tel.: +34-961246706

Simple Summary: Blood sampling from laboratory animals is a routine procedure in biomedical research, and husbandry is known to affect the quality of animal models. The effect of handling required by these techniques can have a major impact on the condition and responses of experimental animals. Comparing three common methods of blood extraction in mice, i.e., saphenous vein phlebotomy, caudal vein phlebotomy and tail cut blood collection, together with the effect of acclimation, we study which technique is less stressful for the animal. Our results suggest that saphenous vein phlebotomy causes less stress even when acclimation is not performed.

Abstract: Blood sampling in rodents is common practice in scientific studies. Some of the refined methods widely used are the puncture of the saphenous vein or tail vein, or even tail docking. The handling needs of these different blood sampling methods are different and can directly affect stress, increasing the variability of the study. Moreover, there is less aversion and stress if the animal is accustomed to the environment, handling and technique. Therefore, our study aimed to assess the influence of these three blood sampling techniques (saphenous puncture, tail vein puncture and tail vein docking) and the use of previous acclimation on different indicators of animal stress, assessing blood glucose concentrations and faecal corticosterone metabolites (FCMs). Twenty-four young adult male and female C57Bl6/J mice were divided in three groups by sampling method: tail docking (TD), saphenous vein puncture (SV) and caudal vein puncture (CV) groups. All mice were studied with and without acclimation, which was performed during 9 consecutive days. The results showed that both males and females present very similar responses to the different handling and sampling methods without significant differences. Nevertheless, acclimation in all sampling methods decreased glucose and FCM levels significantly. The method that obtained the lowest glucose and FCM levels with significance was saphenous vein puncture. Therefore, we can say that it causes less stress when performing prior acclimation, even when this involves greater handling of the animal. Our results contribute to refinement within the 3R concept and could serve researchers to programme and select a good handling technique and a welfare-friendly blood sampling method for their experiments.

Keywords: blood sampling; mice; stress; glucose refinement; 3R; acclimation; animal welfare

Citation: Marin, N.; Moragon, A.; Gil, D.; Garcia-Garcia, F.; Bisbal, V. Acclimation and Blood Sampling: Effects on Stress Markers in C57Bl/6J Mice. *Animals* **2023**, *13*, 2816. https://doi.org/10.3390/ani13182816

Academic Editor: Garikoitz Azkona

Received: 15 July 2023
Revised: 31 August 2023
Accepted: 1 September 2023
Published: 5 September 2023

Copyright: © 2023 by the authors. Licensee MDPI, Basel, Switzerland. This article is an open access article distributed under the terms and conditions of the Creative Commons Attribution (CC BY) license (https://creativecommons.org/licenses/by/4.0/).

1. Introduction

"Good welfare equals good science" was demonstrated by Trevor Poole in 1997 [1], and, according to the principles of replacement, reduction and refinement, adverse effects such as pain, fear and distress should be avoided or minimised [2]. In addition, the refinement of techniques is one of the most important principles in laboratory animal science and must be considered an ethical and legal requirement. Blood sampling is a common procedure in biomedical research and could be a source of stress that can affect the variability of results and compromise animal welfare. Different scientific organisations have published recommendations and guidelines for commonly used blood sampling techniques in laboratory mice [3–5], but the techniques differ in their degree of invasiveness and in the handling duration needed [6,7], which might provoke different grades of distress in the animals. In this sense, any pain or distress has to be an objective to avoid in all experiments, and the optimal method needs to be used for collecting blood to accomplish minimal stress according to principles of replacement, reduction and refinement [8].

The literature describes different methods of blood sampling with more or less restraint techniques in mice. Hurst and West [9] showed that the handling method itself could be critical and can induce fear and anxiety responses to human contact. They found that picking mice up in a handling tunnel or cupping them in the open hand leads to substantial voluntary interaction with the handler and reduces stress and anxiety. Some blood sampling methods used in mice require intense handling (vena facialis puncture, retrobulbar sinus puncture, sublingual vein puncture, saphenous phlebotomy, etc.), and this makes them probably more stressful than those methods needing less handling. Meyer demonstrated that a single blood collection from the vena facialis, retrobulbar sinus or tail vessel led to an acute increase in plasma corticosterone levels, with a strong response when sampling from the facial vein and retrobulbar sinus [10]. It has been described that mice respond negatively when the base of the tail is picked up and do not readily habituate to this widely used method. Furthermore, the use of non-aversive tunnels or cupping methods has shown a reduction in plasma corticosterone and glucose levels [11] and reduced pain grimace scores when compared with tail-handled mice [12]. Some studies have shown the influence of handling on animal welfare and how acclimation reduces stress in mice [13–16].

Several studies compare the quality of blood samples obtained by different blood sampling methods [4,11,17,18]. Facial vein phlebotomy is a common blood sampling method in mice due to its simplicity and good quality of blood but requires high restraint. It allows a maximum allowable sample with minimal trauma [19]. Tail vein sampling is a quick and simple method that can be practised with a bit of restraint or without restraint by senior practitioners but requires some dilatation of blood vessels and does not allow maximum volumes to be obtained. Lateral saphenous vein puncture is a refined method of blood sampling, which is relatively quick but requires a moderate-to-high restraint technique [18,20].

The literature describes different methods of blood sampling with more or less restraint techniques in mice. Manipulation and restraint can have a great impact on animal welfare and could result in different responses of laboratory animals. In fact, handling stress is often pointed out as a potential source of unexplained variability of results [21]. This is due to the influence of management on both behaviour and physiology of animals [22,23], and not considering this might lead to an increase in the number of animals required for experiments. Animals in captivity are very sensitive to interaction with humans, so handling is habitually an unavoidable and variable procedure and animals have to be familiarised to allow them to learn that these interactions are not harmful [24].

Blood sampling remains the most widely used method for the assessment of hypothalamic–pituitary–adrenal (HPA) axis activity, and there are currently no appropriate alternative methods to assess acute changes [4,11,25]. It is essential to minimise the stress associated with these techniques and refine to the maximum the techniques used.

Plasma glucose levels are one of the blood parameters used as indicators of stress in rodents [21,26], since stress-mediated corticosterone production leads to gluconeogenesis and inhibition of insulin secretion [22]. These studies compare the glucose levels associated with the blood sampling technique but do not try to minimise these with prior acclimation of the animals to handling. Corticosterone is the primary glucocorticoid produced and secreted by the adrenal cortex in mice. It is often referred to as the "stress hormone" as it is involved in the stress response and affects blood pressure, blood sugar levels and other actions of stress adaptation [13,27]. The sampling procedure itself can be a source of stress, and it would be ideal if one could measure it with non-invasive techniques. In this way, faecal sampling must be considered a valid non-invasive method for steroid hormone assessment in laboratory mice and rats [14,25,28,29].

This study aims to compare three different blood sampling methods (saphenous vein phlebotomy, caudal vein phlebotomy and tail cut blood collection) and the influence of acclimation to the handling needed in these sampling methods to assess which blood sampling method is less stressful and whether stress can be reduced by acclimation techniques, together with the haematological and faecal detection of "stress markers" (glucose and corticosterone faecal metabolites (FCMs)), with the additional focus on sex. The experiment was performed with C57Bl/6J mice, one of the most widely used inbred strain of mice in biomedical research and a background strain to most genetically modified mouse models [30]. Considering different degrees of handling of the three blood sampling methods, differences in "stress markers" (glucose levels and FCMs) were expected [10,28].

2. Materials and Methods

2.1. Ethical Statement

All procedures were previously approved by the Ethics Committee of the Principe Felipe Research Center according to the National Law for the Protection of laboratory animals (RD 53/2013) and the European Union (European Directive 2010/63/EU).

2.2. Animals and Housing Conditions

Twenty-four SPF (specific pathogen-free) young adult male and female C57Bl/6J mice at the age of 10 to 12 weeks were studied. Mice were obtained from a commercial supplier (Charles River Laboratories, France) and randomly allocated to the different groups. They were housed in SPF conditions according to the FELASA guidelines in pressurised and individually ventilated 1145T (403 × 165 × 174 mm; 435 cm^2 floor area; Tecniplast) cages (70 air changes/h) with irradiated feeding (2014, Envigo, Barcelona, Spain) and autoclaved water. Nesting material and an autoclaved cardboard cylinder were used. Animals were allowed to acclimatise for ten days before experimentation. The light/dark cycle in the animal room consisted of a 12 h/12 h cycle. The temperature was 21 ± 2 °C, with a relative humidity of 50 ± 5% and 15 complete changes of filtered air per hour. Mice were housed in groups of 4 animals to minimise the impact of individual housing. During the sample collection, animals were housed in the same described conditions and all manipulations and sample collections were obtained in diurnal rhythms, between 10:00 and 12:00 in the morning.

2.3. Experimental Design

Mice were randomly housed, separated by sex and in groups of 4 mice per cage. Cages were randomly divided into the 3 experimental groups: tail vein (TV) group, saphenous vein (SV) group and tail cut (TC) group. Blood samples were obtained before and after acclimation: (a) non-acclimation (PRE-acclimation), where measures were taken without a handling routine; and (b) with acclimation (POST-acclimation), where the animals were trained with the habitual handling needed for blood sampling, as described below (Figure 1).

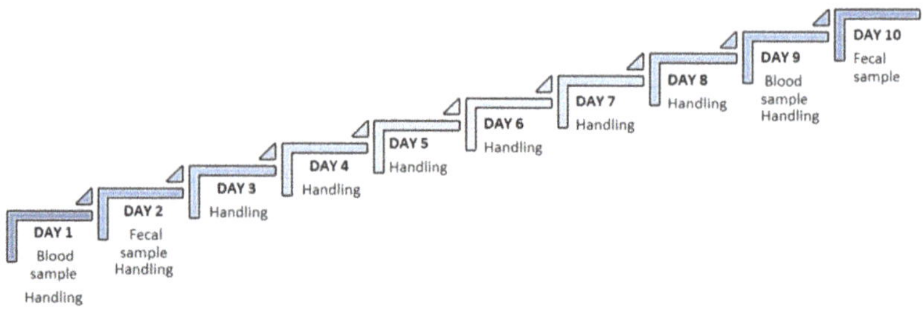

Figure 1. Experimental design timetable. Blood sampling was performed on day 1 and 9. Faecal sampling was performed on day 2 and 10.

2.4. Handling Technique

Handling acclimation was always performed in an adjacent experimental room and the mice were transported in their home cage.

A manipulation routine was established based on the handling needs for each of the three blood sampling methods of the study. Therefore, a specific handling technique was developed for each group, with a duration of 60 s per day for 9 continuous days. Procedures were performed by the same two trained and experienced technicians.

Based on other studies [10–12], the technique has the next sequence:

1. First, animals were placed between the hands of the technician, (as a cave) for about 30 s.
2. Then, the necessary immobilisation for each blood sampling method was performed for another 30 s: the animals from the TC and TV groups were placed on the cage rack and their tails were held, simulating the hold necessary for the blood sampling method, for 30 sec. The animals from the SV group were introduced to a containment "tunnel/tube", used later as a trap for sampling for 30 s [14,15].

2.5. Blood and Faecal Sampling

Blood and faecal sampling were always performed in an adjacent experimental room and the mice were transported in their home cage.

Glucose levels were obtained using a glucometer Contour®XT (Bayer AG, Leverkusen, Germany) according to the manufacturer's instructions, with a range of detection of 10 mg/dL to 600 mg/dL of glucose.

After blood sampling, mice were transferred to clean cages and faeces were collected 24 h after in all groups, following the protocol described by DetectX®. Then, 200 µgr of dried faecal samples were collected into a tube and stored at −21 °C until analysis, as the protocol described. Studies in male and female C53Bl/6 mice report the FCM peak radioactivity to be about 10 h (range 8–12 h) after injection [16]. In addition, corticosteroid hormone secretion is usually pulsatile and influenced by feed intake and environmental factors [17]. In this sense, collection samples from the cage 24 h after blood sampling include the FCM peak, which avoids the problems of circadian variation and timing of metabolism and excretion and is a good method for both chronic and acute studies [16].

Samples were identified by condition, not individually, to avoid the need to individualise animals and minimise this source of stress [31]. The use of this kind of sample that does not require restraining animals is a good method to avoid the effect of hormone secretion. The FCMs were extracted and quantified by enzyme immunometric assay (EIA) according to the manufacturer's instructions using a DetectX® Cortisol Immunoassay kit, with a sensitivity of 27.6 pg/mL and a detection limit of 45.4 pg/mL.

2.5.1. Tail Vein (TV)

The lateral caudal vein was pricked and a 25-gauge needle was used to puncture the vein. A blood drop was applied on to the glucose strip. Bleeding was stopped by applying a slight pressure with fingers.

2.5.2. Saphenous Vein (SV)

The sample was obtained using a similar technique described by Hem and cols. [18]. The mice were restrained, and introduced into a 20 mL falcon. This restraint allowed them to breathe through a falcon pipe hole at the top. The hind leg was externalized to visualise the saphenous vein. Hair over the saphenous vein was shaved and a 25-gauge needle was used to puncture the vein. After the sample was obtained, slight pressure was applied to stop any bleeding.

2.5.3. Tail Cut (TC)

To obtain samples via cutting the tail, mice were located on the rack cage, allowing movement as tail vein groups, and a surgical scalpel was used to obtain the sample. In the cut zone, haemostatic powder (Bioline Pet Styptic Powder) was applied to stop the bleeding that habitually cannot be stopped with finger pressure alone.

2.6. Statistics

The sample size was calculated and recognised as statistically significant if a minimum difference of 20 units between any pair of groups in the 3 groups existed, accepting an alpha risk of 0.05, and a beta risk of 0.2 in a two-sided test. In this condition, a sample size of four animals per group was established.

Data were summarised using mean (standard deviation) and median (1st–3rd quartile) in the case of continuous variables, and by absolute frequencies in the case of categorical variables. The Shapiro–Wilk test confirmed the normal distribution of levels of glucose in all groups. Comparisons between different groups were made using the *t*-test (2 groups) and one-way ANOVA, followed by multiple comparisons (more than two groups). P values lower than 0.05 were considered statistically significant. All analyses were performed using R (version 3.5.2, Foundation for Statistical Computing, Vienna, Austria).

3. Results

3.1. Differences between Sexes

Results obtained analysing FCM levels (Table 1) and glucose levels (Figure 2) show there were no differences between sexes.

Table 1. FCM levels (pg/mL) description by acclimation (PRE, POST) and sex (Female, Male). Descriptives: sample size, minimum, 1st quartile, median, mean, 3rd quartile, maximum, standard deviation, t-statistic, *p* value.

PRE	n	**Min.**	**1st Qu.**	**Median**	**Mean**	**3rd Qu.**	**Max.**	**t**	***p* Value**
Female	12	0.438	0.762	1.086	0.916	1.155	1.223	0.457	0.680
Male	12	0.810	0.959	1.108	1.039	1.153	1.198		
POST	n	**Min.**	**1st Qu.**	**Median**	**Mean**	**3rd Qu.**	**Max.**	**t**	***p* Value**
Female	12	0.532	0.635	0.738	0.708	0.797	0.855	1.221	0.298
Male	12	0.224	0.417	0.610	0.502	0.641	0.672		

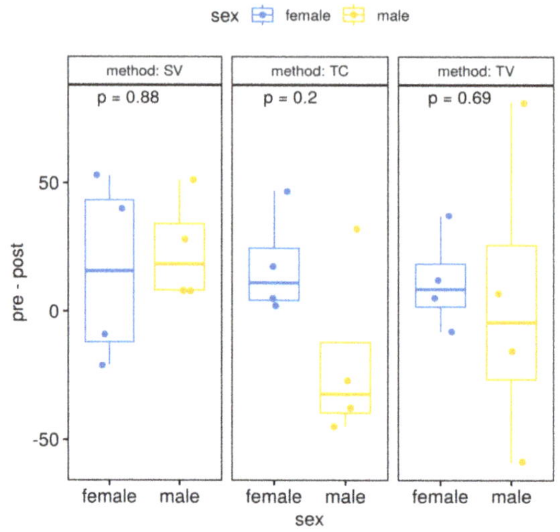

Figure 2. Differences between glucose levels pre and postacclimation between sexes. Results showed no differences by sex in all methods used for sampling (TC group $p = 0.2$; TV group $p = 0.69$; SV group $p = 0.88$).

This is interesting because it is not necessary to use a specific method by sex, and this allows us to analyse all animals by technique, without taking into account this variable, the sex. It also allows us to assess the set of all animals as a single group (instead of separate males and females), so we have a higher sample size for the analysis strategy.

3.2. Differences between Acclimation and Non-Acclimation

Overall, the mean glucose levels were lower in POST- than PRE-acclimation, but these differences were not statistically significant. We see the same trend when analysing the glucose levels by technique (Figure 3).

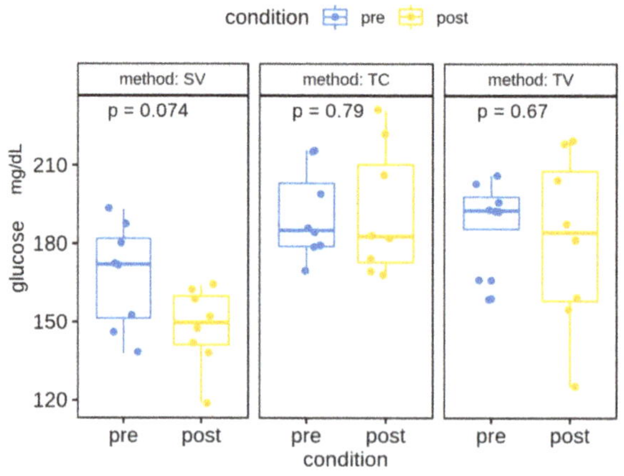

Figure 3. Glucose levels (mg/dL) by technique before (pre-) and after (post-) acclimation.

3.3. Differences between Measurement Methods

Our results show that the mean glucose levels differed in the three evaluated methods: SV, TC and TV before and after acclimation (Table 2, Figures 4 and 5). Specifically, glucose levels were lower in the SV group compared with the other two methods: TV and TC (both in PRE as in POST). These results showed statistically significant differences without acclimation between groups: SF–TC (*p*-value < 0.05) and SV–TV (*p*-value < 0.1). Statistically significant differences were shown post-acclimation between groups: SV–TC (*p*-value < 0.05) and SV–TV (*p*-value < 0.05).

Table 2. Description of glucose levels (mg/dL) by acclimation and sampling methods. Descriptives: sample size, minimum, 1st quartile, median, mean, 3rd quartile, maximum, standard deviation.

		n	Min.	Q1	Median	Mean	Q3	Max.	SD
PRE	TC	8	170	179	185	191	203	216	17.2
	TV	8	159	185.5	192.5	188.4	197.8	206	16.9
	SV	8	138	151.2	172	167.6	181.8	193	19.9
POST	TC	8	168	172.8	182.5	191.9	210	231	24.6
	TV	8	125	158	184	181	207.5	219	33.1
	SV	8	119	141	149.5	147.9	159.8	164	15

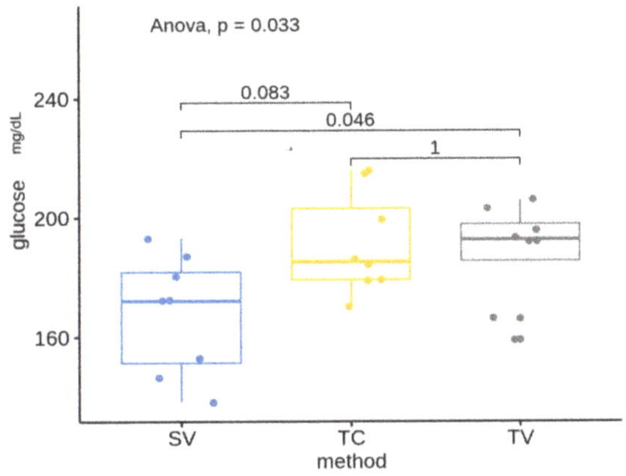

Figure 4. Differences of glucose levels between methods, without acclimation. Results show differences were statistically significant between the SV and TC groups ($p = 0.083$), and between the SV and TV groups ($p = 0.046$).

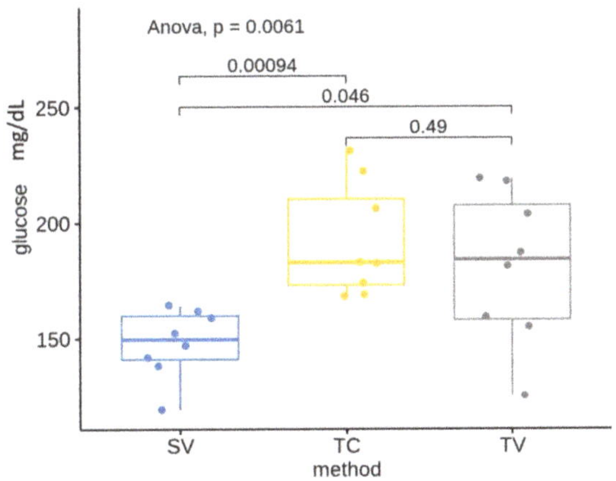

Figure 5. Differences of glucose levels between methods, post-acclimation period. Results show differences were statistically significant between the SV and TC groups ($p = 0.00094$) and between the SV and TV groups ($p = 0.046$).

3.4. Faecal Corticosterone Metabolites Levels

Regarding FCM levels by method, we saw lower FCM levels in the TC groups without these differences being significant (Figure 6).

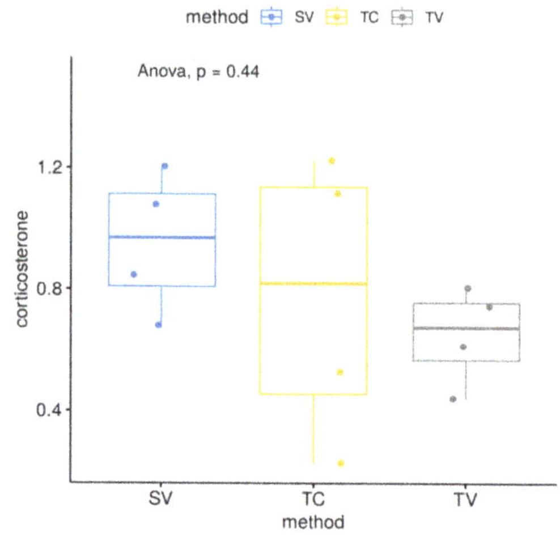

Figure 6. Differences in FCMs (pg/mL) comparing by methods (p = 0.44).

Comparing groups with or without acclimation, FCM levels were significantly higher in measurements without acclimation compared with those with acclimation (Figure 7) but without statistical significance (p = 0.078).

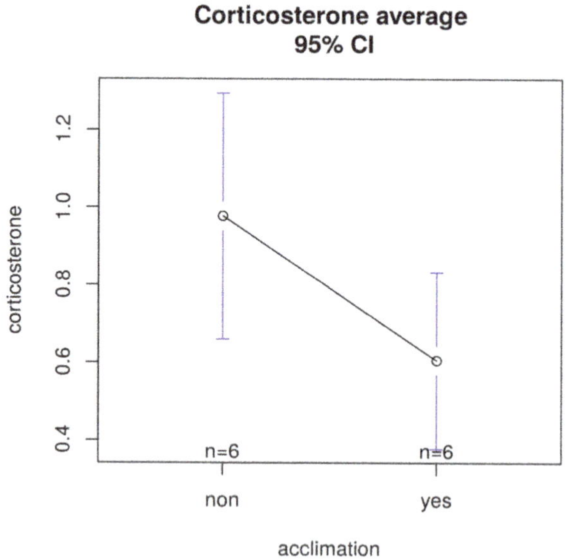

Figure 7. Differences in FCMs (pg/mL) comparing groups with (mean: 0.6052 pg/mL) and without acclimation (mean: 0.9772 pg/mL).

4. Discussion

The study reported here compares the impact of widely used blood sampling methods on glucose and FCM levels in C57Bl/6J mice. Plasma glucose in blood samples obtained from the tail and saphenous vein by different methods was measured. Faecal corticosterone was measured by obtaining faeces directly from cages without any kind of manipulation. In mice, the use of caudal veins in different methods or the use of saphenous phlebotomy are common techniques for blood sampling [17,18,20,21]. Some studies evaluate and compare some of them, analysing the quality of blood samples or how stressful the use of such techniques is for animals [17,26,32]. However, the handling need for these techniques is not considered and it is well known that handling is a source of stress for animals [12,22]. When mice are handled using a home cage or external tunnel, they show less anxiety in an elevated plus maze than those picked up by the tail [9,24]. Further, compared with mice picked up by the tail, mice handled by non-aversive tunnel or cupping methods have reduced plasma corticosterone, reduced blood glucose and improved glucose tolerance [33]. Monitoring endocrine functions in mice is constrained seriously by the adverse effects of blood sampling [34,35]. Therefore, non-invasive techniques to monitor stress hormones are highly demanded in the laboratory as well as in field research [10,29].

Our results show lower glucose levels when saphenous phlebotomy is used comparing groups with or without acclimation (mean PRE: 167.6 mg/dL; mean POST: 147.6 mg/dL). These results are not statistically significant but show an important reduction close to significance. Routine acclimation was obtained based on different studies [9,22]. It has been described that prior manipulation and habituation reduce anxiety and stress in mice, facilitate routine management, improve animal welfare, decrease shortages of data and improve experimental reliability [10]. However, there is no standard and established technique so we have to consider that these results could be due to a short routine acclimation period, and it would be interesting to consider longer periods when blood sampling techniques are needed.

Groups in which the tail needed to be manipulated did not show this downward trend. Despite the beneficial effects of handling being known, the tail-pick-up approach, which is particularly stressful, is still widely used. Some handling procedures such as picking up animals by the tail may actually simulate the act of being captured and provoke stress responses [9,36]. The method in which a saphenous vein was used for sampling (SV) resulted in lower glucose levels. This technique, which apparently needs high manipulation and could be more stressful than TC or TV, without less handling, could be considered as a less stressful method. These results confirm that tail manipulation is a stressful technique, as other studies have shown [32,37,38]. In recent years, less aversive handling methods (for example, tunnel handling or bowl hand) have been shown to mitigate anxiety and depressive behaviours in mice [24,39]. In this sense, the SV sampling method allows the animals to be sheltered (hidden) in the tunnel while the sample is taken, while, in the other two methods, the animals cannot hide, and the tail is manipulated to a greater or lesser extent. The standard handling method of picking up mice by the tail increases behavioural and physiological measures of stress and anxiety, which may explain our results [24,32,35,40].

Comparing groups by technique without acclimation, we confirmed animals manipulated by the tail (TC and TV groups) showed no differences between them, while the differences between the saphenous group and the TC group were statistically significant ($p = 0.046$). Kress and cols. considered the saphenous technique more stressful than the puncture of the facial vein because of the time required for sampling and the increase of corticosterone in urine production, but our results show lower glucose levels, indicating that the SV method is less stressful than the TV or TC methods [20]. Comparing techniques after the handling routine, after the acclimation period was completed, we saw similar results in which the statistically significant differences were between the SV and TV groups, with a p-value = 0.046, and between the SV and TC groups, with a p-value = 0.00094. These results confirm that manipulation of the tail and the handling needed for a tail vein phlebotomy or a tail vein docking is a stressful technique, with the independence of a previous acclimation period.

As an additional measure of stress levels, faecal samples were collected to measure FCMs. Typically, blood serum or plasma is used to measure corticosterone concentrations, with an increase implying an acute stress response [27], but the measurement of FCMs has been proposed due to the advantage of it being a non-invasive technique. Corticosterone in the blood can usually be demonstrated after a few minutes but it requires a quick analysis because of its short half-life in plasma [11,19]. Nevertheless, corticosterone is a stable metabolite to detect in faeces. In fact, measurement of FCMs has become a common approach in evaluating HPA activity because it is non-invasive and provides a relatively more stable and time-integrated picture of HPA activation than do circulating corticosterone levels [41]. Therefore, in order to avoid the activation of the HPA axis, which quickly leads to the secretion of glucocorticoids associated with restraint and manipulation needs to blood sampling, we analysed faecal corticosterone metabolites using DetectX® Cortisol Immunoassay, as the protocol described. To minimise interaction with animals to stop them stressing or influence the results we analysed faecal metabolites by cage 24 h after sampling [37]. It has been demonstrated that 24 h FCM collection avoids the problems of circadian variations and it is a good method for both chronic and acute studies [14]. Our results reflected a reduction in FCMs after acclimation but without statistical significance. However, since we only measured the total amount of FCMs excreted 0–24 h after blood sampling by cage, additional faecal sampling would be necessary to assess whether there is an effect of blood sampling on FCMs.

We know that isolation allows easier collection of faeces and would have increased the number of samples, but it has been demonstrated to induce an increase in corticosterone levels with respect to animals housed in standard cages, in groups of two-to-three per cage [42]. Alterations in neurochemistry, metabolism, growth, reproduction and dopaminergic hyperactivity have been found in shared neural regions, implicated with the performance of stereotypy in isolated mice. Animals habituated to stable groups show less stress than when in individual housing [18,43], but our results do not show differences between methods and between pre- and post-acclimation ($p = 0.078$). The TV method shows lower FCM metabolites than the other methods, and FCM levels are significantly higher without acclimation than those obtained with acclimation.

Interestingly, there are some studies concerning the differences in anxiety and stress responsiveness by sex [42,44], but we found no statistically significant differences in this respect. Hurst and West demonstrated similar responses to the different handling methods in males and females mice. Further, their studied ICR and C57Bl6/J strains, and both strains showed the same general differences in responses on tunnel and tail handling [9]. On the basis of the results obtained in this work, the following studies will be proposed in which some methodological limitations concerning the experimental design or the consideration of multifactorial designs in the analysis will be improved, allowing joint treatment of all the variables of interest, with a global approach to all the possible sources of variability. In light of all these data, we suggest the use of the SV technique for blood sampling as a less stressful method and highlight the importance of acclimation in any manipulation and any technique in which restraint is needed.

5. Conclusions

We conclude that acclimation must be considered a requirement to minimise stress in mice blood sampling techniques. The use of the saphenous vein for blood sampling, despite the handling needed, could be considered a less stressful technique than tail docking or tail phlebotomy.

These results allow researchers to select the most welfare-friendly blood sampling technique objectively from studied techniques for a given experiment and contribute to refinement within the 3R concept, the essential concept on which laboratory animal science is based.

Author Contributions: Conceptualization, V.B. and N.M.; methodology, and acquisition of the data, A.M. and D.G.; analysis and interpretation of the data, F.G.-G.; analysis and interpretation of the data N.M.; writing—review and editing, V.B. All authors have read and agreed to the published version of the manuscript.

Funding: This research received no external funding.

Institutional Review Board Statement: The animal study protocol was approved by the Institutional Ethics Committee of Prince Felipe Research Center (CIPF) and by Valencia Government as national laws required with code 2018/VSC/PEA/0075 on 18 May 2018.

Informed Consent Statement: Not applicable.

Data Availability Statement: All the data of the study can be made available upon request.

Conflicts of Interest: The authors declare no conflict of interest.

References

1. Poole, T. Happy animals make good science. *Lab. Anim.* **1997**, *31*, 116–124. [CrossRef] [PubMed]
2. Prescott, M.J.; Lidster, K. Improving quality of science through better animal welfare: The NC3Rs strategy. *Lab Anim.* **2017**, *46*, 152–156. [CrossRef]
3. National Centre for the Replacement Refinement & Reduction of Animal in Research (NC3Rs). *Blood Sampling Mouse*; NC3Rs: London, UK, 2013.
4. Morton, D.B.; Abbot, D.; Barcley, R.; Close, B.S.; Ewbank, R.; Gask, D.; Heath, M.; Mattic, S.; Poole, T.; Seamer, J.; et al. Removal of blood from laboratory mammals and birds. First report of the BVA/FRAME/RSPCA/UFAW Joint Working Group on Refinement. *Lab. Anim.* **1993**, *27*, 1–22.
5. Dülsner, A.; Hack, R.; Krüger, C.; Pils, M.; Scherer, K.; Schmelting, B.; Schmidt, M.; Weinert, H.; Jourdan, T. *Recommendation for Blood Sampling in Laboratory Animals, Especially Small Laboratory Animals*; Gesellschaft für Versuchstierkunde: Freiburg, Germany, 2017.
6. Hoff, J. Methods of Blood Collection in the Mouse. *Lab. Anim.* **2000**, *29*, 47–53.
7. Forbes, N.; Brayton, C.; Grindle, S.; Shepherd, S.; Tyler, B.; Guarnieri, M. Morbidity and mortality rates associated with serial bleeding from the superficial temporal vein in mice. *Lab. Anim.* **2010**, *39*, 236–240. [CrossRef]
8. Russell, W.M.S.; Burch, R.L. *The Principles of Humane Experimental Technique*; Universities Federation for Animal Welfare: Hertfordshire, UK, 1959.
9. Hurst, J.L.; West, R.S. Taming anxiety in laboratory mice. *Nat. Methods* **2010**, *7*, 825–826. [CrossRef]
10. Meyer, N.; Kröger, M.; Thümmler, J.; Tietze, L.; Palme, R.; Touma, C. Impact of three commonly used blood sampling techniques on the welfare of laboratory mice: Taking the animal's perspective. *PLoS ONE* **2020**, *15*, e0238895. [CrossRef] [PubMed]
11. Balcombe, J.P.; Barnard, N.D.; Sandusky, C. Laboratory routines cause animal stress. *J. Am. Assoc. Lab. Anim. Sci.* **2004**, *43*, 42–51.
12. Roughan, J.V.; Sevenoaks, T. Welfare and Scientific Considerations of Tattooing and Ear Tagging for Mouse Identification. *J. Am. Assoc. Lab. Anim. Sci.* **2019**, *58*, 142–153. [CrossRef]
13. Meijer, M.K.; Sommer, R.; Spruijt, B.M.; van Zutphen, L.F.M.; Baumans, V. Influence of environmental enrichment and handling on the acute stress response in individually housed mice. *Lab. Anim.* **2007**, *41*, 161–173. [CrossRef]
14. Festing, M.F.; Baumans, V.; Combes, R.D.; Haider, M.; Hendriksen, C.F.; Howard, B.R.; Lovell, D.P.; Moore, G.J.; Overend, P.; Wilson, M.S. Reducing the Use of Laboratory Animals in Biomedical Research: Problems and Possible Solutions. *Altern. Lab. Anim.* **1998**, *26*, 283–301. [CrossRef]
15. Howard, B.R. Control of variability. *ILAR J.* **2002**, *43*, 194–201. [CrossRef] [PubMed]
16. Kim, S.; Foong, D.; Cooper, M.S.; Seibel, M.J.; Zhou, H. Comparison of blood sampling methods for plasma corti-costerone measurement in mice with minimal stress-related artefacts. *Steroids* **2018**, *135*, 69–72. [CrossRef]
17. Christensen, S.D.; Mikkelsen, L.F.; Fels, J.J.; Bodvarsdóttir, T.B.; Hansen, A.K. Quality of plasma sampled by different methods for multiple blood sampling in mice. *Lab. Anim.* **2009**, *43*, 65–71. [CrossRef] [PubMed]
18. Aasland, K.E.; Skjerve, E.; Smith, A.J. Quality of blood samples from the saphenous vein compared with the tail vein during multiple blood sampling of mice. *Lab. Anim.* **2010**, *44*, 25–29. [CrossRef]
19. Lambillotte, C.; Gilon, P.; Henquin, J.C. Direct glucocorticoid inhibition of insulin secretion. An in vitro study of dexame-thasone effects in mouse islets. *J. Clin. Invest.* **1997**, *99*, 414–423. [CrossRef]
20. Kress, A.P.A.; Zhang, Y.; Kaiser-Vry, A.R.; Sauer, M.B. A Comparison of Blood Collection Techniques in Mice and their Effects on Welfare. *J. Am. Assoc. Lab. Anim. Sci.* **2022**, *61*, 287–295. [CrossRef] [PubMed]
21. Parasuraman, S.; Raveendran, R.; Kesavan, R. Blood sample collection in small laboratory animals. *J. Pharmacol. Pharmacother.* **2010**, *1*, 87–93. [CrossRef]
22. Gouveia, K.; Hurst, J.L. Reducing Mouse Anxiety during Handling: Effect of Experience with Handling Tunnels. *PLoS ONE* **2013**, *8*, e66401. [CrossRef]

23. Meaney, M.J.; Diorio, J.; Francis, D.; Widdowson, J.; LaPlante, P.; Caldji, C.; Sharma, S.; Seckl, J.R.; Plotsky, P.M. Early environmental regulation of forebrain glucocorticoid receptor gene expression: Implications for adrenocortical responses to stress. *Dev. Neurosci.* **1996**, *18*, 49–72. [CrossRef]
24. Gouveia, K.; Hurst, J.L. Improving the practicality of using non-aversive handling methods to reduce background stress and anxiety in laboratory mice. *Sci. Rep.* **2019**, *9*, 20305. [CrossRef]
25. Mormède, P.; Andanson, S.; Aupérin, B.; Beerda, B.; Guémené, D.; Malmkvist, J.; Manteca, X.; Manteuffel, G.; Prunet, P.; van Reenen, C.G.; et al. Exploration of the hypothalamic–pituitary–adrenal function as a tool to evaluate animal welfare. *Physiol. Behav.* **2007**, *92*, 317–339. [CrossRef] [PubMed]
26. Moore, E.S.; Cleland, T.A.; Williams, W.O.; Peterson, C.M.; Singh, B.; Southard, T.L.; Pasch, B.; Labitt, R.N.; Daugherity, E.K. Comparing Phlebotomy by Tail Tip Amputation, Facial Vein Puncture, and Tail Vein Incision in C57BL/6 Mice by Using Physiologic and Behavioral Metrics of Pain and Distress. *J. Am. Assoc. Lab. Anim. Sci.* **2017**, *56*, 307–317.
27. Thanos, P.K.; Cavigelli, S.A.; Michaelides, M.; Olvet, D.M.; Patel, U.; Diep, M.N.; Volkow, N.D. A non-invasive method for detecting the metabolic stress response in rodents: Characterization and dis-ruption of the circadian corticosterone rhythm. *Physiol. Res.* **2009**, *58*, 219–228. [CrossRef] [PubMed]
28. Palme, R. Non-invasive measurement of glucocorticoids: Advances and problems. *Physiol. Behav.* **2018**, *199*, 229–243. [CrossRef]
29. Touma, C.; Palme, R.; Sachser, N. Analyzing corticosterone metabolites in fecal samples of mice: A noninvasive technique to monitor stress hormones. *Horm. Behav.* **2004**, *45*, 10–22. [CrossRef] [PubMed]
30. Crawley, J.N.; Belknap, J.K.; Collins, A.; Crabbe, J.C.; Frankel, W.; Henderson, N.; Hitzemann, R.J.; Maxson, S.C.; Miner, L.L.; Silva, A.J.; et al. Behavioral phenotypes of inbred mouse strains: Implications and recommendations for molecular studies. *Psychopharmacology* **1997**, *132*, 107–124. [CrossRef]
31. Ortega-Saez, I.; Díez-Solinska, A.; Grífols, R.; Martí, C.; Zamora, C.; Muñoz-Culla, M.; Vegas, O.; Azkona, G. Individualized Housing Modifies the Immune–Endocrine System in CD1 Adult Male Mice. *Animals* **2023**, *13*, 1026. [CrossRef]
32. Dobrakovová, M.; Kvetnanský, R.; Oprsalová, Z.; Jezová, D. Specificity of the effect of repeated handling on sympathetic-adrenomedullary and pitui-tary-adrenocortical activity in rats. *Psychoneuroendocrinology* **1993**, *18*, 163–174. [CrossRef]
33. Ghosal, S.; Nunley, A.; Mahbod, P.; Lewis, A.G.; Smith, E.P.; Tong, J.; D'Alessio, D.A.; Herman, J.P. Mouse handling limits the impact of stress on metabolic endpoints. *Physiol. Behav.* **2015**, *150*, 31–37. [CrossRef]
34. Ayala, J.E.; Bracy, D.P.; McGuinness, O.P.; Wasserman, D.H. Considerations in the Design of Hyperinsulinemic-Euglycemic Clamps in the Conscious Mouse. *Diabetes* **2006**, *55*, 390–397. [CrossRef]
35. Ayala, J.E.; Samuel, V.T.; Morton, G.J.; Obici, S.; Croniger, C.M.; Shulman, G.I.; Wasserman, D.H.; McGuinness, O.P. Standard operating procedures for describing and performing metabolic tests of glucose homeostasis in mice. *Dis. Model. Mech.* **2010**, *3*, 525–534. [CrossRef]
36. Resch, M.; Neels, T.; Tichy, A.; Palme, R.; Rülicke, T. Impact assessment of tail-vein injection in mice using a modified anaesthesia induction chamber versus a common restrainer without anaesthesia. *Lab. Anim.* **2019**, *53*, 190–201. [CrossRef]
37. Harikrishnan, V.; Hansen, A.K.; Abelson, K.S.; Sørensen, D.B. A comparison of various methods of blood sampling in mice and rats: Effects on animal welfare. *Lab. Anim.* **2018**, *52*, 253–264. [CrossRef] [PubMed]
38. Madetoja, J.; Madetoja, M.; Makinen, J.; Riuttala, E.; Jokinen, J. Blood sampling from the tail vein, in comparison with two other techniques, causes less stress to mice. *Scand. J. Lab. Anim. Sci.* **2009**, *36*, 215–221.
39. Marcotte, M.; Bernardo, A.; Linga, N.; Pérez-Romero, C.A.; Guillou, J.-L.; Sibille, E.; Prevot, T.D. Handling Techniques to Reduce Stress in Mice. *J. Vis. Exp.* **2021**, *175*, e62593.
40. Teilmann, A.C.; Kalliokoski, O.; Sørensen, D.B.; Hau, J.; Abelson, K.S.P. Manual versus automated blood sampling: Impact of repeated blood sampling on stress parameters and behavior in male NMRI mice. *Lab. Anim.* **2014**, *48*, 278–291. [CrossRef]
41. Sheriff, M.J.; Dantzer, B.; Delehanty, B.; Palme, R.; Boonstra, R. Measuring stress in wildlife: Techniques for quantifying glucocorticoids. *Oecologia* **2011**, *166*, 869–887. [CrossRef]
42. Gomez-Sanchez, E.P.; Gomez-Sanchez, C.E. The effect of gonadectomy and aromatase inhibition on the excretion of 19-nordeoxycorticosterone in rats. *J. Steroid Biochem. Mol. Biol.* **1991**, *39*, 185–188. [CrossRef]
43. Krohn, T.; Sørensen, D.; Ottesen, J.; Hansen, A. The effects of individual housing on mice and rats: A review. *Anim. Welf.* **2006**, *15*, 343–352. [CrossRef]
44. Touma, C.; Sachser, N.; Möstl, E.; Palme, R. Effects of sex and time of day on metabolism and excretion of corticosterone in urine and feces of mice. *Gen. Comp. Endocrinol.* **2003**, *130*, 267–278. [CrossRef] [PubMed]

Disclaimer/Publisher's Note: The statements, opinions and data contained in all publications are solely those of the individual author(s) and contributor(s) and not of MDPI and/or the editor(s). MDPI and/or the editor(s) disclaim responsibility for any injury to people or property resulting from any ideas, methods, instructions or products referred to in the content.

animals

MDPI

Article

Evaluation of Parameters Which Influence Voluntary Ingestion of Supplements in Rats

Santiago Ruvira [1,2,†], Pilar Rodríguez-Rodríguez [1,2], Silvia Cañas [2,3,4], David Ramiro-Cortijo [1,2], Yolanda Aguilera [2,3,4], David Muñoz-Valverde [5] and Silvia M. Arribas [1,2,*]

1 Department of Physiology, Faculty of Medicine, Universidad Autónoma de Madrid, C/Arzobispo Morcillo 2, 28029 Madrid, Spain
2 Food, Oxidative Stress and Cardiovascular Health (FOSCH) Research Group, Universidad Autónoma de Madrid, Ciudad Universitaria de Cantoblanco, 28049 Madrid, Spain
3 Department of Agricultural Chemistry and Food Science, Faculty of Science, Universidad Autónoma de Madrid, C/Francisco Tomás y Valiente 7, 28049 Madrid, Spain
4 Institute of Food Science Research (CIAL), Universidad Autónoma de Madrid (UAM-CSIC), C/Nicolás Cabrera 9, 28049 Madrid, Spain
5 Animal House Facility, Veterinary Unit, Faculty of Medicine, Universidad Autónoma de Madrid, C/Arzobispo Morcillo 2, 28029 Madrid, Spain
* Correspondence: silvia.arribas@uam.es
† Student of PhD Program in Pharmacology and Physiology, Doctoral School, Universidad Autónoma de Madrid, Ciudad Universitaria de Cantoblanco, 28049 Madrid, Spain.

Simple Summary: Preclinical studies evaluating the safety and efficacy of new drugs require experimental animals, which is important to ensure both adequate dosage and animal welfare. Administration through voluntary ingestion can improve animal wellbeing. We aimed to develop a protocol using small gelatin cubes to train rats, assessing the influence of age, sex, fasting, flavors (vanilla), and sweeteners (sucralose) to accept the new food. We tested the usefulness of the protocol to supplement rats during lactation. We demonstrated that most animals were easily trained to accept the gelatin cube in 2–3 days. However, some rats refused the new food even after 8 days. The proportion of rats that did not train was higher in adult males compared to young males, adults, and young females, suggesting the influence of sex. Four-hour food deprivation reduced the time for acceptance only in females, but flavoring or sweeteners in the gelatin did not modify it. Rats trained prior to gestation remembered training 2 months later and ate a gelatin containing a supplement daily during lactation for 1–5 min, without problems with the pups. We conclude that gelatin-based supplementation can be used for drug studies in rats, ensuring adequate dosage and wellbeing, which is important for the detection of non-trained rats.

Citation: Ruvira, S.; Rodríguez-Rodríguez, P.; Cañas, S.; Ramiro-Cortijo, D.; Aguilera, Y.; Muñoz-Valverde, D.; Arribas, S.M. Evaluation of Parameters Which Influence Voluntary Ingestion of Supplements in Rats. *Animals* **2023**, *13*, 1827. https://doi.org/10.3390/ani13111827

Academic Editor: Garikoitz Azkona

Received: 28 March 2023
Revised: 29 May 2023
Accepted: 30 May 2023
Published: 31 May 2023

Copyright: © 2023 by the authors. Licensee MDPI, Basel, Switzerland. This article is an open access article distributed under the terms and conditions of the Creative Commons Attribution (CC BY) license (https://creativecommons.org/licenses/by/4.0/).

Abstract: Drug safety and efficacy studies frequently use oral gavage, but repetitive usage may cause problems. Administration through voluntary ingestion represents an opportunity for refinement. We aimed to develop a protocol for voluntary ingestion of gelatin-based supplements in rats, assessing the influence of age, sex, fasting (4 h), and additives (vanilla, VF; sucralose, S), and to test it in lactating dams. Three-week-old and 5-month-old Sprague-Dawley rats were placed individually in an empty cage containing a gelatin cube and trained daily (5 days/week), recording the day the whole cube was consumed (latency). Rats trained prior to gestation were offered a gelatin containing 250 mg/kg cocoa shell extract (CSE) during lactation. Rats that did not eat the cube after 8 training days were considered non-habituated, with a proportion similar in young males (7.1%), young females (11.1%), and adult females (10.3%), but significantly higher in adult males (39.3%). Excluding non-habituated rats, latency was 2–3 days, without differences between young and adult rats (p = 0.657) or between males and females (p = 0.189). VF or VF + S in the gelatin did not modify latency, while fasting significantly reduced it in females (p = 0.007) but not in males (p = 0.501). During lactation, trained females ate the CSE-gelatin within 1–5 min without litter problems. Conclusions: Acceptance of a gelatin-based supplement is negatively influenced by male sex, facilitated by fasting, and not modified by additives. Training is remembered after 2 months and does not interfere with lactation. Gelatin-based voluntary ingestion is suitable to administer drugs that need to pass through the digestive system, ensuring adequate dosage, and is important to detect non-habituated rats prior to the study. The current protocol may be implemented by training the rats in their own cage.

Keywords: food supplement; gelatin; voluntary ingestion; habituation; rats; sex; refinement

1. Introduction

Animal experimentation constitutes a relevant aspect of biomedical research and a key piece of preclinical research to evaluate the safety and efficacy of new drugs. This type of research requires repetitive administration of substances, which is a critical component of experimental design and represents a good opportunity for refinement [1].

The development of nutraceuticals or the testing of drugs that need to pass through the digestive system involves the oral administration of selected compounds. The most common form is oral gavage, which consists of force-feeding a substance with a tube inserted in the mouth and then into the stomach. Gavage provides a rapid and efficient means of accurately delivering oral dosing to rodents, and it is applicable to unanesthetized animals. However, oral administration is not exempt from serious complications, including reflux, aspiration, esophageal irritation, and inflammation. The main complication is reflux and respiratory aspiration, which have been reported to cause around 20% mortality in some studies, depending on dosage [2,3]. Besides, the procedure causes stress, as evidenced by the increase in corticosterone levels [4,5]. Even though these alterations may be minimized by appropriate handling by an experienced technician [6], they may interfere with animal welfare and with the interpretation of results. Thus, reducing the stress associated with chronic drug delivery to experimental animals is desirable. Besides, the use of oral gavage may be particularly detrimental in periods such as gestation or lactation since it may influence maternal behavior with detrimental effects on pups [7].

An alternative to oral gavage is the administration of the compound of interest mixed with an attractive vehicle by voluntary ingestion. Among vehicles, jam [8], gelatin [9], cookie dough [10], or nut paste [11] have been successfully used in rodents. One of the challenges is to guarantee adequate dosage, i.e., to ensure full ingestion, particularly if the substance under study is bitter [10], and most of the above-mentioned studies include flavoring substances and sweeteners in the vehicle. It must be noted that for some types of studies, it would be preferable to avoid a vehicle with a high caloric content.

In the voluntary acceptance of supplements, a key issue is the fact that rodents have a natural neophobia, and recent data indicate that this may be modulated by sex and age [12]. Therefore, our objective was to evaluate several factors that may influence the voluntary ingestion of a novel food in rats and refine protocols for supplement administration. Using gelatin cubes (GC), we studied in Sprague-Dawley rats the effect of intrinsic (sex and age) and extrinsic (additives and fasting) factors on (1) the percentage of successfully trained rats, (2) the training time requested for full acceptance (latency to ingestion), and (3) the capacity to recall after a period of training. Besides, we have also evaluated the effectiveness of the training to supplement lactating rats with an extract derived from cocoa shell (CSE), a by-product derived from the chocolate manufacturing industry rich in bioactive compounds with potential to reduce cardiometabolic diseases, which we have previously analyzed in vitro [13]. Based on our findings, we propose some key points that may be useful for researchers willing to use this method of supplementation.

2. Materials and Methods

2.1. Experimental Animals

Sprague-Dawley rats from the colony bread at the animal house facility of the Universidad Autónoma de Madrid (ES-28079-0000097) were used at the following age points: 3-week-old (young rats; $n = 9$ females and $n = 14$ males); 5-month-old (adult rats; $n = 35$ females and $n = 40$ males). We also used 7 lactating rats to evaluate the acceptance of a gelatin containing a compound of interest (CSE) after training. The experimental procedures conformed to the Guidelines for the Care and Use of Laboratory Animals (National Institutes of Health publication no. 85-23, revised in 1996), the Spanish legislation (RD 53/2013), and the Directive 2010/63/EU on the protection of animals, and were approved by the Ethics Review Board of Universidad Autónoma de Madrid and the Regional Committee of Comunidad Autónoma de Madrid (PROEX 19/04; approval date: 20 March 2019).

After weaning, rats from the same sex were housed in groups of 3–5 per box and maintained in type III cages (24 × 19 × 45 cm; length × height × width) or type IV cages (55 × 18 × 32 cm; length × height × width), according to the number and weight of the rats, with poplar bedding. Cellulose nestlets and play tunnels (Index Research S.L., Madrid, Spain) were used for environmental enrichment. The animals were constantly kept under controlled conditions of temperature (22 °C), humidity (40%), and photoperiod of 12 h of darkness and 12 h of light. Rats were fed *ad libitum* (except during fasting periods) with a diet containing 51.7% carbohydrates, 21.4% protein, 5.1% lipids, 3.9% fiber, 5.7% minerals, and 12.2% humidity (SafeA03; Safe Augy, France). Drinking water was also provided *ad libitum*. The animal health was regularly monitored by staff, ensuring rats were free from pathogens that may interact with any of the parameters studied.

2.2. Gelatin Cubes Preparation

The cubes for training were prepared with 100% bovine gelatin (Inkafoods, S.L., Barcelona, Spain) in water at a concentration of 140 g/L. Water was first heated at 50–60 °C in a glass beaker, and the gelatin was slowly added, stirring it until complete dissolution. At this point, the different additives can be incorporated into the mixture: vanilla flavor (VF, 4.8 mL/L; MyProtein, Hut.com Ltd., Manchester, England) as a non-caloric flavoring agent, alone or with sucralose (S, 0.6 g/L; sucralin, sucralose S.L., Barcelona, Spain). The mixture was transferred to a mold, ensuring homogeneous distribution, to prepare 1 cm^3-sized GC, since preliminary results showed that this size is easily handled by rats (Figure S1). The mold was left to cool down to allow solidification, first at room temperature and then in the fridge. Thereafter, the individual cubes were extracted from the mold and stored in the fridge in plastic bags, or, if the gelatin was not going to be consumed immediately, they were stored frozen at −20 °C. Cubes can be defrosted at room temperature without changing their consistency or shape.

Cubes with cocoa shell extract (CSE). Cocoa shell was kindly supplied by Chocolates Santocildes S.A. (https://www.chocolatessantocildes.com/ (accessed on 10 March 2023), Castilla y León, León, Spain). An extract (CSE) rich in phenolic compounds was prepared as previously described [14]. The CSE was incorporated after the gelatin was dissolved. The dose of CSE used for the study was 250 mg/kg. The animals were first weighed, and the cubes were prepared according to the weight of the rat for supplementation during the entire lactation period.

2.3. Administration Protocols

Training protocol. The rat was placed in an individual empty box without bedding with the GC and left for 2 h. At the end of this period, we recorded if the animal ingested part or the whole cube. This procedure was carried out for 5 days/week (Monday to Friday) at the same time (9:00 to 11:59 a.m.). The first day that the rat ingested the entire GC was considered the acceptance day, and the number of days from the first exposure was counted as the latency period. The cubes that were not eaten were discarded.

Fasting protocol. To study the influence of fasting, a group of adult males and females were deprived of feed in their own cage for 4 h prior to the training protocol (from 7:00 to 11:00 a.m.).

Training protocol in their own cage. In a group of rats who did not accept the GC in an empty cage, the cube was presented individually in the usual cage, placing it in the hopper without food.

Recall protocol. A group of adult males and females previously trained were exposed to the NF gelatin 1 month later, and the latency to ingestion was recorded. In these rats, the time needed for complete ingestion was also registered.

Protocol for CSE supplementation in lactating rats. Seven female rats previously trained were mated, and after giving birth, they were weighed to prepare the CSE GC at a dose of 250 mg/kg. From the second day postpartum, the rats were offered the CSE cube 5 days/week during the entire lactation period. For this protocol, once the rat ate the whole cube, it was immediately returned to the cage with the pups.

2.4. Statistical Analysis

Data analysis was performed by GraphPad software (GraphPad Prism, version 8.0, Boston, MA, USA). The distribution of the variables was evaluated using the Kolmogorov–Smirnov test. Since some of the data did not follow a normal distribution and the sample size was small in some protocols, quantitative data were reported as the median and maximum and minimum levels, and statistical analysis was performed by Kruskal–Wallis or Mann–Whitney's U tests. To assess differences in the proportion of non-habituated rats, Fisher's exact test was performed. Significance was established with a *p*-value (p) < 0.05.

3. Results

We evidenced that some rats did not eat the gelatin after 8 days of training. Some preliminary data indicated that they did not eat it even if exposed for 3 weeks. Based on this, and to avoid unnecessary stress, rats that did not eat the gelatin cube in 8 days were counted as "non-habituated".

3.1. Influence of Age and Sex on Habituation and Latency

In adult males, the proportion of non-habituated rats was significantly higher compared to young males ($\chi^2 = 4.728$; $p = 0.03$), while no significant difference was detected between young and adult females ($\chi^2 = 0.004$; $p = 0.948$). In young rats, the proportion of non-habituated rats was similar between sexes ($\chi^2 = 0.109$; $p = 0.742$), while in adult rats it was higher in males compared to females ($\chi^2 = 6.440$; $p = 0.011$) (Figure 1A). We evaluated differences in the latency to ingestion in all rats, considering 8 days of latency for non-habituated rats. Latency was significantly higher in adult males compared to young males and of near statistical significance in adult males compared to adult females ($p = 0.056$). No significant differences were found between young and adult females (Figure 1B).

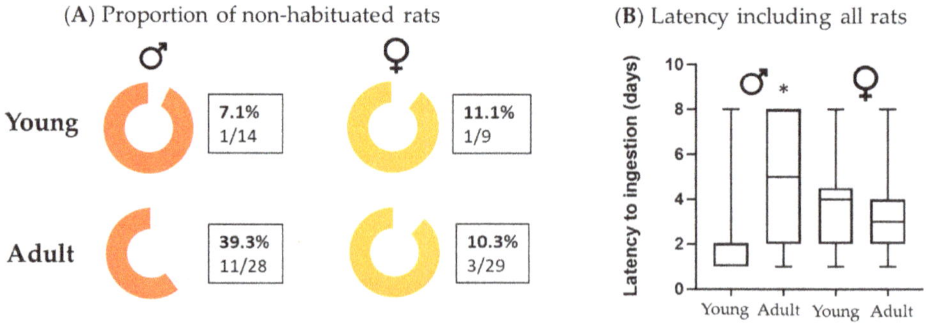

Figure 1. Influence of sex and age on the proportion of non-habituated Sprague-Dawley rats (**A**) and latency to ingestion including all rats (**B**). Young (3-week-old) and adult (5-month-old) rats. Data in (**B**) show the median and maximum and minimum days of latency. Number of rats from each group is shown in (**A**). Statistical analysis was performed by Fisher's exact test (**A**) or Kruskal–Wallis test (**B**); * *p*-value < 0.05 adult males compared to young males.

We assessed the latency to ingestion in the population of rats that accepted the novel food (i.e., excluding non-habituated rats). We did not find statistical differences between groups, suggesting no influence of sex or age on latency (Figure 2).

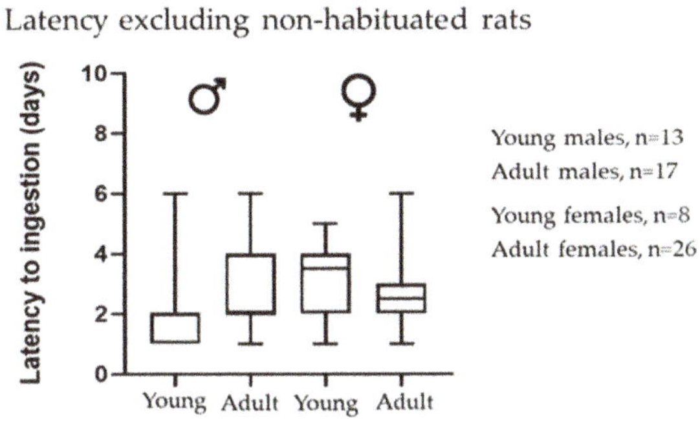

Figure 2. Influence of sex and age on the latency to ingestion excluding non-habituated rats. Young (3-week-old) and adult (5-month-old) rats. Data show the median and maximum and minimum days of latency. Statistical analysis was performed by Kruskal–Wallis test; *n*, number of rats.

In a group of 5 non-habituated males, we evaluated if the presentation of the GC in their own cage would improve acceptance. We observed that they did not try the GC when placed in the empty cage. However, on the same day, all tried the GC if it was placed in the food hopper of their own cage (Video S1). The following day, 4 out of the 5 rats ate the cube when the gelatin was placed in the empty cage.

3.2. Influence of Additives and Fasting on Habituation and Latency

The influence of additives (NF, VF alone, or VF + S) was evaluated in adult rats. Additives did not influence the proportion of non-habituated rats, which was similar in both males ($\chi^2 = 1.504$; $p = 0.471$) and females ($\chi^2 = 0.932$; $p = 0.627$) (Figure 3A). Similarly, the latency to ingestion, including all rats, did not show significant differences between groups (NF, VF, and VF + S) either in males or females (Figure 3B).

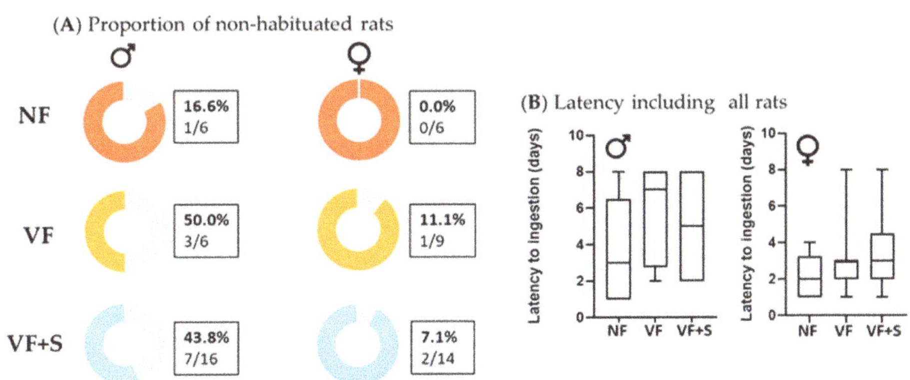

Figure 3. Influence of additives on the proportion of non-habituated Sprague-Dawley rats (**A**) and latency to ingestion excluding non-habituated rats (**B**). Young (3-week-old) and adult (5-month-old) rats. NF, neutral flavor; VF, vanilla flavor; VF + S, vanilla flavor, and sucralose. Data in (**B**) show the median and maximum and minimum levels of latency. Number of rats from each group is shown in figure (**A**). Statistical analysis was performed by Fisher's exact test (**A**) or Kruskal–Wallis test (**B**).

Latency to ingestion, excluding non-habituated rats, was not statistically different between groups (NF, VF, or VF + S) in males or females (Figure 4).

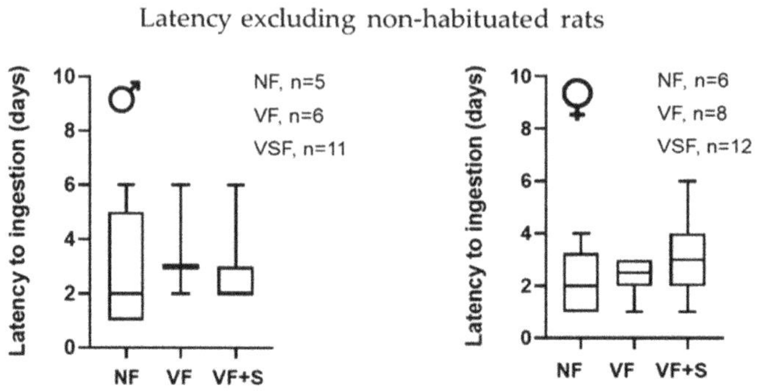

Figure 4. Influence of additives on latency to ingestion excluding non-habituated rats. Young (3-week-old) and adult (5-month-old) rats. NF, neutral flavor; VF, vanilla flavor; VF + S, vanilla flavor, and sucralose. Data show the median and maximum and minimum days of latency. Statistical analysis was performed by Kruskal–Wallis test; *n*, number of rats.

In a group of adult rats trained with NF, we evaluated the time needed for complete ingestion. Female rats were faster (mean time = 1.7 ± 0.2 min, minimum = 1.25 min, and maximum = 2.1 min; $n = 4$) compared to males (mean time = 7.9 ± 1.5, minimum = 4.0 min, and maximum = 11.5 min; $n = 4$; $p = 0.02$).

We also analyzed if short (4 h) fasting affected the latency to ingestion. We observed that 4 h prior fasting did not modify latency in males (Figure 5) but fasting reduced it significantly in females (Figure 5).

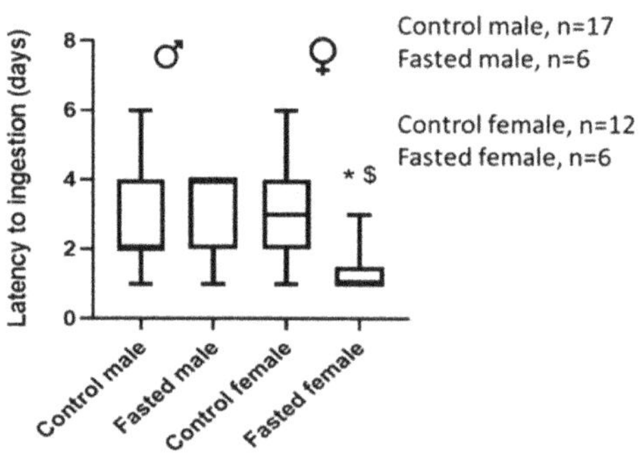

Figure 5. Effect of 4 fasting on latency to ingestion in adult male and female rats excluding non-habituated rats. Data show the median and maximum and minimum days of latency. Statistical analysis was performed by Mann–Whitney's U test; * *p*-value < 0.05 with control females, $, *p*-value < 0.05 with fasted adult males; *n*, number of rats.

3.3. Capacity to Remember Training and Acceptance of GC with a Compound of Interest

In a group of 6 adult males and 6 adult females previously trained with NF gelatin, we evidenced that after 1 month, all rats, except one female, ate the GC.

We also tested the capacity to accept a supplement (CSE) incorporated into the GC in a group of 7 previously trained adult females during the lactation period. All the dams accepted the gelatin and ate it during the 3 weeks of lactation and completed ingestion in a single attempt and between 1 and 5 min. No alterations in dam behavior or problems with the pups were detected, compared to our previous studies in lactating rats [15]. A video showing the ingestion of a CSE cube by a lactating rat is included (Video S2).

4. Discussion

Experimental animals are an essential tool for the development of new drugs; they are commonly used to evaluate efficacy and toxicity. The procedures involve repetitive administration of the compound of interest, ensuring adequate dosage. Some drugs can be injected, while others, such as nutraceuticals, need to pass through the digestive tract to exert their actions. Ingestion is also a method required to evaluate drug activity for the development of orally administered medicines. For these studies, voluntary ingestion methods have several advantages over involuntary procedures (such as oral gavage), the most important being that they are non-invasive, reducing stress and possible complications [2–5], particularly during sensitive periods such as gestation and lactation. One possible way of delivery is to include the substance of interest in the drinking water. Although this method is suitable and easy, some substances cannot be incorporated, i.e., if they are hydrophobic, and daily dosage may be subject to differences in the amount of water each animal drinks. The present study was conducted to implement a protocol for voluntary ingestion of food supplements in rats based on a gelatin matrix, evaluate key factors influencing acceptance, and assess the capacity to remember previous training. Once the training protocol was established, we also evaluated its effectiveness as a supplement with a compound of interest (CSE) during lactation, a critical period when oral gavage should be avoided. Our main results are that training rats for voluntary ingestion of a GC is easy since it requires very basic equipment and does not require specific skills, and it is a suitable method for supplementation during lactation. Regarding factors that affect training, we detected a higher proportion of adult males who did not habituate, although acceptance may be improved if the training is conducted in a familiar environment. Based on our findings, we suggest that it is important to detect animals who refuse to eat the gelatin during training to exclude them from the study, avoiding unnecessary stress and experimental bias. These animals may be useful for other studies that do not require voluntary ingestion of compounds, contributing to the three Rs. In Appendix A, we propose a protocol for voluntary ingestion of supplements in rats.

There are several vehicles to incorporate the compounds of interest for voluntary supplementation in rodents (jams, cookies, paste, and gelatins). We chose gelatin as a vehicle since it is cheap and easy to produce and because we could incorporate all types of compounds into it. Besides, it can be shaped in various sizes, and the rats can handle it easily without breaking into pieces (Video S1), as previously described in mice [16]. Thus, this vehicle ensures complete ingestion and adequate dosage. Besides, gelatin is a non-caloric vehicle and therefore suitable for studies analyzing compounds interfering with metabolic processes. In fact, it has been proposed that it can be used to deliver glucose for tolerance tests as a substitute for oral gavage or injection since it causes less stress and guarantees a better interpretation of results.

Caution is a natural response of animals to unfamiliar objects, and the term "neophobia" is used to characterize fear-like responses based on the novelty of a stimulus, regardless of its modality. Neophobic responses may occur to objects, places, or sounds, and food neophobia applies to the refusal or suppression of the ingestion of a new food. Rodents are neophobic animals, with different degrees depending on the strain of mice or rats and if they are wild or laboratory-bred [17]. When a rat is exposed to a new food, the initial response is avoidance, followed by gradual sampling, and, if there is no harm, consumption increases across successive encounters [17,18]. Our experiments showed that Sprague-Dawley rats took a median of 2–3 days to accept a novel food (a GC), a latency like that reported in mice (ranging from 2–4 days) [19] or in Lister-hooded rats, who needed 3 days of training for full acceptance [20].

It is important to note that we detected a percentage of rats that maintained neophobia over a period of 2 weeks, and we considered these rats non-habituated. The proportion of non-habituated rats was higher in adult males compared to the other groups. Maintained neophobia has been previously detected in male mice, where it was shown that 5% refused to eat a gelatin even after prolonged fasting [16,19], and in Long Evans rats, where it was reported that 4 out of 8 males never ate the novel food [21]. We observed that the proportion of non-habituated rats was larger in adults but not in young males, which, according to our previous data, are still in the prepubertal stage [22]. Therefore, our data indicate an influence of sex and age on the degree of neophobia, with a higher susceptibility among adult males. These difficulties in habituation can be related to a previous high level of stress that drives food avoidance [17,21], for example, due to laboratory noise [16]. Since gonadal hormones play a key role in the regulation of the hypothalamic–pituitary–adrenal axis [23], it is possible that male rats are more susceptible to maintaining neophobia in response to environmental stressors.

In non-neophobic rats, we also analyzed if sex and age affected time to acceptance, i.e., latency. We did not find sex differences, in contrast to the findings of other authors, who reported that Long Evans male rats habituated to a novel food faster than females if exposed to an unfamiliar environment, although no significant differences between sexes were found if the new food was presented in their own cage [21]. This is likely related to the fact that rats respond with neophobia not only to the new object but also to the container [17]. The faster adjustment of males to an unfamiliar environment has been explained by their higher risk-taking behavior compared to females [24]. We did not find significant differences in latency between young and adult rats. Some studies evidence a better acceptance of a new food in adolescent Sprague-Dawley rats, showing that they are more sensitive than adults to the hedonic properties of an appetitive stimulus and less sensitive to aversive stimuli, such as quinine [25]. Even though we did not detect differences in latency, the fact that young males had a significantly lower rate of non-habituation compared to adults suggests the interest in training rats at a young age for supplementation studies. We did not use old rats, but it is possible that they would have a higher latency since neophobia habituation shares neural circuits with mechanisms of declarative memory [18], and studies comparing adult and old rats show increased neophobia in aging together with deficits in spatial learning [26].

Regarding the time the rats need for GC intake once they are trained, our data are in accordance with other studies showing that rodents take an average of 1 to 10 min for complete ingestion. Zhang and co-workers using a gelatin-based supplement reported that most mice ate it within 1 min of presentation and finished the entire piece in a single attempt [19]. Neto et al. using female mice trained with a cookie reported that the animals need 5–10 min [27], and Lister hooded rats can achieve a complete intake within 5 min [20]. We evidenced that male rats took a longer time for complete ingestion, which is in accordance with data from several rat strains showing that females eat faster and approach the food before feeding more frequently than males [17]. However, the study of Teixera-Santos and co-workers in mice showed a lower latency in males, who were able to eat the supplement in 1 min compared to females, who took 5 min [8]. Thus, species differences may occur.

We also analyzed the capacity of external factors (additives and prior fasting), which may facilitate training and can be incorporated into future protocols. Rodents refuse some substances, such as alcohol, due to aversive odors, as well as foods with a bitter or acidic taste. It has been demonstrated that acceptance can be improved if the supplement includes a flavoring agent or a sweetener. In laboratory mice, vanilla flavor increases nicotine consumption [28]. House mice preferred foods with flavors (vanilla, chocolate, hazelnut, or peanut) to natural foods, while rats rejected chocolate flavor and did not show a preference for the other additives [29]. Rats also increase alcohol consumption in the presence of sweeteners (sucralose, saccharine, or maltodextrins) [30,31], and sugar-free fruit juices improve the acceptance of nicotine [20]. In our study, we did not observe a reduction in latency or in the proportion of non-habituated rats using gelatins with vanilla with or without sucralose; we even found worse results than with neutral gelatin. This is in accordance with the study of Sclafani and Clare using the same strain of rats, showing that sucralose reduced food palatability in Sprague-Dawley rats [32]. We conclude that the food additives and sweeteners tested (vanilla and sucralose) do not improve the acceptance of gelatin in Sprague-Dawley rats.

We also tested the influence of prior fasting since it has been shown that the hungrier the rat, the quicker it starts to eat an unfamiliar food [33]. This was confirmed in our study in females, which demonstrated that prior food deprivation reduced latency. However, this was not observed in males. We used a short fasting period of 4 h compared to previous studies using overnight fasting [19], and it would be possible to reduce it since it has been shown that 30 min fasting is sufficient for the acceptance of a jam including a drug in mice [8]. Although it has been demonstrated that fasting for less than 16 h does not represent a stress for the rat [34], and we demonstrated a positive effect on latency, reducing acceptance by 1 day, we think that it is not strictly necessary for the training protocol.

Finally, one of our aims was to evaluate the effectiveness of the training protocol for supplementation in rats during lactation, which, together with gestation, is a very sensitive period where stress induced by oral gavage should be avoided. In fact, stress induced by oral gavage in the rat leads to weight loss, and although this can be minimized if the rat is anesthetized [35], this can be a sign of stress with a negative influence, particularly during gestation and lactation. Rats subjected to prenatal stress around the gestational period have more spontaneous abortions and fewer viable pups [36], and oral gavage during pregnancy alters the behavioral development of the offspring [37]. During lactation, dams exposed to chronic social stress show impaired maternal care and lactation [38]. We demonstrated that 100% of trained female rats prior to gestation remembered the training two months later and ate the gelatin, including the supplement (CSE), on a single attempt in 1–2 min. This procedure greatly reduces the rat's stress since the animal is returned to the cage once the gelatin has been eaten, and we have evidenced no modifications in the dam or pups' behavior. Besides, complete gelation consumption ensured an adequate dosage of the compound of interest. We also have evidence from a metabolomic study that CSE bioactive compounds (caffeine and theobromine) are present in rat plasma after 1 week of supplementation [13]. Therefore, this training protocol is suitable for experiments during gestation and lactation.

Limitations and protocol implementation. The present study aimed to improve methods of supplementation by voluntary ingestion by assessing some factors that may affect neophobia and latency. We evidence the feasibility of training rats to eat a gelatin-based vehicle and the capacity to remember and accept the supplement months later. The protocol is an alternative to oral gavage and improves animal wellbeing. However, it is not devoid of limitations, the main one being the resistance of some rats, mainly adult males, to training; therefore, a pre-screening of animals is required. We also showed that males do not improve latency by short fasting. As indicated above, males may have more neophobia due to the higher influence of stress factors. One of them is isolation in an empty cage, which was used to facilitate the view of the gelatin and avoid hiding or breaking it. However, this is an unfamiliar environment, which may contribute to neophobia [17]. Besides, we exposed the rats to repeated short isolation periods, which have been shown to also be a stressor [39]. Presentation of the novel food in their cage may facilitate acceptance and has been previously used in rats [10] and mice [19], placing the supplement in a glass Petri dish overnight. However, this protocol requires isolating the rat for the whole night. We used an alternative method in the group of adult males who refused to eat the cube with the usual protocol in the empty cage after isolation and observed that if the cube was placed in the hopper, where they usually receive the food, they accepted it. Moreover, after this test, most rats ate the gelatin with the usual protocol in the empty cage in the following days. It has also been shown that acceptance of a novel food does not depend solely on the individual's experience but has a social component. For example, if a mouse has tried a novel food, it will be investigated by the other mice in the cage, which may result in other animals from the colony selecting food with the same scent [40]. We also observed this behavior when presenting the gelatin in the hopper, and it is possible that an initial training in the regular cage may also help acceptance through this social process. However, after training, the supplement should be given individually to ensure dosage. Therefore, we suggest implementing our protocol by training rats with the gelatin in their usual environment, followed by individualization in the empty cage to ensure dosage.

Neophobia has an important genetic component, and studies in KO mice have identified some genes influencing it. Thy-1 (a cell adhesion molecule) KO mice display neophobia, which has been suggested to be mediated by neuronal plasticity regulation [40]. In addition, a study of wild brown rats living in various Tokyo locations demonstrates differences in neophobia and the existence of rats that are indifferent to novel objects, probably due to genetic variations [41]. These data suggest the possibility of obtaining genetically modified animals with lower neophobia through selective breeding.

Regarding the utility of this method, we think it is not restricted to food supplements, but it would be a way of testing the effectiveness of drugs. Some studies using voluntary administration of analgesics with gelatin-based vehicles have found poorer results compared to intraperitoneal administration [42], likely due to modifications of the drug by passage through the gastrointestinal tract. However, it has to be noted that this study evaluated analgesia after a short period (1 h), and it is possible that long-term studies would yield better results. In addition, oral intake of drugs is the preferred route of administration, and voluntary supplementation would be a useful method for this purpose.

5. Conclusions

The present study demonstrates the feasibility of training rats to voluntarily accept a supplement based on a gelatin vehicle that can be used during lactation and gestation. Training is not influenced by additives, but it is improved by fasting in females. Male sex may negatively influence neophobia, making it important to detect non-habituated animals to exclude them from the study and avoid unnecessary stress. It is possible that this neophobia can be reduced by the initial presentation of the novel food in their own cage. We propose some tips for future use of this protocol in Appendix A.

Supplementary Materials: The following supporting information can be downloaded at: https://www.mdpi.com/article/10.3390/ani13111827/s1, Figure S1: Preparation of gelatin cubes: (A) mold to prepare 1 cm^3 cubes; (B) neutral gelatin cubes. Video S1: Rat eating gelatin from hopper; Video S2: Ingestion of a gelatin cube with supplement in a trained Sprague-Dawley lactating rat.

Author Contributions: Conceptualization, S.M.A., Y.A. and D.M.-V.; methodology, S.R., S.C. and P.R.-R.; software, S.R. and D.R.-C.; validation, Y.A., D.M.-V., S.C. and P.R.-R.; formal analysis, D.R.-C. and S.R.; investigation, S.M.A., Y.A. and D.R.-C.; resources, S.M.A.; data curation, S.M.A.; writing—original draft preparation, S.M.A.; writing—review and editing, S.R., P.R.-R., S.C., D.R.-C., Y.A., D.M.-V. and S.M.A.; visualization, S.R. and D.R.-C.; supervision, S.M.A. and D.R.-C.; project administration, S.M.A.; funding acquisition, S.M.A. and D.R.-C. All authors have read and agreed to the published version of the manuscript.

Funding: This research was funded by Ministerio de Ciencia Innovación y Universidades (Spain), grant number RTI2018-097504-B-I00 and The APC was funded by D.R.-C.

Institutional Review Board Statement: The animal study protocol was approved by the Ethics Review Board of Universidad Autónoma de Madrid (Spain) and Regional Committee from Comunidad Autónoma de Madrid (Spain; protocol code PROEX 19/04, approval date: 20 March 2019).

Informed Consent Statement: Not applicable.

Data Availability Statement: The data are available upon request to the corresponding author by institutional email and after ethical evaluation.

Conflicts of Interest: The authors declare no conflict of interest.

Appendix A

Proposed protocol for training and supplementation in rodents based on GC.

1. Preparation of GC:
    - No need for additives. However, they may be used if supplement is unpleasant.
    - GC can be stored in bags in the fridge or frozen.
    - Do not recycle GC since they dry.
    - Prepare the GC with supplement based on the weight of the animal.
2. Training:
    - The first day, present several neutral gelatin cubes in their own cage, in the food hopper. This procedure may reduce the number of non-habituated rats.
    - The second day, see if they eat the cube individually in their own cage.
    - The third day, see if they complete ingestion in an empty cage. Once they eat it completely, the rats can be considered trained and given the supplement.
3. Training tips:
    - Training rats at young age may reduce the number of non-habituated.
    - Fasting reduces latency to acceptance in females. However, it is not strictly necessary.
    - If the supplement is unpleasant, it may be necessary to mask it with a flavoring agent. Sucralose did not improve acceptance in Sprague-Dawley rats.
4. Supplementation:
    - The rats can remember training for at least 2 months.
    - Place the rat with the supplemented gelatin in an empty cage to ensure dosage.
    - Return it to the cage once the rat eats the supplement.

References

1. Turner, P.V.; Brabb, T.; Pekow, C.; Vasbinder, M.A. Administration of Substances to Laboratory Animals: Routes of Administration and Factors to Consider. *J. Am. Assoc. Lab. Anim. Sci.* **2011**, *50*, 600–613. [PubMed]
2. Damsch, S.; Eichenbaum, G.; Tonelli, A.; Lammens, L.; Bulck, K.V.d.; Feyen, B.; Vandenberghe, J.; Megens, A.; Knight, E.; Kelley, M. Gavage-Related Reflux in Rats. *Toxicol. Pathol.* **2011**, *39*, 348–360. [CrossRef] [PubMed]
3. Eichenbaum, G.; Damsch, S.; Looszova, A.; Vandenberghe, J.; Van den Bulck, K.; Roels, K.; Megens, A.; Knight, E.; Hillsamer, V.; Feyen, B.; et al. Impact of Gavage Dosing Procedure and Gastric Content on Adverse Respiratory Effects and Mortality in Rat Toxicity Studies. *J. Appl. Toxicol.* **2011**, *31*, 342–354. [CrossRef] [PubMed]
4. Balcombe, J.P.; Barnard, N.D.; Sandusky, C. Laboratory Routines Cause Animal Stress. *Contemp. Top. Lab. Anim. Sci.* **2004**, *43*, 42–51. [PubMed]
5. Brown, A.P.; Dinger, N.; Levine, B.S. Stress Produced by Gavage Administration in the Rat. *Contemp. Top. Lab. Anim. Sci.* **2000**, *39*, 17–21. [PubMed]
6. Arantes-Rodrigues, R.; Henriques, A.; Pinto-Leite, R.; Faustino-Rocha, A.; Pinho-Oliveira, J.; Teixeira-Guedes, C.; Seixas, F.; Gama, A.; Colaço, B.; Colaço, A.; et al. The Effects of Repeated Oral Gavage on the Health of Male CD-1 Mice. *Lab. Anim.* **2012**, *41*, 129–134. [CrossRef]
7. Branchi, I.; Capone, F.; Vitalone, A.; Madia, F.; Santucci, D.; Alleva, E.; Costa, L.G. Early Developmental Exposure to BDE 99 or Aroclor 1254 Affects Neurobehavioural Profile: Interference from the Administration Route. *Neurotoxicology* **2005**, *26*, 183–192. [CrossRef]
8. Teixeira-Santos, L.; Albino-Teixeira, A.; Pinho, D. An Alternative Method for Oral Drug Administration by Voluntary Intake in Male and Female Mice. *Lab. Anim.* **2021**, *55*, 76–80. [CrossRef]
9. Abraham, A.D.; Leung, E.J.Y.; Wong, B.A.; Rivera, Z.M.G.; Kruse, L.C.; Clark, J.J.; Land, B.B. Orally Consumed Cannabinoids Provide Long-Lasting Relief of Allodynia in a Mouse Model of Chronic Neuropathic Pain. *Neuropsychopharmacology* **2020**, *45*, 1105–1114. [CrossRef]
10. Corbett, A.; McGowin, A.; Sieber, S.; Flannery, T.; Sibbitt, B. A Method for Reliable Voluntary Oral Administration of a Fixed Dosage (Mg/Kg) of Chronic Daily Medication to Rats. *Lab. Anim.* **2012**, *46*, 318–324. [CrossRef]
11. Diogo, L.N.; Faustino, I.V.; Afonso, R.A.; Pereira, S.A.; Monteiro, E.C.; Santos, A.I. Voluntary Oral Administration of Losartan in Rats. *J. Am. Assoc. Lab. Anim. Sci.* **2015**, *54*, 549–556. [PubMed]
12. Angulo, R.; Arévalo-Romero, C.A. Sexual Dimorphism in Classical Conditioning? Sex Differences in Neophobia, Latent Inhibition, Generalization, and Extinction for Rats (Rattus Norvegicus) in a Conditioned Taste Aversion Preparation Irrespective of Housing Conditions. *J. Comp. Psychol.* **2021**, *135*, 315–326. [CrossRef] [PubMed]

13. Rodríguez-Rodríguez, P.; Ragusky, K.; Phuthong, S.; Ruvira, S.; Ramiro-Cortijo, D.; Cañas, S.; Rebollo-Hernanz, M.; Morales, M.D.; López de Pablo, Á.L.; Martín-Cabrejas, M.A.; et al. Vasoactive Properties of a Cocoa Shell Extract: Mechanism of Action and Effect on Endothelial Dysfunction in Aged Rats. *Antioxidants* **2022**, *11*, 429. [CrossRef] [PubMed]
14. Rebollo-Hernanz, M.; Zhang, Q.; Aguilera, Y.; Martín-Cabrejas, M.A.; Mejia, E.G. Cocoa Shell Aqueous Phenolic Extract Preserves Mitochondrial Function and Insulin Sensitivity by Attenuating Inflammation between Macrophages and Adipocytes In Vitro. *Mol. Nutr. Food Res.* **2019**, *63*, 1801413. [CrossRef]
15. Rodríguez-Rodríguez, P.; Monedero-Cobeta, I.; Ramiro-Cortijo, D.; Puthong, S.; Quintana-Villamandos, B.; Gil-Ramírez, A.; Cañas, S.; Ruvira, S.; Arribas, S.M. Slower Growth during Lactation Rescues Early Cardiovascular and Adipose Tissue Hypertrophy Induced by Fetal Undernutrition in Rats. *Biomedicines* **2022**, *10*, 2504. [CrossRef] [PubMed]
16. Zhang, L.; Lee, N.J.; Nguyen, A.D.; Enriquez, R.F.; Riepler, S.J.; Stehrer, B.; Yulyaningsih, E.; Lin, S.; Shi, Y.C.; Baldock, P.A.; et al. Additive Actions of the Cannabinoid and Neuropeptide Y Systems on Adiposity and Lipid Oxidation. *Diabetes Obes. Metab.* **2009**, *12*, 591–603. [CrossRef]
17. Modlinska, K.; Stryjek, R.; Pisula, W. Food Neophobia in Wild and Laboratory Rats (Multi-Strain Comparison). *Behav. Process.* **2015**, *113*, 41–50. [CrossRef]
18. Vicente, L.; De la Casa, L.G. Context Properties Modulate Flavor Neophobia Habituation. *Psicothema* **2021**, *33*, 617–622. [CrossRef]
19. Zhang, L. Method for Voluntary Oral Administration of Drugs in Mice. *STAR Protoc.* **2021**, *2*, 100330. [CrossRef]
20. Dhawan, S.S.; Xia, S.; Tait, D.S.; Bundgaard, C.; Bowman, E.; Brown, V.J. Oral Dosing of Rodents Using a Palatable Tablet. *Psychopharmacology* **2018**, *235*, 1527–1532. [CrossRef]
21. Greiner, E.M.; Petrovich, G.D. The Effects of Novelty on Food Consumption in Male and Female Rats. *Physiol. Behav.* **2020**, *223*, 112970. [CrossRef]
22. Rodríguez-Rodríguez, P.; López de Pablo, A.L.; García-Prieto, C.F.; Somoza, B.; Quintana-Villamandos, B.; Gómez de Diego, J.J.; Gutierrez-Arzapalo, P.Y.; Ramiro-Cortijo, D.; González, M.C.; Arribas, S.M. Long Term Effects of Fetal Undernutrition on Rat Heart. Role of Hypertension and Oxidative Stress. *PLoS ONE* **2017**, *12*, e0171544. [CrossRef]
23. Oyola, M.G.; Handa, R.J. Hypothalamic-Pituitary-Adrenal and Hypothalamic-Pituitary-Gonadal Axes: Sex Differences in Regulation of Stress Responsivity. *Stress* **2017**, *20*, 476–494. [CrossRef] [PubMed]
24. Orsini, C.A.; Blaes, S.L.; Hernandez, C.M.; Betzhold, S.M.; Perera, H.; Wheeler, A.-R.; Ten Eyck, T.W.; Garman, T.S.; Bizon, J.L.; Setlow, B. Regulation of Risky Decision Making by Gonadal Hormones in Males and Females. *Neuropsychopharmacology* **2021**, *46*, 603–613. [CrossRef] [PubMed]
25. Wilmouth, C.E.; Spear, L.P. Hedonic Sensitivity in Adolescent and Adult Rats: Taste Reactivity and Voluntary Sucrose Consumption. *Pharm. Biochem. Behav.* **2009**, *92*, 566–573. [CrossRef] [PubMed]
26. Collier, T.J.; Greene, J.G.; Felten, D.L.; Stevens, S.Y.; Collier, K.S. Reduced Cortical Noradrenergic Neurotransmission Is Associated with Increased Neophobia and Impaired Spatial Memory in Aged Rats. *Neurobiol. Aging* **2004**, *25*, 209–221. [CrossRef]
27. Neto, T.; Faustino-Rocha, A.I.; Gil da Costa, R.M.; Medeiros, R.; Oliveira, P.A. A Quick and Low-Intensity Method for Oral Administration to Large Numbers of Mice: A Possible Alternative to Oral Gavage. *Lab. Anim.* **2022**, *56*, 185–190. [CrossRef]
28. Tannous, S.; Darlot, F.; Cador, M.; Caille, S. Flavor Additives Facilitate Oral Self-Administration of Nicotine Solution in Mice. *Psychopharmacology* **2021**, *238*, 2235–2247. [CrossRef]
29. Schmolz, E.; Kalle, A. Effect of Addition of Artificial Flavour on Rodent Bait Attractiveness. *Proc. Vertebr. Pest Conf.* **2010**, *24*, 371–376. [CrossRef]
30. Sardarian, A.; Liu, S.; Youngentob, S.L.; Glendinning, J.I. Mixtures of Sweeteners and Maltodextrin Enhance Flavor and Intake of Alcohol in Adolescent Rats. *Chem. Senses* **2020**, *45*, 675–685. [CrossRef]
31. Truxell, E.M.; Molina, J.C.; Spear, N.E. Ethanol Intake in the Juvenile, Adolescent, and Adult Rat: Effects of Age and Prior Exposure to Ethanol. *Alcohol. Clin. Exp. Res.* **2007**, *31*, 755–765. [CrossRef]
32. Sclafani, A.; Clare, R.A. Female Rats Show a Bimodal Preference Response to the Artificial Sweetener Sucralose. *Chem. Senses* **2004**, *29*, 523–528. [CrossRef] [PubMed]
33. Barnett, S.A. *The Rat*; Routledge: Oxford, UK, 2017; ISBN 9781315134468.
34. Nowland, M.H.; Hugunin, K.M.S.; Rogers, K.L. Effects of Short-Term Fasting in Male Sprague-Dawley Rats. *Comp. Med.* **2011**, *61*, 138–144. [PubMed]
35. Murphy, S.J.; Smith, P.; Shaivitz, A.B.; Rossberg, M.I.; Hurn, P.D. The Effect of Brief Halothane Anesthesia during Daily Gavage on Complications and Body Weight in Rats. *Contemp. Top. Lab. Anim. Sci.* **2001**, *40*, 9–12. [PubMed]
36. Herrenkohl, L.R. Prenatal Stress Reduces Fertility and Fecundity in Female Offspring. *Science* **1979**, *206*, 1097–1099. [CrossRef]
37. McDonnell-Dowling, K.; Kleefeld, S.; Kelly, J.P. Consequences of Oral Gavage during Gestation and Lactation on Rat Dams and the Neurodevelopment and Behavior of Their Offspring. *J. Am. Assoc. Lab. Anim. Sci.* **2017**, *56*, 79–83.
38. Murgatroyd, C.A.; Taliefar, M.; Bradburn, S.; Carini, L.M.; Babb, J.A.; Nephew, B.C. Social Stress during Lactation, Depressed Maternal Care, and Neuropeptidergic Gene Expression. *Behav. Pharmacol.* **2015**, *26*, 642–653. [CrossRef]
39. Hodges, T.E.; McCormick, C.M. Adolescent and Adult Male Rats Habituate to Repeated Isolation, but Only Adolescents Sensitize to Partner Unfamiliarity. *Horm. Behav.* **2015**, *69*, 16–30. [CrossRef]
40. File, S.E. Factors Controlling Measures of Anxiety and Responses to Novelty in the Mouse. *Behav. Brain Res.* **2001**, *125*, 151–157. [CrossRef]

41. Koizumi, R.; Kiyokawa, Y.; Tanaka, K.D.; Kimura, G.; Tanikawa, T.; Takeuchi, Y. Existence of Wild Brown Rats (Rattus Norvegicus) That Are Indifferent to Novel Objects. *J. Vet. Med. Sci.* **2021**, *83*, 78–83. [CrossRef]
42. Cannon, C.Z.; Kissling, G.E.; Hoenerhoff, M.J.; King-Herbert, A.P.; Blankenship-Paris, T. Evaluation of Dosages and Routes of Administration of Tramadol Analgesia in Rats Using Hot-Plate and Tail-Flick Tests. *Lab. Anim.* **2010**, *39*, 342–351. [CrossRef] [PubMed]

Disclaimer/Publisher's Note: The statements, opinions and data contained in all publications are solely those of the individual author(s) and contributor(s) and not of MDPI and/or the editor(s). MDPI and/or the editor(s) disclaim responsibility for any injury to people or property resulting from any ideas, methods, instructions or products referred to in the content.

animals

MDPI

Review

Ultrasound Guided Surgery as a Refinement Tool in Oncology Research

Juan Antonio Camara Serrano

Preclinical Therapeutics Core, University of California San Francisco, San Francisco, CA 94158, USA; juanantonio.camaraserrano@ucsf.edu; Tel.: +1-628-629-3555

Simple Summary: This review describes the potential use of ultrasound as a refinement tool in research. After a general state-of-the-art discussion, the most frequent organs used as a host for orthotopic models in oncology research are listed, including the thyroid gland, heart, liver, spleen, kidney, pancreas, uterus, and testicles. In each organ and after a short ultrasonography description, the practical protocol for the ultrasound-guided injection is described as well as the main risk of the procedure and technical limitations. The main objective of this work is to help users with the use of ultrasound-guided injection. For this purpose, the descriptions of the protocols are mainly practical with tips which are frequent mistakes carried out during the injections.

Abstract: Refinement is one of the ethical pillars of the use of animals in research. Ultrasonography is currently used in human medicine as a surgical tool for guided biopsies and this idea can be applied to preclinical research thanks to the development of specific instruments. This will eliminate the necessity of a surgical opening for implanting cells in specific organs or taking samples from tissues. The approach for the injection will depend on the target but most of the case is going to be lateral, with the probe in a ventral position and the needle going into from the lateral. This is the situation for the thyroid gland, heart, liver, spleen, kidney, pancreas, uterus, and testicles. Other approaches, such as the dorsal, can be used in the spleen or kidney. The maximum injected volume will depend on the size of the structure. For biopsies, the technical protocol is similar to the injection knowing that in big organs such as the liver, spleen, or kidney we can take several samples moving slightly the needle inside the structure. In all cases, animals must be anesthetized and minimum pain management is required after the intervention.

Keywords: preclinical models; oncology; ultrasounds injections; refinement

Citation: Camara Serrano, J.A. Ultrasound Guided Surgery as a Refinement Tool in Oncology Research. *Animals* **2022**, *12*, 3445. https://doi.org/10.3390/ani12233445

Academic Editor: Garikoitz Azkona

Received: 10 November 2022
Accepted: 1 December 2022
Published: 6 December 2022

Copyright: © 2022 by the author. Licensee MDPI, Basel, Switzerland. This article is an open access article distributed under the terms and conditions of the Creative Commons Attribution (CC BY) license (https://creativecommons.org/licenses/by/4.0/).

1. Introduction

The 3Rs principle was enunciated in the early years of the 60s of the past century by Russell and Burch, two English biologists, in their book "The principle of Humane Experimental Technique". This publication was an outbreak in the manipulation of lab animals, that until this work, were basically considered a research tool such as an Eppendorf tube or a pipette [1]. In their work, Russell and Burch defined the ethical principles that must rule over any experimental work which involves animals, especially the ones with a more developed neurological system, such as mammals or birds. In recent years, these ethical principles cover other animals such as octopuses, which have been raised as conscious and sensible to pain subjects, at a higher level that could be assumed due to their phylogenetic stratum [2].

The concepts included in the 3Rs principle are replacement, reduction, and refinement. The first one refers to the ethical command to use non-in vivo methodologies for research when accessible, for example, using in vitro studies or in silico experiments. The second concept, reduction, commits to using fewer animals as possible. The total number of animals included in an experiment must be justified and statistical analysis must be run to calculate the minimum sample size for every experimental group [3]. The third concept, refinement, stands for the development of less invasive techniques and the reduction in the damage done to the animals during the research work.

Looking at the current scenario, the reduction principle has been propelled in the last years with big data tools and artificial intelligence that make it possible to obtain trustable conclusions from experiments with only a single patient or specimen [4]. In the same way, replacement is taking advantage of the astonishing development of computer science and nowadays we can design algorithms for simulating viral replications, drug effects in cells, or cancer cell mutations [5]. Perhaps, refinement has been the less developed concept since the beginning of the 3Rs principle. We still need to manipulate the animals and in most cases create some sort of disconformity to them. Even so, the minimum required welfare levels are much higher than in the past decades, taking advantage, for example, of the advances in analgesia and anesthesia or the new surgical techniques, less invasive and more refined [6,7].

Oncology research is one of the most important in medicine, due to the global economic impact of these pathologies, as well as the emotional charge that "the big C" generates in the patient and the families. In oncology research, animal models are a basic part of the process, not only for drug discovery but also because we do not still have full knowledge of the process that drives a cell to become a cancerous entity. In both basic research and drug discovery, animal models are a pillar in the projects. In some of these works, the animals get implanted with a sample of human tumor tissue, or tumoral cells, and a surgical procedure is required for this implantation. Frequently, these implantations are performed in the subcutaneous tissue, but in other cases, the implantation is performed in different locations, and we talk about orthotopic models, where the tumor cells are implanted in the physiological host organ [7,8]. A clarifying example can be the pancreatic implantation of a human pancreatic cancer sample, or the intravascular injection of blood cancer cells. These orthotopic models are, in theory, more similar to the clinical scenario than the subcutaneous models because the host tissue is similar to the original one [7].

The use of imaging techniques for preclinical research dated from several decades ago, when lab animals were scanned in clinical devices for obtaining information about the internal structures of the animals, due to the absence of specific technologies designed for preclinical research. However, in recent years, we witnessed the arising of different technology for lab animals, including specific instruments such as microPET (positron emission tomography), microCT (computed tomography), MRI (magnetic resonance imaging), optical imaging, or ultrasonography. All these technologies play a significant role in the application of the 3Rs principle in cancer models. The possibility to look into the animals avoiding their sacrifice represents a dual effect on animal welfare: on one side the necessity to euthanize the animal is eliminated (refinement) and on the other side we can rescan the same animal at different time points during the experiment (reduction). Furthermore, repeating the imaging on the same animal allows to use them as their own control, which increases the statistical significance of the experiments. Resuming, the imaging technologies development had a significant impact on animal welfare in preclinical research [7].

Specifically talking about ultrasonography, new equipment with higher ultrasound frequencies which increase the spatial and temporal resolution make it possible to obtain images that will be not accessible from a clinical device designed for human patients and sizes. Ultrasonography has a significant advantage which is real-time imaging. What we see on the monitor is happening inside the patient. This opens a door to the use of this technology as a surgical tool for accessing internal structures in the animal. We can make injections or take biopsies avoiding conventional surgical procedures, which will reduce the requirements of post-surgical care, due to the minimum damage created and the low pain generated to the patient/animal. Ultrasonography-guided surgery has been performed for years in human medicine and pets and now, with specific equipment for lab animals, it is time for including it as a refinement tool in animal models and preclinical research.

2. State of the Art

When talking about preclinical imaging, it is necessary to make a difference between animal models. In big animals, above the size of a rabbit, clinical imaging devices have been used for research for several decades, due to the similar size of the animals to the human patients and the acceptable quality of the images. This did not happen in lab animals. The maximum spatial resolution of the clinical equipment was not enough to get acceptable images from an animal that could weigh, in the worst case, less than 20 grams. During the last years, a significant improvement in imaging technologies has made an impact on animal research. Specific systems for lab animals have been developed, with technical parameters according to the requirements of lab animals. For example, microCT and preclinical ultrasonography are now able to reach spatial resolutions as 5 um [9]. These new capabilities make these systems accessible for almost any research field involving lab animals such as oncology, infectious diseases, metabolism, orthopedics, and so.

The development of new ultrasound technologies in the last decade made accessible the scanning of lab animals almost independent of the patient size. Even mouse fetuses can be scanned with an acceptable quality in both spatial and temporal resolution. Moreover, new ultrasonographic modes have been developed, in addition to the classic A, B, M, and Doppler. In this way, now we have elastography, contrast, 4D, digital RF, or oxy-hemo modes [10]. All of them are based on the same principles as the former modes, the analysis of ultrasound waves during their travel through the tissues, but from them, we can obtain different and valuable data, such as tissue stiffness, micro-vascularization of the tissues, oxygen and hemoglobin levels, and other physiological data [10].

In this review, we will focus on the use of ultrasonography for interventionism and its potential to reduce the manipulation of animals and improve their welfare. Almost all these procedures are carried out using the B mode (brightness mode, the most typical ultrasound mode), and at some points, the Doppler mode will be useful for distinguishing the regional vascularization in order to avoid accidental punctures of the vessels. The other ultrasonography modes are, in theory, not necessary during the interventionism. The effect of ultrasound-guided surgery on the welfare of the animals could be evaluated by looking at the manipulation required for developing the same cancer model with or without this technology. We can use a pancreatic cancer model as an example. Without using an ultrasound device, we would need to anesthetize the animal, remove the air from the left lateral abdomen, and disinfect the surgical area. This procedure will be the same for the guided injection. But then, without the ultrasound, we would need to make an incision in the skin, followed by the abdominal wall. We would manually move the bowels for reaching the spleen. Later, we would expose this organ followed by the pancreatic tissue. Then, out of the body, we will inject the cells into the tissue and would put the pancreas and spleen back into the abdomen. We would close the abdominal wall and skin with surgical sutures. The whole procedure could take, for an experienced surgeon, at least 5 to 10 min and we would need to use supportive analgesia at least for the following three days. Using ultrasound guidance, after disinfecting the skin, we will scan the abdomen, find the pancreatic area, introduce transabdominally the needle into the pancreas, inject the cells, and remove the needle. Analgesia will be required for 2 days. The time for doing the injection would be 1–2 min for an experienced user. In our opinion, it is clear the effect of this technique on the welfare of the animals and projects.

Because the aim of ultrasound-guided interventionism is to introduce, or remove, material in the tissues, a minimum image quality is required for the procedures. There are no specific values for that and all depends on the target structure. For example, the required spatial resolution for a liver injection is going to be lower than the one required for a thyroid injection in the same animal; basically, because the liver is thousands of times bigger than the thyroid glands and the difficulty of injecting correctly in the organ will be low. For this reason, in this example, we do not need an extremely high spatial resolution, only sufficient to distinguish between abdominal organs. In the same way, works temporal resolution. It is not necessary to have a smooth movement of the organs in the monitor, while injecting or taking a sample, only the minimum is required to synchronize the movement of the needle and its screen visualization.

As we said, there is no optimal system or minimum requirements for doing guided injections. It will depend on the animal model (mice, rats, ferrets, rabbits...) and the structure we need to poke. For example, for injecting into the liver of a rabbit, a clinical ultrasound device with a 7.5 MHz probe will be enough, but for doing the same injection in a 20 gr mouse we will need, at least, 20 MHz, which means that almost none of the clinical devices could be used (there are some dermatology probes that could reach high frequencies such as 20 MHz) [11]. For a mouse thyroid gland or a fetus injection, the requirements reach 35–40 MHz, with values restricted to preclinical devices.

Regarding the anesthesia of the animals, it is mandatory in order to work in proper conditions and follow the minimum ethical requirements. There are several publications that show ultrasonographic examinations with awake and immobilized animals, even trained ones [12], but the interventionism required the introduction of a needle in the body, so some pain is going to be produced. Furthermore, for correct recognition of the target, we will need steady images, without movement artifacts that will complicate the synchronization of the needle movement and the ultrasound images. There are different anesthesia protocols available in the literature, with pros and cons, but for regular injections, an inhaled anesthesia would give an acceptable unconsciousness level, knowing that the pain generated in the procedure will be minimum. This level can be reached using isoflurane with an individual mask and a concentration of 2–3% of isoflurane in fresh air or oxygen. A heating platform will be required in order to compensate for the loss of temperature caused by the anesthesia. A heat light would work too in most cases. In the same way as the temperature, a lubricant must be applied to the eyes for avoiding keratitis due to dry eyes.

An analgesic protocol is required for the procedure. It is recommended to use local anesthesia at the point of injection as well as long-term analgesia during the first 48 to 72 h after the injections. This is especially required when the punctured structure is a hard organ such as the heart, liver, kidney, or spleen. In these cases, the tissular damage caused is higher than the one created when injecting a soft structure such as the pancreas. A standard protocol could be the use of bupivacaine or lidocaine as local anesthetics and buprenorphine as an acute analgesic in the first 24 h, followed by meloxicam or carprofen during the next 48 h [13]. It would be possible to inject local anesthetics in the target organ, but this will require switching syringes between the one containing the anesthetic agent and the one containing the cells. This change could be complicated if we are performing a free hand injection. Using a support for the probe and the syringes could make this change more feasible.

The needle width is a key point during the injection or the extraction of the sample. It should be big enough for the correct internal fluid displacement but the smallest to reduce the damage caused in the tissue. Therefore, the viscosity of the injected substance has an impact on the welfare of the animals and we should try to reduce this viscosity as possible. The injections can be performed manually, with regular needles and syringes, but different microinjectors can be found in the market, specifically designed for ultrasound-guided procedures [14,15]. These systems are able to set low injection volumes, down to 2 nanoliters (Figure 1).

The technical procedure to inject a liquid (drug or cells) and to remove a sample is the same but the final step will be obviously different. While doing an injection, we will push the plunger, during a biopsy we need to pull the same part of the syringe, creating an internal vacuum inside the syringe. We will need to repeat this pulling several times to be sure that we have enough content for the required analysis. This procedure is called Fine needle aspiration or FNA [16]. In hard tissues, multiple punctures can be performed without removing the needle from the organ, only changing slightly its position. With these movements, we will create more damage, but on the other hand, we will be able to sample different parts of the organ and obtain more representative samples.

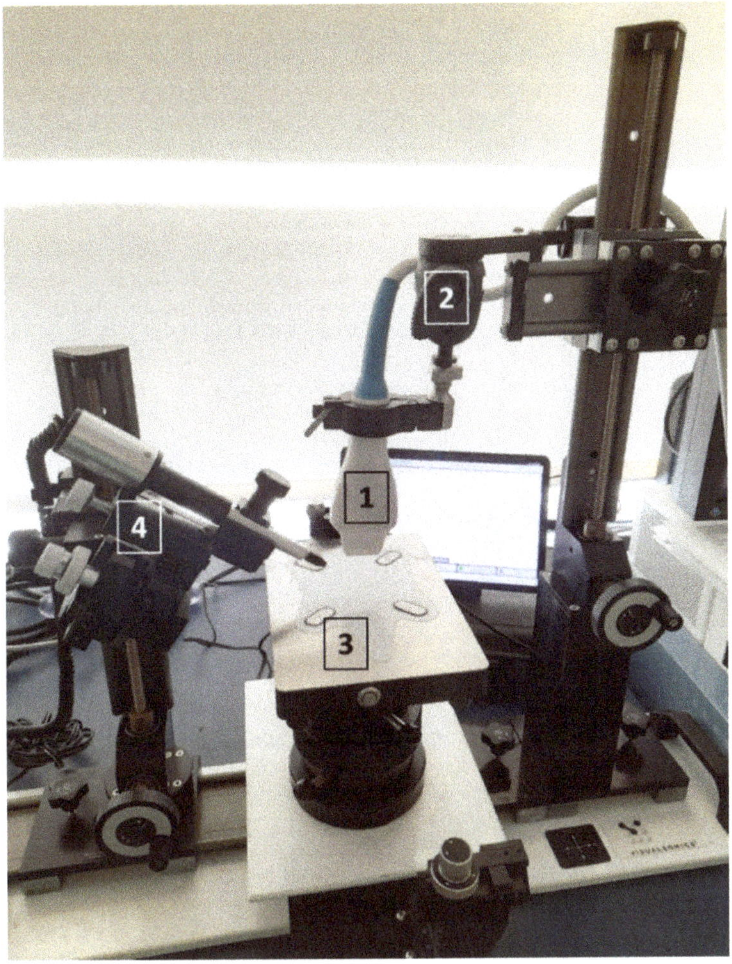

Figure 1. Ultrasound set with the probe (1) placed in the support (2). The heating platform for placing the scanned animal (3) and the microinjector system (4) are displayed too.

3. Ultrasonographic Interventionism

In this section, we will describe the procedural protocols that are applied for the most frequent guided injections or samplings. From cranial to caudal, we will describe the thyroid and intracardiac injections followed by the intrabdominal organs (liver, spleen, pancreas, kidney, uterus) and end with the intratesticular injection. We will try to describe in detail the manipulation of the animal and the probe and syringe, with comments about the injected volumes and potential complications of the procedure.

It is important to say a few words about one of the main limitations of ultrasound-guided injections, and general ultrasound: it is the operator dependency of the technique. Unlike other imaging techniques, in which the analysis of the images can be performed after the acquisition, ultrasonography is a real-time technique. During the exam, we shall decide about different examination aspects, such as specific movements of the probe, increases in the probe pressure, changes in the wave frequency, brightness and contrast of the images, and so. Because of these, the results obtained during the exam are going to be affected by the expertise and experience of the operator. This is even exacerbated in ultrasound-guided surgeries, where we need to synchronize the movements of the probe and the syringe. Even if we fix the probe and needle in the respective supports, the visualization of the needle in the monitor will depend on slight and smooth movements and the probe/needle and this could be hard to achieve without previous experience and practice. For this reason, it is required to make a training before starting the real experiments and procedures with the ultrasound. The practices can be run at the beginning using carcasses for refinement and reduction purposes. As with any other surgical ability, the learning curve will depend on our skills and practice, so it is strongly recommended to repeat the procedures several times before starting the real projects.

Even with the operator-dependency limitation, the reproducibility of the models can be significantly improved using guided surgery. The reduction in the number of animal manipulations as well as the improvement in animal welfare will affect the reproducibility of the whole project, due to the lower side effects we will have from the tumor cells implantation as well as the reduced anesthesia times and improved recovery of the animal. These changes would be reflected in more homogeneous tumor developments through the group of animals. This improvement in reproducibility will have an impact on the quality of the research and animal welfare due to the reduction in outlier animals which will mean a reduction in the total of used animals [17].

3.1. The Thyroid Gland

The thyroid glands are two small ellipsoid and hypoechoic structures located in the ventral aspect of the neck, at both sides of the trachea, and surrounded by the salivary glands and sternohyoideus and sternothyroideus muscles. Other regional structures include the common carotid arteries and the internal jugular veins. These vascular structures make thyroid puncture significantly risky for non-expert practitioners. Even so, this risk can be reduced using the ultrasonographic Doppler mode for distinguishing the vessels from the other structures.

For intrathyroidal injections, a ventral approach is required, positioning the mouse in ventral recumbence and removing the hair of the neck and cranial part of the thorax. The front limbs are fixed with tape in a caudal position, close to the ribs. Intubation of the animal is not required and anesthesia can be supported with a facial mask. After the application of ultrasonographic gel, the scan starts locating the trachea in transversal view at the hyoid bone level. It is recognized due to the acoustic shadow produced by the intratracheal air. Moving the probe caudal, the salivary glands will appear as two superficial, bilateral, hypoechoic, and big structures. At this level, we will need to slightly increase the pressure of the probe against the neck for improving the visualization of deeper structures. Going caudal, a muscular band will appear in the middle line, ventral to the trachea, followed by two bilateral structures, the sternohyoideus and sternothyroideus muscles. The carotid arteries and jugular veins will be visible at this point. The first ones are smaller but have a pulse. We can check the blood flow direction using the Color Doppler mode of the ultrasound system. In a standardized position of the probe (left side of the probe placed over the right side of the animal), the arterial flow should be colored in red and the jugular veins should appear in blue (Figure 2).

The thyroid glands will be located at this level, dorsal and slightly medial to the neck vessels. In a standardized exam, they will appear under the vessels. They are composed of soft tissue, so the echogenicity will be lower than the salivary glands but higher than the vessels. Their shape is irregularly ellipsoidal. The best approach for the puncture is lateral, placing the needle under the ultrasound probe. If the needle is placed correctly in the injection support, we will see it coming from the lateral of the screen (Figure 2). The hardest part of the injection is piercing the skin and for this purpose, we can use forceps for immobilizing the skin. The maximum volume we can inject is low due to the organ size, so more than 10 to 20 microliters in each gland is not recommended [18–20].

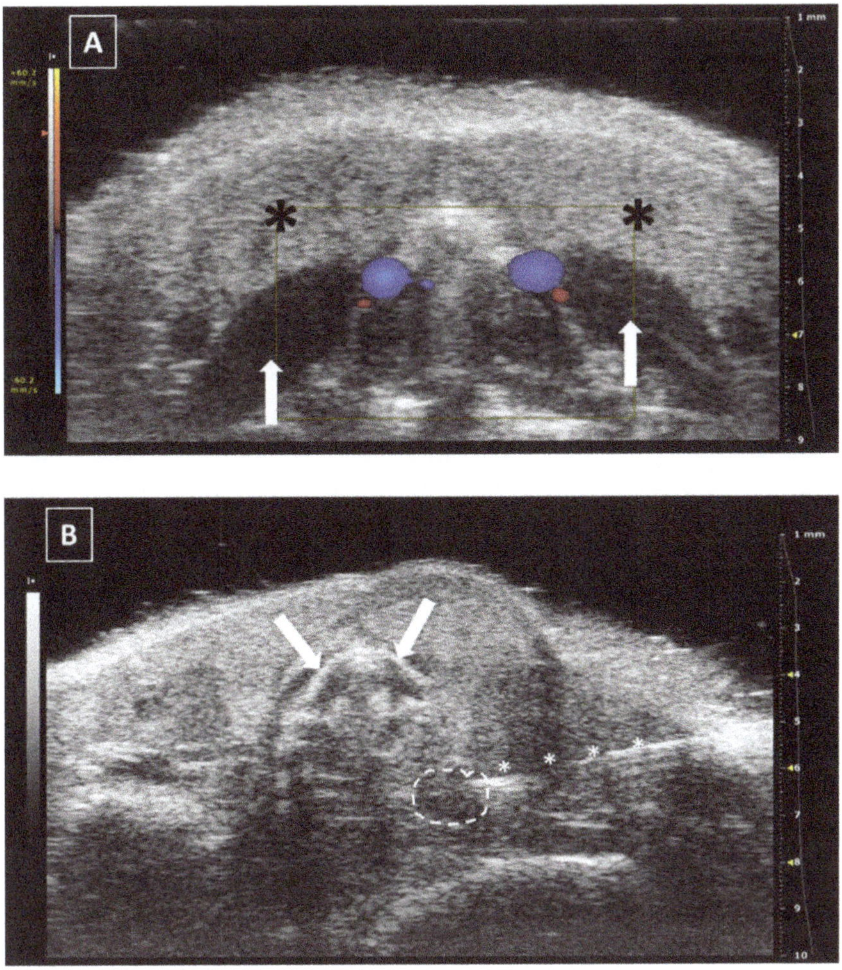

Figure 2. Thyroid gland injection. (**A**) Doppler mode of the medium level of the neck, where the jugular veins are colored in blue while the carotid arteries in red. Salivary glands marked with a black asterisks and neck muscles with a white arrow. (**B**) B mode during the thyroid injection. The needle is marked with white asterisks, and the thyroid gland is surrounded by a dashed line. Trachea cartilage is signaled with white arrows. Images acquired with 40 MHz frequency in B mode and 32 MHz in Doppler mode.

The needle can be slowly removed after the end of the injection and a final revision of the gland is required for confirming the absence of hemorrhages.

The major risk for this technique is the injection in the wrong structure, such as the salivary glands or the regional muscles. In this case, we will see a fluid accumulation in any of the cited structures. Another risk is damage to the carotid or jugular vessels. In this case, we will see an acute hemorrhage in the zone, with a fast separation of the lobes of the salivary glands and a local swelling visually noticeable.

The thyroid biopsy should be performed in the same way as the injection, but the movement for doing multiple samplings is not recommended due to the small size of the organ and the close proximity of relevant structures such as the neck vessels or the trachea, that will have a fatal result if damaged.

3.2. *The Heart*

Intracardiac injection is a frequent task in cancer research for developing general metastatic models. Most of the time it is performed without ultrasonography and there are several manuals that describe the protocol for this injection [21–24]. The advantage of using ultrasonography is to confirm the success of the injection, since we will be able to see the needle coming into the left ventricle and the injected fluid going into the blood torrent.

The animal should be placed in a ventral recumbency and the thorax must be shaved. The probe should be placed in a transversal view over the middle part of the thorax where we will see the heart beating. In a standardized position of the probe, the left ventricle of the heart should be visible on the right side of the monitor and the right ventricle, small in comparison, will be visible on its left, mostly hidden by the acoustic shadow coming from the sternum. The approach for the injection will be from the left side of the animal, where the heart is in close contact with the inner thoracic wall. We will need to avoid the rib bones during the needle introduction into the thorax.

The less traumatic place for injection is the medial part of the left ventricle, right after the papillary muscles which can be observed as pyramidal structures growing inward from the myocardium. At this level, the needle will be more difficult to intrude, due to the muscular thickness, and the damage will be higher. Moving caudal from this point, we will find a bigger region with only the myocardial wall between the outer and inner sides of the heart. Going too caudal will lead us to the apex, where the injection will be more difficult due to its small size and thicker muscular wall.

Once we find the correct spot for injecting, the needle can be introduced from the left side avoiding the rib bones. The movement of the needle needs to be slow, and the thoracic wall can create some resistance that can be exceeded making some external pressure from the contralateral side, pushing with one finger from the right side of the thorax.

The needle can be seen inside the left ventricle as a hyperechoic linear structure with reverberation artifacts in an anechoic background. During the injection, we will be able to see some small hyperechoic dots coming from the needle. These are microbubbles created during the needle filling. These dots will confirm the correct injection in the anechoic ventricular cavity. Once the injection is completed, the needle can be removed. An example of an intracardiac injection can be seen in Figure 3.

The duration of this process depends on the expertise of the user. An experienced ultrasonographer can do the injection in less than a minute. The major risk of this procedure is the incorrect injection in the right ventricle, the lung, or the mediastinum. We will not see the hyperechoic bubbles arising inside the left ventricle. Other less frequent errors can be damaging the aorta the cava vein or any of the cardiac atriums. In these cases, we will see an acute intrathoracic hemorrhage.

For intracardiac blood sampling, the approach and procedure are the same as the injection. Sampling the myocardial tissue will be really challenging due to the thin wall and the continuous movement of the heart. We will need to puncture the cardiac wall without getting into the ventricle and make the pulling of the plunger in synchrony with the movement of the ventricle wall.

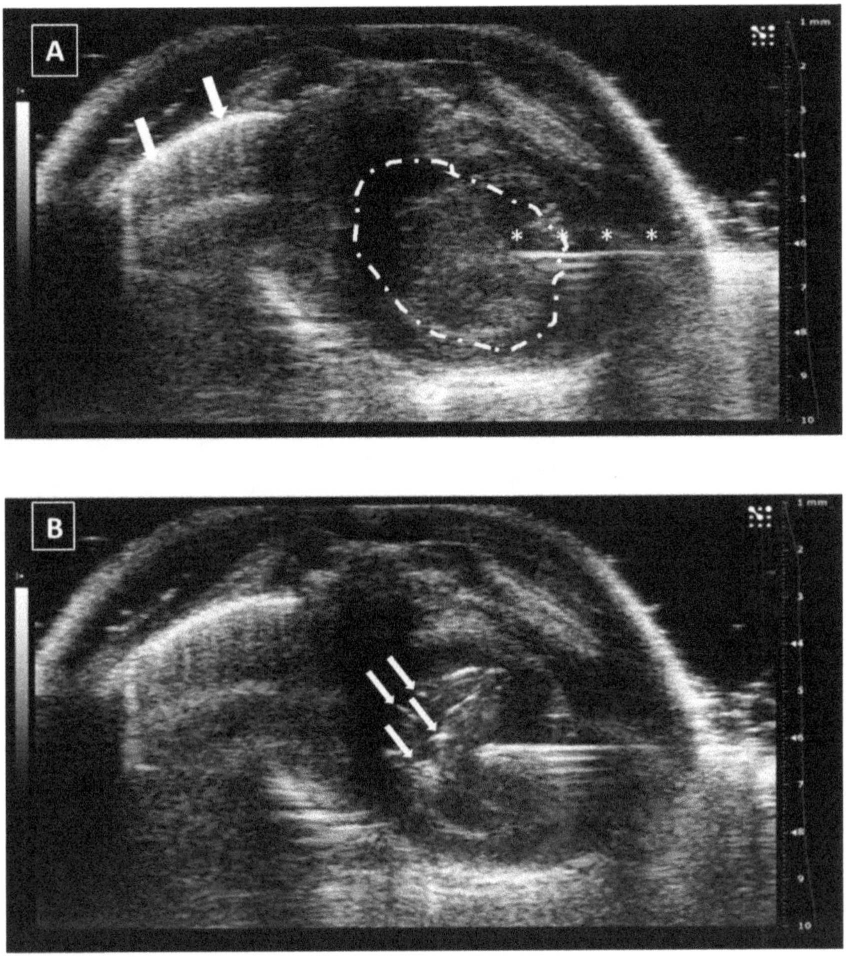

Figure 3. Intracardiac injection. (**A**) Preinjection image. Needle is marked with white asterisks and left ventricle area is surrounded by a dashed line. Lung artifact is labeled with white arrows. (**B**) Injection moment. Multiple white dots (marked with white arrows) inside the left ventricle correspond to microbubbles injected with the suspension. Images obtained with 40 MHz frequency.

3.3. The Liver

The liver is a homogenous organ situated in close contact with the diaphragm, occupying the cranial part of the abdomen. Echographically, it can be described as homogeneous and moderately hyperechoic compared to the spleen [25–27].

The intrahepatic injection is an easy procedure due to the size of the liver, which allows us to inject in both left and right sides of the organ. In our opinion, the right approach is easier due to the presence of the stomach on the left, which reduces the space for maneuvering. The animal is placed in ventral recumbency and hair is shaved in the cranial part of the abdomen. After localizing the desired region of the liver, the needle is moved under and parallel to the probe from the outside and into the abdomen, avoiding the rib bones. We will see a hyperechoic line going into the hypoechoic and homogeneous hepatic tissue. The injection can be confirmed with the appearance of an anechoic structure (the injected fluid) inside the liver tissue. After injecting, the needle should be kept in place for some seconds. Later, it can be removed and the organ should be examined for the presence of hemorrhages. An example of a liver injection is shown in Figure 4.

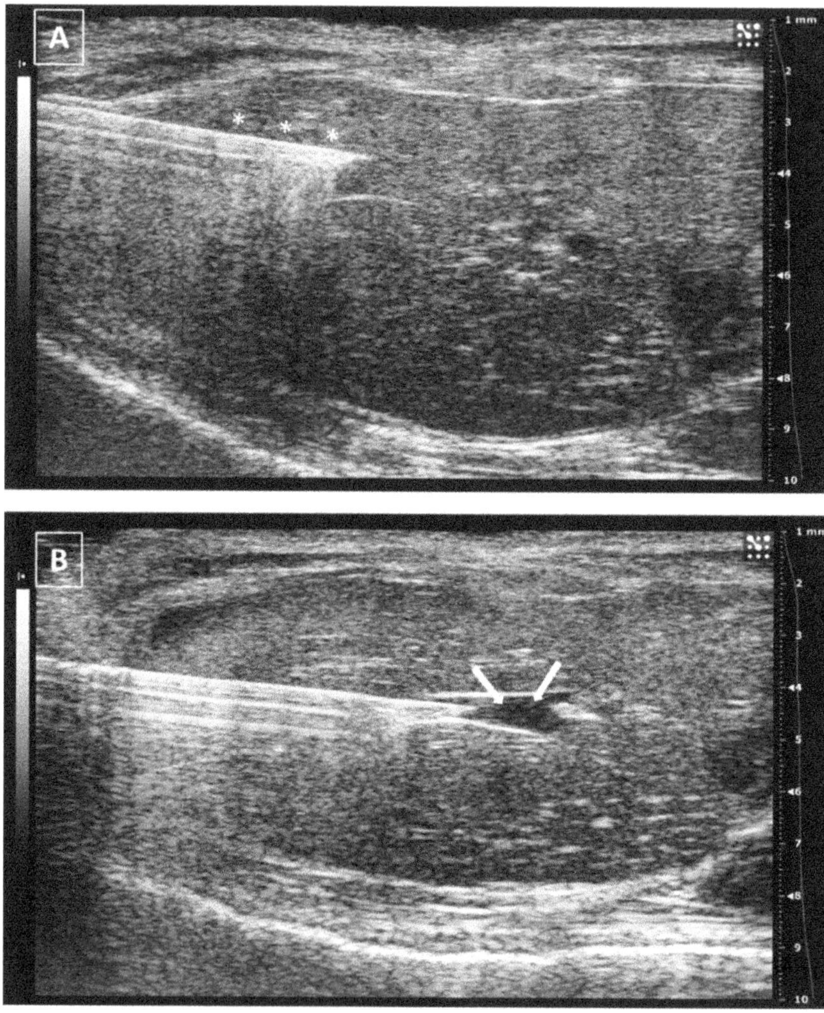

Figure 4. Intrahepatic injection. (**A**) Right side approach. Needle is marked with white asterisks. (**B**) Left side approach. Injected fluid is marked with white arrows. The fluid appears as an anechoic collection inside the homogeneous hypoechoic liver tissue. Images obtained at 40 MHz frequency.

The recommended maximum volume of injection depends on the size of the organ, but in the literature, we can find volumes around 40–50 microliters [28–30]. An excess in the injection volume could lead to a rupture of the hepatic tissue due to a pressure increase in the tissue and this will lead to the appearance of an acute local hemorrhage, or even an hemoabdomen in case the rupture affects the Glisson's capsule.

There are no major risks with this procedure apart from the injection of an excessive volume. Other infrequent problems could be the puncture of the gallbladder or an intrahepatic vessel. In both cases, the probability is significantly low and it could be really hard to confirm any of these situations with the ultrasound. There will be general and unspecific symptoms such as intraabdominal bleeding or peritonitis due to the release of bile into the abdomen but both findings will be delayed in time.

Making a liver biopsy will be as easier as doing an injection, following the same approach and procedure. We will be able to make multiple aspirations of the organ due to its size.

3.4. The Spleen

The spleen is a hypoechoic hematic organ, typically located on the left side of the abdomen, caudal to the stomach and lateral to the left kidney, but it can slide around the cranial part of the abdomen, especially during splenomegaly [25,26]. Due to this reason, the approach for its injection will depend on where it is located. In its usual place, a lateral approach is the best and easiest way of injecting. The animal should be placed in lateral recumbency, with the left side up. After shaving the hair, the scanning probe is placed over the last ribs and slowly displaced caudally. The spleen will appear on top of the screen, just under the skin. We will slightly balance the probe ventrally without losing sight of the structure and will insert the needle from the back of the animal. In a standardized view, the needle will arise from the right side of the screen and go medially. If the pressure from the ultrasound probe is enough, the spleen will be immobilized between it and the needle, and the injection will be performed easily. Similar to the liver, the maximum injected volume depends on the organ size but in previously published work we can find a range from 20 to 50 microliters [31]. After a few seconds, the needle can be removed and a last exam for the absence of bleeding should be performed. A representative image of the injection is shown in Figure 5.

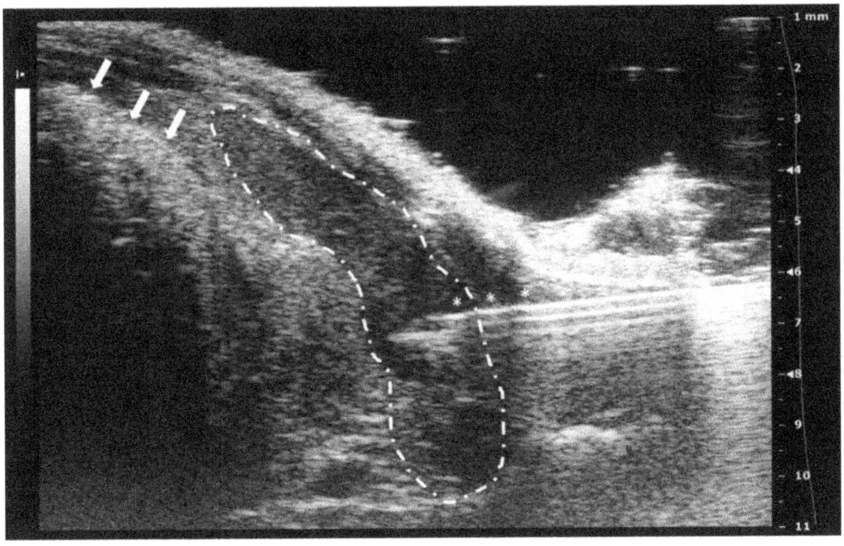

Figure 5. Intrasplenic injection. Needle is marked with white asterisks and the spleen is surrounded by a dashed line. Stomach can be localized due to its typical acoustic shadow. Images obtained at 40 MHz frequency.

When the spleen is located in a different area of the abdomen, we will do a lateral approach for the injection, placing the scanning probe in a ventral view and injecting from the left or right sides depending on the location of the structure.

For a splenic biopsy, the approach is similar and we will be able to make several punctures from the organ due to its size in most of the animal models.

There are no big risks during a splenic injection due to its superficial location and the size of the structure. Only in immunodeficient animals, where the organ is extremely small, the risk of injecting incorrectly in another structure or even freely into the abdominal cavity should be considered. In these cases, increasing the spatial resolution for more detailed visualizations is required. The organ could be found under the last ribs and this will make the injection more complicated.

3.5. The Pancreas

The pancreatic tissue is, in mice, a poorly defined structure divided into three branches. The left branch is located cranial to the left kidney and medial to the spleen. It can be easily found surrounding the principal splenic vessels. The middle branch is located between the caudal aspect of the stomach and the cranial aspect of the transverse colon. The right pancreatic branch is limited by the lateral aspect of the right kidney and the medial aspect of the duodenum. The ultrasonographic image of the pancreas is as an isoechoic tissue compared to the liver, with small parallel hyperechoic lines [25,26].

For pancreatic injections, frequently performed in oncology for developing pancreatic tumors, the best area for injecting is the left branch. The middle one is smaller and the right branch is more complicated to access due to the presence of the duodenum and the right kidney. For injecting in the left branch, we need to localize the primary splenic vein that runs cranially to the cranial pole of the left kidney, from its origin in the spleen to its insertion in the cava vein. The pancreatic area can be recognized as a poorly defined isoechoic area with internal parallel hyperechoic lines around the vein. In this place, the needle can be introduced from the lateral side while we keep the scanning probe in a medial position. In a standardized image, the needle will appear from the right side of the screen. As happens with other lateral approaches, sometimes we will need to make some pressure from the contralateral side of the animal to overcome the skin and abdominal wall resistance to the puncture. Once these two structures are pierced, the injection can be performed without difficulties. A small anechoic bubble arising in the pancreatic area can be observed if the injection is performed correctly. We should wait a few seconds before removing the needle as we do in other procedures. An example of a pancreatic injection is displayed in Figure 6.

The major risk of this injection will not affect the health of the animal but the success of the model. Several times we can observe, after the injection, that the fluid from the syringe is not creating an anechoic bubble but moving freely to the ventral wall of the abdomen (the top of the monitor because we positioned the animal in ventral recumbency). This happens when the injection is not correctly performed in the pancreatic tissue but in the peritoneum. In this situation, we should stop injecting and relocate the tip of the needle to another area. Otherwise, we will have an abdominal disseminated model of pancreatic tumors.

There is no limitation on the volume injected in the pancreas due to the minimum stiffness of the tissue, but in most of the publications, the injected volume range goes from 20 to 50 microliters [32–34].

The guided biopsy of the pancreas is extremely difficult because, as we said before, the pancreas in mice is a membranous structure and there is not a defined tissue for aspirate. In the case of a pancreatic tumor biopsy, for example, the approach will be similar, but we should find the mass prior to the introduction of the needle. In the cells injection was correctly performed, the pancreatic tumor should be located in the same region.

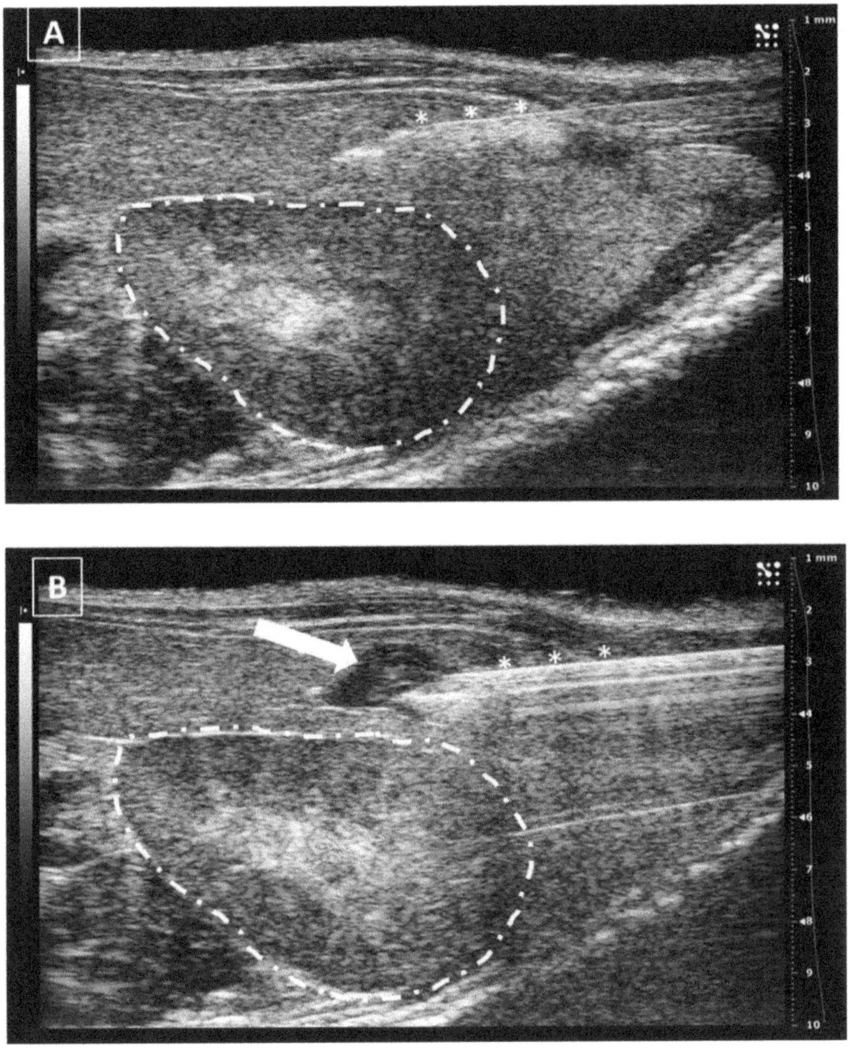

Figure 6. Intrapancreatic injection. (**A**) The needle is marked with white asterisks and injected in the pancreatic area. The left kidney is surrounded with a dashed line. (**B**) Same area after injection. The fluid collection is signaled with a white arrow. Images obtained at 40 MHz frequency.

3.6. The Kidneys

The kidneys are two ellipsoid hyperechoic structures that can be found on both sides of the abdomen. The left kidney is more caudal than the right and is anatomically related to the spleen and stomach, while the right one is close to the liver and duodenum. Their echogenicity is higher than the spleen and similar to the liver, but these ratios can change between animal strains or even individuals [25,26].

The intrarenal injection can be performed in both organs and a lateral approach is recommended, with the animal in ventral recumbency and the probe placed in the middle line of the abdomen. This layout will give us access to the lateral aspect of the kidney, safe from the ilium where the renal artery and vein are located. Depending on the injection depth, we will make a cortical (more superficial) or a medullar (inner part) injection. In both cases, the needle needs to be introduced from the lateral side and will appear in the monitor from the right side (if we inject the left kidney) or the left side (if we inject the right one) as far as we follow the scanning standardization.

A different approach for the injection can be performed with the animal in lateral recumbency, placing the probe in the ventral aspect of the abdomen and injecting from the dorsum. This will create a compression between the needle and the probe, immobilizing the organ in the middle.

The injected fluid will be observed as a hypoechoic accumulation inside the echoic renal tissue. The maximum volume that we can inject is limited to 20–50 microliters if we attend to the published works [35,36]. The renal tissue is rigid and fragile and does not accept significant increments in the tissular pressure.

In the same way as the liver and spleen, the kidney biopsy can be performed following the same approach as the injection but pulling the needle plunger instead of pushing it down. Multiple samples can be collected in both the cortical and medullar areas of the organ. An example of a renal injection is shown in Figure 7.

The major risk during a renal injection is the incorrect settlement of the fluid. If the injection is performed too deeply into the renal tissue, the tip of the needle could reach the pelvic zone. In this case, the injected fluid will be released directly into the pelvic area and moved to the ureters. Another risk of incorrect injections could be injecting into the renal vessels, especially the renal vein, bigger than the artery. But this possibility is low if we make a safe approach from the lateral or dorsum.

3.7. The Uterus and Fetus

The uterus is a structure located in the caudal region of the female abdomen. It's divided into two parts: the neck and the horns, all with a similar ultrasonographic image: tubular shape with middle echogenicity [25,26]. The neck is anatomically related to the urinary bladder and the rectum, while the horns run from their division in the cranial aspect of the neck until they reach the ovaries, caudal and lateral to the kidneys. Recognizing the uterine horns can be challenging for non-experienced users, especially in non-gravid animals when the image can be mixed up with the bowels. During the pregnancy, the uterine horns increase exponentially their size to host the developing fetus and they are easily recognized.

The uterine injection using ultrasonography is one of the most difficult techniques, especially when the organ is in repose (not gravid). The structure is long, thin, and mobile, and the uterine wall is hard. All these characteristics make the intrauterine injection a challenging process. The less mobile part of the uterus is the neck which stays anatomically fixed in place. In this part of the organ, the injection could be feasible. On the other hand, for injecting the uterine horns they should be externally exposed with a surgical opening of the abdomen. For intrauterine injections, the previously published works never went over 30 microliters [37–40]. Like in the other injections, we should keep the needle in place for some seconds before removing the needle. Figure 8 shows a representative image of a uterine injection.

The injection in the gravid uterus is easier due to the increased size of the organ but can be challenging depending on our target. The myometrium will be more complicated to reach due to the reduction in its thickness during pregnancy, as well as the injection into the uterine cavity will be challenging too because of the presence of multiple vesicles corresponding to the amniotic sacs.

If we want to inject into the amniotic sacs, the feasibility of the procedure will depend on the stage of pregnancy and fetal development. During the earliest days, recognizing the sacs will be challenging due to their small size. With the progression of fetal development, the amniotic sacs become bigger and the injection will be easier. At the end of the pregnancy, the amniotic fluid is significantly reduced and almost the whole volume of the sac corresponds to the fetus [41–43].

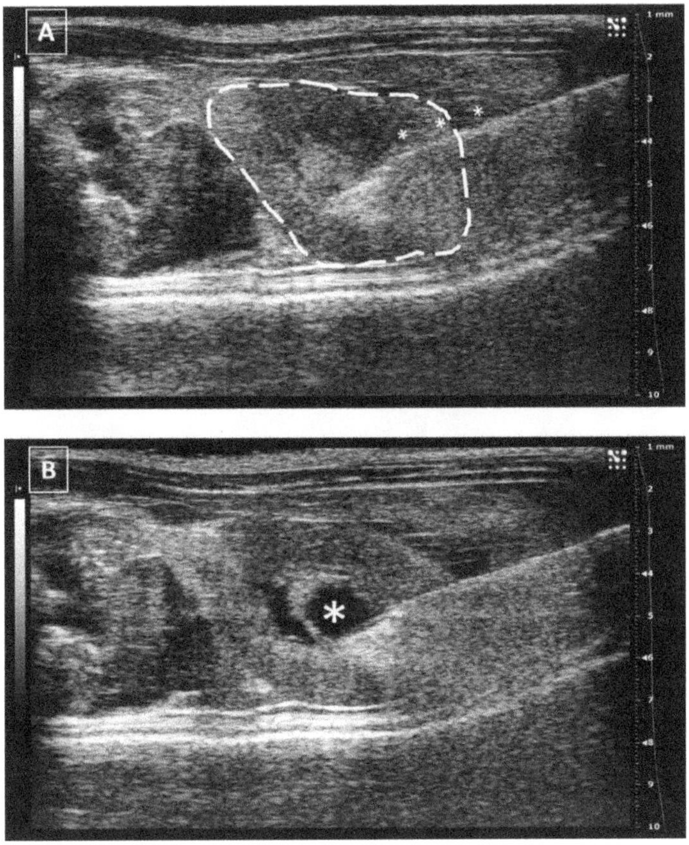

Figure 7. Intrarenal injection. (**A**) Injection of the needle into the kidney. Needle is marked with white asterisks and the kidney is surrounded by a dashed line. The injection is performed in the medullar zone of the organ. (**B**) Administration of the fluid, that is marked with a white asterisk. Images obtained at 40 MHz frequency.

Injecting a fetus could be challenging depending on the organ or structure we want to pierce. The fetuses are mobile and they will be hard to immobilize. For this reason, most of the published works make a surgical opening of the maternal abdomen, exposing the uterine horns for ultrasonography and fetal-guided injection. Once the uterine horns are externalized, the injection procedure would be similar to the one performed in a mature animal, but the recognition of the organs could be more challenging because in some cases their echogenicity is different, such as what happens with the lungs or the kidneys. Even so, some organs, such as the testicles, are located in a different place compared to a born specimen [43].

The biopsy of the non-gravid uterus will be as challenging as the injection, due to the same reasons: its size and the almost free movement of the horns. It could be easier to make in the uterine neck. On the other hand, amniocentesis can be performed in the gravid uterus if we are able to immobilize the uterine horn, but we should consider the potential damage to the fetus due to the reduction in the amnios volume which could be a significant secondary effect of the procedure.

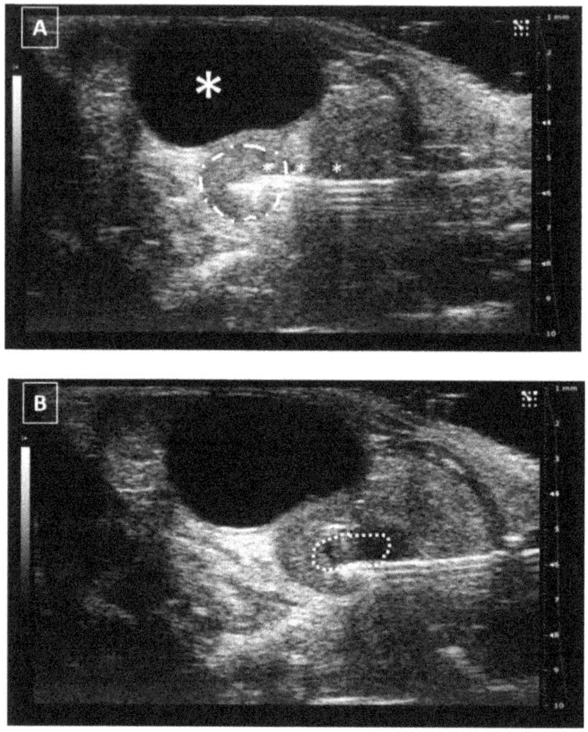

Figure 8. Intrauterine injection. (**A**) The uterus is punched but no fluid is administered. The needle is marked with white asterisks. The uterus is surrounded by a dashed line. The urinary bladder is marked with a big white asterisk. (**B**) Same structure after administration. The fluid collection is marked with a dotted line. Images obtained at 40 MHz frequency.

The main risk of failure during a uterine puncture is the incorrect injection in the abdominal cavity because of an incorrect approach to the organ. Other problems will be more related to the gravid situation, such as uterine wall ruptures if we try to inject an excessive volume into the amniotic sac or damage the fetus during the injection.

3.8. *The Testicles*

The testicles are two mobile hyperechoic structures that can be found randomly inside the scrotum, in the abdominal cavity, or in the inguinal channels [25,26]. For the intratesticular injection, we should move the organs into the scrotum, where the injection can be performed easily. A constant pressure in the caudal region of the abdomen is required to keep the testicles inside the scrotum.

Once the testicle is fixed into the scrotum, the lateral approach is the best for the injection. Placing the scanning probe in a ventral position over the scrotum, the structure can be defined echographically as hyperechoic, homogenous and circular in shape. The lateral approach is the best for injecting into the testicles, placing the probe in the middle line of the body over the scrotum. For reducing the potential movement of the organ, we can use forceps for fixing the skin while piercing with the needle. Introducing the needle in the testicular stroma is easy due to its softness and it can be seen as a hyperechoic line with several comet tails artifacts in a hyperechoic background. The injected fluid will arise as an anechoic accumulation in the testicular stroma. After the injection, we should keep the needle in place for some seconds before removing it. An example of a testicular injection is represented in Figure 9.

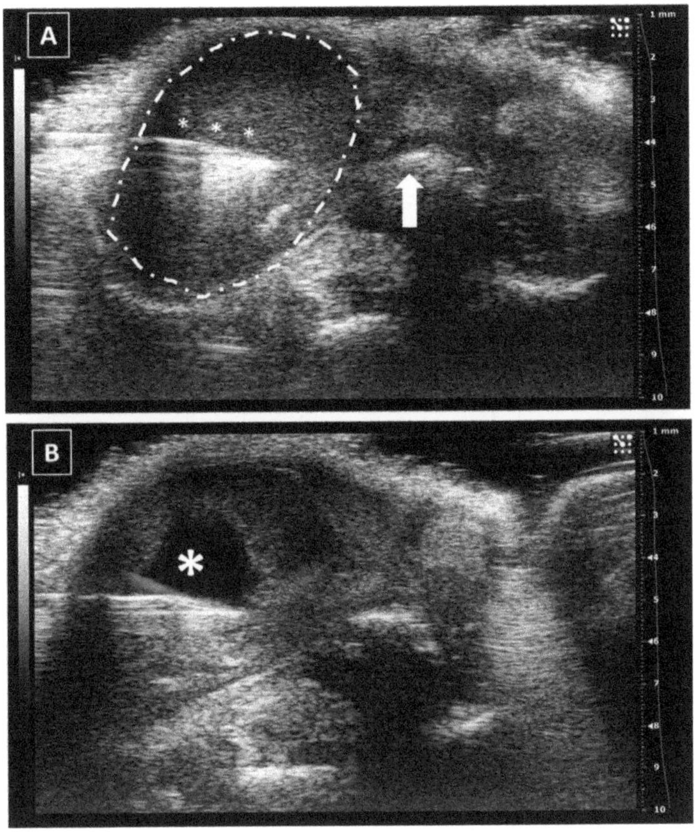

Figure 9. Intratesticular injection. (**A**) Injection moment. The needle is marked with white asterisks and the testicle is surrounded by a dashed line. The penis bone is marked with a white arrow. (**B**) Administration moment. The fluid collection is marked with a white asterisk. Images obtained at 40 MHz frequency.

The maximum injected volume ranges from 20 to 50 microliters if we attend to the previous publications [44,45]. The testicle is a soft and flexible tissue with a significant capability to increase its volume so the possibility of a tissular rupture can be considered low but not inexistent.

The testicle biopsy will be as easier as the injection following the same protocol. We will need to take care of the biopsy point, because a puncture too close to the epididymis will give us a sperm sample instead of real tissue.

The are no major risks during a testicle injection. Only damage to the local vessels could create a negative effect on the viability of the organ. For avoiding this, the color Doppler mode can be used for recognizing the testicular vascular network prior to the injection.

3.9. Other Organs

There are other structures and organs that can be a target for orthotopic cancer models, such as the different sections of the gastrointestinal system or the urinary bowel. In all of them, we will face an additional complication to the procedure and it is that all of these structures are hollow. This means that the injection should be performed in the wall of an empty structure and in most situations, we will not have the technical precision required to make the injection or aspiration. In addition, as we described in the uterine horns, most of these structures are mobile and they will slide with the pressure of the needle tip. The only fixed structures in this group could be the stomach and the urinary bladder. For the first, injecting in the gastric wall could be feasible but the visualization of the injection will be difficult due to all the gas present in this organ, especially in rodents, that will create an acoustic shadow making it almost impossible to visualize the gastric wall. Regarding the urinary bladder, the main difficulty of its injection will be the thin wall of the bladder, especially if there is urine inside. This thin wall makes the exact injection extremely difficult. On the other hand, cystocentesis (sterile urine collection) is an easy procedure that can be performed with the help of an ultrasound and only requires the presence of urine in the bladder. The approach is ventral and we will find the bladder as a complete anechoic structure, round in shape, in the caudal part of the abdomen. Once we find it, the procedure requires introducing the needle from a lateral aspect and almost perpendicular to the probe, firmly and deep into the abdomen. If we succeed in getting into the bladder, a hyperechoic spot will be visible in the anechoic structure of the bladder. An example of a cystocentesis can be seen in Figure 10.

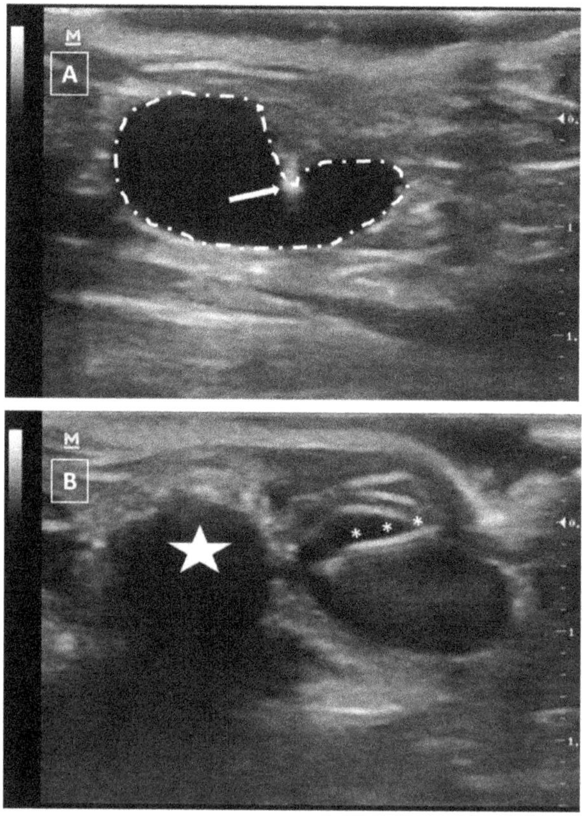

Figure 10. Cystocentesis. (**A**) Moment of the bladder wall rupture in transversal view. The wall is still presenting resistance and the shape of the bladder is not round due to this. The bladder is surrounded by a dotted line. The tip of the needle is marked with a white arrow. (**B**) Collection of urine. Longitudinal view. The bladder has been punctured and its wall recovered the tension, getting back the round shaper of the organ. The needle is marked with white asterisks. Cranial to the bladder, an acoustic shadow indicates the presence of feces in the rectum (white start). Images obtained at 25MHz frequency.

4. Conclusions

Ultrasonography is a potent and trustable technique that can help investigators in developing different cancer models. Ultrasound-guided injection is a significant tool for ethical refinement which should be considered during the design of animal experiments for an improvement in animal welfare and the quality of the research.

For the implementation of the guided injection or sampling, instruments with specific requirements are necessary, depending on the animal model and target structure, as well as a deep knowledge of the procedures by the operator is required.

Funding: This research received no external funding.

Institutional Review Board Statement: Not applicable.

Acknowledgments: The author wants to thank the Preclinical Therapeutics Core at the UCSF for their support during the writing of this review as well as to Marina Ferrer Clotas for her help during the whole process.

Conflicts of Interest: The authors declare no conflict of interest.

References

1. Russell, W.M.S.; Burch, R.L. *The Principles of Humane Experimental Technique*; Methuen & Co.: London, UK, 1959.
2. Fiorito, G.; Affuso, A.; Basil, J.; Cole, A.; de Girolamo, P.; D'Angelo, L.; Dickel, L.; Gestal, C.; Grasso, F.; Kuba, M.; et al. Guidelines for the Care and Welfare of Cephalopods in Research—A consensus based on an initiative by CephRes, FELASA and the Boyd Group. *Lab. Anim.* **2015**, *49* (Suppl. S2), 1–90. [CrossRef] [PubMed]
3. Charan, J.; Kantharia, N.D. How to calculate sample size in animal studies? *J. Pharmacol. Pharmacother.* **2013**, *4*, 303–306. [CrossRef] [PubMed]
4. Smith, J.D. Single-Case Experimental Designs: A Systematic Review of Published Research and Current Standards. *Psychol. Methods* **2012**, *17*, 510. [CrossRef] [PubMed]
5. Piñero, J.; Furlong, L.I. In silico models in drug development: Where we are. *Curr. Opin. Pharmacol.* **2018**, *42*, 111–121. [CrossRef]
6. Carbone, L. Ethical and IACUC Considerations Regarding Analgesia and Pain Management in Laboratory Rodents. *Comp. Med.* **2019**, *69*, 443–450. [CrossRef]
7. Workman, P.; Aboagye, E.O.; Balkwill, F.; Balmain, A.; Bruder, G.; Chaplin, D.J.; Double, J.A.; Everitt, J.; Farningham, D.A.; Glennie, M.J.; et al. Guidelines for the welfare and use of animals in cancer research. *Br. J. Cancer* **2010**, *102*, 1555–1577. [CrossRef]
8. Onaciu, A.; Munteanu, R.; Munteanu, V.C.; Gulei, D.; Raduly, L.; Feder, R.; Pirlog, R.; Atanasov, A.G.; Korban, S.S.; Irimie, A.; et al. Spontaneous and Induced Animal Models for Cancer. *Res. Diagn.* **2020**, *10*, 660. [CrossRef]
9. Plessis, A.d.; Broeckhoven, C.; Guelpa, A.; le Roux, S.G. Laboratory x-ray micro-computed tomography: A user guideline for biological samples. *Gigascience* **2017**, *6*, 1–11. [CrossRef]
10. Moran, C.M.; Thomson, A.J.W. Preclinical Ultrasound Imaging—A Review of Techniques and Imaging Applications. *Front.Phys.* **2020**, *8*, 124. [CrossRef]
11. Hewick, S.A.; Fairhead, A.C.; Culy, J.C.; Atta, H.R. A comparison of 10 MHz and 20 MHz ultrasound probes in imaging the eye and orbit. *Br. J. Ophthalmol.* **2004**, *88*, 551–555. [CrossRef] [PubMed]
12. Brunner, C.; Grillet, M.; Urban, A.; Roska, B.; Montaldo, G.; Macé, E. Whole-brain functional ultrasound imaging in awake head-fixed mice. *Nat. Protoc.* **2021**, *16*, 3547–3571. [CrossRef]
13. Foley, P.L.; Kendall, L.V.; Turner, P.V. Clinical Management of Pain in Rodents. *Comp. Med.* **2019**, *69*, 468–489. [CrossRef]
14. Slevin, J.C.; Byers, L.; Gertsenstein, M.; Qu, D.; Mu, J.; Sunn, N.; Kingdom, J.C.; Rossant, J.; Adamson, S.L. High resolution ultrasound-guided microinjection for interventional studies of early embryonic and placental development in vivo in mice. *BMC Dev. Biol.* **2006**, *6*, 10. [CrossRef] [PubMed]
15. Available online: https://www.visualsonics.com/application/preclinical/image-guided-injection (accessed on 19 November 2022).
16. Available online: https://www.cancer.net/blog/2021-10/fine-needle-aspiration-biopsy-how-prepare-and-what-expect (accessed on 19 November 2022).
17. Voelkl, B.; Altman, N.S.; Forsman, A.; Forstmeier, W.; Gurevitch, J.; Jaric, I.; Karp, N.A.; Kas, M.J.; Schielzeth, H.; Van de Casteele, T.; et al. Reproducibility of animal research in light of biological variation. *Nat. Rev. Neurosci.* **2020**, *21*, 384–393. [CrossRef] [PubMed]
18. Sewell, W.; Reeb, A.; Lin, R.Y. An orthotopic mouse model of anaplastic thyroid carcinoma. *J. Vis. Exp.* **2013**, *74*, 50097. [CrossRef]
19. Mancini, M.; Vergara, E.; Salvatore, G.; Greco, A.; Troncone, G.; Affuso, A.; Liuzzi, R.; Salerno, P.; Scotto di Santolo, M.; Santoro, M.; et al. Morphological ultrasound microimaging of thyroid in living mice. *Endocrinology* **2009**, *150*, 4810–4815. [CrossRef] [PubMed]

20. Greco, A.; Albanese, S.; Auletta, L.; Mirabelli, P.; Zannetti, A.; D'Alterio, C.; Di Maro, G.; Orlandella, F.M.; Salvatore, G.; Soricelli, A.; et al. High-Frequency ultrasound-guided injection for the generation of a novel orthotopic mouse model of human thyroid carcinoma. *Thyroid* **2016**, *26*, 552–558. [CrossRef] [PubMed]
21. Zhou, H.; Zhao, D. Ultrasound imaging-guided intracardiac injection to develop a mouse model of breast cancer brain metastases followed by longitudinal MRI. *J. Vis. Exp.* **2014**, *85*, 51146. [CrossRef] [PubMed]
22. Available online: https://star-protocols.cell.com/protocols/1600 (accessed on 20 November 2022).
23. Toivonen, R.; Koskenvuo, J.; Merentie, M.; Söderström, M.; Ylä-Herttuala, S.; Savontaus, M. Intracardiac injection of a capsid-modified Ad5/35 results in decreased heart toxicity when compared to standard Ad5. *Virol. J.* **2012**, *9*, 296. [CrossRef]
24. Campbell, J.P.; Merkel, A.R.; Masood-Campbell, S.K.; Elefteriou, F.; Sterling, J.A. Models of bone metastasis. *J. Vis. Exp.* **2012**, *67*, e4260. [CrossRef]
25. Available online: https://www.scintica.com/wp-content/uploads/2019/02/The-Mouse-abdominal-Imaging-Guide (accessed on 20 November 2022).
26. Available online: https://www.visualsonics.com/application/preclinical/gastrointestinal (accessed on 19 November 2022).
27. Fernández-Domínguez, I.; Echevarria-Uraga, J.J.; Gómez, N.; Luka, Z.; Wagner, C.; Lu, S.C.; Mato, J.M.; Martínez-Chantar, M.L.; Rodríguez-Cuesta, J. High-frequency ultrasound imaging for longitudinal evaluation of non-alcoholic fatty liver disease progression in mice. *Ultrasound Med. Biol.* **2011**, *37*, 1161–1169. [CrossRef] [PubMed]
28. McVeigh, L.E.; Wijetunga, I.; Ingram, N.; Marston, G.; Prasad, R.; Markham, A.F.; Coletta, P.L. Development of orthotopic tumour models using ultrasound-guided intrahepatic injection. *Sci. Rep.* **2019**, *9*, 9904. [CrossRef] [PubMed]
29. Bou About, G.; Thiebault, E.; Wattenhofer-Donzé, M.; Jacobs, H.; Guimond, A.; Sorg, T.; Robinet, E.; Baumert, T.F.; Monassier, L.; Herault, Y. Ultrasound-Guided Approaches to Improve Orthotopic Mouse Xenograft Models for Hepatocellular Carcinoma. *Curr. Protoc. Mouse Biol.* **2019**, *9*, e62. [CrossRef] [PubMed]
30. Sugase, T.; Lam, B.Q.; Danielson, M.; Terai, M.; Aplin, A.E.; Gutkind, J.S.; Sato, T. Development and optimization of orthotopic liver metastasis xenograft mouse models in uveal melanoma. *J. Transl. Med.* **2020**, *18*, 208. [CrossRef] [PubMed]
31. Takada, M.; Smyth, L.A.; Hix, J.M.; Corner, S.M.; Kiupel, M.; Yuzbasiyan-Gurkan, V. Development of an orthotopic intrasplenic xenograft mouse model of canine histiocytic sarcoma and its use in evaluating the efficacy of treatment with dasatinib. *Comp. Med.* **2019**, *69*, 22–28. [CrossRef]
32. Hay, C.A.; Sor, R.; Flowers, A.J.; Clendenin, C.; Byrne, K.T. Ultrasound-Guided Orthotopic Implantation of Murine Pancreatic Ductal Adenocarcinoma. *J. Vis. Exp.* **2019**, 153. [CrossRef]
33. Moreno, J.A.; Sanchez, A.; Hoffman, R.M.; Nur, S.; Lambros, M.P. Fluorescent Orthotopic Mouse Model of Pancreatic Cancer. *J. Vis. Exp.* **2016**, *115*, 54337. [CrossRef]
34. Huynh, A.S.; Abrahams, D.F.; Torres, M.S.; Baldwin, M.K.; Gillies, R.J.; Morse, D.L. Development of an Orthotopic Human Pancreatic Cancer Xenograft Model Using Ultrasound Guided Injection of Cells. *PLoS ONE* **2011**, *6*, e20330. [CrossRef]
35. Pavathuparambil Abdul Manaph, N.; Al-Hawwas, M.; Zhou, X.F. A direct and non-invasive method for kidney delivery of therapeutics in mice. *Methods X* **2018**, *5*, 1440–1446. [CrossRef]
36. Bukulmez, H.; Horkayne-Szakaly, I.; Bilgin, A.; Baker, T.P.; Caplan, A.I.; Jones, O.Y. Intrarenal injection of mesenchymal stem cell for treatment of lupus nephritis in mice—A pilot study. *Lupus* **2021**, *30*, 52–60. [CrossRef]
37. Rinaldi, S.F.; Makieva, S.; Frew, L.; Wade, J.; Thomson, A.J.; Moran, C.M.; Norman, J.E.; Stock, S.J. Ultrasound-guided intrauterine injection of lipopolysaccharide as a novel model of preterm birth in the mouse. *Am. J. Pathol.* **2015**, *185*, 1201–1206. [CrossRef] [PubMed]
38. Greaves, E.; Cousins, F.L.; Murray, A.; Esnal-Zufiaurre, A.; Fassbender, A.; Horne, A.W.; Saunders, P.T.K. A novel mouse model of endometriosis mimics human phenotype and reveals insights into the inflammatory contribution of shed endometrium. *Am. J. Pathol.* **2014**, *184*, 1930–1939. [CrossRef] [PubMed]
39. Konings, G.F.; Saarinen, N.; Delvoux, B.; Kooreman, L.; Koskimies, P.; Krakstad, C.; Fasmer, K.E.; Haldorsen, I.S.; Zaffagnini, A.; Häkkinen, M.R.; et al. Development of an Image-Guided Orthotopic Xenograft Mouse Model of Endometrial Cancer with Controllable Estrogen Exposure. *Int. J. Mol. Sci.* **2018**, *19*, 2547. [CrossRef] [PubMed]
40. Wang, C.; Wang, Z.; Yang, Y.; Zhu, C.; Wu, G.; Yu, G.; Jian, T.; Yang, N.; Shi, H.; Tang, M.; et al. Pirt Contributes to Uterine Contraction-Induced Pain in Mice. *Mol. Pain* **2015**, *11*, 57. [CrossRef]
41. Greco, A.; Ragucci, M.; Coda, A.R.; Rosa, A.; Gargiulo, S.; Liuzzi, R.; Gramanzini, M.; Albanese, S.; Pappatà, S.; Mancini, M.; et al. High Frequency Ultrasound for In Vivo Pregnancy Diagnosis and Staging of Placental and Fetal Development in Mice. *PLoS ONE* **2013**, *8*, e77205. [CrossRef]
42. Cheung, C.Y.; Brace, R.A. Amniotic fluid volume and composition in mouse pregnancy. *J. Soc. Gynecol. Investig.* **2005**, *12*, 558–562. [CrossRef]
43. Phoon, C. Imaging Tools for the Developmental Biologist: Ultrasound Biomicroscopy of Mouse Embryonic Development. *Pediatr. Res.* **2006**, *60*, 14–21. [CrossRef]
44. Michaelis, M.; Sobczak, A.; Weitzel, J.M. In vivo microinjection and electroporation of mouse testis. *J. Vis. Exp.* **2014**, *90*, 51802. [CrossRef]
45. Penny, G.M.; Cochran, R.B.; Pihlajoki, M.; Kyrönlahti, A.; Schrade, A.; Häkkinen, M.; Toppari, J.; Heikinheimo, M.; Wilson, D.B. Probing GATA factor function in mouse Leydig cells via testicular injection of adenoviral vectors. *Reproduction* **2017**, *154*, 455–467. [CrossRef]

animals MDPI

Article

Camera-Based Respiration Monitoring of Unconstrained Rodents

Lukas Breuer [1,*], Lucas Mösch [1], Janosch Kunczik [1], Verena Buchecker [2], Heidrun Potschka [2], Michael Czaplik [1] and Carina Barbosa Pereira [1]

1 Department of Anesthesiology, Faculty of Medicine, RWTH Aachen University, Pauwelsstraße 30, 52074 Aachen, Germany
2 Institute of Pharmacology, Toxicology, and Pharmacy, Ludwig-Maximilians-University of Munich, Königinstraße 16, 80539 München, Germany
* Correspondence: lubreuer@ukaachen.de

Simple Summary: Monitoring vitals sign such as the respiratory rate, heart rate, or temperature is of high importance to medical and biological research. Using camera-based methods, we monitored the respiratory rate of unconstrained laboratory rats by analyzing the visible breathing movement in the thorax. We hope this is a further step to enabling the non-invasive monitoring of rodent in an experimental environment without using implanted sensors, reducing the stress and pain within an otherwise unneeded operation.

Abstract: Animal research has always been crucial for various medical and scientific breakthroughs, providing information on disease mechanisms, genetic predisposition to diseases, and pharmacological treatment. However, the use of animals in medical research is a source of great controversy and ongoing debate in modern science. To ensure a high level of bioethics, new guidelines have been adopted by the EU, implementing the 3R principles to replace animal testing wherever possible, reduce the number of animals per experiment, and refine procedures to minimize stress and pain. Supporting these guidelines, this article proposes an improved approach for unobtrusive, continuous, and automated monitoring of the respiratory rate of laboratory rats. It uses the cyclical expansion and contraction of the rats' thorax/abdominal region to determine this physiological parameter. In contrast to previous work, the focus is on unconstrained animals, which requires the algorithms to be especially robust to motion artifacts. To test the feasibility of the proposed approach, video material of multiple rats was recorded and evaluated. High agreement was obtained between RGB imaging and the reference method (respiratory rate derived from electrocardiography), which was reflected in a relative error of 5.46%. The current work shows that camera-based technologies are promising and relevant alternatives for monitoring the respiratory rate of unconstrained rats, contributing to the development of new alternatives for a continuous and objective assessment of animal welfare, and hereby guiding the way to modern and bioethical research.

Keywords: respiration; automatic monitoring; rodent; rat; animal welfare; refinement; 3R; laboratory animals; camera-based monitoring; breathing

Citation: Breuer, L.; Mösch, L.; Kunczik, J.; Buchecker, V.; Potschka, H.; Czaplik, M.; Pereira, C.B. Camera-Based Respiration Monitoring of Unconstrained Rodents. *Animals* **2023**, *13*, 1901. https://doi.org/10.3390/ani13121901

Academic Editor: Garikoitz Azkona

Received: 24 May 2023
Revised: 3 June 2023
Accepted: 5 June 2023
Published: 7 June 2023

Copyright: © 2023 by the authors. Licensee MDPI, Basel, Switzerland. This article is an open access article distributed under the terms and conditions of the Creative Commons Attribution (CC BY) license (https://creativecommons.org/licenses/by/4.0/).

1. Introduction

Animal research has played a major role in many scientific breakthroughs for centuries, even though it has been a source of various ethical debates [1]. This caused governing bodies to implement laws and other regulatory means to safeguard animals in experimental settings. The European Union (EU) requires member states by its Directive 2010/63/EU [2] to apply the 3R principles proposed by Russell et al. [3] in 1959. These principles refer to reduction, refinement and replacement as a mean to minimize the use of animals in scientific studies, while maximizing animal welfare. The term reduction refers to reducing the number of animals used in a study, while still providing the scientific significance needed. Refinement refers to minimizing the pain, suffering, or distress introduced by animal trials. This can be achieved by using less invasive methods or improving the living conditions in terms of housing and care. Replacement refers to finding alternatives to animal testing which are similar or more effective, thus making the animal trial needless. Feasible alternatives could be using cell cultures, simulations, or human studies.

However, reality shows that not all experiments with living animals can be replaced. In 2019, the EU reported that 10.61 million animals were still used in animal trials [4], showing the great need for further refinement methods. Of these, 72% were used for research, 17% to satisfy regulatory requirements and another 6% for routine production. Most of the animals were used to enhance the understanding of the nervous system or finding treatments for diseases such as cancer. Until today, research has not been able to find adequate replacements for these kinds of animal testing, which makes the refinement and improvement of these experiments crucial.

Due to their high anatomical, physiological, and genetic similarity to humans, while being small and easy to maintain, mice and other rodents are most used in research [5] and represent about half of all trial animals [4]. Cardiovascular, pharmacological, and toxicological research requires vital parameters such as the heart rate (HR) or respiratory rate (RR) to assess a given theory. Currently, implanted radio transponders are the only methods to monitor these for unrestrained mice or rats [6]. This can be ECG sensors, piezoelectric sensors, implanted catheter, or other implanted devices. Despite its ability to generate highly precise data, there are several significant drawbacks associated with this methodology. First, it requires an initial implantation surgery, which is invasive and time-consuming. The recovery time for animals to regain their normal circadian rhythms can take up to five to seven days, according to Braga and Burmeister [7]. Second, the implanted device may cause distress and discomfort, especially in small species. Braga and Burmeister also noted that the implanted device could have adverse physiological effects, such as an increased volume in abdominal viscera, which can potentially compromise the movement of the diaphragm and alter breathing patterns in terms of depth and rhythm. Therefore, there is a great need for contactless and unobtrusive monitoring of techniques, which, on the one hand, permit continuously monitoring the laboratory animals and on the other hand obtaining objective parameters for welfare assessment.

There are numerous examples of the application of RR monitoring for rodents, including toxicity studies in drug development [8], anesthesia monitoring [9], respiratory disease research [10], stress and pain assessment [11], sleep research [12] and many more. Ohtani et al. [8] compared the analgesic and respiratory effects of norbuprenorphine (NBN) and buprenorphine (BN), finding that BN had a lower concentration for an analgesic effect without inducing respiratory depression compared to NBN. Tsukamoto et al. [9] studied the effect of multiple anesthetics on vital signs such as temperature, heart rate, respiratory rate and $SP0_2$. Card et al. [10] identified the differences in respiratory physiology depending on the sex, using a seral model of respiratory diseases. Schöner et al. [11] found out that an increased respiratory rate might occur in a model of PTSD in rats. Mendelson et al. [12] investigated sleep apnea in rats.

Over the years, numerous researchers have explored monitoring RR remotely. In 2019, Kunczik et al. [13] showed that the monitoring of mice and rats can be achieved using an RGB camera while undergoing anesthesia. In this approach, RR is measured by tracking the movement of the abdominal areas, while HR is measured using a DistancePPG, as proposed by Kumar et al. [14]. Another approach was presented by Takahashi et al. [15], using the camera recordings of mice from below a see-through acrylic glass, monitoring and tracking hairless areas. Both approaches lack the possibility of long-term monitoring as we would like to see, due to the animals being restrained or in a specialized cage with no possibility of litter or enrichment materials such as nesting pads.

The current paper presents an improved approach for respiratory rate monitoring in rodents by using visual imaging from above. In contrast to other publications that use videos of anaesthetized animals to estimate this vital parameter, our focus here is to demonstrate the capability of the presented algorithm in extracting this from moving animals.

2. Materials and Methods

The proposed algorithm is a multi-step approach for monitoring respiration in an RGB video of unconstrained rats, as illustrated in Figure 1. This paragraph provides a brief overview of all steps, which will be described in detail in the following sections, along with the experimental protocol. During the first step, segmentation masks of images are computed from video recordings using a deep learning algorithm to detect the respiration-associated movement. In the second step, the preprocessing of the segmented regions is carried out. In the third step, the signal is extracted. Last, the actual computation of the respiratory rate is carried out. As a reference, respiration signals were extracted from electrocardiography (ECG) data and used to compare the camera-based signal.

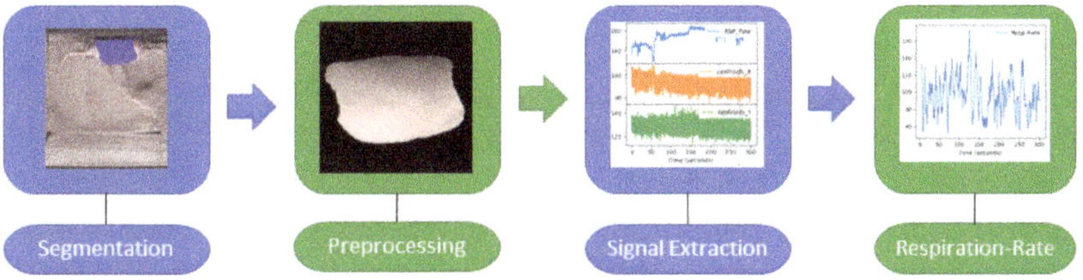

Figure 1. Key stages involved in extracting the RR from the RGB videos of rats: Video preprocessing (segmentation, preprocessing), signal extraction, and RR calculation.

2.1. Experimental Protocol

The data used in this work are part of a larger study that adhered to the 3R principles (replacement, refinement and reduction) to ensure the ethical treatment of animals. The study followed the approved experimental protocol of the governmental animal care and used the institution "Regierung von Oberbayern" (Germany, ROB-55.2-2532.Vet_02-16-105), and was conducted in compliance with the German Animal Welfare Law. All animals received humane care in accordance with the principles outlined in the "Guide for the Care and Use of Laboratory Animals" (8th edition, NIH Publication, 2011, USA).

Three male albino Sprague Dawley rats (360–375 g; 9–11 weeks; Envigo, Horst, The Netherlands) were included in this study. They were subjected to an operation in which ECG and EEG transponders (DSI-HDX02, Data Sciences International, Inc., New Brighton, MN, USA) were implanted. A detailed description about the surgical procedure was already published in 2019 by Seiffert et al. [16]. Prior to and following the operation, the rats were placed into an open glass cage, measuring approximately 0.30 m × 0.30 m, and recorded using two cameras (Cam1 and Cam2). The cage was bedded with a white textile sheet and no additional illumination was provided. The cameras were mounted above the cage on a tripod at about 1.5 m above the bottom of the cage. The distance was selected so that both cameras could acquire the complete bottom of the cage. The experimental setup is depicted in Figure 2.

Cam1 is a long-wave infrared thermal camera (Infratec VarioCAM HD head 820, InfraTec GmbH, Dresden, Germany) with a resolution of 640 × 480 pixels, a thermal resolution of up to 20 mK, a frame rate of 60 FPS and a dynamic range of 16 bit. Cam2 is an RGB camera (Allied Vision Mako G-223C, Allied Vision Technologies GmbH, Stadtrova, Germany) with a resolution of 1368 × 640 pixels and a framerate of 60 FPS, resulting in 18,000 images for a 5 min recording per modality.

The experiment was conducted over five consecutive days, as shown in the experiment schedule displayed in Figure 3. At each measurement time (MT), two 5 min videos were recorded with a parallel ECG recording:

- Day 1: One video recording was obtained for establishing a baseline and the rats allowed to acclimate to the environment. For this recording, no ECG was recorded.

- Day 2: This was surgery day where the EEG and ECG transponders were implanted. Two recordings with all three rats were carried out: the first directly after the surgical procedure and the second approximately two hours later.
- Days 3 to 5 followed a similar schedule, with recordings starting at 9 am, 11 am, 1 pm and 3 pm. On day 5, only the first two video acquisitions were made.

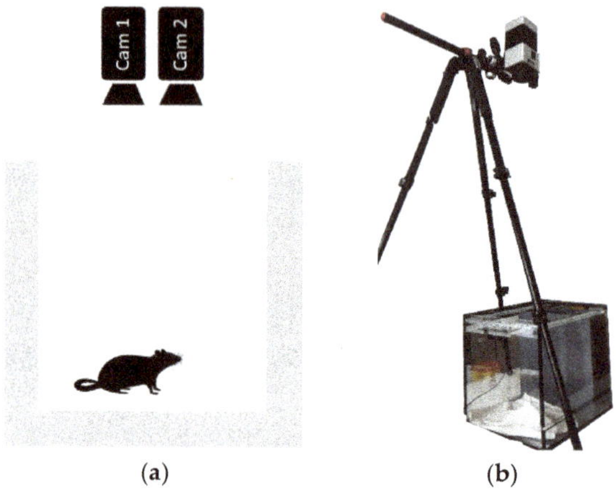

Figure 2. Recording setup. (**a**) Schematic view with both the RGB and thermal camera, which are recording the rat from above. (**b**) Picture of the recording setup. Both cameras were mounted using a tripod 1.5 m above the monitoring cage.

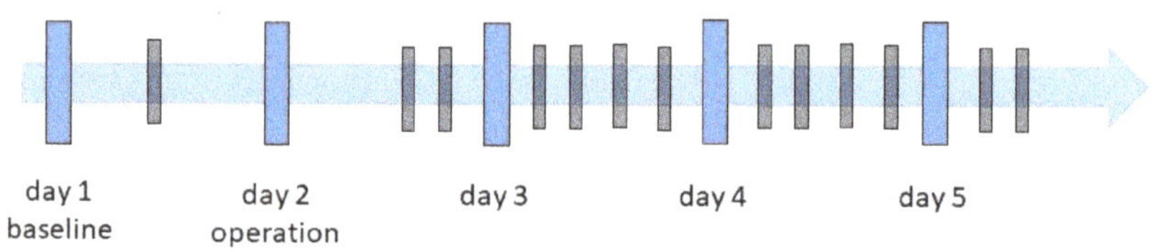

Figure 3. Experiment schedule: The blue bars correspond to the five measurement days. The black bars indicate the times at which the recordings were made.

For every recording, the ECG transponder had to be activated using a magnetic switch. Shortly afterward, the camera recordings were started simultaneously for both cameras. After 5 min of recording time, the cameras switched off automatically, followed by activating the magnetic switch again to turn off the transponder. This allowed for the recording of 13 videos for each rat, 5 min each, totaling to 39 videos (in total 195 min of video recordings). All videos were captured in raw format, without any compression. During the recording, the rats were allowed to move freely, resulting in occasional sections of heavy movements, while most of the videos are made up of minor movements such as sniffing, and fur care is present in all of the videos.

After the experiment, the animals were euthanized with an intraperitoneal sodium pentobarbital injection (600 mg/kg Narcoren®, Merial GmbH, Hallbergmoos, Germany).

2.2. Segmentation

For assessing the heart rate, a target RoI must be defined. In contrast to previous works, which mostly monitored anesthetized animals using on the upper abdomen as the region for signal extraction [8], our goal was to monitor unconstrained animals. This means that the RoI must be detected and tracked over time. Thus, the RoI was set to cover the entire chest and abdomen, and was bounded by the connecting line between both upper and lower legs, which can be recorded by cameras when they are mounted above the cage.

In 2019, Wu et al. [17] published the detectron2 framework for image segmentation and object detection, which was customized for segmenting the RoI in rats in this work. Such supervised deep learning approaches need annotated image data before the training process of the neural network can be started. Therefore, images from our study (described in detail in Section 2.1) were selected, such that 50 images that were automatically extracted from each of the 39 recorded videos, beginning with images with little-to-no movement and then randomly sampling until the required number (50) was reached. These images were annotated using LabelMe, a project created by the MIT Computer Science and Artificial Intelligence Laboratory (Cambridge, MA, USA), which provides an annotation tool to build image databases for computer vision research. An example of an annotation can be seen in Figure 4, which was applied in RGB images. Along with the detectron2 framework, Wu et al. [17] also published pretrained models on various datasets. To begin training our network, the Mask-RCNN-R50-FPN architecture, was chosen, which was pretrained on the CoCo-Dataset [18] (referenced as model-ID: 137849600). Mask-R-CNN-R50-FPN references a deep learning model, for instance segmentation. As a backbone, a ResNet-50 is used, consisting of 50 convolutional layers to extract the features from the input image. These features are then used in a feature pyramid network (FPN) to build a multi-scale feature pyramid for improved object detection and segmentation.

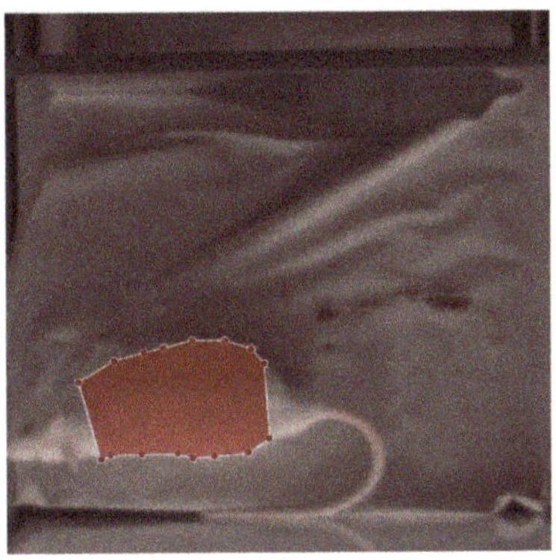

Figure 4. Annotated rat images: The red area corresponds to the desired RoI (thorax and abdomen), which should be automatically identified and segmented.

To adapt Mask-R-CNN R50 FPN to the current data, minor changes were made to its architecture. Appendix A provides a complete set of the changed parameters of the model architecture. The feature extraction layers of the network were frozen, and the number of RoI-heads was set to 128 to enable a batch size of 8 during training. Training was performed using a GeForce RTX 2080 Super (NVIDIA Corporation, Santa Clara, CA, USA). To evaluate the neural network properly, the dataset was divided into three parts (training, validation, test), with each part containing data from a single rat. For each rat, a network was trained on the 650 annotated images per rat, validated on a second rat, and tested on a third rat. This is done to ensure that the neural network had not been exposed to any images of the animals included in the test data, and thus prevent any bias during the evaluation caused by any animal-specific visible features. During training, several augmentations were applied (see Appendix B for a complete set of augmentations). Applying the segmentation network to each frame of the video results in two different outputs: a binary mask, and a certainty score between 0 and 1. Detections which are exceeding a score 0.99 were defined as valid segmentations.

2.3. Preprocessing and Signal Extraction

For an RR assessment from the segmented images, several steps of preprocessing were performed. Based upon the binary masks, from the segmentation step, the centers of the mass were computed, and each image was cropped to the extent of the bounding box of the segmentation mask, after nullifying every pixel outside the segmented area. Subsequent to obtaining all masked images of a given video, the images were shifted so that the centers of mass are overlapping for each frame in a video. The preliminary respiration signal R was obtained by computing the area of the segmentation in each image. To extract the signal, R was denoised using a linear denoising algorithm according to Nowara et al. [19], which was originally developed for denoising remote photoplethysmography signals, but should be also applicable for respiration signals due to a similar temporal profile.

The noise signals include the linear detrended center-of-mass coordinates over time for both X- and Y-coordinates, as well as their first derivatives. The algorithm uses the disturbed signal R projected onto the noise subspace Q to compute the denoised signal Z with $Z = R - \frac{QQ^T}{Q^TQ}R$. Furthermore, the resulting signal was preprocessed with a second-order Butterworth bandpass filter, with a lower and upper cutoff frequency of 1 Hz (60 breaths/min) and 3.3 Hz (200 breaths/min), respectively, and clipped wherever the gradient exceeded 1.5. The clipped values were then filled by interpolating the two neighboring values of the respiration signal.

2.4. RR Computation

Once the filtered respiration signal has been acquired, a peak detection is carried out to determine both in- and exhale cycles, which can later be used to compute the RR. An algorithm developed for electrical impedance tomography (EIT) by Khodadad et al. [20] was adapted for this purpose. First, the signal was detrended by subtracting the means of a best-fit line, and zero crossings in the signal were found. Second, a separate search for extreme points at both rising and falling zero crossings was performed. Third, an outlier detection algorithm was applied to identify the valid peaks based on their distance from the neighboring peaks. Once the peaks have been computed, the instantaneous RR (f_{RR}) can be calculated as the inverse of the distance between two consecutive peaks, using the equation: $f_{RR} = 60/d_{peak}$, where d_{peak} corresponds to the number of sampling points divided by the sampling rate and the respiration signal f_{RR} is given in breaths per minute (breaths/min). Figure 5 illustrates the algorithm, showing two signals an ECG-derived-respiration signal at the top and the corresponding computed RR at the bottom.

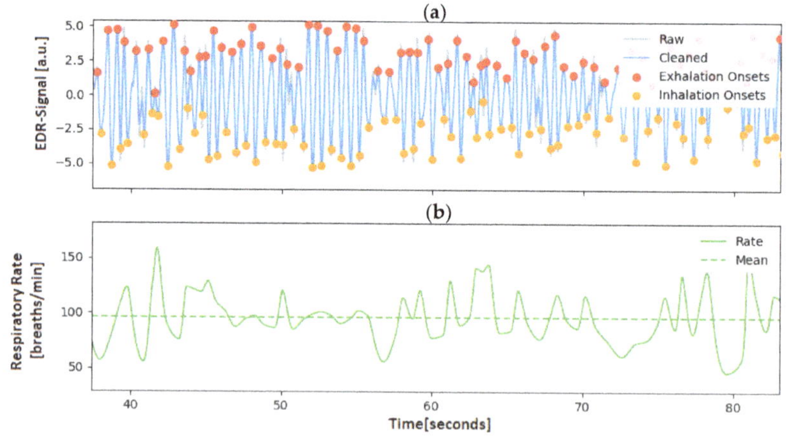

Figure 5. Example of ECG-derived respiration signal and the rate extracted from a rat ECG; (**a**) EDR signal: The blue line corresponds to the EDR signal, on which the red dots represent the maximum and the yellow dots the minimum of the breathing signal. (**b**) The EDR rate is the corresponding instantaneous respiratory rate, with its mean value denoted as a dashed line.

2.5. ECG Analysis and ECG-Derived Respiration

The results were validated using ECG as the ground truth, since the radio transponder employed in the animal trial allowed for the extraction of this parameter. ECG-derived respiration (EDR) describes the process of extracting the respiration signal from a given ECG signal. However, to obtain an EDR signal of interest, the processing of the raw ECG signal was required.

Several methods were proposed for peak detection in an ECG signal, including by Pan et al. [21], Vuong et al. [22], Kalidas et al. [23], Koka et al. [24] and Makowski et al. [25]. Most of these methods focus on detecting the QRS complexes of a given ECG as it is the most prominent feature. The peak detection method used was proposed by Makowski et al. [25], who used the gradients' steepness to detect QRS complexes, followed by searching the local maxima within the detected region to find the R-peak. Customization was required to enable the computation of the HR of rats, as their ECGs have a morphology that is vastly different from that of humans. The schematic ECG of a normal human is shown in Figure 6, along with the recorded an ECG of a rat.

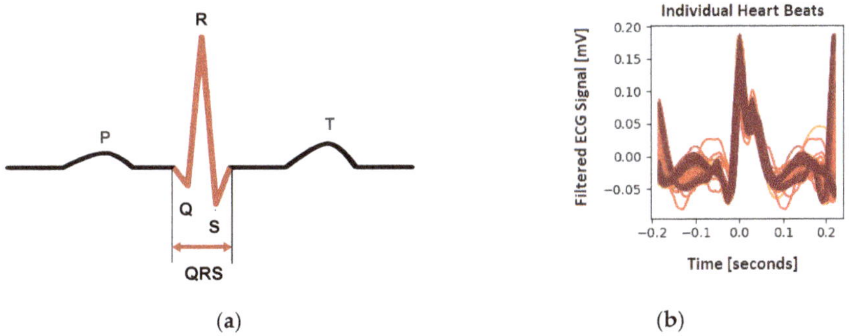

Figure 6. Heartbeat in ECG signals; (**a**) Schematic diagram of an ECG of a human. (**b**) Showcase of individual heart beats by ECG of the captured rats in the experiment.

The customization involves filtering the signal with a Butterworth low-pass filter, with a cutoff at 4 Hz, and discarding possible artifacts resulting from a 50 Hz powerline frequency. To apply the peak detection method to rats, the kernel size for smoothing and averaging was reduced by factors of two and four (smoothwindow = 0.05 s; avgwindow = 0.1875 s), respectively. Additionally, the minimum delay between two different peaks was set to 0.1 s. The threshold for discarding a QRS complex because it is too short was set to 0.1 s. An exemplary detection of the resulting R-peaks can be seen in Figure 7.

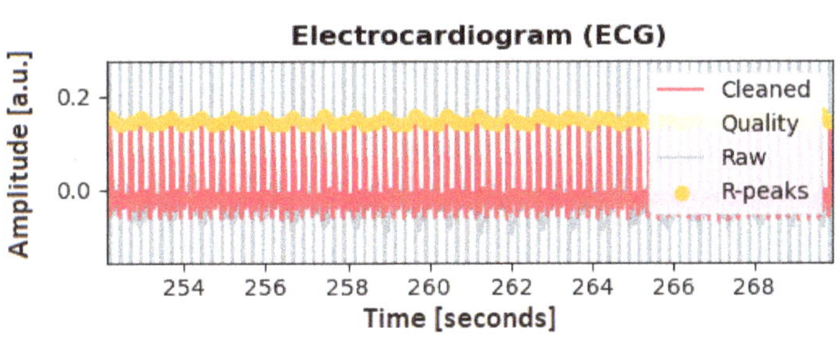

Figure 7. ECG signal of a rat, including the utilized peak detection, as denoted by the yellow markers.

Many methods have been proposed to extract the EDR from an ECG signal. Sarkar et al. [26], Charlton et al. [27] and van Gent et al. [28] used simple filtering to reconstruct the respiratory signal, while Kontaxis et al. [29] computed the respiratory signal from the difference between the maximum and the minimum slopes in the QRS complex. Langley et al. [30], in turn, computed the EDR signal by applying principal component analysis of the global amplitude variation of the QRS complex. To receive the respiratory signal from our data, the approach from van Gent et al. [28] was used, as it was most robust, especially when used on noisy signals. An EDR signal computed with this method can be seen in Figure 8, along with its respiratory rate. Figure 9, in turn, shows the spectrum of a processed ECG spectrum, clearly showing the respiratory rate and the first harmonic.

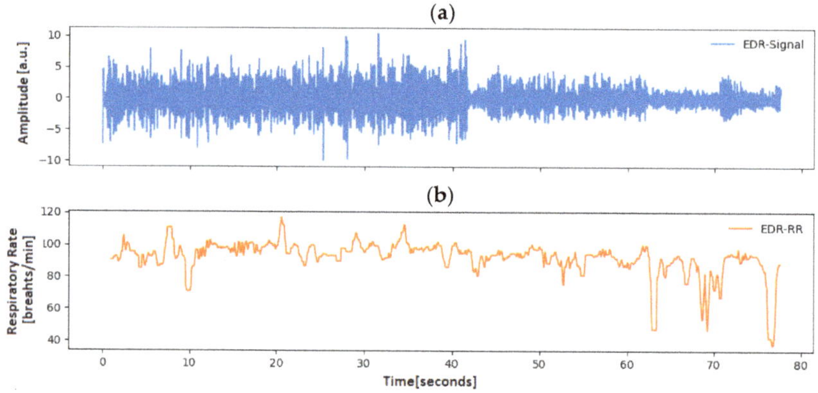

Figure 8. ECG-derived respiration in rats. (**a**) ECG-derived respiratory waveform after applying the approach proposed by van Gent et al. [22]. (**b**) Respiratory rate of the animal computed according to Khodadad et al. [23].

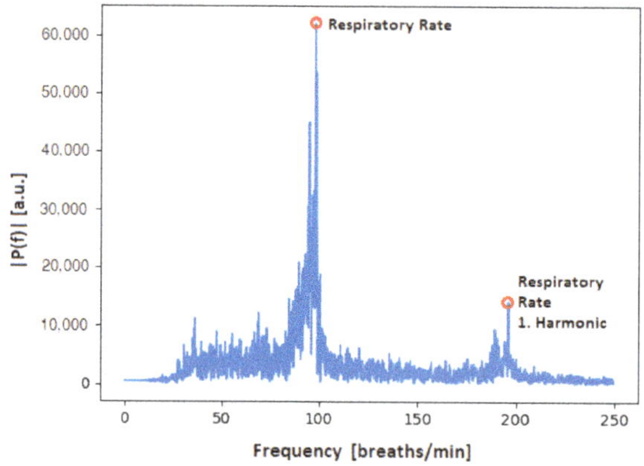

Figure 9. Frequency spectrum of a rat's respiratory signal. The highest peak visible at around 100 breaths/min corresponds to the respiratory rate of the animal. Additionally noticeable is the first harmonic, around approximately 200 breaths/min.

3. Results

3.1. *Reference Respiratory Rate*

Figure 10 shows the RR derived from the ECG for each measurement time point, as well as a box plot diagram showing the variation of the ECG-derived RR for each animal. Looking at the results, it can be observed that the RR ranges from 79.08 breaths/min to 98.87 breaths/min. On average, 92.09 breaths/min was recorded, with a standard deviation of 4.23 breaths/min. A detailed list of respiratory rates for all measurement time points is reported in Table 1.

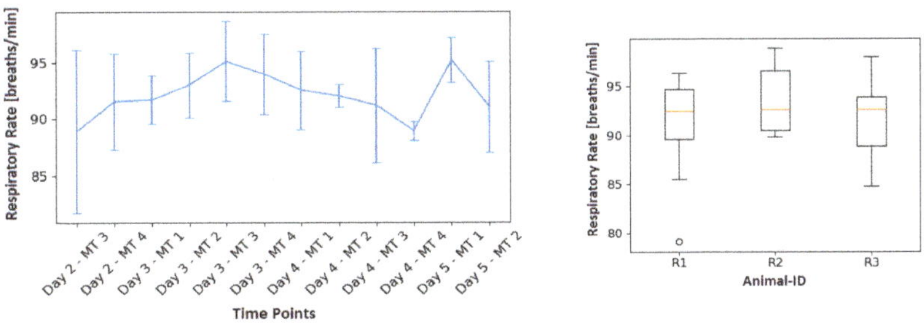

(a) (b)

Figure 10. EDR results: (**a**) Illustration of the temporal aspect of the RR by grouping measurements for the boxplot by measurement time. (**b**) Boxplot of all measurements split by the different animals.

Table 1. RR from camera-based respiration compared to the EDR. For each day and time of the measurement, the table shows the EDR rate, the camera-based RR (Rrcam), as average over the whole measurement. Additionally, the resulting relative error and the absolute error are listed. The last row lists the average of all recorded values.

Day	MT	Rat ID	Mean EDR [Breaths/min]	Mean RR_{cam} [Breaths/min]	Rel. Error [%]	Abs. Error [Breaths/min]
Day 2	MT3	R1	96.28	99.56	3.41	3.28
		R2	79.08	98.63	24.72	19.55
		R3	91.34	103.23	13.02	11.89
	MT4	R1	94.05	80.29	14.63	13.76
		R2	85.55	97.73	14.24	12.18
		R3	94.97	96.83	1.96	1.86
Day 3	MT1	R1	94.61	92.88	1.83	1.73
		R2	90.69	82.72	8.79	7.97
		R3	89.70	91.45	1.95	1.75
	MT2	R1	96.28	91.64	4.82	4.64
		R2	93.45	90.37	3.30	3.08
		R3	89.27	87.77	1.68	1.5
	MT3	R1	98.73	99,01	0.28	0.28
		R2	96.32	103.9	7.87	7.58
		R3	90.27	91.15	0.97	0.88
	MT4	R1	98.87	89.18	9.80	9.69
		R2	92.41	88.86	3.84	3.55
		R3	90.60	95.97	5.93	5.37
Day 4	MT1	R1	97.40	90.03	7.57	7.37
		R2	89.8	87.75	2.33	2.09
		R3	90.34	98.22	8.72	7.88
	MT2	R1	92.79	92.75	0.04	0.04
		R2	92.74	91.49	1.35	1.25
		R3	90.55	92.68	2.35	2.13
	MT3	R1	97.29	97.04	0.26	0.25
		R2	84.8	93.48	10.24	8.68
		R3	91.48	97.03	6.07	5.55
	MT4	R1	89.24	88.32	1.03	0.92
		R2	89.74	87.41	2.60	2.33
		R3	87.79	93.49	6.49	5.7
Day 5	MT1	R1	98.03	93.67	4.45	4.36
		R2	93.69	86.25	7.94	7.44
		R3	93.89	98.2	4.59	4.31
	MT2	R1	93.86	91.41	2.61	2.45
		R2	85.3	86.09	0.93	0.79
		R3	93.95	89.95	4.26	4
	Ø		92.09	92.67	5.47	4.94

3.2. Segmentation

The neural networks were trained on the images of one rat each, over the time of 100,000 iterations, thus leaving the images of the other two rats for validation and testing. Throughout the training process, the weights of the neural network were saved periodically every 10,000 steps and validated on the validation set, as is shown in Figure 11. The figure is split into three parts, showing the validation losses, intersection over union (IoU) for the detected bounding boxes and the IoU of the segmentation masks for each of the three trained networks over time. At the end of the training process, the network-weights with the smallest validation loss were selected for the evaluation of the test set.

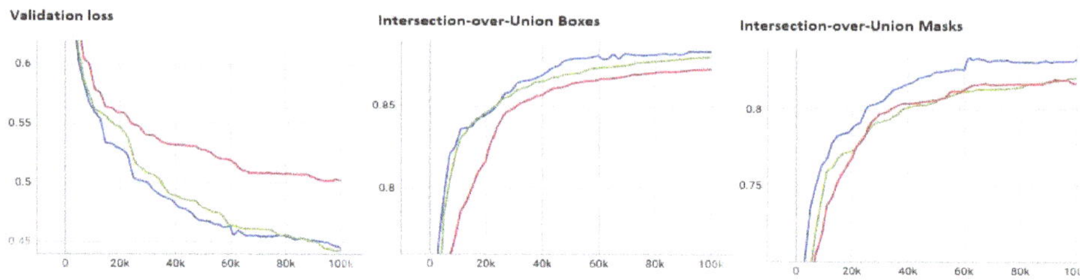

(a) (b) (c)

Figure 11. Validation loss (**a**) and intersection-over-union (**b**,**c**) for the trained networks. Blue: Network trained on R1, validated on R2, tested on R3. Green: Network trained on R2, validated on R1, tested on R2. Pink: Network trained on R3, validated on R1, tested on R2.

Intersection over union is defined as the area of overlap divided by the area of union $IoU = A_{intersection}/A_{union}$. Overall, the segmentation on the test data resulted in an average IoU of 87.75% ± 5.04% for the segmentation masks, and an IoU of 82.52% ± 6.69% for the bounding boxes. Even though the networks were trained on different animals, only small differences can be seen in the IoU scores. Table 2 shows the detailed results for the two IoUs, along with the subjective certainty score computed by the network for all three rats, along with the average.

Table 2. IoU segmentation algorithm: The table shows the results for all three trained networks, Rat ID denotes the rat on which the evaluation was performed. N describes the number of images which were annotated for the corresponding rat and used for testing. IoU is the percentage of the intersection of both annotated and detected RoIs, once computed with the rectangle (IoU Box) around the RoI, and once with the pixelwise-mask of the segmented area (IoU-Mask). The certainty score is the computed certainty that a rat was found in the segmented area.

Rat ID	N	IoU Box [%]	IoU Mask [%]	Certainty Score [%]
R1	637	82.27 ± 7.73	86.86 ± 6.18	99.84 ± 0.40
R2	654	82.85 ± 6.01	88.28 ± 4.61	99.90 ± 0.26
R3	659	82.42 ± 6.37	88.09 ± 4.39	99.80 ± 1.69
Ø	650	82.52 ± 6.69	87.75 ± 5.04	99.85 ± 0.79

3.3. Respiratory Rate

In the left part of Figure 12, the EDR (blue) can be seen together with the RR computed from the RGB videos (orange) for each measurement time point. In turn, the right part is showing the variation of the EDR and camera-based RR for each animal. In addition, Table 1 shows the RR for each video that was analyzed and an average RR of the reference. As can be observed in the table, the relative error averaged 5.47%, while the absolute error was 4.95 breaths/min.

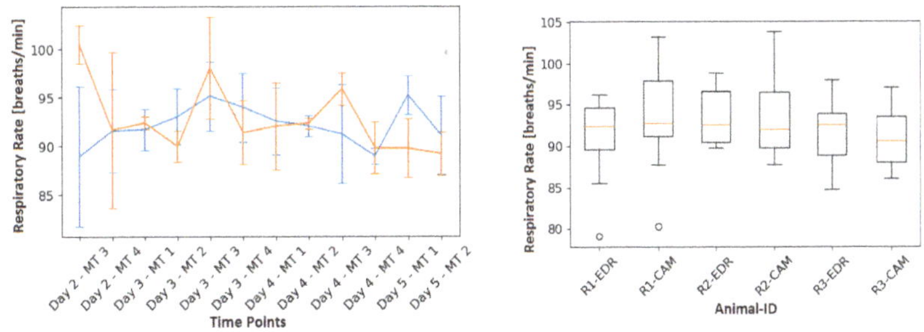

(a) (b)

Figure 12. EDR ref vs. camera-based RR: (**a**) RR over time for each MT and its variation as a boxplot. EDR rate is shown in blue, while the orange curve is the camera-based RR. (**b**) Boxplot of all results grouped by animal and modularity. R1-EDR is the EDR rate of R1 and R1-CAM is the camera-based RR for R1.

4. Discussion

The aim of this research paper is to assess the feasibility, and accuracy of monitoring RR in unrestrained, awake laboratory rats using visible imaging. This is of particular interest considering that previous approaches have only been performed with sedated animals, which does not correspond to reality for most respiratory monitoring applications. Drug development and toxicity studies could especially benefit from the possibility of long-term respiration monitoring, allowing for the assessment of the side-effects with only little interaction with the animal care takeover, while also minimizing the cost and labor for telemetry implants and evaluation. Furthermore, recovery after transmitter implantation would not be needed. Anesthesia monitoring could also profit from the proposed methods, even though it is not of much benefit to replace the current monitoring devices during an operation, as automatic respiration monitoring could be used to ensure a safe recovery from anesthesia without the need of care-takers to be present. For most respiratory diseases, it is necessary to monitor the actual breaths rather than the respiratory rate. Since the outliers are removed to enable monitoring with movement present in the signal, our methods might not be suitable for signal extraction when they are later used for the classification of complex breathing patterns. Stress and pain assessment might be one of the most interesting perspectives for this method, since pain and stress assessment is becoming more and more important in animal experiments.

The results confirm the successful performance of the segmentation and tracking algorithm; it accurately identified the thorax and abdominal area as the RoI and effectively tracked them, achieving an IoU of the segmentation mask of 87.74% on average. Unfortunately, due to the absence of enrichments in the open glass cage, image occlusion testing could not be carried out. However, based on the inherent nature of the algorithm, we have strong confidence in its ability to perform effectively, even when the animal is occluded and reappears in the image. The respiratory waveforms were extracted by leveraging the cyclical changes in the size of the area of the RoI caused by the expansion and contraction of the thorax during the respiratory cycle. Despite the presence of challenging conditions, such as motion artifacts caused by the animal's movement in the cage, the RR could still be extracted with a high degree of accuracy from the videos, with the absolute error averaging 4.95 breaths/min, providing a fist proof of the concept which has to be validated further in future studies with more animals. Nevertheless, the error could be further minimized by reducing the overall coverage. In this work, all available video sequences were used for RR estimation and evaluation. Therefore, animal movement leads to movement artifacts and thus higher errors between the reference and RR computed from visual imaging. Additionally, an ECG-derived RR rate is not the most accurate ground truth as it is very prone to motion artifacts. Other sensors, such as an implanted subcutaneous piezoelectric, may provide a more accurate reference. Varon et al. [31] also reported that EDR is quite prone to errors from noisy ECG signals. This is caused by faulty peak detection propagating into the respiration signal. Nonetheless, alternative gold-standard methods, such as respiratory belt transducers, require the animal to be restrained during the RR measurements.

There are other studies in the literature that aimed to extract the respiratory waveform/RR from rats noninvasively. Wang et al. [32] and Guan et al. [33] used humidity sensors to evaluate the RR of rodents, but both methods require the animal to be restrained. These studies primarily focus on describing the sensors themselves and the extracted respiratory waveform, but lack comprehensive investigations and comparisons with a reference/ground truth. Esquivelzeta Rabell et al. [34] and Kurnikova et al. [35] used camera-based methods to monitor respiration, namely thermal and visual imaging. In these studies, the focus was not on the RR itself, but rather the waveform of the respiratory curve extracted from the temperature variation around the nostrils, to analyze exploratory sniffing. As a result, the parameter RR was not calculated further. The algorithms used required a close-up view of the animal's nostrils, with minimal motion involved. In 2019, Kunczik et al. [13] extracted the RR from six anesthetized laboratory rats. The results have demonstrated excellent algorithm performance, with a root-mean-square error of 0.32 breaths/min. It is worth highlighting that the animals were under anesthesia during the study, and thus the influence of motion artifacts on the algorithm performance was not tested. In a study by Anishchenko et al. [36], the RR of laboratory rats during sleep was remotely measured using a radar, webcam and thermal camera, yet no reference for validation purposes was acquired, which makes a direct comparison with the present approach unfeasible.

While the tests in this study were conducted on rats, the algorithm developed can potentially be applied to other rodents such as mice and hamsters, though retraining the tracking algorithm would be necessary, along with minor adjustments, such as modifying the parameters of the temporal filter to adapt to the expected RR range of the specific animal species.

In relation to the presented study, there are some limitations that should be discussed as they may have influenced the results. First, the similar colors of animals and background (both white) might have impaired the algorithm and most probably decreased the overall accuracy, as the contrast between both is very low. Moreover, when considering the approach for denoising the respiration signal, it solely focuses on the general relative movements and does not consider movements such as scratching or sniffing during the denoising process, which could potentially affect the accuracy of the results. Inaccuracies of the tracking might have also contributed to more noise in the respiration signal, and thus a smaller signal-to-noise ratio. To further enhance the results, a dynamic assessment of the exposure time setting for the camera depending on the illumination of the RoI could be beneficial. This assessment would involve adjusting the exposure time based on the RoI rather than the overall lighting environment. By tailoring the exposure time to the specific RoI, more accurate and precise measurements could be obtained, leading to improved outcomes. Another possibility to improve the results, and thus the overall accuracy, would be to decrease the coverage of the algorithm by considering only those videos sequences in the extraction where no movement is present. However, this would imply that continuous monitoring would no longer be possible. In this context, the question arises as to whether continuous monitoring is really indispensable in laboratory research or whether fewer measurements, for example one measurement per hour, would be sufficient. Obtaining a short video sequence (e.g., 10–20 s) of motionless animals could potentially be adequate for this purpose. This could potentially minimize the monitoring burden, while still providing sufficient data for analysis, depending on the specific research objectives and requirements. Further investigation and validation would be necessary to determine the optimal frequency and duration of measurements for the specific research context. Another potential limitation is that the current segmentation algorithm is not real-time capable, but this could be improved with a different architecture. If continuous monitoring is not necessary, then the algorithm does not necessarily need to be real-time capable.

Overall, the proposed algorithm can evaluate the RR of unconstrained rodents properly. Further studies will focus on the application of the developed methods in a home cage scenario, to assess the feasibility of continuous long-term monitoring and the robustness over a wider range of respiratory rates.

5. Conclusions

Until today, it was not possible to replace animal research entirely in medical and biological science. Therefore, the need for the further refinement of experiments is significant. Vitals signs, such as the respiratory rate, are mostly monitored by using ECG implants. Until now, camera-based methods only allowed for monitoring the respiratory rate in anesthetized animals, thus a new method was proposed for unconstrained and moving animals. The respiratory rate was analyzed through the cyclical expansion and contraction of the rats' thorax/abdominal region. Compared to the EDR, a relative error of 5.47% could be achieved, while the IoU of the segmentation mask of the thorax region averaged 87.74%.

Improvements and further experiments are still needed to evaluate the performance of the algorithm when animals are occluded; furthermore, a higher range of respiratory rates is needed to evaluate the robustness of this approach. This could enable a fully automatic camera-based monitoring of rodents, reducing the need for implanted transmitters and thereby surgeries in animal experiments.

Author Contributions: Conceptualization, C.B.P., L.B. and H.P.; methodology, L.B. and C.B.P.; software, L.B. and L.M.; validation, L.B.; formal analysis, L.B.; investigation, L.B., J.K., V.B. and C.B.P.; resources, C.B.P., M.C. and H.P; data curation, C.B.P. and L.B.; writing—original draft preparation, L.B.; writing—review and editing, L.B., C.B.P., H.P., V.B., J.K. and L.M.; visualization, L.B.; supervision, C.B.P.; project administration, C.B.P.; funding acquisition, C.B.P., H.P. and M.C. All authors have read and agreed to the published version of the manuscript.

Funding: This work was supported by the German Research association under grant DFG-FOR2591 (GZ: BA 7115/1-2, CZ 215/3-2, PO 681/9-1 and PO 681/9-2).

Institutional Review Board Statement: The study followed the approved experimental protocol of the governmental animal care and use institution "Regierung von Oberbayern" (Germany, ROB-55.2-2532.Vet_02-16-105), and was conducted in compliance with the German Animal Welfare Law. All animals received humane care in accordance with the principles outlined in the "Guide for the Care and Use of Laboratory Animals" (8th edition, NIH Publication, 2011, USA).

Informed Consent Statement: Not applicable.

Data Availability Statement: The data presented in this study are available on request from the corresponding author. The data are not publicly available due to the file size of the raw videos.

Conflicts of Interest: The authors declare no conflict of interest.

Appendix A

Table A1. Network parameters. Table of parameters, changed from default parameters in the Detectron 2 RGB model.

Parameter	Value	Description
INPUT.MIN SIZE TRAIN	(480, 512, 544, 576, 608, 640)	Size of short edge for rescale
INPUT.MIN SIZE TEST	(480)	Size of short edge for rescale
SOLVER.IMS PER BATCH	8	Batch size
SOLVER.BASE LR	0.0001	Learning rate
SOLVER.MAX ITER	100,000	Number of training iterations
MODEL.ROI HEADS.BATCH SIZE PER IMAGE	128	Number of regions of interest heads
MODEL.SEM SEG HEAD.LOSS WEIGHT	2	Weight for segmentation loss
MODEL.ROI HEADS.NUM CLASSES	1	Number of classes

Appendix B

Table A2. Image augmentations. Table of applied image augmentations in Detectron 2.

	Augmentation
1.	RandomBrightness (intensity min = 0.5, intensity max = 2)
2.	RandomContrast (intensity min = 0.5, intensity max = 2)
3.	RandomSaturation (intensity min = 0.5, intensity max = 2)
4.	RandomFlip (prob = 0.5)
5.	RandomFlip (prob = 0.5, horizontal=False, vertical=True)
6.	RandomExtent (scale range = (0.8, 1.2), shift range = (0.05, 0.05))
7.	RandomRotation (expand = False, angle = [−15, 15], interp = BILINEAR)
8.	ResizeShortestEdge (short edge length = INPUT.MIN SIZE TRAIN, sample style = "choice", max size = 1368)

References

1. Franco, N.H. Animal experiments in biomedical research: A historical perspective. *Animals* **2013**, *3*, 238–273. [CrossRef]
2. Parliament, E. Directive 2010/63/EU of the European parliament and of the council of 22 September 2010 on the protection of animals used for scientific purposes. *Off. J. Eur. Union* **2010**, *276*, 33–79.
3. Russell, W.M.S.; Burch, R.L. *The Principles of Humane Experimental Technique*; Johns Hopkins University: London, UK, 1959.
4. European Commission. *2019 Report on the Statistics on the Use of Animals for Scientific Purposes in the Member States of the European Union in 2015–2017*; European Commission: Luxembourg, Brussels, 2020.
5. Bryda, E.C. The mighty mouse: The impact of rodents on advances in biomedical research. *Mo. Med.* **2013**, *110*, 207. [PubMed]
6. Cesarovic, N.; Jirkof, P.; Rettich, A.; Arras, M. Implantation of radiotelemetry transmitters yielding data on ecg, heart rate, core body temperature and activity in free-moving laboratory mice. *JoVE (J. Vis. Exp.)* **2011**, *57*, e3260. [CrossRef]
7. Braga, V.A.; Burmeister, M.A. Applications of telemetry in small laboratory animals for studying cardiovascular diseases. In *Modern Telemetry*, 1st ed.; InTech: Rijeka, Croatia, 2011; pp. 183–196. [CrossRef]
8. Ohtani, M.; Kotaki, H.; Nishitateno, K.; Sawada, Y.; Iga, T. Kinetics of respiratory depression in rats induced by buprenorphine and its metabolite, norbuprenorphine. *J. Pharmacol. Exp. Ther.* **1997**, *281*, 428–433.
9. Tsukamoto, A.; Serizawa, K.; Sato, R.; Yamazaki, J.; Inomata, T. Vital signs monitoring during injectable and inhalant anesthesia in mice. *Exp. Anim.* **2015**, *64*, 57–64. [CrossRef]
10. Card, J.W.; Zeldin, D.C. Hormonal influences on lung function and response to environmental agents: Lessons from animal models of respiratory disease. *Proc. Am. Thorac. Soc.* **2009**, *6*, 588–595. [CrossRef]
11. Schöner, J.; Heinz, A.; Endres, M.; Gertz, K.; Kronenberg, G. Post-traumatic stress disorder and beyond: An overview of rodent stress models. *J. Cell. Mol. Med.* **2017**, *21*, 2248–2256. [CrossRef]
12. Mendelson, W.B.; Martin, J.V.; Perlis, M.; Giesen, H.; Wagner, R.; Rapoport, S.I. Periodic cessation of respiratory effort during sleep in adult rats. *Physiol. Behav.* **1998**, *43*, 229–234. [CrossRef]
13. Kunczik, J.; Pereira, C.B.; Zieglowski, L.; Tolba, R.; Wassermann, L.; Häger, C.; Bleich, A.; Janssen, H.; Thum, T.; Czaplik, M. Remote vitals monitoring in rodents using video recordings. *Biomed. Opt. Express* **2019**, *10*, 4422–4436. [CrossRef]
14. Kumar, M.; Veeraraghavan, A.; Sabharwal, A. DistancePPG: Robust non-contact vital signs monitoring using a camera. *Biomed. Opt. Express* **2015**, *6*, 1565–1588. [CrossRef] [PubMed]
15. Takahashi, M.; Yamaguchi, T.; Takahashi, R.; Ogawa-Ochiai, K.; Tsumura, N.; Iijima, N. Non-contact measurement of pulse wave in rats using an RGB camera. In *Optical Diagnostics and Sensing XXI: Toward Point-of-Care Diagnostics*; SPIE: Bellingham, DC, USA, 2021; Volume 11651, pp. 46–53. [CrossRef]
16. Seiffert, I.; van Dijk, R.M.; Koska, I.; Di Liberto, V.; Möller, C.; Palme, R.; Hellweg, R.; Potschka, H. Toward evidence-based severity assessment in rat models with repeated seizures: III. Electrical post-status epilepticus model. *Epilepsia* **2019**, *60*, 1539–1551. [CrossRef] [PubMed]
17. Wu, Y.; Kirillov, A.; Massa, F.; Lo, W.; Girshick, R. Detectron2. 2019. Available online: https://github.com/facebookresearch/detectron2 (accessed on 1 April 2023).
18. Lin, T.-Y.; Maire, M.; Belongie, S.; Bourdev, L.; Girshick, R.; Hays, J.; Perona, P.; Zitnick, C.L.; Dollár, P. Microsoft coco: Common objects in context. *arXiv* **2015**. [CrossRef]
19. Nowara, E.M.; Marks, T.K.; Mansour, H.; Veeraraghavan, A. Near-Infrared Imaging Photoplethysmography During Driving. *IEEE Trans. Intell. Transp. Syst.* **2020**, *23*, 3589–3600. [CrossRef]
20. Khodadad, D.; Nordebo, S.; Müller, B.; Waldmann, A.; Yerworth, R.; Becher, T.; Frerichs, I.; Sophocleous, L.; van Kaam, A.; Miedema, M.; et al. Optimized breath detection algorithm in electrical impedance tomography. *Physiol. Meas.* **2018**, *39*, 094001. [CrossRef] [PubMed]
21. Pan, J.; Tompkins, W.J. A real-time QRS detection algorithm. *IEEE Trans. Biomed. Eng.* **1985**, *3*, 230–236. [CrossRef]

22. Vuong, N.; Nguyen, T.; Tran, L.D.; Nordebo, S.; Müller, B.; Waldmann, A.; Yerworth, R.; Becher, T.; Frerichs, I.; Sophocleous, L.; et al. Detect QRS complex in ECG. In Proceedings of the 2017 12th IEEE Conference on Industrial Electronics and Applications (ICIEA), Siem Reap, Cambodia, 18–20 June 2017; pp. 2022–2027. [CrossRef]
23. Kalidas, V.; Tamil, L. Real-time QRS detector using stationary wavelet transform for automated ECG analysis. In Proceedings of the 2017 IEEE 17th International Conference on Bioinformatics and Bioengineering (BIBE), Washington, DC, USA, 23–25 October 2017; pp. 457–461. [CrossRef]
24. Koka, T.; Muma, M. Fast and Sample Accurate R-Peak Detection for Noisy ECG Using Visibility Graphs. In Proceedings of the 2022 44th Annual International Conference of the IEEE Engineering in Medicine & Biology Society (EMBC), Glasgow, UK, 11–15 July 2022; pp. 121–126. [CrossRef]
25. Makowski, D.; Pham, T.; Lau, Z.J.; Brammer, J.C.; Lespinasse, F.; Pham, H.; Schölzel, C.; Chen, S.H.A. NeuroKit2: A python toolbox for neurophysiological signal processing. *Behav. Res. Methods* **2021**, *53*, 1689–1696. [CrossRef]
26. Sarkar, S.; Bhattacherjee, S.; Pal, S. Extraction of respiration signal from ECG for respiratory rate estimation. In Proceedings of the Michael Faraday IET International Summit 2015, Kolkata, India, 12–13 September 2015; pp. 336–340. [CrossRef]
27. Charlton, P.H.; Bonnici, T.; Tarassenko, L.; Clifton, D.A.; Beale, R.; Watkinson, P.J. An assessment of algorithms to estimate respiratory rate from the electrocardiogram and photoplethysmogram. *Physiol. Meas.* **2016**, *37*, 610. [CrossRef]
28. van Gent, P.; Farah, H.; van Nes, N.; van Arem, B. Heartpy: A novel heart rate algorithm for the analysis of noisy signals. *Transp. Res. Part F Traffic Psychol. Behav.* **2019**, *66*, 368–378. [CrossRef]
29. Kontaxis, S.; Lazaro, J.; Corino, V.D.; Sandberg, F.; Bailon, R.; Laguna, P.; Sornmo, L. ECG-derived respiratory rate in atrial fibrillation. *IEEE Trans. Biomed. Eng.* **2019**, *67*, 905–914. [CrossRef]
30. Langley, P.; Bowers, E.J.; Murray, A. Principal component analysis as a tool for analyzing beat-to-beat changes in ECG features: Application to ECG-derived respiration. *IEEE Trans. Biomed. Eng.* **2009**, *57*, 821–829. [CrossRef] [PubMed]
31. Varon, C.; Morales, J.; Lazaro, J.; Orini, M.; Deviaene, M.; Kontaxis, S.; Testelmans, D.; Buyse, B.; Borzée, P.; Sörnmo, L.; et al. A Comparative Study of ECG-derived Respiration in Ambulatory Monitoring using the Single-lead ECG. *Sci. Rep.* **2020**, *10*, 5704. [CrossRef] [PubMed]
32. Wang, G.; Zhang, Y.; Yang, H.; Wang, W.; Dai, Y.-Z.; Niu, L.-G.; Lv, C.; Xia, H.; Liu, T. Fast-response humidity sensor based on laser printing for respiration monitoring. *RSC Adv.* **2020**, *10*, 8910–8916. [CrossRef] [PubMed]
33. Guan, Y.; Le, X.; Hu, M.; Liu, W.; Xie, J. A noninvasive method for monitoring respiratory rate of rats based on a microcantilever resonant humidity sensor. *J. Micromech. Microeng.* **2019**, *29*, 125001. [CrossRef]
34. Rabell, J.E.; Mutlu, K.; Noutel, J.; del Olmo, P.M.; Haesler, S. Spontaneous rapid odor source localization behavior requires interhemispheric communication. *Curr. Biol.* **2017**, *27*, 1542–1548.e4. [CrossRef]
35. Kurnikova, A.; Moore, J.D.; Liao, S.-M.; Deschênes, M.; Kleinfeld, D. Coordination of orofacial motor actions into exploratory behavior by rat. *Curr. Biol.* **2017**, *27*, 688–696. [CrossRef]
36. Anishchenko, L.; Gaysina, E. Comparison of 4 GHz and 14 GHz SFCW radars in measuring of small laboratory animals vital signs. In Proceedings of the 2015 IEEE International Conference on Microwaves, Communications, Antennas and Electronic Systems (COMCAS), Tel Aviv, Israel, 6–8 November 2015; pp. 1–3. [CrossRef]

Disclaimer/Publisher's Note: The statements, opinions and data contained in all publications are solely those of the individual author(s) and contributor(s) and not of MDPI and/or the editor(s). MDPI and/or the editor(s) disclaim responsibility for any injury to people or property resulting from any ideas, methods, instructions or products referred to in the content.

animals

MDPI

Article

Longitudinal Studies on Alzheimer Disease Mouse Models with Multiple Tracer PET/CT: Application of Reduction and Refinement Principles in Daily Practice to Safeguard Animal Welfare during Progressive Aging

Giovanna Palumbo [1,*], Lea Helena Kunze [1,2], Rosel Oos [1], Karin Wind-Mark [1,2], Simon Lindner [1], Barbara von Ungern-Sternberg [1], Peter Bartenstein [1], Sibylle Ziegler [1] and Matthias Brendel [1,2,3]

1 Department of Nuclear Medicine, University Hospital LMU Munich, Marchioninis Strasse 15, 81377 Munich, Germany
2 German Center for Neurodegenerative Diseases (DZNE), Feodor-Lynen-Strasse 17, 81377 Munich, Germany
3 Munich Cluster for Systems Neurology (SyNergy), 81377 Munich, Germany
* Correspondence: giovanna.palumbo@med.uni-muenchen.de

Simple Summary: Animal models continue to be necessary in many research fields, accompanied by ongoing ethical discussions regarding animal welfare. Therefore, we describe, in detail, our daily practice focused on the improvement of animal welfare (such as handling, enriched environment, study design, and experimental procedures), which results in a weight gain over time that has been shown to be an indicator of well-being. We also describe the reduction in the number of animals needed for our projects, thanks to the establishment of longitudinal studies.

Abstract: Longitudinal studies on mouse models related to Alzheimer disease (AD) pathology play an important role in the investigation of therapeutic targets to help pharmaceutical research in the development of new drugs and in the attempt of an early diagnosis that can contribute to improving people's quality of life. There are several advantages to enriching longitudinal studies in AD models with Positron Emission Tomography (PET); among these advantages, the possibility of following the principle of the 3Rs of animal welfare is fundamental. In this manuscript, good daily experimental practice focusing on animal welfare is described and commented upon, based on the experience attained from studies conducted in our Nuclear Medicine department.

Keywords: refinement; enrichment environment; reduction; small animal positron emission tomography; longitudinal study; Alzheimer disease

Citation: Palumbo, G.; Kunze, L.H.; Oos, R.; Wind-Mark, K.; Lindner, S.; von Ungern-Sternberg, B.; Bartenstein, P.; Ziegler, S.; Brendel, M. Longitudinal Studies on Alzheimer Disease Mouse Models with Multiple Tracer PET/CT: Application of Reduction and Refinement Principles in Daily Practice to Safeguard Animal Welfare during Progressive Aging. *Animals* **2023**, *13*, 1812. https://doi.org/10.3390/ani13111812

Academic Editor: Garikoitz Azkona

Received: 14 April 2023
Revised: 26 May 2023
Accepted: 27 May 2023
Published: 30 May 2023

Copyright: © 2023 by the authors. Licensee MDPI, Basel, Switzerland. This article is an open access article distributed under the terms and conditions of the Creative Commons Attribution (CC BY) license (https://creativecommons.org/licenses/by/4.0/).

1. Introduction

Alzheimer disease (AD) is a debilitating disease that causes progressive decline in cognitive and motor function, significantly reducing quality of life [1]. It is characterized by an early decrease in brain glucose metabolism [2], as well as the presence of amyloid plaques and neurofibrillary tangles. In addition, histological studies demonstrate that neuroinflammation is also a key feature of the AD brain [3]. Amyloid plaques are surrounded by activated astrocytes that produce reactive oxygen and nitrogen species, which may contribute to AD [4]. Animal models of AD are useful for studying these changes and the progression and development of the disease, with the goal of finding new diagnostic and treatment strategies. As the factors involved in the development and progression of the disease are different depending on the time during which the disease starts to manifest, different animal models for longitudinal studies are available.

Each model reflects biological features related to AD. Involvement and the effects of all these factors during the time of disease development are of high interest. Positron emission tomography (PET) is a molecular imaging method offering a large variety of radioactively labelled substances targeting different biological structures or processes. It offers non-invasive measurement of the radiotracer concentration in tissue and is clearly translational, since the exact same tracers and imaging technology can be applied in animal models, as well as in research and clinical use in humans. For example, PET remains one of the few methods to allow direct assessment of the human central nervous system (CNS) pharmacology, providing information on target engagement and supporting dose selection [5]. Thus, it has an increasing role in studying the biochemical and physiological dynamics of the CNS. The goal of the 3Rs Principle [6] is to avoid animal experiments (Replacement), to limit the number of animals (Reduction), and to limit their suffering in tests to an absolute minimum (Refinement). From the perspective of the design and development of a longitudinal study, the principle of the 3Rs plays a fundamental role in the research of all the strategies aimed at maintaining animal welfare for the entire time of the experiment. Due to their shorter lifespan, life-course results are obtained much more quickly in animal models, which is especially important when studying aging and transgenerational disease transmission [7]. Unfortunately, the complex interactions between organs and cells within their regular environment still necessitate the use of animal models in some cases. For example, in order to establish cause–effect relationships, connections between systems and the onset of age-related pathologies need to be studied. Since animal experiments cannot be completely replaced in our research on the development of AD, we implemented different strategies, following the principles of Reduction and Refinement. Improving the quality of life for experimental animals and reducing their stress and pain are among the goals that must be considered when it is not possible to replace animals, according to the principle of the 3Rs. EU Directive 2010/63 states that the animals should have "space of sufficient complexity to allow expression of a wide range of normal behavior" [8] (Annex III, Section A, paragraph 3.1.). For this reason, we aim for the enrichment environment we use to reflect the normal living conditions of the animals as much as possible. Currently, the term enrichment environment (EE) refers to various objects, or a combination of them, that can be added to the bedding and nest material (which are now considered basic components). These components play the roles of cognitive, sensory, social, and motor stimulators, which promote the interaction of animals with objects and with each other [9]. The scientific literature is full of examples of benefits due to EE. For example, it can prevent barbering [10], it reduces the likelihood of alopecia [11], and it decreases the expression of abnormal repetitive behaviors and anxiety [12], as well as the development of depressive-like phenotypes [13]. We can also assume that EE prevents or mitigates the onset of boredom-like symptoms in mice, since, in humans, this can be triggered by predictability, monotony, and confinement, and similar phenomena in rodents indicate that boredom in laboratory animals is real [14]. Considering the gender difference, groups of males in an environment with excessive EE demonstrated a higher occurrence of aggression [15]. On the other hand, without adequate EE, the conditions of their natural habitat would not be reproduced (spread out, keep away, escape, hide, and predict the occurrence of aggressive encounters). In order to prevent aggressive events in male mice, in our facility, they are always located in a cage with littermates with the same genotype, or with around one week of age difference with the same genotype, as it is known that this can prevent aggression and, consequentially, the necessity of single housing [16]. In this report, we summarize methods of ensuring animal welfare in longitudinal studies characterizing age-related processes in mouse models of AD. In addition to the non-invasiveness of PET imaging, a number of established techniques have been adapted, which contribute to the 3Rs in repeated measurements.

2. Materials and Methods

2.1. Animal Models

In our longitudinal studies over recent years, we used the following mouse models:

- APPSL70 mice, transgene of the Amyloid Precursor Protein (APP) in Swedish and London mutations [17] (Figures 5–8 Sections 3.1–3.3) (n = 92, half group was treated);

- APP-NL-GF mice knock-in model for the APP gene in Swedish, London, Arctic, and Iberian mutations [18] (Section 3.2) (n = 79);
- APPPS1 and PS2APP mice, mutated, respectively, for the APP gene and for the genes of presenilin 1 (PS1) and 2 (PS2) [19] (Section 3.2) (n = 46 and n = 36);
- APPPS1 X Trem 2 mice, transgene of APP, as well as of PS1 for triggering receptor expressed on myeloid cells type 2 (TREM2) [20] (Section 3.2) (n = 33);
- P301S mice, mutated for the microtubule-associated protein (TAU), which is abnormally aggregated in neuronal and glial cells in AD [21] (Section 3.2) (n = 46);
- C57BL/6J wild–type mice (WT) as control group (Sections 3.2 and 3.3) (n = 46).

In the present study concerning the principle of the 3Rs applied to longitudinal studies in AD research, the APPSL70 is the mouse line used the most for examples.

2.2. Study Design

We performed longitudinal studies in AD mouse models, in which each animal acted as its own control. The models consisted of PET/CT scans from 3 to 5 time points, with 2 to 4 different radiotracers characterizing changes in biological targets up to the age of 12 to 18 months. Furthermore, a Morris water maze (MWM) test was integrated before or after the last PET/CT time point in order to study spatial learning. At the end of each study, an intracardial perfusion with PBS was performed after intraperitoneal injection of a solution of 300–500 µL (weight-dependent) ketamine/xylazin (4 mL Ketamine 10%/1 mL Xylazine 20 mg/mL from Serumwerk Bernburg Germany, up to 24 mL with NaCl 0.9%), for the preparation of samples for histological and biochemical analyses (Figure 1).

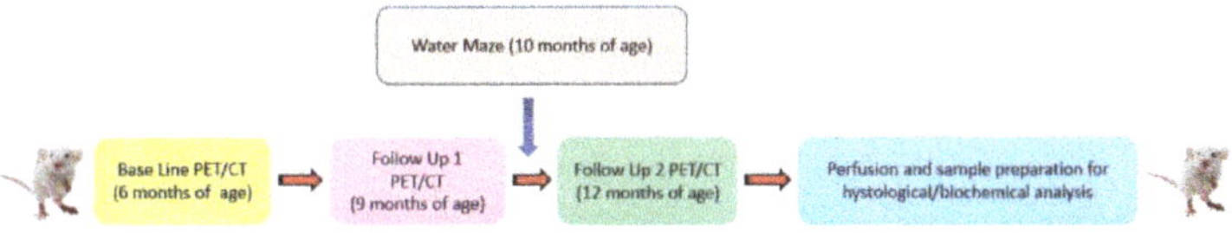

Figure 1. Study design of longitudinal AD studies.

2.3. Positron Emission Tomography

In each animal, we used 2 to 4 of the following ^{18}F labelled tracers: [^{18}F] D2-Deprenyl for reactive astrocytes which surround the Beta Amyloid plaques [3]; [18F] Florbetaben for beta amyloid accumulation; [^{18}F] Ge-180 for the 18-kDa translocator protein (TSPO), as its local upregulation is a sensitive marker for the microglial activation in AD brains [22]; and [^{18}F] UCB-H for synaptic density (synaptic loss or synaptic sprouting) [23]. PET imaging was performed under constant anesthesia with isoflurane (1.5% at 1.5L oxygen flow per minute) with a Nanoscan PET/CT (Mediso Ltd., Budapest, Hungary). For anatomical information, the system was equipped with an X-ray Computed Tomography System (CT) in line with the PET scanner. After induction of anesthesia with isoflurane, the eyes were protected from drying out by topical application of an eye ointment (Bepanthen, Bayer AG, Leverkusen, Germany). In order to judge the depth of anesthesia before the radiotracer´s injection, surgical tweezers were used to check whether the inter-toe reflex could still be triggered. If it could no longer be triggered, a micro-catheter was inserted into the lateral tail vein (30 G needle, 7 cm plastic tube, 30 G attachment, flushing with 0.9% isotonic saline solution); the correct position of the indwelling venous cannula for the application of the radiotracer was checked by administering a small amount of isotonic saline solution (20–30 µL) into the catheter. The injected radioligand solution consisted of a total of 150 µL, and contained approximately 20 MBq of radioactive tracer. PET/CT measurement was carried out up to 60 min post injection: for [^{18}F] Ge-180 and [^{18}F] Florbetaben, the measurement was carried out for 30 min., 60 and 30 min. after radiotracer uptake, respectively (static scan); those for [^{18}F] D2-Deprenyl and [^{18}F] UCB-H were carried out immediately post injection, for 60 min (dynamic scan). Up to four animals were measured in parallel, and image data were generated for each animal from head to tail. After the experiment, the mice were placed in a fresh temporary cage with food and water, warmed by heating mats. The animals were returned to their home cages only when they were fully awake.

The radioactive waste coming from the animals (feces, urine) and all the equipment used on the workspace (e.g., syringes, paper tissue, paper mats, and gloves) were collected in black boxes, upon which were written the isotope used and the date and time of the experiment. These boxes were then collected by the radiation safety personnel of the department.

2.4. Water Maze

The Morris Water Maze (MWM) is useful to test hippocampal-dependent learning, including acquisition of short- and long-term spatial memory [24]. Typically, it consists of a six-day trial, and it has to be conducted by the same operator in the same room in order to reduce odor trail interferences [25]. In brief, the first test day served for acclimatization to the visible platform (5 min per mouse). Thereafter, the mice underwent five training days, during which each mouse had to perform four trials per day, with the platform visible on the first training day, and the platform hidden under water for all other training days. After the trial, the mice were placed in a heated box to dry. The test day entailed a single trial with complete removal of the platform. The trial length on all training and test days was set to a maximum of 70 s. The video tracking software EthoVision®XT (Noldus) was used for analyses of escape latency, platform frequency, and attendance on the platform quadrant during the trial. In our longitudinal studies, the test was conducted in a room adjacent to both the cage facility and the room where the preparation and PET scans took place, and the mice resided in a cabinet in the same behavioral room for the duration of the trial. This minimized the stress caused by this additional procedure and by moving animals to other, more distant, locations. The personnel who carried out the MWM test received appropriate training from a veterinarian before the experiment and was also part of the team that regularly deals with animal welfare, as well as with the preparation and execution of PET scans.

2.5. Enrichment Environment

In our facility, the male mice are housed (in groups of 3–4 siblings, with one week age difference and same genotype) with the same EE as the females. They are more intensively monitored (for too much grooming, self-isolation attributable to the attempt to escape from the attacks of the other mice, and the presence of wounds) so as to prevent the onset of high levels of aggression, but still keeping them in a condition to be able to maintain a good level of play and sociopositive behavior. The female mice are housed in groups of 3 to 5. All of the mice are housed in large IVC-Techniplast cages (425 mm × 266 mm × 185 mm), with a 12:12 h light: dark cycle, humidity of 45–65%, and temperature of 23–26 °C. All toys are changed once a week. The food (standard diet from Sniff Spezialdiäten GmbH) is placed ad libitum, as is water. Mice were allowed to acclimatize to their environment for 7 days before any experimental procedure. The EE can be divided into categories (Figure 2) [26]. In our studies, we used the basic nest (2A), structural EE (2B), foraging EE (2C) and housing EE (2D):

2.6. Handling

We use a handling method that reduces stress, and it can be used both during experiments and in daily practices (such as weighing, animal checks, and changing cages) and does not need any additional material (for example, tunnels). Once the animal is picked up by the tail from the cage, with a quick but gentle twist of the hand, it is immediately placed in the cupped hand (Figure 3).

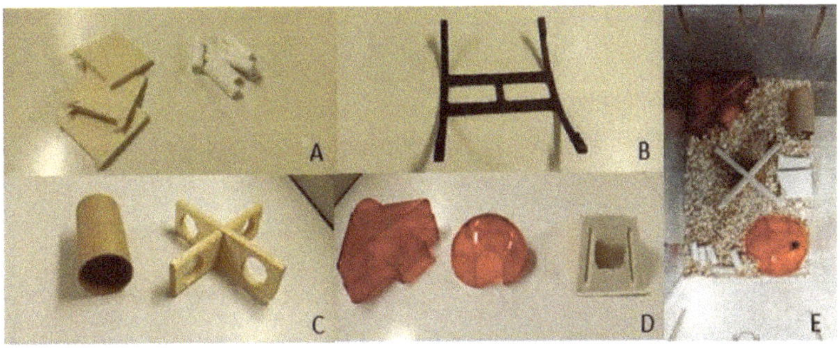

Figure 2. Basic nest (**A**), double mouse swing (**B**), paper or plastic tunnel and wooden cross with holes (**C**), and mouse angle house, igloo, and paper angle (**D**). Example of cage with EE used in our studies (**E**).

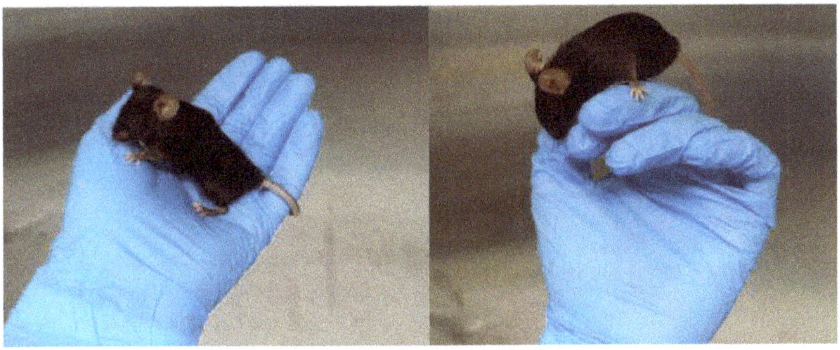

Figure 3. Gentle handling applied for all experimental procedures. As can be seen, the mouse shows normal exploration behavior.

2.7. Monitoring Signs of Stress

Animals are monitored for any sign of stress using the Mouse Grimace Scale. It consists of a standardized behavioral coding system with high accuracy and reliability in which no-pain pictures and descriptions are compared with pictures of moderate and severe distress via mouse facial expression (e.g., orbital tightening, nose bulge, cheek bulge, ear position, and whisker change) [27]. In our longitudinal studies, the mice are scanned two or, maximum, three times a week for each time point, and they are always observed in the postanesthesia phase, until they have completely recovered. They are also checked on the day after the procedures according to the Mice Grimace Scale, in order to identify the onset of visible postprocedural stress symptoms. It is known that this scoring system allows for the identification of the degree of pain in mice, but also shows that they manifest their pain using facial expressions [26]. Furthermore, the MGS measures not only pain, but also distress, fear, and discomfort, which can be detected by the ear and eye score [28]. Mice hide signs of pain and suffering to avoid becoming prey, and, generally, they show only subtle signs of suffering and pain, such as weight loss [29]. In addition to the parameters shown in Figure 4, the weight of each mouse is monitored up to once per week during all the experimental phases, as it has been demonstrated that mice gain weight when the stress level is lower, and this can be achieved with appropriate EE and enough ventilation (in IVC cages) [30].

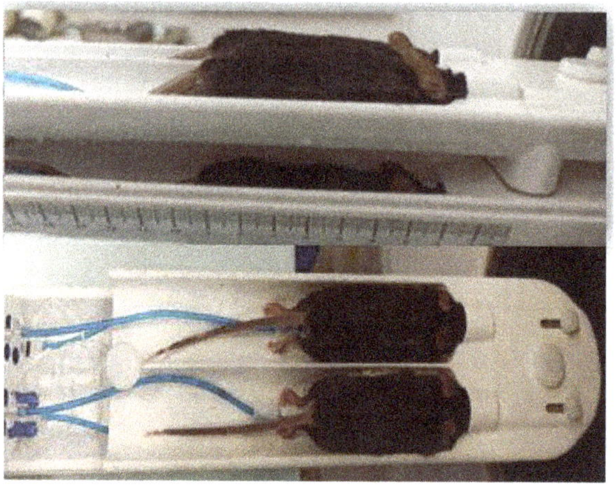

Figure 4. Four-mouse Hotel Bed (Mediso, Budapest, Hungary), in which two mice are positioned on the lower part and two on the upper part. The parts are connected, and the isoflurane and oxygen can easily and continuously be administered to the mice for anesthesia. Underneath the mice, there are small pillows (at the end of the blue tubes, under the mice's chests) that monitor the respiratory rate.

2.8. Anesthesia

In contrast to human studies, the imaging of small animals generally requires anesthesia [31]. In our longitudinal studies, we use isoflurane because it is well suited for animal PET scans for up to 6h of measurement time [32]. Repeated isoflurane anesthesia causes only mild short-term distress and impairment of well-being, mainly in the immediate postanesthetic period [33]. This can be kept under control by monitoring the respiratory rate and heart rate, and keeping the body temperature of the animals constant throughout the process in order to prevent the onset of hypothermia. Maintaining the temperature and anesthesia at a constant level during the experiment minimizes any potential pain, suffering, distress, or lasting harm to the animal [34]. In a study conducted by Baier J et al. in 2020 [35] on repeated MRI scans, using isoflurane for general anesthesia and its maintenance three times a week for four weeks, the mice showed no alterations in animal welfare due to the repeated procedures. Another study [36] demonstrated that nest-building activity is not altered by a second exposure to isoflurane.

2.9. Mouse Hotel Bed

Using a four-mouse bed (Mediso Ltd., Budapest, Hungary) it was possible to scan multiple mice (up to four) simultaneously (Figure 4). The use of this chamber is advantageous in many aspects:

1. The chance to perform more scans in the same day using radioactive isotopes with a short half-life (^{18}F: 109, 7 min). This is especially important in longitudinal studies with a large number of animals because all of the scans have to be performed respective of the age-related time point.
2. There is a lower injected volume. This is important, as the ^{18}F-based tracers decay quickly. Since the activity has to be the same for each animal, the volume of the injection needs to be increased according to the radioactive decay in sequential scans. Thus, the volume to be injected would be above the maximum limit allowed for these animal experiments. Using the 4-mouse bed, this problem is circumvented, since four mice are injected and scanned at the same time (instead of four individual scans one after the other).
3. This particular bed for four animals is equipped with a heating system that keeps the temperature of the bed constant at about 37 °C, which is very important during

anesthesia, as hypothermia can arise, and also because the temperature can have an effect on the biodistribution of the injected radiotracer. In addition, there is a monitoring system for cardiac and respiratory function that allows continuous monitoring in each animal.

3. Results

3.1. Longitudinal Studies Enable the Analysis of the Same Animal with Different Radioligands during Progressive Aging

Figure 5 shows an example of PET, CT, and Fusion PET/CT of four mice imaged simultaneously.

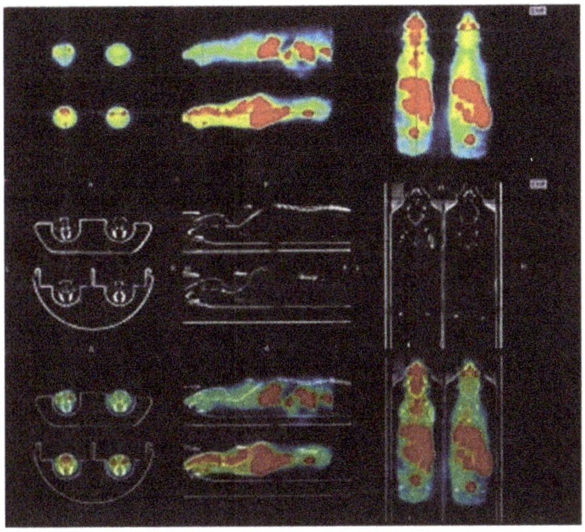

Figure 5. From top: example of PET, CT, and FUSION PET/CT of four mice imaged simultaneously using the four-mouse bed. The four APPSL70 mice were scanned for 30 min after 30 min uptake with 20 MBq [^{18}F] Florbetaben tracer for beta amyloid accumulation. Axial, sagittal, and coronal slices.

Figure 6 shows an example of beta amyloid in a transgenic mouse brain over time with the same radiotracer ([^{18}F] Florbetaben): the same mouse was scanned at 6, 9, and 12 months of age. Coronal slices of the brain from the same animal after the injection of four different radioligands at 12 months of age with a reference MRI (Magnetic Resonance Imaging) used as anatomical reference are also included.

3.2. The Longitudinal PET Studies follow the Principle of Reduction by Russell and Burch

The graph in Figure 7 shows the number of mice used by our research group in longitudinal studies in recent years, and the number of mice that would have been necessary, in theory, for the same projects using separate groups for each time point.

3.3. Gentle Handling and EE Reduce Stress in Longitudinal Studies

For all the experiments, the MGS resulted in a score of zero.

We observed that, during the entire stay of the animals in the facility (9 months or more), the nest material was regularly used, as were the other components of the EE. The combination of gentle handling and EE makes the animals more docile and less stressed, even during the time before/during/after the experimental procedures. The graph in Figure 8 shows the development of the weight in a female AD mouse model (APPSL70, n = 28) and in wild-type controls (n = 17), each receiving PET scans with four different radiotracers at three different time points, with a MWM test before the last time point. We observed an average of 13% mortality during this experiment of up to six months, independent of the mouse line.

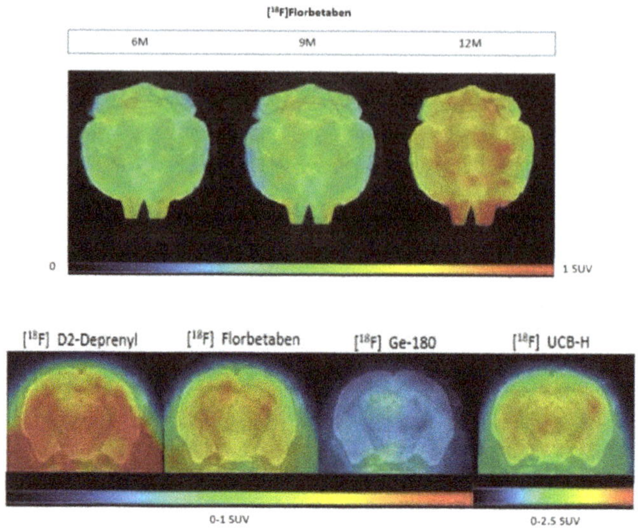

Figure 6. Top panel: Example of an APPSL70 mouse brain in standard uptake value (SUV) scale. PET projection view and MRI reference with the same ^{18}F radiotracer ([^{18}F] Florbetaben) at 6, 9, and 12 months of age (left to right). **Lower panel:** Coronal slices of PET scans of a transgenic AD-related mouse brain at 12 months of age after injection of four different radioligands.

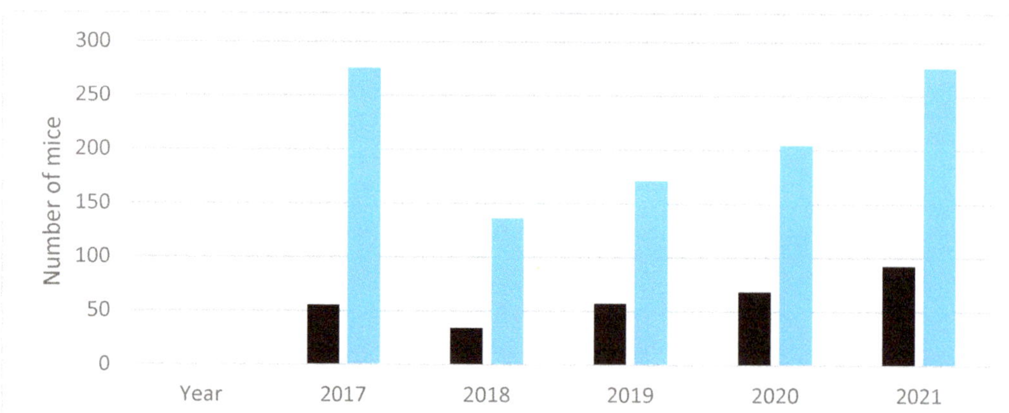

Figure 7. Difference between number of mice used (in black) and the number that would have been necessary, in theory, if separate groups had been used in longitudinal studies (in blue). The data reported include the AD models APPPS1, APPPSL70, P301S, APPPS1XTrem2, APP-NL-GF, PS2APP, and C57BL/6 control mice.

While the body weight of wild-type mice increased continuously, the body weight in the APPSL70 AD model did not increase further after ~10 months of age. A very similar weight gain (9.6% mean percentage of gain weight from 3 to 6 months of age and 11.8% from 6 to 9 months of age) was observed in the APPPS1 group, which consisted of 11 animals. No mouse, either wild-type or transgenic, had to be excluded from the trials due to stress/fear factors, and no animal developed signs of anxiety/stress over six months of study (up to one year of age for mice).

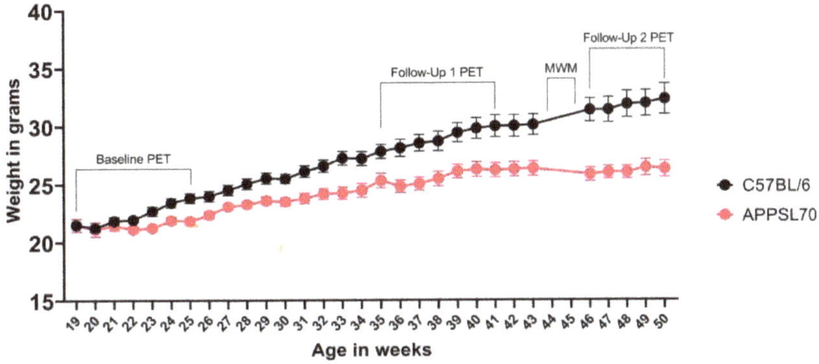

Figure 8. Weight/time graph during the longitudinal study with mice up to one year of age. Shown is the average weight (±STD) of C57BL/6J wild-type and APPSL70 AD mice. The time points within which the PET scans and MWM were performed are marked (APPSL70 n = 28, WT n = 17). These animals did not receive any pharmacological treatment.

4. Discussion

We report on aspects of implementation of the 3Rs Principle in longitudinal mouse studies for AD research. Of note, the reported methods and findings concerning handling and housing are not restricted to AD research. Repeated PET/CT measurements, combined with a behavioral test, proved feasible in AD mouse models and wild-type mice without losing animals due to increased stress, up to 12–18 months of age. Reduction in the number of animals used for experimentation is achieved by longitudinal, multi-tracer PET scans in the same animal; thus, each mouse is its own control, and the results of behavior tests can also be correlated for each individual animal. No episodes of severe weight loss were detected during the study (according to our approved scoring system). The weight loss observed in the APPSL70 AD model is compatible with the findings in previous studies on different AD mouse models, showing that weight loss is connected to the reduction in body adiposity, an increase in energy expenditure, and a decrease in food efficiency connected to the progression of AD [37,38]. It is known that stress has a negative effect on the results of the MWM [39,40], and in some studies, it has been hypothesized that the stress from less frequent handling before the test can be responsible for the failure of some mice in finding the platform [41]. It is also known that the induction of stress produces learning deficits in the MWM [42,43]. In this context, considering that routine handling influences animal welfare and the relationship that is established between the handler and the animal, we observed that the non-stressful handling method assures low variability in the results obtained from the test. In all procedures, we adhered to the same animal handling procedure. The standard method, picking up the mice by the tail, induces more anxiety and stress compared to more gentle methods, such as cupping and tunnel handling [44,45]; it was also previously demonstrated that gentle handling can foster a better relationship between the handlers and the rodents, and, if implemented as the standard of care, handling can reduce depressive symptoms in mice, producing data that are more reliable and, in general, improve the animals' well-being [46]. Ueno et al. [47], in 2020, demonstrated that repeated exposure of the mice to the experimenter´s hand before conducting behavioral tests allows them to become accustomed to it and to reduce anxiety about high altitudes. In addition, it has been previously shown that reducing mouse anxiety due to handling contributes to a reduction in the number of animals required for experiments [47]. The same handling method (described in Methods 2.6) is used to put the mice under general anesthesia before the radiotracer injection. Thus, no immobilization method is necessary. How much our long-term gentle handling (up to six months stay in the facility) effectively reduced stress in both male and female mice will need to be shown in a further study including non-gentle handling control mice. While there are several advantages to using the four-mouse hotel chamber for PET imaging, it has the limitation of keeping one single flow rate for all animals. Thus, individual adaption is not possible, but could be improved in a new design. Providing enriched environments in sufficiently large cages also contributes to decreasing mortality, as has been demonstrated by comparison to conventional housing [48]. We used established, qualitative measures of stress, as well as animal weight, as a surrogate of well-being. Quantitative measures, such as corticosterone level, could not be integrated into the study protocol, since the mice are sacrificed only at the end of the project, in order to follow the Reduction Principle.

5. Conclusions

In the context of longitudinal studies, animal welfare assumes particular relevance, since the animals reside for long periods in the facility during their progressive aging and, at the same time, are involved in all of the experimental phases. Non-invasive imaging using PET allows repeated measurements of multiple biologically relevant tracer distributions over time in the same animal, thus significantly reducing the number of animals required. Reduction is achieved with the establishment of longitudinal studies that use radioactively labelled molecules in very small-volume solutions without pharmacological effect. Refinement is ensured by the possibility of conducting PET/CT scans with four animals at a time, a varied and stimulating enriched environment, and a standardized handling method during all phases of the experiment with a minimal number of experimenters.

Author Contributions: Conceptualization, M.B. and S.Z.; methodology, G.P. and R.O.; formal analysis, L.H.K.; investigation, G.P. and L.H.K.; resources, S.L.; writing—original draft preparation, G.P.; writing—review and editing, S.Z., M.B. and P.B.; visualization, L.H.K.; supervision, B.v.U.-S.; project administration, K.W.-M.; funding acquisition, M.B. All authors have read and agreed to the published version of the manuscript.

Funding: This work was supported by grants from the Deutsche Forschungsgemeinschaft (DFG, German Research Foundation) under Germany's Excellence Strategy within the framework of the Munich Cluster for Systems Neurology (EXC 2145 SyNergy—ID 390857198).

Institutional Review Board Statement: All experiments were carried out in compliance with the National Guidelines for Animal Protection, Germany, with the approval ROB-55.2-2532.Vet_02-19-26 from 01.07.2019 of the regional Animal Care Committee of the Government of Oberbayern (Regierung Oberbayern), and were overseen by a veterinarian. Animal experiments were conducted in accordance with the guidelines of EU Directive 2010/63/EU for animal experiments.

Informed Consent Statement: Not applicable.

Data Availability Statement: Data presented here will be provided by the authors upon request.

Conflicts of Interest: The authors declare no conflict of interest.

References

1. Gurdon, B.; Kaczorowski, C. Pursuit of precision medicine: Systems biology approaches in Alzheimer's disease mouse models. *Neurobiol. Dis.* **2021**, *161*, 105558. [CrossRef]
2. Bouter, C.; Henniges, P.; Franke, T.N.; Irwin, C.; Sahlmann, C.O.; Sichler, M.E.; Beindorff, N.; Bayer, T.A.; Bouter, Y. (18)F-FDG-PET Detects Drastic Changes in Brain Metabolism in the Tg4-42 Model of Alzheimer's Disease. *Front. Aging Neurosci.* **2018**, *10*, 425. [CrossRef]
3. Spangenberg, E.E.; Lee, R.J.; Najafi, A.R.; Rice, R.A.; Elmore, M.R.; Blurton-Jones, M.; West, B.L.; Green, K.N. Eliminating microglia in Alzheimer's mice prevents neuronal loss without modulating amyloid-beta pathology. *Brain* **2016**, *139*, 1265–1281. [CrossRef] [PubMed]
4. Scholl, M.; Carter, S.F.; Westman, E.; Rodriguez-Vieitez, E.; Almkvist, O.; Thordardottir, S.; Wall, A.; Graff, C.; Langstrom, B.; Nordberg, A. Early astrocytosis in autosomal dominant Alzheimer's disease measured in vivo by multi-tracer positron emission tomography. *Sci. Rep.* **2015**, *5*, 16404. [CrossRef]
5. Honer, M.; Gobbi, L.; Martarello, L.; Comley, R.A. Radioligand development for molecular imaging of the central nervous system with positron emission tomography. *Drug Discov. Today* **2014**, *19*, 1936–1944. [CrossRef] [PubMed]
6. Russell, W.M.S.; Burch, R.L. *The Principles of Humane Experimental Technique*; Methuen & Co., Ltd.: London, UK, 1959.

7. Diaz, L.; Zambrano, E.; Flores, M.E.; Contreras, M.; Crispin, J.C.; Aleman, G.; Bravo, C.; Armenta, A.; Valdes, V.J.; Tovar, A.; et al. Ethical Considerations in Animal Research: The Principle of 3R's. *Rev. Investig. Clin.* **2020**, *73*, 199–209. [CrossRef]
8. Directive 2010/63/EU of the European Parliament and of the Council of 22 September 2010 on the Protection of Animals Used for Scientific Purposes; OJ L 276/33. 2010. Available online: https://eur-lex.europa.eu/LexUriServ/LexUriServ.do?uri=OJ:L:2010:276:0033:0079:en:PDF (accessed on 13 April 2023).
9. Slater, A.M.; Cao, L. A Protocol for Housing Mice in an Enriched Environment. *J. Vis. Exp.* **2015**, *100*, e52874. [CrossRef]
10. Moody, C.M.; Paterson, E.A.; Leroux-Petersen, D.; Turner, P.V. Using Paper Nest Pucks to Prevent Barbering in C57BL/6 Mice. *J. Am. Assoc. Lab. Anim. Sci.* **2021**, *60*, 133–138. [CrossRef] [PubMed]
11. Bechard, A.; Meagher, R.; Mason, G. Environmental enrichment reduces the likelihood of alopecia in adult C57BL/6J mice. *J. Am. Assoc. Lab. Anim. Sci.* **2011**, *50*, 171–174.
12. Bailoo, J.D.; Murphy, E.; Boada-Sana, M.; Varholick, J.A.; Hintze, S.; Baussiere, C.; Hahn, K.C.; Gopfert, C.; Palme, R.; Voelkl, B.; et al. Effects of Cage Enrichment on Behavior, Welfare and Outcome Variability in Female Mice. *Front. Behav. Neurosci.* **2018**, *12*, 232. [CrossRef] [PubMed]
13. Chourbaji, S.; Zacher, C.; Sanchis-Segura, C.; Spanagel, R.; Gass, P. Social and structural housing conditions influence the development of a depressive-like phenotype in the learned helplessness paradigm in male mice. *Behav. Brain Res.* **2005**, *164*, 100–106. [CrossRef] [PubMed]
14. Mieske, P.; Hobbiesiefken, U.; Fischer-Tenhagen, C.; Heinl, C.; Hohlbaum, K.; Kahnau, P.; Meier, J.; Wilzopolski, J.; Butzke, D.; Rudeck, J.; et al. Bored at home?—A systematic review on the effect of environmental enrichment on the welfare of laboratory rats and mice. *Front. Vet. Sci.* **2022**, *9*, 899219. [CrossRef]
15. Marashi, V.; Barnekow, A.; Ossendorf, E.; Sachser, N. Effects of different forms of environmental enrichment on behavioral, endocrinological, and immunological parameters in male mice. *Horm. Behav.* **2003**, *43*, 281–292. [CrossRef]
16. Azkona, G.; Caballero, J.M. Implementing strategies to reduce singly housed male mice. *Lab. Anim.* **2019**, *53*, 508–510. [CrossRef] [PubMed]
17. Shin, R.W.; Ogino, K.; Shimabuku, A.; Taki, T.; Nakashima, H.; Ishihara, T.; Kitamoto, T. Amyloid precursor protein cytoplasmic domain with phospho-Thr668 accumulates in Alzheimer's disease and its transgenic models: A role to mediate interaction of Abeta and tau. *Acta Neuropathol.* **2007**, *113*, 627–636. [CrossRef]
18. Nilsson, P.; Saito, T.; Saido, T.C. New mouse model of Alzheimer's. *ACS Chem. Neurosci.* **2014**, *5*, 499–502. [CrossRef]
19. Puzzo, D.; Gulisano, W.; Palmeri, A.; Arancio, O. Rodent models for Alzheimer's disease drug discovery. *Expert Opin. Drug Discov.* **2015**, *10*, 703–711. [CrossRef] [PubMed]
20. Xue, F.; Du, H. TREM2 Mediates Microglial Anti-Inflammatory Activations in Alzheimer's Disease: Lessons Learned from Transcriptomics. *Cells* **2021**, *10*, 321. [CrossRef]
21. Chung, D.C.; Roemer, S.; Petrucelli, L.; Dickson, D.W. Cellular and pathological heterogeneity of primary tauopathies. *Mol. Neurodegener.* **2021**, *16*, 57. [CrossRef]
22. Blume, T.; Focke, C.; Peters, F.; Deussing, M.; Albert, N.L.; Lindner, S.; Gildehaus, F.J.; von Ungern-Sternberg, B.; Ozmen, L.; Baumann, K.; et al. Microglial response to increasing amyloid load saturates with aging: A longitudinal dual tracer in vivo muPET-study. *J. Neuroinflammation* **2018**, *15*, 307. [CrossRef]
23. Warnock, G.I.; Aerts, J.; Bahri, M.A.; Bretin, F.; Lemaire, C.; Giacomelli, F.; Mievis, F.; Mestdagh, N.; Buchanan, T.; Valade, A.; et al. Evaluation of 18F-UCB-H as a novel PET tracer for synaptic vesicle protein 2A in the brain. *J. Nucl. Med.* **2014**, *55*, 1336–1341. [CrossRef] [PubMed]
24. Morris, R. Developments of a water-maze procedure for studying spatial learning in the rat. *J. Neurosci. Methods* **1984**, *11*, 47–60. [CrossRef]
25. Block, F. Global ischemia and behavioural deficits. *Prog. Neurobiol.* **1999**, *58*, 279–295. [CrossRef] [PubMed]
26. Hobbiesiefken, U.; Mieske, P.; Lewejohann, L.; Diederich, K. Evaluation of different types of enrichment—Their usage and effect on home cage behavior in female mice. *PLoS ONE* **2021**, *16*, e0261876. [CrossRef]
27. Leach, M.C.; Klaus, K.; Miller, A.L.; Scotto di Perrotolo, M.; Sotocinal, S.G.; Flecknell, P.A. The assessment of post-vasectomy pain in mice using behaviour and the Mouse Grimace Scale. *PLoS ONE* **2012**, *7*, e35656. [CrossRef]
28. Swan, J.; Boyer, S.; Westlund, K.; Bengtsson, C.; Nordahl, G.; Tornqvist, E. Decreased levels of discomfort in repeatedly handled mice during experimental procedures, assessed by facial expressions. *Front. Behav. Neurosci.* **2023**, *17*, 1109886. [CrossRef] [PubMed]
29. Langford, D.J.; Bailey, A.L.; Chanda, M.L.; Clarke, S.E.; Drummond, T.E.; Echols, S.; Glick, S.; Ingrao, J.; Klassen-Ross, T.; Lacroix-Fralish, M.L.; et al. Coding of facial expressions of pain in the laboratory mouse. *Nat. Methods* **2010**, *7*, 447–449. [CrossRef]
30. Lee, G.H.; Kim, K.; Jo, W. Stress Evaluation of Mouse Husbandry Environments for Improving Laboratory Animal Welfare. *Animals* **2023**, *13*, 249. [CrossRef] [PubMed]
31. Hildebrandt, I.J.; Su, H.; Weber, W.A. Anesthesia and other considerations for in vivo imaging of small animals. *ILAR J.* **2008**, *49*, 17–26. [CrossRef]
32. Alstrup, A.K.; Smith, D.F. Anaesthesia for positron emission tomography scanning of animal brains. *Lab. Anim.* **2013**, *47*, 12–18. [CrossRef]
33. Hohlbaum, K.; Bert, B.; Dietze, S.; Palme, R.; Fink, H.; Thone-Reineke, C. Severity classification of repeated isoflurane anesthesia in C57BL/6JRj mice-Assessing the degree of distress. *PLoS ONE* **2017**, *12*, e0179588. [CrossRef] [PubMed]

34. Greenwood, H.E.; Nyitrai, Z.; Mocsai, G.; Hobor, S.; Witney, T.H. High-Throughput PET/CT Imaging Using a Multiple-Mouse Imaging System. *J. Nucl. Med.* **2020**, *61*, 292–297. [CrossRef] [PubMed]
35. Baier, J.; Rix, A.; Drude, N.I.; Darguzyte, M.; Baues, M.; May, J.N.; Schipper, S.; Mockel, D.; Palme, R.; Tolba, R.; et al. Influence of MRI Examinations on Animal Welfare and Study Results. *Investig. Radiol.* **2020**, *55*, 507–514. [CrossRef]
36. Gjendal, K.; Ottesen, J.L.; Olsson, I.A.S.; Sorensen, D.B. Effect of Repeated Exposure to Isoflurane on Nest Building and Burrowing in Mice. *J. Am. Assoc. Lab. Anim. Sci.* **2020**, *59*, 30–36. [CrossRef]
37. Joly-Amado, A.; Serraneau, K.S.; Brownlow, M.; de Evsikova, C.M.; Speakman, J.R.; Gordon, M.N.; Morgan, D. Metabolic changes over the course of aging in a mouse model of tau deposition. *Neurobiol. Aging* **2016**, *44*, 62–73. [CrossRef]
38. Ishii, M.; Wang, G.; Racchumi, G.; Dyke, J.P.; Iadecola, C. Transgenic mice overexpressing amyloid precursor protein exhibit early metabolic deficits and a pathologically low leptin state associated with hypothalamic dysfunction in arcuate neuropeptide Y neurons. *J. Neurosci.* **2014**, *34*, 9096–9106. [CrossRef] [PubMed]
39. Holscher, C. Stress impairs performance in spatial water maze learning tasks. *Behav. Brain Res.* **1999**, *100*, 225–235. [CrossRef]
40. Harrison, F.E.; Hosseini, A.H.; McDonald, M.P. Endogenous anxiety and stress responses in water maze and Barnes maze spatial memory tasks. *Behav. Brain Res.* **2009**, *198*, 247–251. [CrossRef]
41. Fridgeirsdottir, G.A.; Hillered, L.; Clausen, F. Escalated handling of young C57BL/6 mice results in altered Morris water maze performance. *Ups. J. Med. Sci.* **2014**, *119*, 1–9. [CrossRef]
42. Bisaz, R.; Schachner, M.; Sandi, C. Causal evidence for the involvement of the neural cell adhesion molecule, NCAM, in chronic stress-induced cognitive impairments. *Hippocampus* **2011**, *21*, 56–71. [CrossRef] [PubMed]
43. Chesler, E.J.; Wilson, S.G.; Lariviere, W.R.; Rodriguez-Zas, S.L.; Mogil, J.S. Identification and ranking of genetic and laboratory environment factors influencing a behavioral trait, thermal nociception, via computational analysis of a large data archive. *Neurosci. Biobehav. Rev.* **2002**, *26*, 907–923. [CrossRef] [PubMed]
44. Hurst, J.L.; West, R.S. Taming anxiety in laboratory mice. *Nat. Methods* **2010**, *7*, 825–826. [CrossRef] [PubMed]
45. Gouveia, K.; Hurst, J.L. Reducing mouse anxiety during handling: Effect of experience with handling tunnels. *PLoS ONE* **2013**, *8*, e66401. [CrossRef] [PubMed]
46. Neely, C.; Lane, C.; Torres, J.; Flinn, J. The Effect of Gentle Handling on Depressive-Like Behavior in Adult Male Mice: Considerations for Human and Rodent Interactions in the Laboratory. *Behav. Neurol.* **2018**, *2018*, 2976014. [CrossRef]
47. Ueno, H.; Takahashi, Y.; Suemitsu, S.; Murakami, S.; Kitamura, N.; Wani, K.; Matsumoto, Y.; Okamoto, M.; Ishihara, T. Effects of repetitive gentle handling of male C57BL/6NCrl mice on comparative behavioural test results. *Sci. Rep.* **2020**, *10*, 3509. [CrossRef]
48. Cait, J.; Cait, A.; Scott, R.W.; Winder, C.B.; Mason, G.J. Conventional laboratory housing increases morbidity and mortality in research rodents: Results of a meta-analysis. *BMC Biol.* **2022**, *20*, 15. [CrossRef]

Disclaimer/Publisher's Note: The statements, opinions and data contained in all publications are solely those of the individual author(s) and contributor(s) and not of MDPI and/or the editor(s). MDPI and/or the editor(s) disclaim responsibility for any injury to people or property resulting from any ideas, methods, instructions or products referred to in the content.

animals

Article

Refining Stereotaxic Neurosurgery Techniques and Welfare Assessment for Long-Term Intracerebroventricular Device Implantation in Rodents

Ester Pérez-Martín [1,*,†], Almudena Coto-Vilcapoma [2,3,†], Juan Castilla-Silgado [2,3,†], María Rodríguez-Cañón [1], Catuxa Prado [1], Gabriel Álvarez [1], Marco Antonio Álvarez-Vega [4,5], Benjamín Fernández-García [2,3,6], Manuel Menéndez-González [3,7,8] and Cristina Tomás-Zapico [2,3,*]

1 Neuroscience Innovative Technologies S.L., Neurostech, 33428 Llanera, Spain; catuxa.prado@neurostech.com (C.P.)
2 Departamento de Biología Funcional, Área de Fisiología, Universidad de Oviedo, 33006 Oviedo, Spain
3 Instituto de Investigación Sanitaria del Principado de Asturias (ISPA), 33011 Oviedo, Spain
4 Departamento de Cirugía, Área de Cirugía, Universidad de Oviedo, 33006 Oviedo, Spain
5 Servicio de Neurocirugía, Hospital Universitario Central de Asturias, 33011 Oviedo, Spain
6 Departamento de Morfología y Biología Celular, Área de Anatomía, Universidad de Oviedo, 33006 Oviedo, Spain
7 Servicio de Neurología, Hospital Universitario Central de Asturias, 33011 Oviedo, Spain
8 Departamento de Medicina, Universidad de Oviedo, 33011 Oviedo, Spain
* Correspondence: ester.perez@neurostech.com (E.P.-M.); tomascristina@uniovi.es (C.T.-Z.)
† These authors contributed equally to this work.

Simple Summary: The development of innovative therapeutic strategies involving chronic drug delivery to specific brain regions has been crucial in preclinical neuroscience. However, these strategies involve complex surgeries due to the need for implanting a drug storage system connected to the brain through a cannula while ensuring animal welfare, which is a challenge, especially for long-term studies in rodents. In this study, we propose an optimized method with three main refinements: (i) modifying the dimensions of the implantable devices, (ii) using a combination of adhesive tissue and UV light-curing resin, and (iii) implementing a customized scoresheet to closely monitor animal welfare throughout the experiment. Overall, the proposed refinements significantly improved animal welfare, reduced complications related to surgery, increased animal survival, and ensured safe long-term implantations.

Abstract: Stereotaxic surgeries enable precise access to specific brain regions, being of particular interest for chronic intracerebroventricular drug delivery. However, the challenge of long-term studies at this level is to allow the implantation of drug storage devices and their correct intrathecal connection while guaranteeing animal welfare during the entire study period. In this study, we propose an optimized method for safe intrathecal device implantation, focusing on preoperative, intraoperative, and postoperative procedures, following the 3Rs principle and animal welfare regulations. Our optimized protocol introduces three main refinements. Firstly, we modify the dimensions of the implantable devices, notably diminishing the device-to-mouse weight ratio. Secondly, we use a combination of cyanoacrylate tissue adhesive and UV light-curing resin, which decreases surgery time, improves healing, and notably minimizes cannula detachment or adverse effects. Thirdly, we develop a customized welfare assessment scoresheet to accurately monitor animal well-being during long-term implantations. Taken together, these refinements positively impacted animal welfare by minimizing the negative effects on body weight, surgery-related complications, and anxiety-like behaviors. Overall, the proposed refinements have the potential to reduce animal use, enhance experimental data quality, and improve reproducibility. Additionally, these improvements can be extended to other neurosurgical techniques, thereby advancing neuroscience research, and benefiting the scientific community.

Citation: Pérez-Martín, E.; Coto-Vilcapoma, A.; Castilla-Silgado, J.; Rodríguez-Cañón, M.; Prado, C.; Álvarez, G.; Álvarez-Vega, M.A.; Fernández-García, B.; Menéndez-González, M.; Tomás-Zapico, C. Refining Stereotaxic Neurosurgery Techniques and Welfare Assessment for Long-Term Intracerebroventricular Device Implantation in Rodents. *Animals* **2023**, *13*, 2627. https://doi.org/10.3390/ani13162627

Academic Editor: Garikoitz Azkona

Received: 25 July 2023
Revised: 7 August 2023
Accepted: 8 August 2023
Published: 14 August 2023

Copyright: © 2023 by the authors. Licensee MDPI, Basel, Switzerland. This article is an open access article distributed under the terms and conditions of the Creative Commons Attribution (CC BY) license (https://creativecommons.org/licenses/by/4.0/).

Keywords: 3Rs; animal research; animal welfare; cannula fixation; intrathecal implantation; preclinical neurosurgery; reduction; refinement; stereotaxic surgery

1. Introduction

Stereotaxic surgeries, in combination with other cutting-edge approaches, have revolutionized preclinical neuroscience research [1]. These techniques have enabled precise and controlled access to specific regions of the nervous system, offering unprecedented opportunities to study brain function and evaluate potential therapeutic approaches, including chronic intracerebral or intraventricular administration of drugs, in vivo optogenetic manipulations for activating or inactivating brain structures or systems, and in vivo electrophysiological recordings in awake behaving rodents (Refs. [1–7], among others). However, the success of these techniques relies heavily on complex, highly precise, and time-consuming surgical procedures, the failure of which not only entails a significant loss in economic and human resources, but also compromises the welfare of the animals involved [8]. In fact, we have recently conducted a proof-of-concept study to assess the feasibility and safety of implanting a device for continuous cerebrospinal fluid (CSF) apheresis [2]. In this study, more than 30% of the operated mice had to be euthanized following humane criteria. Retrospective analysis of the causes of this high mortality rate led us to identify two factors that may have the largest effect at this level: (i) the size and weight of the device, which accounted for more than 65% and 10% of the operated mice's size and body weight, respectively [2]; and (ii) the correct fixation of the cannula, allowing freedom of movement of the animals without detachment from the skull. Based on our experience, this last aspect stands out as the most critical one since it entails the formation of wounds that do not heal and undergo necrosis, becoming the main reason for euthanasia of the affected animals.

Although adjusting the size of implantable devices would be one of the simplest refinement procedures, this does not apply to cannula fixation. In fact, one of the main challenges and limiting factors of these procedures in small rodents is the secure fixation of cannulas, guides, or electrodes at the specific stereotaxic coordinates on the skull surface for long-term studies. Traditionally, three methods have been employed: (i) small anchoring screws combined with dental cement (zinc-polycarboxylate), (ii) a combination of white and pink dental cement (zinc-polycarboxylate and methyl-methacrylate, respectively), or (iii) cyanoacrylate adhesive gel [1,9,10]. Despite the benefits of the aforementioned options, all of them pose drawbacks not only for experimental rodents, but also for researchers as some of these compounds contain respiratory tract irritants [11]. Regarding the reported complications in small rodents, the use of dental cements or cyanoacrylate has been associated with an increased incidence of surgical problems, such as skin necrosis, brain damage and trauma, infection, and, most frequently, cannula detachment due to the round shape of mouse skull. To address the last limitation, Sike et al. proposed in 2017 an improved method involving the development of a custom skull-shaped silicone spacer as a fixation adapter to be used in combination with cyanoacrylate tissue adhesive [10]. This method reduced the time of surgery compared to dental cement and minimized adverse effects, although it significantly increased the time required for preoperative procedures. In addition, it required specific materials and equipment, such as a 3D printer or microCT scanner, as the silicone spacer was reconstructed and molded from an original mouse skull, making it a non-universal method and not affordable for all researchers.

Following the European Directive 2010/63/EU (for review [12]) for the protection of laboratory animals and adhering to the 3Rs principle (replacement, reduction, and refinement) from *The Principles of Humane Experimental Technique* [13], in this work, we propose a new refined method to securely performed stereotaxic surgeries involving intracerebroventricular or intrathecal device implantation in long-term studies. These refinements focused on the main critical steps of intrathecal implantation in rodents (i.e., preoperative, intraoperative, and postoperative cares and procedures [14]) and were based on our experience in different studies conducted in our laboratory. Our proposed protocol introduces three significant changes. Firstly, we have modified the dimensions of the implantable devices we have previously used [2], while ensuring the functionality of the device. This allowed us to significantly reduce the ratio between the weight of the devices and the body weight of the mice. Secondly, we use a combination of cyanoacrylate tissue adhesive and UV light-curing resin, which significantly reduces surgery time, improves wound healing, and reduces the postoperative recovery period, with a near 100% success rate. Thirdly, we have designed a customized welfare assessment scoresheet to monitor animals undergoing long-term cannula implantation, including indicators that accurately and effectively reflect animals' well-being for this particular surgery [15–23]. By implementing all the improvements, we evidenced a positive impact on animal welfare, the reduction in the number of animals used, and the quality of the experimental data, all of which align with the "refinement" and "reduction" principles of the 3Rs [13]. Furthermore, these advances hold the potential to be applied in other neurosurgical techniques that require long-term implantation, including optic fiber fixation for optogenetics or electrode placement for in vivo electrophysiological recordings, among others, which may greatly benefit the neuroscience community and research.

2. Materials and Methods

2.1. Animals

Seven/eight-month-old male transgenic APP/PS1 (n = 40, hereafter referred to as "APP") of the 129Sv strain background mice expressing the chimeric mouse/human amyloid precursor protein (Mo/HuAPP695swe) and the mutant human presenilin 1 (PS1-dE9; [2,24]) and their non-transgenic wild-type (WT) littermates (n = 32) were used in this study. The animals were randomly assigned to one of four experimental groups depending on the intrathecal device implanted and the neurosurgery protocol followed (Figure 1a): (i) naïve (n = 10 for each genotype); (ii) group implanted with the original device using traditional surgical techniques (n WT = 9; n APP = 10) from [2]; (iii) group implanted with the miniaturized device using optimized surgical techniques (n WT = 3; n APP = 10); and (iv) group implanted with a commercial osmotic pump using optimized surgical techniques (n = 10 for each genotype; see specifications of the different implantable devices in Section 2.2).

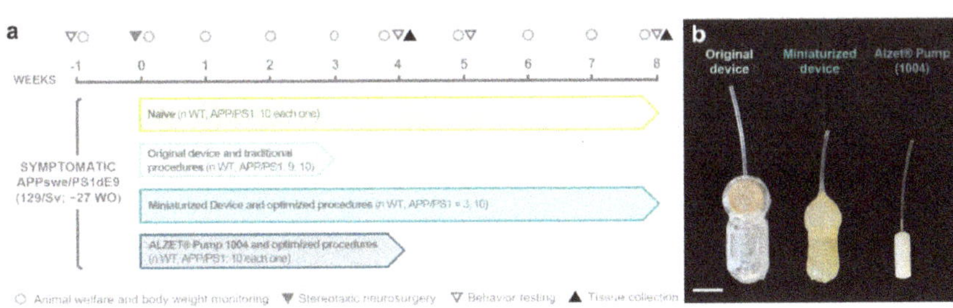

Figure 1. Experimental design and devices used in this study. (**a**) Schematic timeline of the experimental procedures and groups included in the study. (**b**) Photograph of the different implantable devices employed (from left to right): original device, miniaturized device, and a commercial ALZET® Micro-Osmotic Pump 1004 (Cupertino, CA, USA). Note the difference in dimensions among the implantable devices. The original and miniaturized devices are prototypes of a medical device designed to selectively filter cerebrospinal fluid. Scale bar: 1 cm. WO, weeks old; WT, wild-type mice.

Considering that the experiment end times varied among the different experimental groups, which are part of larger studies, our primary focus in this work was to analyze the parameters at timepoints shared by most of the groups, mainly week (W)-1, W3, and W8.

The exact sample sizes for each parameter and timepoint analyzed are provided in Table 1. In addition, the experimental groups, design, and timeline are depicted in Figure 1a and will be described in detail in the following sections.

Table 1. Specific sample sizes of each experimental group for different parameters at different timepoints.

Parameter and Time Analyzed	Naïve		Original Device		Miniaturized Device		Alzet Pump [a]	
	WT	APP	WT	APP	WT	APP	WT	APP
Recovery time after surgery	10	10	9	10	3	10	10	10
Humane endpoint application	10	10	9	10	3	10	10	10
Body weight at W1	10	10	9	10	3	10	10	10
Body weight at W3	10	10	7	7	3	9	10	9
Body weight at W8	5	5	-	-	3	9	-	-
% Change in body weight at W0	10	10	9	10	3	10	10	10
% Change in body weight at W3	10	10	7	7	3	9	10	9
% Change in body weight at W8	5	5	-	-	3	9	-	-
Welfare assessment score at W3	10	10	7	7	3	9	10	9
Welfare assessment score at W8	5	5	-	-	3	9	-	-
Behavioral testing at W1	10	10	-	-	3	9	9	10
Behavioral testing at W4/5	10	10	-	-	3	9	9	9
Behavioral testing at W8	6	5	-	-	3	9	-	-

Legend: W, week. [a] "Alzet pump" refers to micro-osmotic pump model 1004 (ALZET®, Cupertino, CA, USA).

The mice were housed under a 12/12 light/dark cycle (with lights on at 8:00 a.m.) at both constant room temperature (22 ± 2 °C) and relative humidity (55 ± 7%) and provided with ad libitum access to water and rodent global diet chow (A40; SAFE®, Rosenberg, Germany) at the animal facilities of Universidad de Oviedo (Spain). All animal procedures were performed during the light period (8:30–15:00 a.m.) and were approved by the Research Ethics Committee of the University of Oviedo (PROAE IDs 32/2020; 05/2022 and 06/2022) in compliance with European (Directive 2010/63/UE) and Spanish (RD118/2021, Law 32/2007) legislation. Every possible effort was made to ensure animal welfare, minimize pain and distress, and employ the smallest sample size necessary to achieve statistically relevant results, in accordance with the ARRIVE guidelines developed by the NC3R [25].

2.2. *Implantable Devices*

In our study, we employed different intrathecal implantable devices, namely, the original device [2], miniaturized device, and commercial pump (Figure 1b, Table 2). Both the original and miniaturized devices correspond to two prototypes of a medical device designed and manufactured by Neuroscience Innovative Technologies S.L. (https://neurostech.com/; accessed on 31 May 2023) for continuous and selective apheresis of CSF (in-depth review of this innovative therapeutic strategy [26–28]). Briefly, these devices comprise two main components: a subcutaneous reservoir and an apheresis module with selective nanoporous membranes connected to an intrathecal brain infusion cannula (Brain Infusion Kit 3; ALZET®, Cupertino, CA, USA). Their unique design allows the clearing of toxic molecules from CSF while avoiding the side effects caused by the direct contact between the therapeutic agent and the brain parenchyma [2,27,28]. The main and pivotal distinctions between the original and miniaturized devices lie in the dimensions of the device components and the change in the coating material, as outlined in Tables 2 and 3, without compromising its functionality. These modifications result in the following benefits: (i) a reduction in total weight and volume by 47.55% and 57.09%, respectively; (ii) a decrease in length, width, and height by 17.44%, 13.33%, and 40%, respectively; and (iii) a drop in the device weight-to-animal body weight ratio from 14.16 ± 1.35% to 6.93 ± 0.46% (equating to a 51.05% reduction). Additionally, we included the ALZET® 1004 osmotic pump (referred to as "Alzet pump" hereafter; Figure 1b) as a commercial implantable device with smaller dimensions than the miniaturized device for comparative purposes (Tables 2 and 3).

Table 2. Comparison of design parameters among devices and the commercial micro-osmotic pump used in this study.

Parameter	Original Device	Miniaturized Device	ALZET® Pump 1004 [a]
Weight (filled; g)	4.29 ± 0.41	2.25 ± 0.15	0.56 ± 0.06
% Weight (with respect to animal body weight [b]; %)	14.16 ± 1.35	6.93 ± 0.46	1.85 ± 0.19
Total Volume (cm^3)	5.92	2.54	0.54
Maximum L (cm)	3.5	2.9	1.5
Maximum W (cm)	1.5	1.3	0.6
Maximum H (cm)	1.5	0.9	0.6
Coated Material	Polyetheretherketone (PEEK; ISO 13485 [29])	LOCTITE® SI 5248 silicone (Henkel, Düsseldorf, Germany; ISO 10993 [30])	Cellulose ester blend
Catheter Tube Length and Outside Diameter (cm)	3 ± 0.2, 0.11 ± 0.008	3 ± 0.2, 0.11 ± 0.008	3 ± 0.2, 0.11 ± 0.008
Catheter Tube Material and Configuration	Polyvinyl chloride (MG), pre-attached	Polyvinyl chloride (MG), pre-attached	Polyvinyl chloride (MG), attachable
Brain Infusion III Cap Cannula Dimensions (L, W, H; cm) [c]	0.59, 0.59, 0.2	0.59, 0.59, 0.2	0.59, 0.59, 0.2
Brain Infusion III Cap Material [c]	Polycarbonate	Polycarbonate	Polycarbonate
Brain Infusion III Cannula Dimensions (L, W, H; cm) [c]	0.3, 0.031, 0.031	0.3, 0.031, 0.031	0.3, 0.031, 0.031
Brain Infusion III Cannula Material [c]	Stainless steel	Stainless steel	Stainless steel

Legend: H, height; L, length; MG, medical grade; W, width; ISO, International Organization for Standardization. [a] Data obtained from Micro-osmotic Pump Model 1004 (ALZET®, Cupertino, CA, USA) Datasheet; [b] Mean ± SEM Body weight of 7-months old male mouse (30.30 ± 0.16); [c] Data obtained from Brain Infusion Kit III 1–3 mm (ALZET®) Datasheet.

Table 3. Detailed dimensions of the different device components.

Parameter	Original Device	Miniaturized Device	ALZET® Pump 1004 [a]
Apheresis Module Dimensions (L, W, H; cm)	1.4, 1.5, 1.5	1.3, 1.3, 0.9	-
Apheresis Module External Volume (cm^3)	3.150	1.521	-
Apheresis Module Internal Volume (μL)	235.5	235.5	-
Reservoir Dimensions (L, W, H; cm)	2.1, 1.2, 1.1	1.6, 0.8, 0.8	1.5, 0.6, 0.6
Reservoir External Volume (cm^3)	2.772	1.024	0.540
Reservoir Internal Volume (μL)	100	100	100
Total Volume (cm^3)	5.922	2.545	0.540

Legend: H, height; L, length; W, width. [a] Data obtained from micro-osmotic pump model 1004 (ALZET®, Cupertino, CA, USA) datasheet.

Preparation of the implantable devices was conducted prior to surgery. To begin with, the original and miniaturized devices were sterilized by exposure to 25 kGy gamma irradiation from a ^{60}Co source, following a similar sterilization process as that specified in ALZET® products and other medical instruments. On the day of surgery, the filling and priming of implantable devices were performed under sterile conditions as previously described [2] and adhering to the manufacturer's instructions. The original and miniaturized devices were loaded with a solution containing non-conjugated anti-β-Amyloid antibody or artificial CSF (aCSF; Tocris Bioscience, Bristol, UK). Alzet pumps were filled with a recombinant mouse factor or aCSF.

Since no significant differences in body weight and behavior were observed between animals receiving the therapeutic agent and those receiving the vehicle at the analyzed timepoints employed in this study, all animals were included in the same experimental group regardless of the treatment.

2.3. Intrathecal Implantable Device Surgery

The following sections will describe the optimized protocol for intrathecal device implantation surgery, with a particular emphasis on the refined steps that have been proven to be essential in ensuring and enhancing the welfare of mice during long-term implantations compared to existing protocols [2,4,5,7,8,10].

2.3.1. Preoperative Care and Treatment

Regardless of the implanted device and surgical protocol followed, all animals received a presurgical analgesic and antibiotic cocktail 15–20 min prior to the start of the surgery. This treatment consisted of buprenorphine (0.05 mg/kg, subcutaneous injection (SC), Bupaq; Richter Pharma, Wels, Austria), enrofloxacin (10 mg/kg, SC, Syvaquinol; Syva Laboratories, Leon, Spain). Moreover, sterile saline solution was administered intraperitoneally (IP) or SC (1 mL of 0.9% (*w*/*v*) NaCl; B. Braun, Melsungen, Germany) to prevent dehydration associated with the surgical procedure itself.

To minimize the high risk of aggression and potential wound opening/infection in male mice sharing housing, all animals were individually housed. In the case of animals implanted with the original device, immediate isolation following surgery was carried out, which may have contributed to increased distress. To address this concern, in the refined protocol (i.e., animals implanted with the miniaturized device and Alzet pump) individual isolation was conducted one week prior to surgery (W1). Additionally, a diverse range of cage enrichments, such as a red polycarbonate arch (Bio-Serv, Flemington, NJ, USA), sizzle-nest (Datesand, Bredbury, UK), disposable cardboard tunnels, cocoon (Datesand), or paper wool (Datesand) were simultaneously provided. Moreover, palatable gel diet energy and transport food (SAFE®, Rosenberg, Germany) in a Petri dish was placed in the cage bedding. This approach allowed the animals to adapt to their new housing conditions for a 7-day acclimation period prior to surgery, minimizing possible stress and anxiety associated with novel food and environments [31,32].

On the day of surgery, animals implanted with miniaturized and Alzet pumps were placed inside a veterinary intensive care unit (S50 Advance Series II Model; Vetario, Westen-Super-Mare, UK) and maintained at a controlled temperature of 33–36 °C for 30 min, as pre-anesthetic warming has been shown to prevent hypothermia-related complications and to shorten recovery from anesthesia in rodents [33–35].

2.3.2. Stereotaxic Surgery Procedure

Depending on the experimental group, the protocol employed for implantation of the intrathecal device varied. For animals implanted with the original device, traditional protocols were followed (as detailed in [2]). However, for animals implanted with the miniaturized device or Alzet pump, a refined technique was used. In this section, we will primarily focus on the modifications made to the stereotaxic neurosurgery protocol for long-term intrathecal device implantation compared to the available traditional protocols [2,4,5,7,8]. A detailed step-by-step protocol is provided in Supplementary Information S1.

A crucial aspect of all surgeries is the meticulous preparation of the operating room and surgical instruments, which involves maintaining "clean" and "dirty" areas clearly separated (see Figure 2). Every effort was made to employ strict aseptic techniques and to ensure a sterile surgical field throughout the entire surgical process. For instance, the working space was first cleaned with soap and water, dried, and then wiped with 70% ethanol. The surface was covered with protective paper (Nalgene™ Super Versi-Dry™ Surface Protectors, Thermo Scientific™, Waltman, MA, USA), allowing the sterilized material to be placed on it, organized as shown in Figure 2. Prior to each procedure and between surgeries on different animals, all instruments underwent a thorough cleaning with an enzymatic detergent (Helizyme; B. Braun), followed by sterilization both with 70% (v/v) ethanol and by heating at 230 °C for 1 min in a microbead sterilizer (B1305-E-FIS; Fisherbrand, Waltman, MA, USA). In addition, all the researchers involved in the surgery and in assisting the surgery changed gloves between mice and avoided touching other areas of the operating room that may have been dirty. On the other hand, the larger equipment (isoflurane delivery system, surface of the induction chamber, the stereotaxic frame and its digital controller, and the stereo microscope), as well as the containers that carried the medication or products to be used during surgery, were cleaned with 70% ethanol. All the surgeries were performed from 08:30 to 15:00 by the same surgeons (E.P.-M. and M.A.Á.-V.) and surgery assistants (A.C.-V.; J.C.-S. and B.F.-G.).

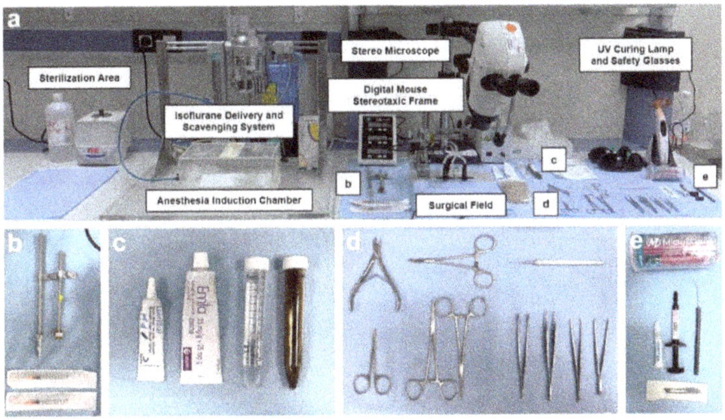

Figure 2. Operating room setup and surgical instrument arrangement. (**a**) Overview of the operating room showing the main equipment and instruments required. (**b–e**) Close-up views highlighting specific instruments and components used during surgery: cannula and stereotaxic needle holders (**b**); ocular and topical anesthetic ointments, 70% (v/v) ethanol, and 1% (v/v) povidone-iodine solution (**c**); surgical instruments (**d**); and cyanoacrylate tissue adhesive, UV light-curing resin, and applicators (**e**).

After pre-anesthetic warming the animals and administering the presurgical analgesic and antibiotic cocktail (see details in the previous section), anesthesia was induced with 4% (v/v) isoflurane (IsoFlo; Zoetis, Parsippany, NJ, USA) and 2.5% (v/v) O_2. The animals were gently shaved on top of the head (Aesculap Isis Rodent Shaver; Aesculap Schermaschinen, Buschbach, Germany) and a topical anesthetic ointment (lidocaine 25 mg/g and prilocaine 25 mg/g, EMLA; AstraZeneca, Cambridge, UK) was applied around the ear canal (Figure 3). Thereafter, the animal was securely placed in a digital stereotaxic frame (Harvard Apparatus, Holliston, MA, USA) with a heating pad underneath to prevent hypothermia during the procedure. Temperature monitoring using a rectal thermometer connected to a physiological monitoring system (Harvard Apparatus) is strongly recommended. Once the absence of withdrawal reflex was confirmed and ocular ointment (Lubrithal; Dechra Pharmaceuticals, Northwich, UK) was applied, a U-shaped incision was made in the specific area between the two auditory pavilions, and the upper skin was removed rostrally by using two hemostats (see Figure 3). To facilitate visualization of the craniometric points bregma and lambda on the cranial surface and to enhance posterior cannula fixation, the periosteum was scraped using a microcurette (Model 10080-05; Fine Surgical Tools F.S.T., Chandler, AZ, USA) and dried with a cotton swab moistened with 70% (v/v) ethanol. The coordinates of the bregma and lambda landmarks were then measured to accurately position them in the dorso-ventral (DV), medio-lateral (ML) and antero-posterior (AP) planes [1,36,37] and readjusted whenever the difference between coordinates exceeded 0.05 mm. Once a "flat-skull" position was reached, a 0.5 mm diameter craniotomy was drilled in the skull at the specific coordinates AP= −0.70 mm and ML= −1.26 mm (Figure 3). Subsequently, a subcutaneous pocket was created with a hemostat by dissecting from the original U-incision towards the mouse's flanks, and the implantable device was carefully inserted. The intraventricular cannula was slowly inserted into the craniotomy until DV= −2.5 mm using a cannula holder (Stoelting Co., Wood Dale, IL, USA) and was glued to the skull surface with cyanoacrylate adhesive tissue (Cicastick; Sutuvet, Santiago de Surco, Peru). To secure the cannula in place, a thin annular layer of UV light-curing resin (Transbond XT 4; 3M, Saint Paul, MN, USA) was applied around the base of the brain cannula and polymerized in 3 cycles of 20 s using a portable UV lamp (Lux E; Bestdent, Gansu, China). After the resin was fully hardened (1 min), the upper tip of the cannula was removed using a bone nipper (Model 16102-11, F.S.T.; Figure 3). The previously removed skin was rehydrated with sterile saline solution and the incision was closed with single stitches using 5-0 monofilament non-resorbable polypropylene suture (Atramat®, Mexico City, Mexico). It was crucial to increase the density of stitches at the position of maximum tension of the U-shaped suture, corresponding to the portion where the catheter was positioned beneath the skin (Figure 3). Finally, a 1% (*v*/*v*) povidone-iodine solution (Betadine; Mylan, Canonsburg, PA, USA) and topical antibiotic ointment (Fucidine 20 mg/g; LEO Pharma, Ballerup, Denmark) were applied using a sterile cotton swab. The isoflurane anesthetic was gradually reduced, and the animal was carefully removed from the stereotaxic frame. Recovery time from anesthesia was recorded and classified into the following intervals: >20 min, 5–10 min, or <5 min.

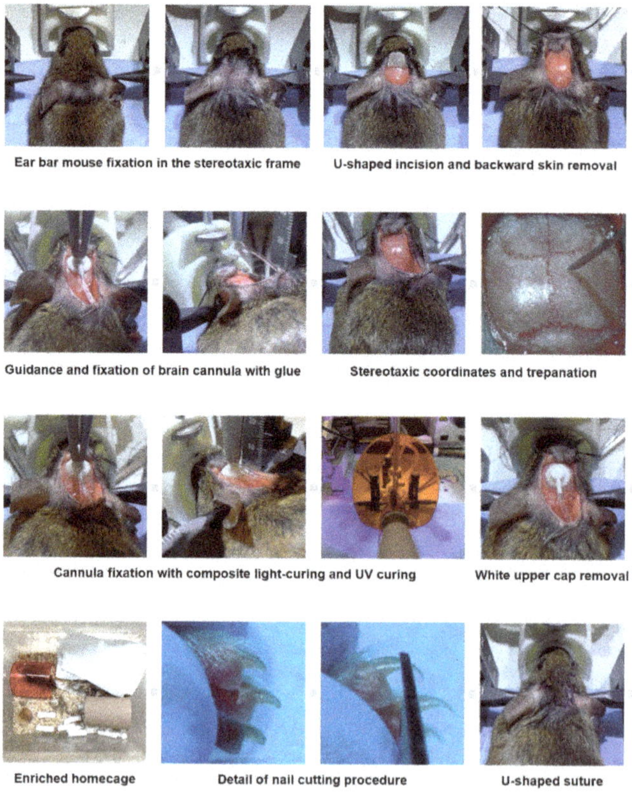

Figure 3. Roadmap of key steps in the refined neurosurgical protocol for long-term implantation of intrathecal cannula in mice.

2.3.3. Postoperative Care

Immediately after surgery, while the animals were still recovering from anesthesia, the nails of the front and rear feet were carefully trimmed to prevent scratching that could worsen incision healing or lead to wound reopening (Figure 3). It is important to note that only the most distal portion of the nail plate, identifiable by its "pearly" coloration, should be trimmed to avoid bleeding. Subsequently, anti-inflammatory treatment was administered using meloxicam (2 mg/kg, SC, Metacam; Boehringer Ingelheim, Rhein, Germany), and rehydration was ensured by supplying sterile saline solution (1 mL, SC). If necessary, sterile saline with 5% (*w*/*v*) glucose (B. Braun) can be SC administered. The mice were then returned to a clean and enriched cage (Figure 3) located within the veterinary intensive care unit, and received a postoperative regimen of analgesic, anti-inflammatory, and antibiotic treatment for two weeks following surgery [38], as described in Supplementary Information S2.

2.4. Body Weight Monitoring and Welfare Assessment

Body weight was measured weekly between 09:00 and 13:00 p.m., starting one week before surgery (referred to as "basal body weight" or W1) and continuing until the end of the experiment, using a portable balance (CB 501; Adam Equipment, Milton Keynes, UK), as the body weight reduction is widely used as a humane endpoint (Refs. [39,40], among others). From the week of surgical implantation onwards, the actual body weight of the animals was inferred by subtracting the weight of the implanted device from the total body weight measured. Then, the percentage change in body weight for subsequent weeks was calculated as follows:

$$\%\text{ Change in Body weight} = \left(\frac{\text{Basal W} - 1\text{ Body weight} - \text{Real Body weight}}{\text{Basal W} - 1\text{ Body weight}}\right) \times 100 \quad (1)$$

Regarding welfare evaluation, all implanted animals were closely and intensively monitored for 24 h, which is considered the critical period. For the remainder of the experiment, daily monitoring and welfare assessment were performed every morning (9:00–13:00 p.m.). All observations were made and diligently documented on a shared data sheet by blinded scorers/observers (E.P.-M. and A.C.-V. or J.C.-S.; an example of the "postoperative monitoring form" is provided in Supplementary Information S3). To ensure consistency and reproducibility, a scoresheet was designed, including several indicators related to general condition, nutritional and hydration status, spontaneous behavior, and surgery-related parameters, as well as compensatory measures, as detailed in Table 4 [15,17,20–22,39,40]. Overall, these welfare indicators were carefully chosen based on our experience with these specific surgical procedures and their relevance, ease of recognition, reliability, and effectiveness in providing accurate assessment of welfare [17].

As indicated in Table 4, if compensatory measures failed to alleviate complications or welfare scores reached values above 12 points, a humane endpoint was implemented, leading to the euthanasia of animals. Complications and the percentage of animals reaching the humane endpoint were recorded for each experimental group (see details in Table S1).

Table 4. Scoresheet for animal welfare monitoring criteria in mice undergoing intrathecal catheter surgery.

Parameters		Score
General Condition, Physical Appearance, and Posture		
Smooth and shiny fur, clean forelimbs, and nose		0
Presence of piloerection, unkempt fur		1
Abnormal posture (abdominal curvature and/or kyphosis, increased muscle tone)		2
Skin lesions unrelated to surgery (ear dermatitis, scratches, excessive barbering)		2
Nutritional and Hydration Status		
Unaffected or increased		0
Dehydration signs	<2 s after pinching the back skin	1
	>2 s after pinching the back skin	2
Animal Body weight	1 ≤ 10% weight loss	1
	10–20% weight loss	4
	>20% weight loss (duration >4 postoperative days)	8
Spontaneous Behavior		
Normal behavior (sleeping, exploration, grooming, nesting, interaction with environmental enrichment objects)		0
No use of enrichment objects, no nesting behavior		1
Impairment of motor function (hypo-locomotion)		4
Lethargy and/or slight loss of balance		8
Hind-limbs paralysis, tremors and/or signs of vestibulocochlear dysfunction		12
Surgery-specific Parameters		
Clean and dry surgical incision, no signs of infection, no pain or signs of distress		0
Scratches around the scar or slight redness		1
Redness and/or necrosis of the skin around the scar		3
Grimace scale	Moderate	4
	Severe	6
Suture opening Grade I (loose stitches with closed and healed wound)		3
Suture opening Grade II (loose stitches with unhealed wound)		6
Suture opening Grade III (open unhealed wound)		10
Suture opening Grade IV (*Brain Infusion Kit cannula* disconnected)		12

Score 0: normal physical condition, no pain. Scores 1–6: presence of discomfort indicating need for sustained observation and increased vigilance, measures such as providing moist and palatable food, administering analgesia, antibiotics, and subcutaneous saline or glucose may be necessary. Scores 6–10: significant suffering and pain indicating need for compensatory measures, subcutaneous glucose, seeking veterinary advice, re-operating scar with recovery in intensive care unit, and considering euthanasia if maintained for four consecutive days. Scores 10–12: moderate to severe suffering and pain, consider a humane endpoint if compensatory measures do not resolve complications and scores are maintained for 24 h. Scores > 12: severe suffering indicative of a humane endpoint application and termination of the experiment.

2.5. Behavioral Analysis

To explore general and anxiety-like behaviors [41], the elevated zero maze was performed both before (W1) and after surgery (W4/5 and W8, according to experimental groups; for specific sample size see Table 1) between 09:00 and 11:00 am. The apparatus (UGO Basile, Gemonio, Italy) consists of an annular gray platform 60 cm in diameter, positioned 60 cm above the floor. It features two closed corridors of 15 cm in length. For the test, a single mouse was placed in the center of one of the closed areas, and 5 min video recordings were captured using a Basler ace acA1300-60gm GigE camera (Basler, Ahrensburg, Germany). The acquired videos were automatically analyzed using the Ethovision XT version 16 software for Windows (Noldus, Wageningen, The Netherlands).

The rodents' instinctive tendency to explore novel environments and to avoid unprotected open and elevated spaces were analyzed based on three different measures: total distance traveled, and the percentage of time spent exploring the open areas (TO) or closed areas (TC), calculated as follows:

$$\%\ \text{Open areas} = \frac{T_O}{T_O + T_C},\quad \%\ \text{Closed areas} = \frac{T_C}{T_O + T_C} \tag{2}$$

Following each trial, the elevated zero maze apparatus was cleaned with a 70% (*v*/*v*) ethanol solution to avoid the influence of odors.

To ensure minimal impact on the overall animal well-being, behavioral testing was exclusively conducted in the following experimental groups: naïve, animals implanted with the miniaturized device, and animals implanted with Alzet pumps, as these animals showed higher welfare assessment scores (see Section 3). This approach aimed to minimize any potential additional adverse effect or stress on animals, which could potentially worsen their welfare.

2.6. *Catheter Placement Validation*

After the experiments, correct catheter placement at the specific stereotaxic coordinates in the right lateral ventricle was confirmed by two different methods detailed below.

2.6.1. In Vivo Dye Infusion and Visualization

Firstly, a solution of 5% (*w*/*v*) blue dextran (5 kDa; TdB Labs, Ultuna, Arlanda) dye diluted in aCSF was infused by percutaneous access through the reservoir of one of the animals implanted with the miniaturized device. A total of 200 µL of tracer solution was infused in two 24 h interval sessions, at a flow rate of 5 µL/min, using a programmable syringe pump (Remote Infuse/Withdraw Pump 11 Elite Nanomite Programmable Syringe Pump; Harvard Apparatus). Following CSF tracer infusion, while the mouse was under anesthesia with 1.5–2% (*v*/*v*) isoflurane and 2.5% (*v*/*v*) O_2, the cisterna magna was exposed using a customized protocol based on a previous study [42]. A stereomicroscope with integrated camera (S9i; Leica Microsystems, Wetzlar, Germany) was employed to visualize in vivo the CSF color in the mouse ventricular system. Subsequently, the mouse was euthanized by decapitation, and the brain immediately removed and embedded in Tissue-Tek® O.C.T. (Sakura Finetek Europe, Alphen aan den Rijn, The Netherlands) before being frozen at −80 °C. The brain was then sliced into 80 µm thick transversal sections using a cryostat (CM1900; Leica Microsystems). Images were captured using a digital camera to observe the visualization of the blue/dark color tracer within the mouse ventricular system.

2.6.2. Histological Analysis

To verify the specific AP, ML, and DV stereotaxic coordinates at which the intracerebroventricular catheter was placed, histological analysis was also performed. Briefly, an animal implanted with the miniaturized device was deeply anesthetized with 4% (*v*/*v*) isoflurane, and both the device and intracerebroventricular cannula were carefully removed. After exsanguination, the mouse was perfused with 40 mL of sterile ice-cold phosphate buffer saline (PBS, pH 7.4; Corning Incorporated, Corning, NY, USA) to clean tissues. The brain was immediately extracted and fixed in 4% (*v*/*v*) paraformaldehyde solution (Electron Microscopy Science, Hatfield, PA, USA) diluted in Sorensen's phosphate buffer overnight at 4 °C with continuous shaking. After cryoprotection in a solution of 30% (*w*/*v*) sucrose (Panreac AppliChem, Darmstadt, Germany) in PBS for 24 h, the brain was frozen in Tissue-Tek® O.C.T. and stored at −80 °C until further use. The brain was sectioned into 30 µm thick coronal slices using a cryostat (CM1900; Leica Microsystems), then washed in PBS and cryoprotected in a solution containing 30% (*v*/*v*) glycerol (VWR Chemicals, Radnor, PA, USA) and 30% (*v*/*v*) ethylene glycol (Sigma-Aldrich, St. Louis, MO, USA) in 0.02 M phosphate buffer, pH 7.2 at −20 °C until used for histological staining. Three non-consecutive coronal sections were washed in PBS (3 × 10 min) and stained with 0.5% (*w*/*v*) Toluidine blue O (Sigma-Aldrich) in distilled water for 1 min at room temperature. Next, the slices were washed in PBS (3 × 10 min), dried overnight at room temperature, and mounted on gelatin-coated slides (VWR). Finally, the sections were covered with an aqueous mounting medium (Aquatex®, Merck, Rhaway, NJ, USA). Visualization and image acquisition were performed using the integrated camera of a stereomicroscope (S9i; Leica Microsystems) and compared with Paxinos and Franklin's Mouse Atlas [37] and the Allen Brain Mouse Atlas (http://mouse.brain-map.org/; accessed on 1 July 2023) for reference.

2.7. Statistics

The data are presented as individual data points and as the median ± interquartile range (IQR), unless otherwise specified. The normality of the data was assessed before conducting statistical analyses using both the Kolmogorov–Smirnov and the Shapiro–Wilk tests. Due to the non-normal distribution of most variables, as indicated by *p*-values < 0.05 in the normality tests, and the unequal sample size of the experimental groups (Table 1), the following non-parametric tests were employed in this work. For independent or dependent quantitative two-samples comparisons, the Mann–Whitney U test or the Wilcoxon test was carried out, respectively. For independent or repeated-measures quantitative multiple comparisons, the Kruskal–Wallis test or the Friedman test were, respectively, employed, followed by Dunn's post hoc test when required. Finally, for qualitative analyses (i.e., "range of recovery time after anesthesia" and "humane endpoint"), variables were properly dichotomized when necessary and the Fisher exact test was performed. A *p*-value < 0.05 was set as the minimum level of statistical significance. All statistical analyses were conducted with SPSS Statistics version 26 for Windows (IBM, Armonk, NY, USA), and graphical representations were generated with GraphPad Prism version 9.0.2 for Windows (GraphPad Dotmatics, San Diego, CA, USA).

3. Results

3.1. The Optimized Protocol of Stereotaxic Surgeries Shortens the Recovery Time Required after Surgery and Significantly Minimizes Humane Endpoint Application

Recovery time was measured as the time interval between the end of stereotaxic surgery (i.e., withdrawal of inhaled anesthesia) and complete recovery of the animal (i.e., interaction with food or enrichment objects) and was classified in three different ranges: higher than 20 min, between 5 and 10 min, and less than 5 min, based on the data obtained (Table 5 and Figure 4a). Regarding the original device and traditional protocols, 100% (n = 20 out of 20) of the animals required more than 20 min to fully recover from anesthesia (Figure 4a,b). For animals implanted with the miniaturized device and following the optimized protocol, 30.76% (n = 3 out of 13) and 69.24% (n = 10 out of 13) of mice needed between 5 and 10 and 5 min to recover, respectively (Figure 4a,b). Finally, 100% (n = 20 out of 20) of the Alzet-pump-implanted mice following the optimized protocol recovered in less than 5 min (Figure 4a,b). These results indicate a positive impact of both the changes in the device dimensions and the implementation of a refined protocol in reducing the time required to fully recover from stereotaxic surgery (Figure 4a,b). Furthermore, the Fisher exact test revealed a clear relationship between the implanted device and recovery time range for most dichotomized variables and comparisons carried out (specific *p*-values are provided in Table 5).

Table 5. Comparisons and specific *p*-values for statistical qualitative analyses of recovery time and humane endpoint application parameters.

Parameter	Comparisons and *p*-Values			
Recovery time		Original Device	Miniaturized Device	Alzet Pump
>20 min		**<0.001**	**0.002**	**<0.001**
5–10 min		0.284	**0.003**	0.151
<5 min		**<0.001**	0.341	**<0.001**
	Naïve	Original Device	Miniaturized Device	Alzet Pump
Humane endpoint	**0.029**	**<0.001**	0.681	0.159

Bold values indicate statistical significance at the $p < 0.05$ in Fisher exact test, i.e., compared variables are related.

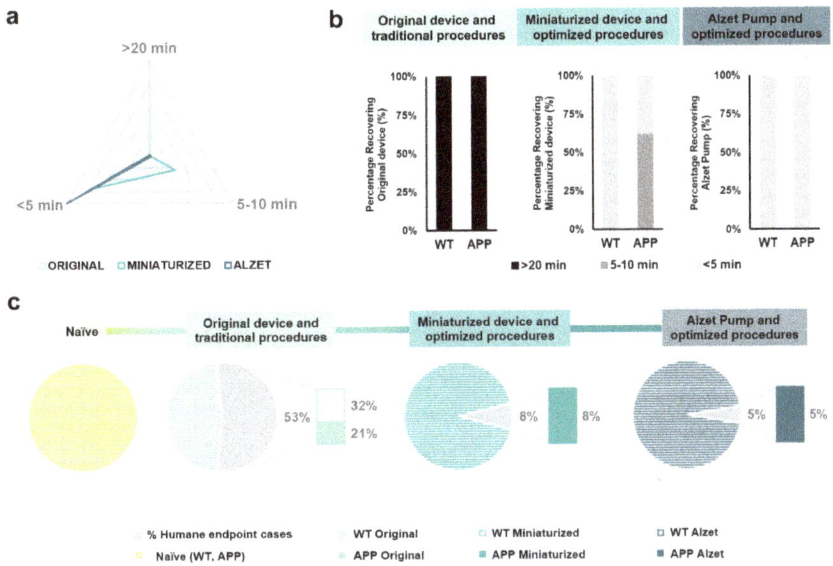

Figure 4. Qualitative analyses of the effect of changes in implantable device dimensions and stereotaxic protocol followed on both the recovery time after surgery and the percentage of humane endpoint application. (**a**) Prism representation of the percentage of animals in each range of recovery time (i.e., >20 min, 5–10 min and <5 min, individually plotted in each vertex) for each experimental group, regardless of genotype. Each inner grey prism line represents 25%. (**b**) Quantification of the percentage of animals in each range of recovery time separated both by implanted device and protocol, and by genotype. Note that 100% of the animals implanted with the original device and Alzet pump required more than 20 min and less than 5 min to recover from surgery, respectively; whereas all animals implanted with the miniaturized device needed less than 10 min. (**c**) Quantification of the percentage of humane endpoint application for each experimental group. Note that changes in the implanted device and refinement of the protocol significantly reduced the application of humane endpoint from 53% in animals implanted with the original device to 5% in those implanted with Alzet pumps. Specific sample sizes and Fisher exact test *p*-values are provided in Tables 1 and 5, respectively.

Regarding the application of humane endpoint based on welfare assessment scores, changes in the device dimensions and the implementation of refined steps in the protocol strongly decreased the percentage of application, this value being 53% for mice implanted with the original device, 8% for animals implanted with the miniaturized device, and 5% for Alzet-implanted mice (Figure 4c). In addition, the Fisher exact test demonstrated a relationship between the device implanted in the experimental group and the percentage of humane endpoint application, particularly in the naïve group and in those implanted with the original device (specific *p*-values in Table 5; Figure 4c).

3.2. Improved Protocol Reduces the Drop in Body Weight Observed in Mice from the First Postoperative Surgery Onwards and Normalizes It over Time

Body weight monitoring was performed throughout the experiment, starting at W1 and lasting for the remainder (Figure 5), as a broadly general humane endpoint ([39,40], among others). Overall, one week before surgery (W1), a higher body weight was evidenced in APP mice compared to their WT counterparts in the naïve group and in animals that were to be implanted with the original device and Alzet pumps (p naïve= 0.007; p original = 0.043; p Alzet pump = 0.004; Figure 5a,b and Figure S1). Moreover, we also detected differences among animals planned to be implanted with different devices, with body weight being higher in those mice that would receive the original device compared to the rest of the mice (p WT naïve vs. original = 0.043; p WT original vs. Alzet = 0.001; p APP naïve vs. Alzet = 0.043; p APP original vs. miniaturized = 0.028; p APP original vs. Alzet = 0.002; Figure 5b). It is worth mentioning that the original device weighed 4.29 ± 0.41 g (see Table 2 for specifications). Therefore, mice with higher body weight values were selected to be implanted with this device in order to ensure their welfare and the feasibility of the surgery. In addition, the reduced dimensions of the miniaturized device and Alzet pump enabled us to perform the implantation surgery on animals of a lower weight (Figure 5b), which would not have been possible with the original one and explains the initial differences observed among experimental groups at W1 (Figure 5b).

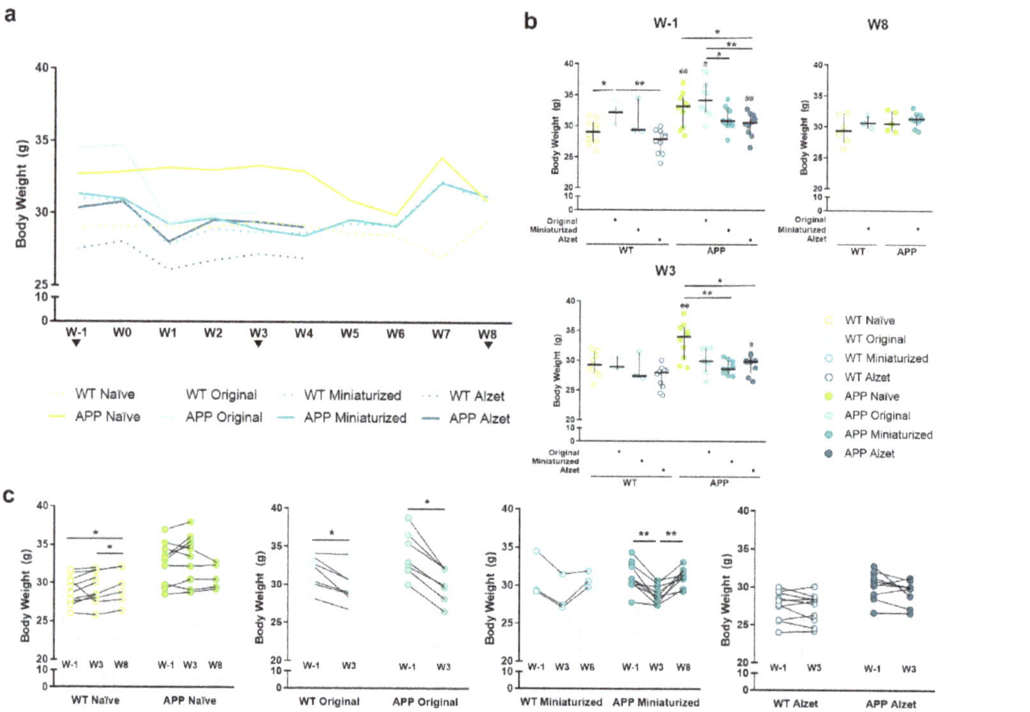

Figure 5. Effect of changes in the dimensions of implantable device and the refined stereotaxic protocol on the body weight at different timepoints. (**a**) Representation of the median body weight

throughout the duration of the experiment of the different experimental groups included in this work (based on the implanted device and genotype). Note the striking reduction in body weight in all the animals undergoing stereotaxic surgery from the first postoperative week, particularly in those implanted with the original device; and the partial recovery in mice implanted with the miniaturized device from postoperative W6 onwards. No statistical test has been performed for these data over time due to the incompatibility with the wide variety of experiment terminations. (**b**) Quantification of animal body weight at timepoints shared by most experimental groups, i.e., W1, W3, and W8. (**c**) Pre- and post-representations of body weight at W1 versus (vs.) W3 vs. W8 or W1 vs. W3, for each experimental group. Note that animals implanted with the miniaturized device regain the body weight loss observed at W3 over time. Specific sample size is provided in Table 1. Data expressed as median and IQR. Kruskal–Wallis test for multiple comparisons (W1 and W3 in (**b**)); Mann–Whitney U test for independent pairwise comparisons (WT vs. APP comparisons in b; W8 in (**b**)); Friedman test for repeated-measures multiple comparisons (naïve and miniaturized in **c**); and Wilcoxon test for dependent pairwise comparisons (original and Alzet in (**c**)). * $p < 0.05$; ** $p < 0.01$, for differences between experimental groups based on the implanted device and multiple comparisons; # $p < 0.05$; ## < 0.01, for differences between WT and APP genotypes in b (see Figure S1 for detailed analysis). W, week.

Regarding the differences among the animals with different devices and protocols after surgery, a significant overall decrease in body weight was observed from the first post-operative week onwards, regardless of the implanted device (Figure 5a,b). However, in the particular case of W3 the decrease was only found in APP animals implanted with the miniaturized and Alzet pumps compared to naïve APP (p APP naïve vs. miniaturized = 0.001; p APP naïve vs. Alzet = 0.013; Figure 5b). Finally, at W8 no differences were detected between animals implanted with the miniaturized device compared to naïve, regardless of the genotype ($p > 0.05$ for all comparisons; Figure 5b), indicating a tendency for body weight to normalize over time.

Taking the evolution of body weight for each experimental group individually (Figure 5c), we can notice a striking drop at W3 compared to W1 in the animals implanted with the original device (p WT = 0.018; p APP = 0.018; Figure 5c), a moderate decrease in those with the miniaturized device (p WT = 0.097; p APP < 0.001; Figure 5c), and no differences in those with the Alzet pump (p WT = 0.202; p APP = 0.086; Figure 5c). Interestingly, animals implanted with the miniaturized device regained basal body weight at W8 ($p = 0.480$; Figure 5c), results that are consistent with those described above.

To better understand and visualize the impact of the different implantable devices and protocols employed on animal body weight, we also included the analysis of the specific percentage change at different timepoints compared to W1 or basal body weight (Figure 6), where the significant loss in body weight at W3 is even more prominent than in the previous analysis and representations (p WT W0 = 0.035; p WT W3 = 0.002; p APP W0 = 0.049; p APP W3 < 0.001; Figure 6). In this regard, and in agreement with the aforementioned results, the percentage change in body weight decreased in all animals undergoing stereotaxic surgery, regardless of the implanted device at W3 (p WT naïve vs. original= 0.001; p WT naïve vs. miniaturized= 0.004; p WT original vs. Alzet = 0.045; p WT miniaturized vs. Alzet = 0.044; p APP naïve vs. original < 0.001; p APP naïve vs. miniaturized = 0.002; p APP original vs. Alzet = 0.006; Figure 6). Furthermore, this plot evidenced the remarkable drop in APP mice implanted with the original device, which largely contributed to us considering and applying a humane endpoint in these animals and terminating the experiment (Figure 6).

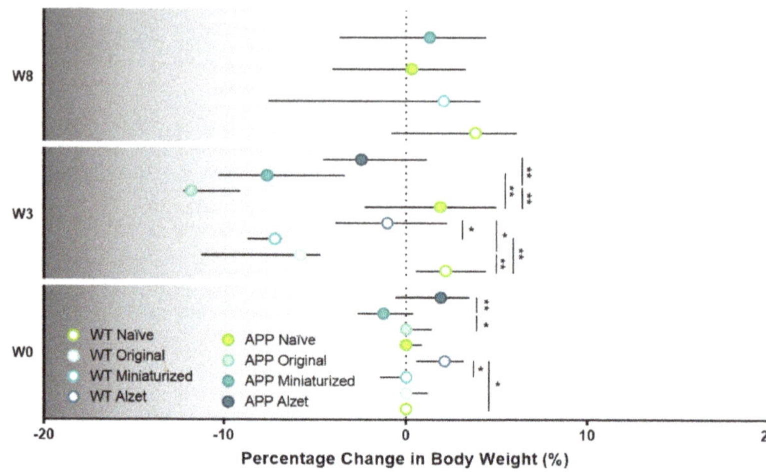

Figure 6. Effect of the different implantable devices and the refined stereotaxic protocol on the percentage change in animal body weight at different timepoints. Note the striking drop in APP mice implanted with the original device at W3 leading to us considering and applying a humane endpoint in this experimental group. Specific sample size is provided in Table 1. Data expressed as median and IQR. Kruskall–Wallis test for multiple comparisons (W0 and W3) and Mann–Whitney U test for independent pairwise comparisons (W8). * $p < 0.05$; ** $p < 0.01$. W, week.

3.3. Optimized Protocol and Implantation of Smaller Devices Significantly Improve Animal Welfare and Enable Longer and Safer Experimental Periods

Animal welfare was monitored intensively after surgery and recorded weekly during the remainder of the experiment, following a customized scoresheet for intrathecal device implantation (Table 4, Figure 7). Among others, the welfare monitoring sheet included parameters of general condition, physical appearance and posture, nutritional and hydration status, spontaneous behavior, and surgery-related indicators (see details in Table 4).

Regarding the overall welfare assessment scores of each experimental group (Figure 7a), higher scores were observed in animals undergoing stereotaxic surgery in the weeks following surgery compared to naïve animals, reaching the highest values at postoperative W3 (Figure 3a). These values were markedly elevated in both WT and APP mice implanted with the original device and following the traditional protocols (Figure 7a). In fact, significant statistical differences were detected among animals implanted with the original device with respect to the rest of the experimental groups at W3 (p WT naïve vs. original < 0.001; p WT naïve vs. Alzet = 0.006; p WT original vs. miniaturized = 0.038; p WT original vs. Alzet = 0.017; p APP naïve vs. original < 0.001; p APP naïve vs. miniaturized = 0.002; p APP original vs. Alzet = 0.006; Figure 7b). These evaluations reached values above 10 points in some animals, representing moderate to severe suffering and pain and indicating a humane endpoint and termination of the experiment at this timepoint. For animals implanted with the miniaturized device, only significant statistical differences were detected compared to the original device at W3 (p WT original vs. miniaturized = 0.038; p APP original vs. miniaturized = 0.004; Figure 7b), but significant differences were found at W8 compared to naïve mice ($p < 0.001$ for all comparisons, Figure 7b). In this case, the values were below three points, indicating the presence of some discomfort and the need for increased vigilance and compensatory measures, but without compromising animal welfare. Finally, for animals implanted with Alzet pumps, differences in welfare scores were detected at W3 compared to naïve in both genotypes (p WT naïve vs. Alzet = 0.006; p APP naïve vs. Alzet < 0.001; Figure 7b) and compared to the original device only in WT mice (p WT =0.017; Figure 7b). This experimental group shared similar welfare values to those obtained in mice implanted with the miniaturized device ($p > 0.05$ for all comparisons; Figure 7b), however, termination of the experiment was conducted at W4 due to the experimental design of this particular project, but not because of humane endpoint considerations.

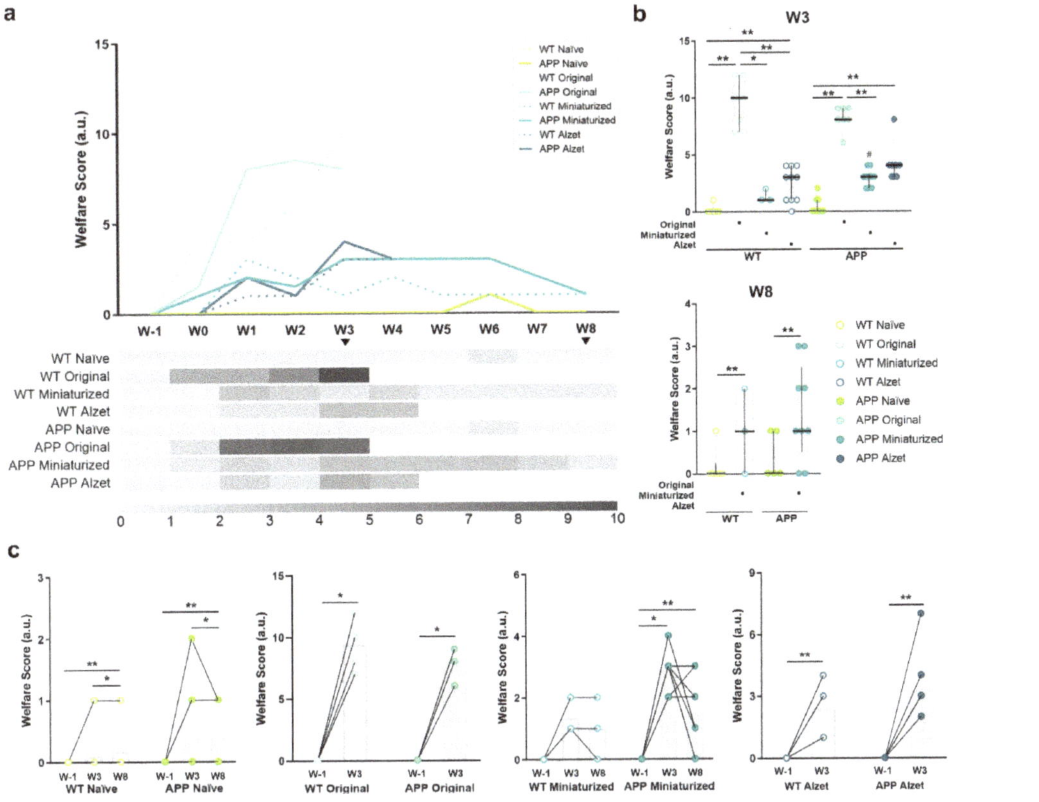

Figure 7. Effect of changes in the dimensions of implanted device and the refined stereotaxic protocol on animal welfare throughout the experiment. (**a**) Representation (top) and heatmap (bottom) of the median welfare assessment scores throughout the duration of the experiment for each experimental group. The higher the scores, the more compromised the welfare of animals. It is important to point out the elevated scores obtained in both WT and APP mice implanted with the original device and following the traditional protocol at W2 and W3, which resulted in the application of humane endpoint and termination of the experiment for this experimental group at this timepoint. No statistical test has been performed for these data over time due to the incompatibility with the wide variety of experiment termination. (**b**) Quantification of animal welfare at timepoints shared by most experimental groups, i.e., W3 and W8. (**c**) Pre- and post-representations of animal welfare scores at W1 vs. W3 vs. W8 or W1 vs. W3, for each experimental group individually. Specific sample size is provided in Table 1. Data expressed as median and IQR. Kruskal–Wallis test for multiple comparisons (W3 in (**b**)); Mann–Whitney U test for independent pairwise comparisons (WT vs. APP comparisons in b; W8 in (**b**)); Friedman test for multiple comparisons with repeated measures (naïve and miniaturized in (**c**)); and Wilcoxon test for dependent pairwise comparisons (original and Alzet in (**c**)). * $p < 0.05$; ** $p < 0.01$, for differences between experimental groups based on the implanted device and multiple comparisons; # $p < 0.05$; for differences between WT and APP genotypes in (**b**) (see Figure S2 for detailed analysis).

As described above, when the pre- and post-score values of welfare assessment were plotted and compared individually for each experimental group, an increase was detected in all the experimental groups at W3 and W8 compared to baseline, including both WT and APP naïve mice (p WT naïve = 0.003; p APP naïve = 0.010; p WT original = 0.017; p APP original = 0.016; p APP miniaturized = 0.001; p WT Alzet = 0.007; p APP Alzet = 0.006; Figure 7c). In the latter, this increase was mainly driven by variations in the body weight of the animals, indicating that these fluctuations are a normal situation even in healthy mice (Figure 7c). Finally, it is noteworthy that no differences in animal welfare were detected in WT animals implanted with the miniaturized device over time (p = 0.097; Figure 7c). In addition, a trend towards improved welfare was observed in some of their APP counterparts implanted with the miniaturized device over time, although no statistical differences were found between W3 and W8 (p = 0.157; Figure 7c).

3.4. *Optimized Intrathecal Implantation Procedures Do Not Appear to Negatively Affect General and Anxiety-like Behaviors for at Least Two Months after Surgery*

To minimize any additional stress on animals undergoing surgery that negatively influenced their well-being, behavioral testing was conducted exclusively in the experimental groups with the best welfare status (Figure 8a,b), i.e., naïve, animals implanted with the miniaturized device and animals implanted with Alzet pumps, at W1, W4/5, and W8 (depending on the experimental design of each group), as previously described (see corresponding Material and Methods section). The elevated zero maze was used to analyze general and anxiety-like behavior (Figure 8c), assessing the total distance traveled and the percentage of time spent in the open and closed areas of the maze [41].

Regarding total distance traveled (Figure 8d), no statistical differences were detected between genotypes (i.e., WT vs. APP) in most of the experimental groups ($p > 0.05$ for all comparisons; p Alzet = 0.029; Figure 8d). In addition, a longer distance traveled was only evidenced in APP mice implanted with the Alzet pump compared to those implanted with the miniaturized device or naïve ones at W1 and W4/5 (W1: p naïve vs. Alzet = 0.001, p miniaturized vs. Alzet = 0.004; W4/5: p naïve vs. Alzet = 0.006, p miniaturized vs. Alzet = 0.004; Figure 8d), whereas no differences were found in the rest of the comparisons at any of the timepoints analyzed ($p > 0.05$ for the remainder comparisons; Figure 8d). These results clearly indicate that optimized intrathecal implantation does not limit or reduce animal movement.

In terms of percentage of time spent in open and closed areas, most of the experimental groups had a strong preference for closed areas within maze corridors except WT mice implanted with the miniaturized device (p WT naïve W1 = 0.008, p WT naïve W4/5 = 0.015, p WT naïve W8 = 0.028, p APP naïve W1 = 0.013, p APP naïve W2 = 0.005, p APP naïve W8 = 0.042, p WT miniaturized W1, W4/5 and W8 = 0.109, p APP miniaturized W1 = 0.005, p APP miniaturized W4/5 = 0.008, p APP miniaturized W8 = 0.011, p WT Alzet W1 = 0.005, p WT Alzet W4/5 = 0.005, p APP Alzet W1 = 0.005 and p APP Alzet W4/5 = 0.005; Figure 8e and Figure S3). This result aligns with normal unprotected and elevated space avoidance behavior reported in mice [41]. Regarding the analysis of novel environment exploration, inferred as the percentage of time spent in open areas (Figure 8d), differences were only found between APP mice implanted with Alzet pumps compared to their counterparts implanted with the original device or the naïve ones at W1 (p naïve vs. Alzet = 0.007; p miniaturized vs. Alzet= 0.005; Figure 8d), with the percentage being always lower in the Alzet-implanted mice (Figure 8d). In addition, WT animals implanted with the miniaturized device also showed a lower percentage of time spent in open areas at W8 compared with WT naïve (p = 0.024; Figure 8d). These findings could indicate a possible unfavorable effect of the intrathecal implantation surgery on the particular behavior of novel environment exploration, although further investigation is needed (Figure 8d). Nevertheless, the overall results demonstrated that the refined intrathecal implantation procedures do not adversely influence the neurobehavioral functions analyzed in the implanted animals, even after two months of implantation.

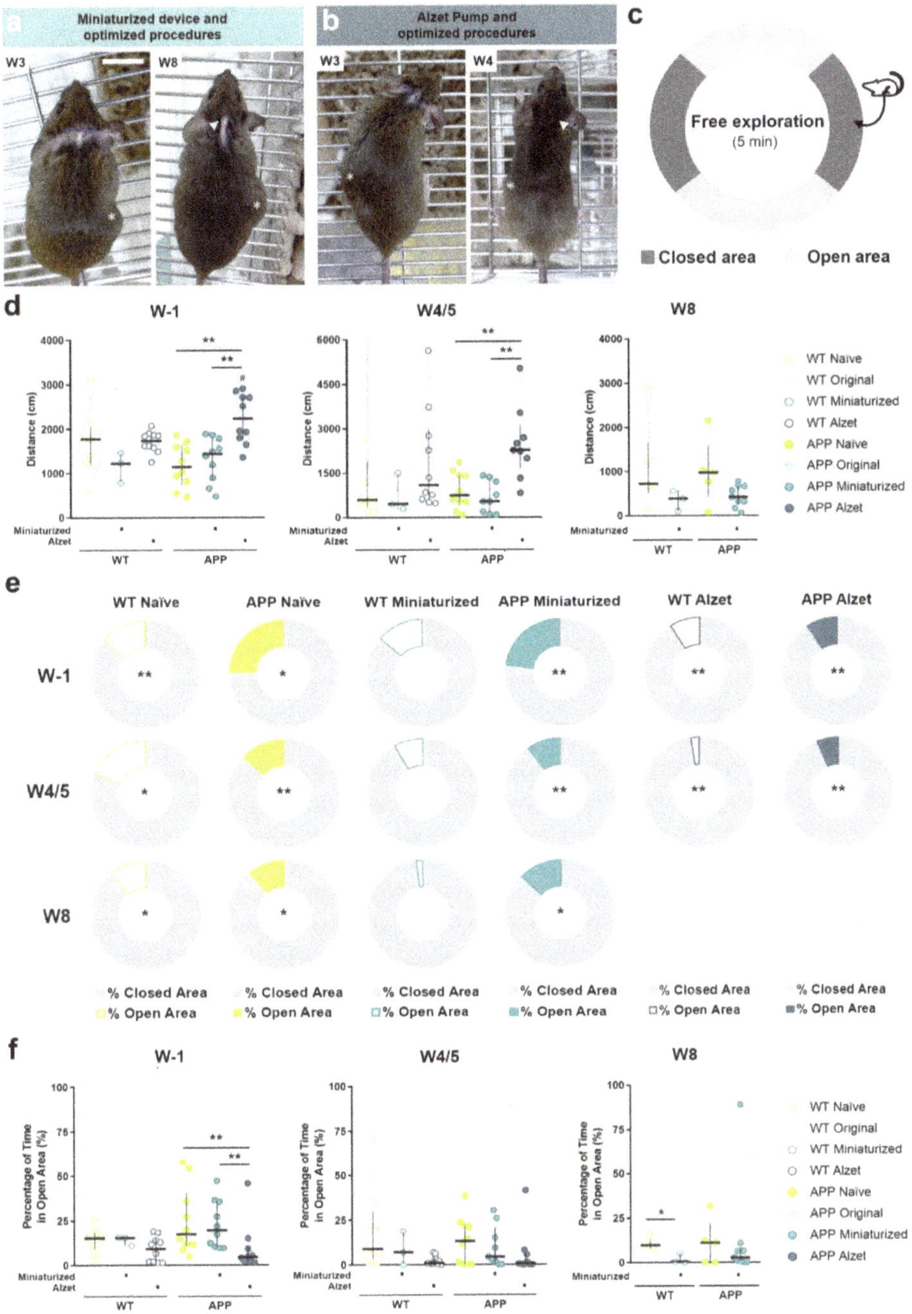

Figure 8. Effect of implantation of smaller devices following refined intrathecal implantation procedures on general and anxiety-like behaviors. (**a**,**b**) Dorsal views of two different mice implanted with the miniaturized device at W3 and W8 (**a**), and with the Alzet pump at W3 and W4 (**b**), following in both cases the optimized surgery protocol. Note the good general appearance of the animals and of

the healed wound, especially at the end of the experiment where it is almost unnoticeable (arrowheaded). The asterisks point out the subcutaneous pocket where device is implanted. (**c**) Schematic representation of the elevated zero maze apparatus. (**d**) Quantification and comparisons of the total distance traveled in the maze at W1, W4/5 and W8. (**e**) Quantification and ring plot of the median percentage of time spent in closed vs. open areas individually for each experimental group and timepoint (see Figure S3 for detailed analysis). (**f**) Quantification and comparisons of the percentage of time spent in open area among the different experimental groups at W1, W4/5 and W8. Specific sample size is provided in Table 1. Data expressed as median and IQR. Kruskal–Wallis test for multiple comparisons (W1, W4/5 in (**d**) and (**f**)); Mann–Whitney U test or Wilcoxon test for independent (W8 in d and W8 in (**f**)) and dependent (% of open vs. % of closed in (**e**)) pairwise comparisons; respectively. * $p < 0.05$; ** $p < 0.01$, for differences between experimental groups based on the implanted device and multiple comparisons; # $p < 0.05$, for differences between WT and APP genotypes in (**d**).

3.5. Validation of Cannula Placement Is Required after Experiment Termination to Confirm Correct Implantation at the Stereotaxic Coordinates of Interest

To verify the correct placement of the intrathecal cannula at the specific AP, ML, and DV stereotaxic coordinates, two different approaches were performed. First, in vivo infusion and visualization of blue dextran dye in the mouse cisterna magna (Figure 9a,b) demonstrated both the correct positioning of the intracerebroventricular cannula in the mouse right ventricle and the flow of the dye solution through the device system and cannula into the ventricular system (Figure 9). In addition, post mortem dark blue visualization of blue dextran within the entire anatomy of the mouse ventricular system also verified the correct intrathecal implantation and infusion of the cannula (Figure 9c–j). Finally, histological analysis with Toluidine blue staining in coronal brain sections also confirmed the correct placement of the intracerebroventricular cannula in the stereotaxic coordinates of interest, compared to and with reference to the most widely employed mouse brain atlases: Paxinos and Franklin [37] and Allen Brain Mouse Atlas (Figure 10).

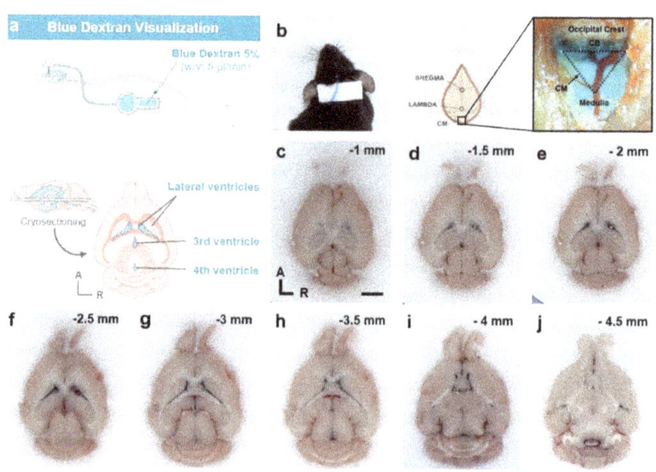

Figure 9. Verification of intracerebroventricular cannula implantation by infusion and visualization of blue dextran. (**a**) Experimental design of blue dextran (5%, v/v in aCSF) study. (**b**) Dorsal view of in vivo visualization of blue dextran in the connecting tube between the cannula and the device (left) and in the mouse cisterna magna (right). (**c–j**) Transversal cryosections (spaced 0.5 mm apart) of mouse brain infused with blue dextran. Note that blue color of dextran is visualized throughout the anatomy of the mouse ventricular system, indicating correct placement and flow of the cannula. Scale bar (c–j): 2 mm. A, anterior; CB, cerebellum; CM, cisterna magna; R, right.

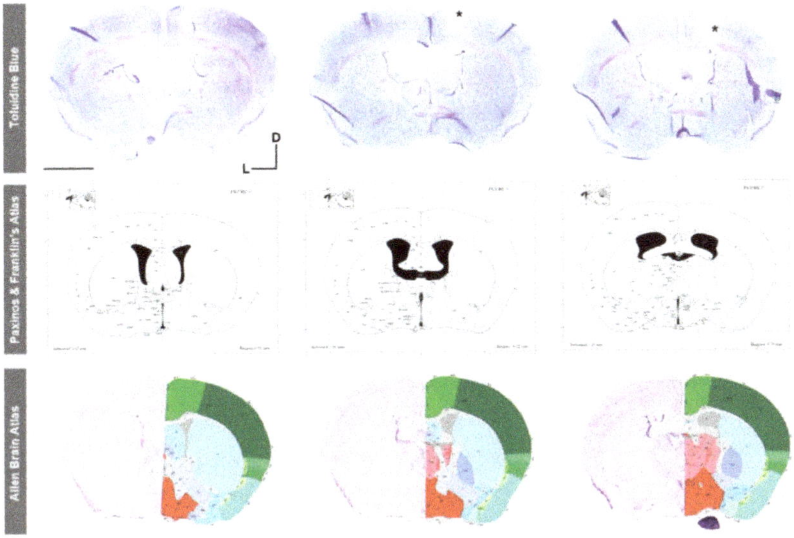

Figure 10. Histological validation of the specific stereotaxic coordinates (AP = −0.70 mm; ML = −1.26 mm; and DV = −2.5 mm) used for intracerebroventricular cannula implantation with Toluidine blue staining and reference to mouse brain atlases. Images of middle and bottom rows obtained from [37] and http://mouse.brain-map.org/ (accessed on 1 July 2023), respectively. Scale bar: 2 mm. D, dorsal; L, left; *, cannula placement.

4. Discussion

Over the past decades, significant advances have been made in stereotaxic neurosurgery in rodents, driven by the need to adhere to higher ethical standards. This progress requires not only a high level of knowledge and technical skills, but also deep understanding and awareness of ethics, animal welfare, and the importance of minimizing the number of animals used to obtain robust conclusions, as outlined in the 3Rs principles of "refinement" and "reduction" [13].

While certain general recommendations can be universally applied to all stereotaxic surgeries, such as pain management during and after surgery, correct determination of stereotaxic coordinates, and using appropriate aseptic techniques [1], other considerations must be tailored on a case-by-case basis. In the context of our study, which involves the long-term implantation of intracerebroventricular or intrathecal cannula in small rodents, two critical steps are pivotal for successful experimentation. The first step involves securely fixing and stabilizing the cannula on the surface of skull. Previous existing protocols described several methods employing dental cement, screws, and cyanoacrylate (Refs. [1,9,10], among others). However, based on our experience and the literature, these fixation methods have proven to be less satisfactory due to the mismatch between the relatively large and flat surface of the cannula and the round-shaped skull surface of mice [10]. To address this issue, we proposed a refined stereotaxic surgery protocol that uses a UV light-curing resin that can be molded and individually adapted to each mouse in combination with cyanoacrylate adhesive tissue. The use of this resin offered several advantages, including shorter surgical intervention times due to the rapid curing time, lower incidence of surgical problems such as skin necrosis or infection (reported with the use of dental cement), and stable and secure attachment of the cannula to the skull for prolonged periods. Importantly, these improvements contribute to minimizing humane endpoint applications to very low levels, in line with the "reduction" principle of the 3Rs [13]. Additionally, our findings suggest that modifying the dimensions of the implantable devices may also be contributing to improve surgical outcomes. Therefore, careful consideration of device dimensions is crucial before designing and performing specific surgeries, especially in mice, where size constraints may arise. Importantly, a limitation of our study is that we cannot attribute the observed improvements in long-term implantation exclusively to the refined stereotaxic protocol, since both modifications (surgical refinement and changes in device dimensions) were implemented simultaneously. However, in view of the fact that with the optimized protocol we successfully overcame the most significant and critical surgical complication reported [2], which involved cannula detachment triggering the application of humane endpoint, we strongly believe that protocol refinement is the key factor mostly influencing the observed improvements. Nevertheless, additional research would clarify the level of contribution of each modification.

The second crucial step for the success of long-term intrathecal implantation studies involves careful and thorough assessment and recording of animal welfare during the immediate and long-term postoperative periods. To our knowledge, no studies have proposed or described a specific protocol to accurately monitor the animal welfare of mice undergoing such stereotaxic surgeries. Previous works have relied on individual parameters such as changes in body weight, physical appearance, or behavior related to the specific brain structure in which the cannula was implanted (Refs. [39,40], among others). In our study, we designed a standardized scoresheet for effective and accurate follow-up of the animal's overall well-being, covering a wide range of indicators related to general condition, physical appearance and posture, nutritional and hydration status, spontaneous behavior, and the surgical procedure itself [15–23]. Our results corroborate the fact that relying on a single parameter is insufficient in most cases. Conversely, the use of diverse indicators provides a holistic view of the animal's actual well-being [16,17,43]. Additionally, close monitoring has enabled us to promptly apply compensatory measures whenever necessary, thereby alleviating any discomfort or pain experienced by the animals in a timely manner and allowing studies to be extended for a longer period of time.

Although the proportion of drop-out animals due to the application of humane endpoint has reached a remarkably low level (below 5% of cases) in our studies, and we have significantly enhanced the welfare of animals undergoing long-term intrathecal implantation, we remain committed to refining our procedures further with the aim of improving animal welfare and increasing reproducibility. It is important to emphasize that maximizing animal welfare in research is crucial not only from ethical and legal perspectives, but also because it positively impacts the quality of scientific outcomes derived from these animals, thus benefiting the scientific community.

5. Conclusions

In summary, our study provides a refined protocol for safe intracerebroventricular cannula fixation over extended experimental periods, as well as a customized protocol to accurately monitor animal welfare in mice undergoing long-term intrathecal device implantation, among other notable findings. The inclusion of these refined steps helps to improve the safety and reproducibility of long-term experiments involving intracerebroventricular cannula fixation, providing a valuable contribution to the neuroscience community and research.

Supplementary Materials: The supporting information can be downloaded at: https://www.mdpi.com/article/10.3390/ani13162627/s1. Table S1: Overview of complications observed in the animals employed in our study.; Figure S1: WT versus APP comparisons for body weight parameter at W-1, W3 and W8.; Figure S2: WT versus APP comparisons for welfare assessment score parameter at W-1, W3 and W8; Figure S3: Detail of the percentage spent in open versus closed area in Elevated Zero Maze at W-1, W3 and W8; Supplementary Information S1: Step-by-step protocol for stereotaxic surgery procedure; Supplementary Information S2: Postoperative regimen of analgesic, anti-inflammatory and antibiotic treatment; Supplementary Information S3: Procedure and Postoperative monitoring form.

Author Contributions: Conceptualization, E.P.-M., M.M.-G. and C.T.-Z. manufacturing the prototypes of medical devices named as "original" and "miniaturized" in this work: C.P., G.Á. and M.R.-C.; device dimension adjustments, M.R.-C.; surgeries, E.P.-M. and M.A.Á.-V. surgery mouse monitoring, A.C.-V., J.C.-S. and B.F.-G. mouse welfare supervision and behavioral testing, E.P.-M., A.C.-V. and J.C.-S.; dextran study, E.P.-M. and A.C.-V.; histological analysis, E.P.-M.; data analysis, E.P.-M. and C.T.-Z.; original draft and figure preparation, E.P.-M.; review and editing, E.P.-M., M.M.-G. and C.T.-Z. All authors revised the manuscript critically. All authors have read and agreed to the published version of the manuscript.

Funding: This work was supported by the Fundación para el Fomento en Asturias de la Investigación Científica Aplicada y la Tecnología (FICYT; Plan de Ciencia, Tecnología e Innovación 2018-2022 del Principado de Asturias), cofunded by the Fondo Europeo de Desarrollo Regional (FEDER), under grant AYUD/2021/57540 to C.T.-Z.; the Centro para el Desarrollo Tecnológico Industrial (CDTI), Ministerio de Ciencia e Innovación of the Spanish Government, NextGenerationEU European Union funding, Plan de Recuperación, Transformación y Resiliencia, under grant SNEO-20211309 to C.P.; and the Fundación Alimerka (Convocatoria de ayudas a la investigación en Ciencias de la Salud I edición, 2022) to M.M.-G.

Institutional Review Board Statement: Animal studies included in this work were approved by the Research Ethics Committee of the University of Oviedo (PROAE IDs 32/2020; 05/2022 and 06/2022) in compliance with European (Directive 2010/63/UE Reference) and Spanish (RD118/2021, Law 32/2007 Reference) legislation.

Informed Consent Statement: Not applicable.

Data Availability Statement: The data that support the findings of this study are available from the corresponding authors upon reasonable request.

Acknowledgments: The authors would like to thank José MP Freije and Olaya Santiago-Fernández for giving us access to their colony of APP/PS1 mice at the University of Oviedo. We would also like to thank the technical support provided by the Scientific and Technological Resources of the University of Oviedo, especially the veterinarians Teresa Sánchez Álvarez and Agustín Brea Pastor, as well as María del Mar Rodríguez Santamaría (ISPA); and to Claudia Roces Llorián (Neuroscience Innovative Technologies S.L.) for technical support in the manufacture of devices. A.C.-V. is supported by a contract associated with grant AYUD/2021/57540 from FICYT. J.C.-S is supported by a contract associated with grant AC20/00017 from Instituto de Salud Carlos III.

Conflicts of Interest: M.M.-G. is the inventor of the prototypes designated as "original" and "miniaturized", which are protected by the International Patent No. WO2019/158791A1 (application number: PCT/ES2019/070038), titled "Device for the selective removal of molecules from tissues or fluids". Patent proprietor is Fundación de Neurociencias.

References

1. Ferry, B.; Gervasoni, D. Improving Stereotaxic Neurosurgery Techniques and Procedures Greatly Reduces the Number of Rats Used per Experimental Group—A Practice Report. *Animals* **2021**, *11*, 2662. [CrossRef] [PubMed]
2. Coto-Vilcapoma, M.A.; Castilla-Silgado, J.; Fernández-García, B.; Pinto-Hernández, P.; Cipriani, R.; Capetillo-Zarate, E.; Menéndez-González, M.; Álvarez-Vega, M.; Tomás-Zapico, C. New, Fully Implantable Device for Selective Clearance of CSF-Target Molecules: Proof of Concept in a Murine Model of Alzheimer's Disease. *Int. J. Mol. Sci.* **2022**, *23*, 9256. [CrossRef] [PubMed]
3. Chen, S.Y.; Kuo, H.Y.; Liu, F.C. Stereotaxic Surgery for Genetic Manipulation in Striatal Cells of Neonatal Mouse Brains. *J. Vis. Exp.* **2018**, *137*, e57270. [CrossRef] [PubMed]
4. DeVos, S.L.; Miller, T.M. Direct Intraventricular Delivery of Drugs to the Rodent Central Nervous System. *J. Vis. Exp.* **2013**, *75*, e50326. [CrossRef] [PubMed]
5. Geiger, B.M.; Frank, L.E.; Caldera-Siu, A.D.; Pothos, E.N. Survivable Stereotaxic Surgery in Rodents. *J. Vis. Exp.* **2008**, *20*, 880. [CrossRef] [PubMed]
6. Harvey, K.; Madsen, P.J.; Smith, T.; Griffin, C.; Patterson, L.; Vitanza, N.A.; Storm, P.B.; Resnick, A.C.; Foster, J.B. Intracranial Cannula Implantation for Serial Locoregional Chimeric Antigen Receptor (CAR) T Cell Infusions in Mice. *J. Vis. Exp.* **2023**, *192*, e64886. [CrossRef] [PubMed]
7. Poole, E.I.; McGavin, J.J.; Cochkanoff, N.L.; Crosby, K.M. Stereotaxic Surgery for Implantation of Guide Cannulas for Microinjection into the Dorsomedial Hypothalamus in Young Rats. *MethodsX* **2019**, *6*, 1652–1659. [CrossRef] [PubMed]

8. Fornari, R.V.; Wichmann, R.; Atsak, P.; Atucha, E.; Barsegyan, A.; Beldjoud, H.; Messanvi, F.; Thuring, C.M.A.; Roozendaal, B. Rodent Stereotaxic Surgery and Animal Welfare Outcome Improvements for Behavioral Neuroscience. *J. Vis. Exp.* **2012**, *59*, e3528. [CrossRef] [PubMed]
9. Criado, A.; Barford, J.; Parker, F.; Bate, S.; Whelan, G. Use of Cyanoacrylate Gel as a Substitute for Dental Cement in Intracerebroventricular Cannulations in Rats. *Contemp. Top. Lab. Anim. Sci.* **2003**, *42*, 13–16.
10. Sike, Á.; Wengenroth, J.; Upīte, J.; Brüning, T.; Eiriz, I.; Sántha, P.; Biverstål, H.; Jansone, B.; Haugen, H.J.; Krohn, M.; et al. Improved Method for Cannula Fixation for Long-Term Intracerebral Brain Infusion. *J. Neurosci. Methods* **2017**, *290*, 145–150. [CrossRef] [PubMed]
11. Leggat, P.A.; Smith, D.R.; Kedjarune, U. Surgical Applications of Methyl Methacrylate: A Review of Toxicity. *Arch. Environ. Occup. Health* **2009**, *64*, 207–212. [CrossRef] [PubMed]
12. Díez-Solinska, A.; Vegas, O.; Azkona, G. Refinement in the European Union: A Systematic Review. *Animals* **2022**, *12*, 3263. [CrossRef]
13. Russel, W.M.S.; Burch, R.L. *The Principles of Humane Experimental Technique*; Methuen & Co., Ltd.: London, UK, 1959.
14. Arman, A.; Hutchinson, M.R. Intrathecal Implantation Surgical Considerations in Rodents; a review. *J. Neurosci. Methods* **2021**, *363*, 109327. [CrossRef] [PubMed]
15. Aulehner, K.; Leenaars, C.; Buchecker, V.; Stirling, H.; Schönhoff, K.; King, H.; Häger, C.; Koska, I.; Jirkof, P.; Bleich, A.; et al. Grimace Scale, Burrowing, and Nest Building for the Assessment of Post-Surgical Pain in Mice and Rats—A Systematic Review. *Front. Vet. Sci.* **2022**, *9*, 930005. [CrossRef] [PubMed]
16. Giménez-Llort, L.; Torres-Lista, V. Social Nesting, Animal Welfare, and Disease Monitoring. *Animals* **2021**, *11*, 1079. [CrossRef] [PubMed]
17. Hawkins, P.; Morton, D.B.; Burman, O.; Dennison, N.; Honess, P.; Jennings, M.; Lane, S.; Middleton, V.; Roughan, J.V.; Wells, S.; et al. A Guide to Defining and Implementing Protocols for the Welfare Assessment of Laboratory Animals: Eleventh Report of the BVAAWF/FRAME/RSPCA/UFAW Joint Working Group on Refinement. *Lab. Anim.* **2011**, *45*, 1–13. [CrossRef] [PubMed]
18. Van der Meer, M.; Rolls, A.; Baumans, V.; Olivier, B.; van Zutphen, L.F.M. Use of Score Sheets for Welfare Assessment of Transgenic Mice. *Lab. Anim.* **2001**, *35*, 379–389. [CrossRef]
19. Tappe-Theodor, A.; Pitzer, C.; Lewejohann, L.; Jirkof, P.; Siegeler, K.; Segelcke, A.; Drude, N.; Pradier, B.; Pogatzki-Zahn, E.; Hollinderbäumer, B.; et al. The "WWHow" Concept for Prospective Categorization of Post-Operative Severity Assessment in Mice and Rats. *Front. Vet. Sci.* **2022**, *9*, 841431. [CrossRef]
20. Burkholder, T.; Foltz, C.; Karlsson, E.; Linton, C.G.; Smith, J.M. Health Evaluation of Experimental Laboratory Mice. *Curr. Protoc. Mouse Biol.* **2012**, *2*, 145–165. [CrossRef]
21. Jirkof, P.; Rudeck, J.; Lewejohann, L. Assessing Affective State in Laboratory Rodents to Promote Animal Welfare—What Is the Progress in Applied Refinement Research? *Animals* **2019**, *9*, 1026. [CrossRef]
22. FELASA Working Group; Mähler, M.; Berard, M.; Feinstein, R.; Gallagher, A.; Illgen-Wilcke, B.; Pritchett-Corning, K.; Raspa, M. FELASA Recommendations for the Health Monitoring of Mouse, Rat, Hamster, Guinea Pig and Rabbit Colonies in Breeding and Experimental Units. *Lab. Anim.* **2014**, *48*, 178–192. [CrossRef]
23. National Research Council (US) Committee on Recognition and Alleviation of Pain in Laboratory Animals. *Recognition and Alleviation of Pain in Laboratory Animals*; National Academies Press (US): Washington, DC, USA, 2009. Available online: https://www.ncbi.nlm.nih.gov/books/NBK32658/ (accessed on 27 June 2023). [CrossRef] [PubMed]
24. Jankowsky, J.L.; Fadale, D.J.; Anderson, J.; Xu, G.M.; Gonzales, V.; Jenkins, N.A.; Copeland, N.G.; Lee, M.K.; Younkin, L.H.; Wagner, S.L.; et al. Mutant Presenilins Specifically Elevate the Levels of the 42 Residue β-Amyloid Peptide in Vivo: Evidence for Augmentation of a 42-Specific γ Secretase. *Hum. Mol. Genet.* **2004**, *13*, 159–170. [CrossRef]
25. Kilkenny, C.; Browne, W.J.; Cuthill, I.C.; Emerson, M.; Altman, D.G. Improving Bioscience Research Reporting: The Arrive Guidelines for Reporting Animal Research. *PLoS Biol.* **2010**, *8*, e1000412. [CrossRef]
26. Schreiner, T.G.; Menéndez-González, M.; Popescu, B.O. The "Cerebrospinal Fluid Sink Therapeutic Strategy" in Alzheimer's Disease—From Theory to Design of Applied Systems. *Biomedicines* **2022**, *10*, 1509. [CrossRef]
27. Menéndez-González, M.; Tamba, B.-I.; Leclere, M.; Mabrouk, M.; Schreiner, T.-G.; Ciobanu, R.; Cristina, T.-Z. Intrathecal Pseudodelivery of Drugs in the Therapy of Neurodegenerative Diseases: Rationale, Basis and Potential Applications. *Pharmaceutics* **2023**, *15*, 768. [CrossRef]
28. Menéndez-González, M. Mechanical Filtration of the Cerebrospinal Fluid: Procedures, Systems, and Applications. *Expert Rev. Med. Devices* **2023**, *20*, 199–207. [CrossRef]
29. ISO Standard. Available online: https://www.iso.org/iso-13485-medical-devices.html (accessed on 6 August 2023).
30. ISO Standard. Available online: https://www.iso.org/standard/68936.html (accessed on 6 August 2023).
31. Gurfein, B.T.; Stamm, A.W.; Bacchetti, P.; Dallman, M.F.; Nadkarni, N.A.; Milush, J.M.; Touma, C.; Palme, R.; di Borgo, C.P.; Fromentin, G.; et al. The Calm Mouse: An Animal Model of Stress Reduction. *Mol. Med.* **2012**, *18*, 606–617. [CrossRef]
32. Lee, G.H.; Kim, K.S.; Jo, W. Stress Evaluation of Mouse Husbandry Environments for Improving Laboratory Animal Welfare. *Animals* **2023**, *13*, 249. [CrossRef] [PubMed]
33. Rufiange, M.; Leung, V.S.; Simpson, K.; Pang, D.S. Prewarming Followed by Active Warming Is Superior to Passive Warming in Preventing Hypothermia for Short Procedures in Adult Rats (Rattus Norvegicus) under Isoflurane Anesthesia. *J. Am. Assoc. Lab. Anim. Sci.* **2020**, *59*, 377–383. [CrossRef]

34. Bailey, K.T.; Jantre, S.R.; Lawrence, F.R.; Hankenson, F.C.; Del Valle, J.M. Evaluation of Active Warming and Surgical Draping for Perioperative Thermal Support in Laboratory Mice. *J. Am. Assoc. Lab. Anim. Sci.* **2022**, *61*, 482–494. [CrossRef]
35. Schuster, C.J.; Pang, D.S.J. Forced-Air Pre-Warming Prevents Peri-Anaesthetic Hypothermia and Shortens Recovery in Adult Rats. *Lab. Anim.* **2018**, *52*, 142–151. [CrossRef] [PubMed]
36. Ferry, B.; Gervasoni, D.; Vogt, C. *Stereotaxic Neurosurgery in Laboratory Rodent*; Springer: Paris, France, 2014; ISBN 978-2-8178-0471-2. Available online: https://link.springer.com/book/10.1007/978-2-8178-0472-9 (accessed on 27 June 2023).
37. Paxinos, G.; Franklin, K.B.J. *The Mouse Brain in Stereotaxic Coordinates*; Academic Press: San Diego, CA, USA, 2001.
38. Foley, P.L.; Kendall, L.V.; Turner, P.V. Clinical Management of Pain in Rodents. *Comp. Med.* **2019**, *69*, 468–489. [CrossRef] [PubMed]
39. Ullman-Culleré, M.H.; Foltz, C.J. Body Condition Scoring: A Rapid and Accurate Method for Assessing Health Status in Mice. *Lab. Anim. Sci.* **1999**, *49*, 319–323. [PubMed]
40. Talbot, S.R.; Biernot, S.; Bleich, A.; van Dijk, R.M.; Ernst, L.; Häger, C.; Helgers, S.O.A.; Koegel, B.; Koska, I.; Kuhla, A.; et al. Defining Body-Weight Reduction as a Humane Endpoint: A Critical Appraisal. *Lab. Anim.* **2020**, *54*, 99–110. [CrossRef] [PubMed]
41. Tucker, L.B.; McCabe, J.T. Behavior of Male and Female C57Bl/6J Mice Is More Consistent with Repeated Trials in the Elevated Zero Maze than in the Elevated plus Maze. *Front. Behav. Neurosci.* **2017**, *11*, 13. [CrossRef] [PubMed]
42. Šakić, B. Cerebrospinal Fluid Collection in Laboratory Mice: Literature Review and Modified Cisternal Puncture Method. *J. Neurosci. Methods* **2019**, *311*, 402–407. [CrossRef]
43. Talbot, S.R.; Struve, B.; Wassermann, L.; Heider, M.; Weegh, N.; Knape, T.; Hofmann, M.C.J.; von Knethen, A.; Jirkof, P.; Häger, C.; et al. RELSA—A Multidimensional Procedure for the Comparative Assessment of Well-Being and the Quantitative Determination of Severity in Experimental Procedures. *Front. Vet. Sci.* **2022**, *9*, 937711. [CrossRef]

Disclaimer/Publisher's Note: The statements, opinions and data contained in all publications are solely those of the individual author(s) and contributor(s) and not of MDPI and/or the editor(s). MDPI and/or the editor(s) disclaim responsibility for any injury to people or property resulting from any ideas, methods, instructions or products referred to in the content.

animals

MDPI

Article

Studying and Analyzing Humane Endpoints in the Fructose-Fed and Streptozotocin-Injected Rat Model of Diabetes

Rita Silva-Reis [1,2], Ana I. Faustino-Rocha [2,3,4], Jéssica Silva [2], Abigaël Valada [2], Tiago Azevedo [2], Lara Anjos [2], Lio Gonçalves [5,6], Maria de Lurdes Pinto [7,8], Rita Ferreira [1], Artur M. S. Silva [1], Susana M. Cardoso [1] and Paula A. Oliveira [2,8,*]

1 LAQV-REQUIMTE, Department of Chemistry, University of Aveiro, 3810-193 Aveiro, Portugal
2 CITAB Inov4Agro—Institute for Innovation, Capacity Building and Sustainability of Agri-Food Production, University of Trás-Os-Montes and Alto Douro, 5000-801 Vila Real, Portugal
3 Department of Zootechnics, School of Sciences and Technology, University of Évora, 7000-812 Évora, Portugal
4 Comprehensive Health Research Center, 7000-812 Évora, Portugal
5 Engineering Department, University of Trás-Os-Montes and Alto Douro, 5000-801 Vila Real, Portugal
6 Institute for Systems and Computer Engineering, Technology and Science, 4200-465 Porto, Portugal
7 Animal and Veterinary Research Center (CECAV), University of Trás-Os-Montes and Alto Douro, 5000-801 Vila Real, Portugal
8 Department of Veterinary Sciences, University of Trás-Os-Montes and Alto Douro, 5000-801 Vila Real, Portugal
* Correspondence: pamo@utad.pt

Simple Summary: This study assessed a humane endpoint scoring system to detect animal suffering in a rat model of type 2 diabetes. Sprague-Dawley male rats were divided into control and induced (fructose-fed and streptozotocin (STZ) administration) groups. Induced animals drank 10% fructose for 14 days, then received STZ (40 mg/kg) intraperitoneally. Weekly monitoring of body weight, water, and food consumption, 14 parameters of animal welfare, and blood glucose levels were conducted. Results showed weight loss, polyuria, polyphagia, and polydipsia, as well as lack of grooming, narrowing of the orbital area, curved posture, liquid/pasty diarrhea, and abdominal distension. The most useful parameters to evaluate humane endpoints in this type 2 diabetes rat induction model were dehydration, absence of grooming, the posture of the animals, abdominal visualization and palpation, and fecal appearance. The glycemia was significantly higher in the induced group, validating the animal model of diabetes. The humane endpoints table was suitable for monitoring animal welfare.

Abstract: This work aimed to define a humane endpoint scoring system able to objectively identify signs of animal suffering in a rat model of type 2 diabetes. Sprague-Dawley male rats were divided into control and induced group. The induced animals drink a 10% fructose solution for 14 days. Then, received an administration of streptozotocin (40 mg/kg). Animals' body weight, water and food consumption were recorded weekly. To evaluate animal welfare, a score sheet with 14 parameters was employed. Blood glucose levels were measured at three time points. After seven weeks of initiating the protocol, the rats were euthanized. The induced animals showed weight loss, polyuria, polyphagia, and polydipsia. According to our humane endpoints table, changes in animal welfare became noticeable after the STZ administration. None of the animals hit the critical score limit (four). Data showed that the most effective parameters to assess welfare in this type 2 diabetes rat induction model were dehydration, grooming, posture, abdominal visualization, and stool appearance. The glycemia was significantly higher in the induced group when compared to the controls ($p < 0.01$). Induced animals' murinometric and nutritional parameters were significantly lower than the controls ($p < 0.01$). Our findings suggest that in this rat model of type 2 diabetes with STZ-induced following fructose consumption, our list of humane endpoints is suitable for monitoring the animals' welfare.

Citation: Silva-Reis, R.; Faustino-Rocha, A.I.; Silva, J.; Valada, A.; Azevedo, T.; Anjos, L.; Gonçalves, L.; Pinto, M.d.L.; Ferreira, R.; Silva, A.M.S.; et al. Studying and Analyzing Humane Endpoints in the Fructose-Fed and Streptozotocin-Injected Rat Model of Diabetes. *Animals* **2023**, *13*, 1397. https://doi.org/10.3390/ani13081397

Academic Editor: Garikoitz Azkona

Received: 21 February 2023
Revised: 10 April 2023
Accepted: 15 April 2023
Published: 18 April 2023

Copyright: © 2023 by the authors. Licensee MDPI, Basel, Switzerland. This article is an open access article distributed under the terms and conditions of the Creative Commons Attribution (CC BY) license (https://creativecommons.org/licenses/by/4.0/).

Keywords: rat model; welfare; hyperglycemia; polyuria; polyphagia; polydipsia

1. Introduction

Several animal models have been used to better understand the biopathology of chronic diseases, such as diabetes [1]. The use of laboratory animals raises many ethical concerns, particularly when stress and pain are induced in the animals. Humane endpoints have become a major element in experiments involving laboratory animals to overdue these ethical issues, applying the 3R's principles (Replacement, Reduction and Refinement) [2]. According to the Canadian Council on Animal Care (CCAC), humane endpoints are physiological or behavioral signs that define the critical stage at which the animal's pain and/or stress must be finished or mitigated through decisions such as no application of a painful procedure, treatment to alleviate pain and/or distress, or ultimately the animals' sacrifice [3]. The assessment of the animals' pain and stress can become subjective, and the researchers may incorrectly interpret certain signs as indicators of pain even when they are not. To reduce misinterpretations, scoring systems have been developed, thus allowing the assessment of animal welfare through the attribution of a score to each parameter. It is worth noting that the parameters examined in a certain study must be adjusted to the study's goal. Normally, at the end of the evaluation, the total score of each animal is recorded and determines whether an animal should or not be removed from the protocols [4–6]. The evolution of chromodacryorrhea [7], the animals' grooming habits (self-grooming) [8], the Rat Grimace scale (which allows the assessment of pain and stress in animals through an analysis of the animals' expressions) [9], body weight [10], and body temperature using infrared thermometers [11] are some of the parameters generally used to define humane endpoints in animal models of disease. However, a systematic approach published by David Morton [4] suggested additional parameters that should be included in the evaluation of humane endpoints in a rat model of type 2 diabetes, the most prevalent type of diabetes worldwide [12]. This study utilized a non-genetic animal model that mimics the pathogenesis of type 2 diabetes by administering STZ to animals that had been fed with fructose in their drinking water [13]. Various parameters were compiled to assess the welfare of animals throughout experimental trials, incorporating those proposed by David Morton [12] and additional ones, such as signs of dehydration, diarrhea, abdominal distension, and posture. Moreover, this work aimed to clearly define a humane endpoint scoring system able to objectively identify signs of animal suffering before the development of severe disease in this model of type 2 diabetes, mitigating all ethical and animal welfare concerns.

2. Materials and Methods

2.1. Ethical Statements

The experimental protocol was approved by the University of Trás-os-Montes and Alto Douro Ethics Committee ("ORBEA—Órgão Responsável pelo Bem-Estar e Ética Animal; approval no 852-e-CITAB-2020 and A_1-e-CITAB-2021). All the experiments performed on the animals were carried out under the European (European Directive 2010/63/EU) and national legislation (Decree-Law no. 113/2013) on the protection of animals used for scientific purposes.

2.2. Animals and Chemicals

Twenty-four Sprague-Dawley male rats (*Rattus norvegicus*) at three weeks of age and weighing 134.24 ± 10.11 g were obtained from Envigo (Barcelona, Spain). The animals were kept in polycarbonate cages with smooth surfaces and rounded edges (1500 U Eurostandard Type IV S, Tecniplast, Buguggiate, Italy), three to five animals per cage, with appropriate identification. Animals were negative for viruses, bacteria, mycoplasma, fungi, parasites, and pathological lesions. The bedding for the animals was prepared from corncob

(Ultragene, Santa Comba Dão, Portugal), and it was changed every week. Following the introduction of fructose and the administration of STZ, the animals urinate more frequently, demanding daily bedding changes. Polyvinyl chloride tubes were placed in each cage to promote the rat's environmental enrichment. With these materials, animals can hide and mimic a natural environment. This research was conducted in a controlled environment with a 12-h light/12-h dark cycle at a temperature of 20 ± 2 °C and relative humidity of 50 ± 10%. Fructose, acquired from BioPortugal-Químico, Farmacêutica, Lda (Porto, Portugal), was prepared daily in the drinking water at a concentration of 10% immediately before being provided to the animals to avoid fermenting and placed in opaque bottles. STZ was obtained from BioPortugal-Químico, Farmacêutica, Lda (Porto, Portugal) and prepared before its use in a 0.1 M citrate buffer solution (pH = 4.3).

2.3. Experimental Protocol

After one week of acclimatization, the rats were randomly assigned to two experimental groups: group control (n = 8) and group induced (n = 16). The experimental design was done without "nuisances", with no need for randomized block designs. After reviewing various published papers concerning animal models of diabetes [14–16], we identified a mortality rate between 20–40% and a rate of disease induction from 58% to 70%. So, to ensure the necessary sampling to have a statistical analysis power of the results, the number of animals differs between groups. This number intends to be conservative and not compromise the analysis of the results, considering Russell and Burch's 3R principles. Thus, the number of animals per group was calculated by the Experimental Design Assistant (https://efa.nc3rs.org.uk (accessed on 1 May 2021)). A higher number of animals was used in the group induced due to the higher mortality rate expected in the animals with diabetes when compared with those animals from group control [13–16]. At 4 weeks of age (beginning of week 1), the animals from the induced group began to drink a 10% fructose solution prepared as indicated above for 14 consecutive days (two weeks). After this, all animals from both groups were exposed to a 12 h overnight fast. Then, the animals from the group induced received a single intraperitoneal administration of STZ (40 mg/kg), while the animals from group control received an injection of the vehicle (0.1 M citrate buffer). Throughout the trial, the animals had unlimited access to tap water via capped bottles and a standard laboratory diet (SAFE® Custom Diets, Augy, France). Only the bottles with the 10% (*m/v*) fructose solution were available to the induced animals during fructose administration (Figure 1).

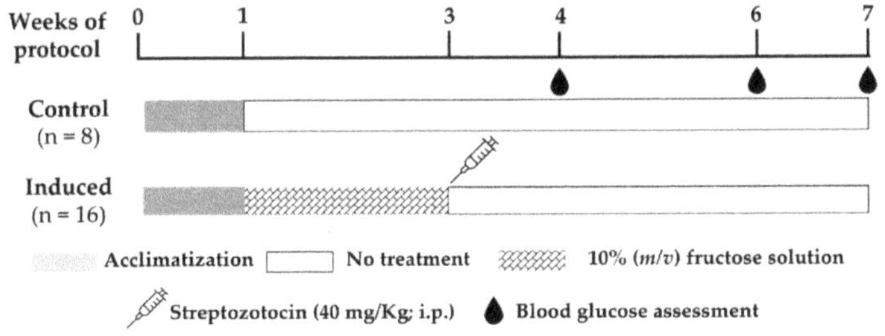

Figure 1. Experimental design.

2.4. Body Weight, Food, and Water Consumption

Weekly, individual rats' weight, as well as food and water consumption, were determined, using a top-loading scale (Mettler PM4000, LabWrench, Midland, ON, Canada).

The body weight gain (BWG) for each rat (%) was also determined using the following formula [17]:

$$BWG = \frac{\text{Body weight at a certain point} - \text{Initial body weight}}{\text{Body weight at a certain point}} \times 100, \quad (1)$$

The following formula was used to calculate the mean daily food consumption for each rat (g) [10]:

$$\text{Daily food consumption} = \frac{\text{Food weight (beggining of the week)} - \text{Food weight (end of the week)}}{\text{No. of animals in the cage} \times \text{No. of days}} \quad (2)$$

Initially, water consumption was determined similarly to food consumption. However, with the disease development, the animals from the group induced started to drink more water and, together with the weekly water losses, it became difficult to assess the real water consumption. In this way, the research team decided to calculate the water consumption for just 24 h, with three bottles available per cage. Thus, water consumption from week 4 onwards was calculated by dividing the total cage water consumption by the number of animals in the cage.

2.5. Scoring Sheet

To monitor the animals' welfare, the registration of humane endpoints was implemented during the experimental protocol, according to Table 1. The parameters were defined in accordance with CCAC guidelines [3] and updated based on earlier works performed by our team [10,17,18]. A score from zero to three was given for each parameter listed in the table. The animals were observed daily, and the scores were recorded once a week by three independent observers to avoid bias. The data were recorded independently and only cross-checked at the beginning and end of the experiment. The score for each animal was calculated by adding the scores assigned to each parameter. It was determined that if the animals' humane endpoints score reached the critical level (total score of four), they should be re-evaluated and, if necessary, removed from the study and euthanized. Moreover, a score of three in some parameters, like body weight and mental status, was an indicator of euthanasia. Animals were handled regularly before and during the trial to reduce stress during manipulation and avoid any interference with the animal welfare assessment.

When the animals did not clean themselves, a lack of grooming (i.e., the presence of dirt on the hair) was registered. According to Mason et al., chromodacryorrhea can be an indicator of stress and has been described in several studies with laboratory animals [8]. Harderian glands are secretory pigment lacrimal glands located posterior to the ocular globes. These glands release a reddish lipid- and porphyrin-rich material that lubricates the eyes and eyelids [19]. Healthy animals perform grooming frequently, so the deposition of this secretion around the eyes is reduced. However, in discomfort and disease, the animals reduce grooming, and this secretion accumulates around the rat's eyes and nose. Chromodacryorrhea changes the color of the hair and is also characterized by the presence of excessive reddish secretions [20]. The Grimace scale was used to assess the animals' pain based on their facial expressions [9]. The animals' reaction to external stimuli was evaluated by observing their response to hand clapping above their cages. To assess the animals' hydration status, a skin pinch test was performed. To make this test, the animal was placed on the researcher's forearm, and the pinch test was made on the back region, more precisely on the lumbar region. If the skin took more than about two seconds to normalize, the rat was considered dehydrated [10]. According to our score sheet, animals with a sum of scores that reach 4 should be sacrificed. However, animals with the following changes, including 20% of body weight loss, severe anemia, stupor and coma, skin necrosis and walking on the tip of the extremities, must be re-evaluated and, if necessary, should be sacrificed. When an animal approaches the critical score level, a veterinarian should conduct a comprehensive health assessment, and if the animal's condition has deteriorated, it would be necessary to euthanize it.

Table 1. Humane endpoints applied to the rat model of chemically induced type 2 diabetes. Adapted from Silva-Reis et al., Faustino-Rocha et al. and CCAC guidelines [3,10,17].

Score		0	1	2	3
Body Condition/Weight		Normal	Altered body condition/weight loss of <10%	Emaciated/weight loss of 10–20%	Weight loss of >20% (Euthanasia)
Posture		Normal	Curved	—	—
Hair/tail appearance and grooming		Normal	Lack of grooming	Very bad-looking hair, dirty tail, severe chromodacryorrhea	—
Grimace scale	Eyes and extremities	Normal	Narrowing of the orbital area	Moderate anemia/Cataracts, Eyes fully closed	Severe anemia/Corneal ulcers (Euthanasia)
	Position of ears and whiskers	Normal	Dropped ears, forward whiskers (Stiffened)	Floppy and curved ears, whiskers turned forward and crumpled (clusters)	—
	Nose/cheeks	Normal	Flattening and elongating the tip of the nose	Flattening of the cheeks (cheek appearance)	—
Walk		Normal	—	—	Walking on the tip of the extremities (Euthanasia)
Skin		Absence	Skin injuries/infections <5 mm	Skin injuries/infections >5 mm	Presence of necrosis (Euthanasia)
Mental status		Normal	Inactive	Moribund	Stupor/coma (Euthanasia)
Response to external stimuli		Normal	Moderate	Moderate with vocalization	Violent
Hydration status		Normal	Abnormal skin pinch test (>2 s)	—	—
Stool appearance		Solid	Pasty diarrhea	Liquid diarrhea	—
Convulsions		Absence	—	—	Presence
Visualization and response to mild abdominal palpation		Normal	Vocalization	Distended	—

Recommendation: A total score of 4 or a score of 3 in some parameters, like body weight, were indicators for euthanasia. When an animal's score exceeds a critical level, it was re-evaluated to decide if it should be removed from the protocol.

2.6. Body Temperature

Additionally, to the parameters mentioned above, the animals' body temperature was recorded weekly with a non-contact infrared forehead thermometer (Andon iHealth PT2L, Paris, France). For each animal, the temperature was measured on the forehead, always at the same location, at approximately 3 cm. All readings were performed by the same researcher when the animals stood relaxed in a black-lined cage. To avoid any temperature increase due to stress, all temperature readings were recorded in under one minute.

2.7. Blood Glucose Assessed during Protocol

One week after STZ or vehicle administration (3rd week of protocol, see Figure 1), blood glucose level was measured in all animals using a GlucoMen® Areo 2K and blood test strips (A. Menarini Diagnostics, Florence, Italy). The measurements were taken after 12 h of fasting and 2 h after feeding. Blood was taken from the tail by gently "milking" it after a small cut [21]. To promote hemostasis and to perform asepsis at the puncture site, we used a sterile gauze with 3% hydrogen peroxide [22]. A new blood glucose measurement was conducted two weeks later in all animals, after 12 h fasting and 2 h after feeding.

2.8. Animals' Euthanasia and Samples Processing

At the end of week 7, the animals were euthanized, after 12 h fasting, by intraperitoneal administration of ketamine (75 mg/kg, Rompun® 2%, Bayer Healthcare SA, Kiel, Germany) and xylazine (10 mg/kg, Clorketam 1000, Vetoquinol, Barcarena, Oeiras, Portugal), followed by exsanguination by cardiac puncture. Blood samples were collected directly from the heart and centrifuged at 3000 g/10 min for serum separation. Glucose, albumin, triglycerides and cholesterol levels were determined in an autoanalyzer (Prestige 24i, Cornay PZ, Lomianki, Poland). A complete necropsy was performed in all animals, and the internal organs were macroscopically observed, collected, weighed, and preserved in 10% buffered formalin for 24 h for histopathological analysis. A veterinary pathologist blindly examined 4-μm sections of paraffin-embedded kidneys, stained with hematoxylin and eosin (H&E) under a light microscope.

2.9. Murinometric and Nutritional Measurements

The animals' body weight was determined immediately before euthanasia (after 12 h fasting) to determine the Lee index and body weight index (BWI) [23]. After the anesthesia for euthanasia, the abdominal perimeter (AP), the thoracic perimeter (TC) and the nasal-anal length (NAL) of all rats were measured, and the AP/TC ratio was determined [24]. The following formulas were applied:

$$\text{Lee index} = \sqrt[3]{\frac{\text{Final body weight (g)}}{\text{NAL (cm)}}}, \tag{3}$$

$$\text{BMI} = \frac{\text{Final body weight (g)}}{\text{NAL}^2\ (\text{cm}^2)}, \tag{4}$$

Moreover, the specific rate of weight gain (SRWG) and food efficiency coefficient (FEC) were calculated using the weight of the animals before fasting at the end of the experiment:

$$\text{SRWG} = \frac{\text{Final body weight (g)} - \text{Initial body weight (g)}}{\text{Initial body weight (g)} \times \text{Days of the experiment}}, \tag{5}$$

$$\text{FEC} = \frac{\text{Final body weight (g)} - \text{Initial body weight (g)}}{\text{Animal's total food consumption (g)}}, \tag{6}$$

The food consumption of the animals was measured as the mean per cage, and the mean consumption per animal was considered.

2.10. Statistical Analysis

All data were analyzed using GraphPad Prism® software for Windows (version 8.0.1, San Diego, CA, USA). The mean and the standard deviation (SD) were calculated for each parameter. Body weight, food and water consumption, glucose levels during the protocol and temperature were compared using the Ordinary two-way ANOVA with Šidák correction for multiple comparisons. For post-mortem glucose, albumin, cholesterol, and triglycerides levels, kidneys' relative weight, and murinometric and nutritional parameters, the differences between groups were assessed using Student's t-distribution. Since the values in Table S1, Supplementary Material, do not follow a normal distribution, a non-parametric Mann-Whitney test was performed. *p*-values lower than 0.05 were considered statistically significant. Boxplots were performed to identify outliers. To avoid bias, data analysis was carried out by both researchers related and unrelated to those conducting the animal experiment.

3. Results

3.1. General Observations

No deaths were recorded during the experimental protocol. The animals' mean body weight (Figure 2a), after one week of introducing fructose in the animals' drinking water, was always statistically lower in the induced groups when compared to the controls (week 2, $p < 0.05$; weeks 3–7, $p < 0.0001$). When looking at the body weight gain, some differences were observed between control and induced animals. In fact, a decrease in weight gain over time was observed in both groups, justified by the natural growth of the animals. Throughout the experiment, the weight gain presented several variations in the induced group and was negative in week 3 of the trial (Figure 2b); that was the week after the STZ injection and in the last week. After the STZ injection (week 3), 81.25% of the animals in the induced group showed weight loss. At the end of the protocol, the percentage of animals that lost weight increased to 93.75%. During the consumption of fructose solution, over the first three weeks of the experiment, the animals from the induced group consumed less food when compared with animals from the control group ($p < 0.0001$). However, this was inverted after STZ administration, with the average food consumption in the induced animals higher during the remaining weeks of the experiment (Figure 2c, $p < 0.0001$). Concerning water consumption, induced animals drank more water over time than control animals, and the water intake also increased after STZ injection (Figure 2d, $p < 0.001$).

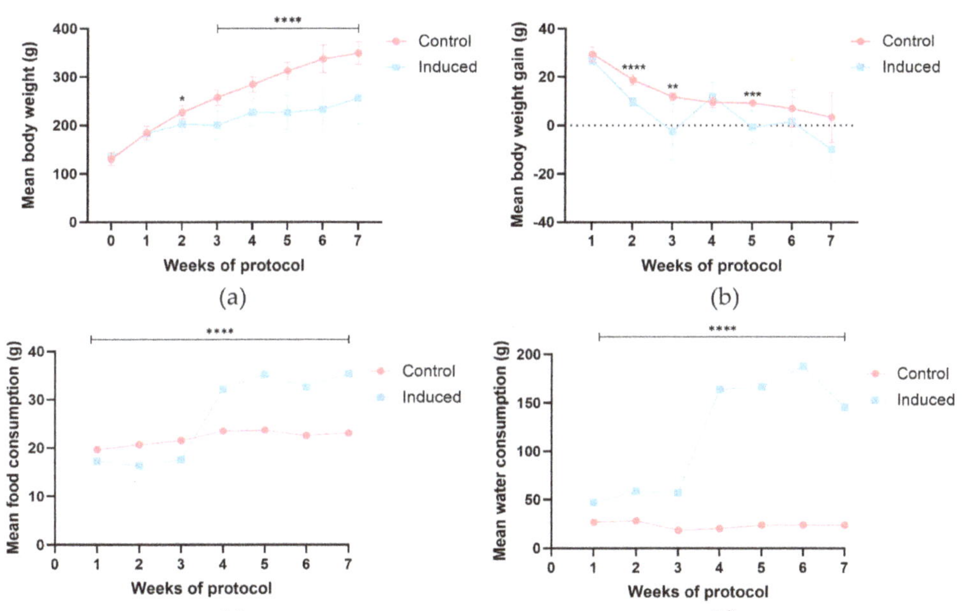

Figure 2. (**a**) Mean body weight of each group, (**b**) mean body weight gain, (**c**) mean food and (**d**) water consumption per day throughout the trial in both experimental groups. * Statistically different from control group ($p < 0.05$); ** $p < 0.01$; *** $p < 0.001$; **** $p < 0.0001$. The group control (n = 8) was intraperitoneally administrated with 0.1 M citrate buffer, while the induced group (n = 16) was intraperitoneally administrated with STZ diluted in 0.1 M citrate buffer after fructose feeding. Data are presented as mean ± standard deviation (SD).

3.2. Animal Welfare

Regarding the assessment of animals' health status by using the parameters stated in the table of humane endpoints, changes were only observed in the induced group (Figure 3).

The first changes were observed in the second week of the experimental protocol, when the animals were being supplemented with fructose solution, with six animals showing dehydration, detected by the skin pinch test as shown in Figure 4a. Changes in animal welfare became increasingly obvious after the STZ administration. Lack of grooming (weeks 3, 4, 6, and 7; Figure 4b), narrowing of the orbital area (week 3; Figure 4c), curved posture and removal of the remaining animals from the cage (weeks 5, 6 and 7; Figure 4d), liquid and pasty diarrhea (week 7; Figure 4e), and distension of the abdomen (Figure 4f) were observed. It is important to note that these changes were not consistent; for example, some animals displayed changes one week, recovered the following week, and the alterations reappeared later. The highest number of alterations in the animals was identified in the last week of the protocol, shortly before they were euthanized. Despite these alterations, none of the animals hit the critical score limit of 4, which implied its removal from the experiment. Nevertheless, the maximum score values assigned to each animal in the induced group were significantly higher when compared to controls ($p < 0.01$; Supplementary Material, Table S1).

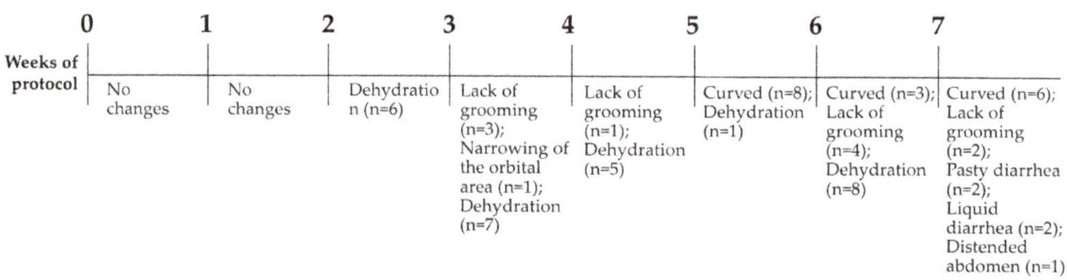

Figure 3. Changes observed in the animals' welfare over time according to previously established humane endpoints. Animals from the control group did not show any change.

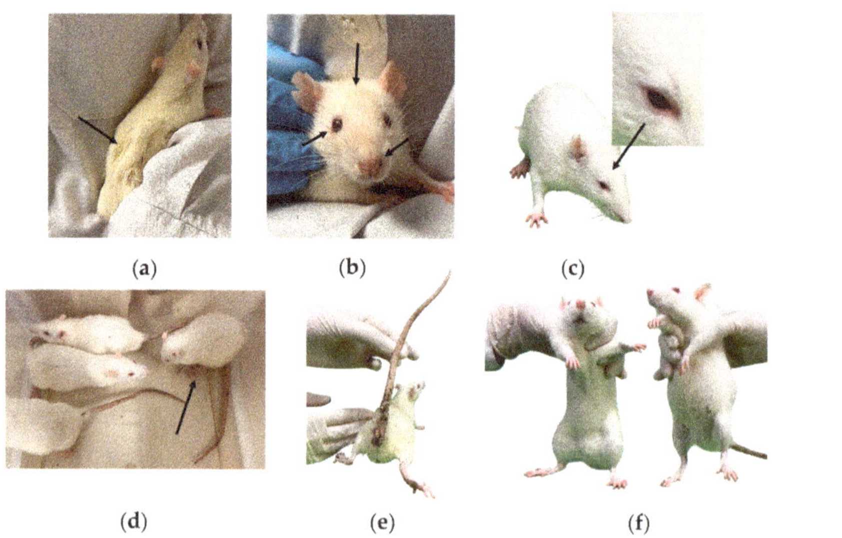

Figure 4. Welfare parameters changed in the induced animals throughout the experimental protocol. (**a**) Dehydration detected by the skin pinch test (black arrow); (**b**) lack of grooming (black arrow); (**c**) narrowing of the orbital area (evaluation of animal expression using the Grimace scale) (black arrow); (**d**) curved posture and isolation from the other animals in the cage (black arrow); (**e**) presence of pasty diarrhea; (**f**) difference between distended abdomen (on the right) and normal abdomen (on the left).

3.3. Body Temperature

No statistically significant variations in the animals' body temperature were reported during the trial ($p > 0.05$). The mean body temperature of animals from both groups ranged from 35.9 to 36.1 °C (Figure 5).

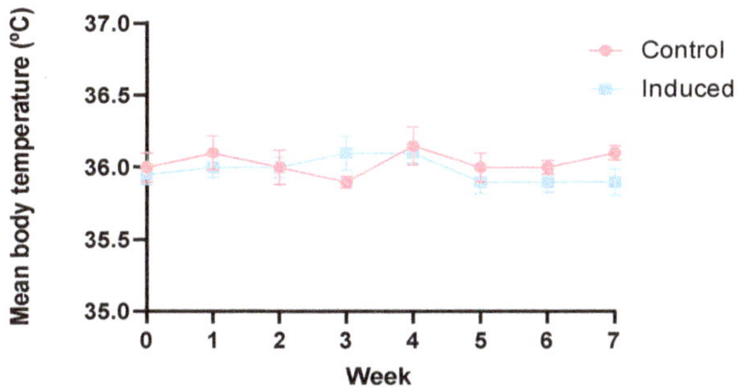

Figure 5. Mean body temperature (°C) recorded weekly per group throughout the protocol. No statistically significant differences were found between the groups ($p > 0.05$). The group control (n = 8) was intraperitoneally administrated with 0.1 M citrate buffer, while the induced group (n = 16) was intraperitoneally administrated with STZ diluted in 0.1 M citrate buffer after fructose feeding. Data are presented as mean ± SD.

3.4. Kidney Analysis

Animals from the induced group exhibited higher kidneys relative weight (animal kidney weight/animal body weight) when compared to controls, as it is shown in Table 2 ($p < 0.01$, right kidneys; $p < 0.001$, left kidneys).

Table 2. Mean relative weight of rats' kidneys after euthanasia.

Relative Weight (g/kg)	Control (n = 6)	Induced (n = 16)
Right kidney	4.053 ± 0.165	4.952 ± 0.676 [a]
Left kidney	3.861 ± 0.266	4.802 ± 0.593 [b]

[a] Statistically different from the control group ($p < 0.01$), [b] $p < 0.001$. Data are presented as mean ± SD. The group control was intraperitoneally administrated with 0.1 M citrate buffer, while the induced group was intraperitoneally administrated with STZ diluted in 0.1 M citrate buffer after fructose feeding.

To gain a better understanding of the increase in fructose-fed and STZ-injected rat kidneys' relative weights, a histological analysis of this organ was conducted. As illustrated in Figure 6a,b, the histology of the control animals' kidneys showed normal glomeruli and cortical tubules. However, the histological sections of the induced animals revealed focal cell necrosis, characterized by a darkly stained cytoplasm and sloughed necrotic cells in the lumen of the cortical tubules (Figure 6c,d).

3.5. Blood Parameters

When animals presented blood glucose levels higher than the maximum value read by the glucometer (600 mg/dL), the maximum value was attributed to them for the assessment of glycemia in each group. In the first evaluation, 56.25% of the rats exhibited feeding glucose readings higher than 600 mg/dL. This percentage increased to 81.25% in the second measurement (performed in week 6). Three animals from the group induced exhibited blood glucose levels <250 mg/dL 2 h after feeding, which were considered outliers by the statistical analysis performed.

In general, the blood glucose levels were higher after feeding, when compared with those measurements taken after fasting, in both experimental groups, reaching a statistically significant difference in the group induced at the 4th and 6th weeks of the experiment ($p < 0.0001$) (Table 3).

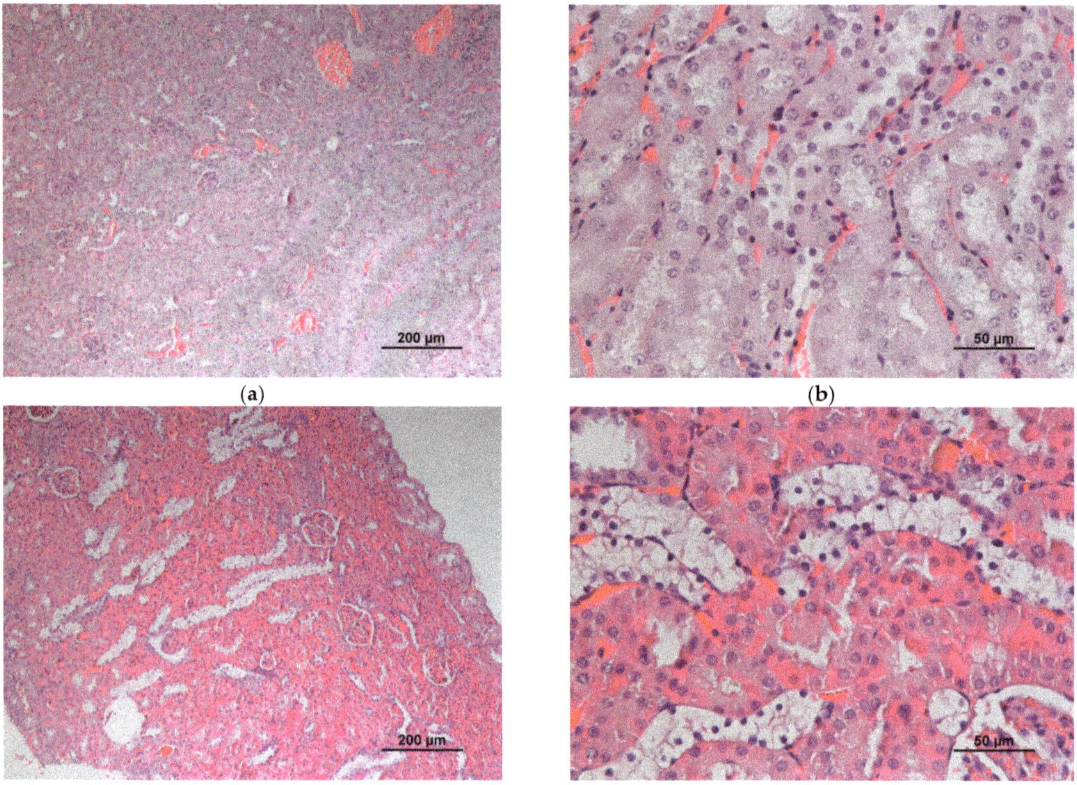

(**a**) (**b**) (**c**) (**d**)

Figure 6. Kidney sections from the animals under study. Images (**a**,**b**) display normal glomeruli and cortical tubules from an animal of the control group. Notice the vacuolation of the cytoplasm and pyknosis of the epithelial cells of cortical tubules. Focal cell necrosis, marked by a darkly stained cytoplasm and sloughed necrotic cells in the lumen of cortical tubules, are also observed in images (**c**,**d**) from fructose-fed and STZ-injected animals.

Table 3. Mean blood glucose levels per group after 12 h fasting and 2 h after feeding in both experimental groups at the 4th and 6th weeks of the experiment.

		Glucose Levels (mg/dL)	
Week of Protocol	**Timepoint**	**Control (n = 8)**	**Induced (n = 16)**
4	After fasting (12 h)	67.00 ± 8.97	65.06 ± 19.06
	After feeding (2 h)	141.00 ± 15.22	568.92 ± 71.43 [a]
6	After fasting (12 h)	94.63 ± 8.35	110.63 ± 30.25
	After feeding (2 h)	141.63 ± 9.55	600.00 ± 0.00 [a]

[a] Statistically different from the control group ($p < 0.0001$). Data are presented as mean ± SD. The group control was intraperitoneally administrated with 0.1 M citrate buffer, while the induced group was intraperitoneally administrated with STZ + 0.1 M citrate buffer.

Regarding albumin, total cholesterol, and triglycerides levels, no statistically significant differences were found ($p > 0.05$; Table 4), supporting the absence of systemic inflammation and lipid metabolism alterations, respectively, in fructose-fed and STZ-injected rats. Fasting glucose levels measured in serum samples obtained from the blood samples collected at necropsy were significantly higher in the induced group when compared to the control group ($p < 0.05$; Table 4), suggesting insulin resistance in fructose-fed and STZ-injected rats.

Table 4. Albumin, total cholesterol, glucose, and triglycerides serum levels per group after animals' euthanasia.

	Control (n = 6)	Induced (n = 16)
Albumin (g/dL)	2.28 ± 0.66	1.81 ± 0.84
Cholesterol (mg/dL)	47.75 ± 10.24	52.19 ± 21.17
Glucose (mg/dL)	107.87 ± 25.23	251.73 ± 136.84 [a]
Triglycerides (mg/dL)	36.77 ± 12.88	60.52 ± 45.23

[a] Statistically different from the control group ($p < 0.05$). The group control was intraperitoneally administrated with 0.1 M citrate buffer, while the induced group was intraperitoneally administrated with STZ diluted in 0.1 M citrate buffer after fructose feeding.

3.6. Murinometric and Nutritional Measurements

Except for the initial body weight, the murinometric and nutritional parameters of induced animals were significantly lower when compared with control animals ($p < 0.01$; Table 5). In fact, the trend of the average weight of the induced animals remains the same as the average weight measured during normal feeding of the animals. Mean nasal-anal length, thoracic perimeter, and abdominal perimeter were significantly lower in induced animals ($p < 0.01$) than in control animals, suggesting that the animals' normal growth was retarded after exposure. The Lee index, BMI, specific rate of body weight gain, and food efficiency coefficient, factors related to an animal's body weight and food intake, were also significantly lower in chemically induced animals ($p < 0.001$, $p < 0.01$, $p < 0.0001$), respectively), revealing the same tendency as the mean body weight.

Table 5. Murinometric and nutritional parameters of animals from both groups control and induced.

Parameter	Control (n = 8)	Induced (n = 16)
Initial body weight (g)	130.57 ± 12.29	133.75 ± 10.92
Final body weight (g)	349.33 ± 21.76	256.07 ± 51.99 [c]
Final fasting body weight (g)	328.75 ± 18.94	235.01 ± 51.41 [c]
Nasal-anal length (cm)	23.50 ± 0.79	21.03 ± 1.39 [b]
Thoracic perimeter (cm)	15.13 ± 0.60	13.22 ± 1.22 [b]
Abdominal perimeter (cm)	15.88 ± 0.70	14.38 ± 1.08 [a]
Lee index	2.41 ± 0.04	2.22 ± 0.11 [b]
BMI (g/cm^2)	0.60 ± 0.03	0.53 ± 0.06 [a]
Specific rate of body weight gain (g/g)	0.03 ± 0.00	0.02 ± 0.0 [c]
Food efficiency coefficient (g/g)	1.41 ± 0.14	0.66 ± 0.28 [c]

[a] Statistically different from group control ($p < 0.01$); [b] $p < 0.001$; [c] $p < 0.0001$. The group control was intraperitoneally administrated with 0.1 M citrate buffer, while the induced group was intraperitoneally administrated with STZ diluted in 0.1 M citrate buffer after fructose feeding. Data are presented as mean ± SD.

4. Discussion

In general, and following the EU directive, to perform an experimental protocol in European countries is necessary to request authorization from the institutional authorities in which the experimental procedures will be carried out, as a first phase, followed by an authorization request to the national authority responsible for the supervising of the use of animals for scientific purposes. To avoid animal suffering, the analysis of the humane endpoints is required by both bodies. However, the bibliographic references related to the analysis of the humane endpoints in the most diverse research with laboratory animals are scarce and, according to our experience, often difficult to publish. To overcome this problem, our team has collected as much information as possible on this subject over the years to expand the available knowledge. So, envisioning to contribute to this, this research had as a main goal to provide a list of humane endpoints to the rat model of fructose-fed and STZ-induced type 2 diabetes in male Sprague-Dawley rats, as well as to validate this induction model. STZ, together with fructose, is widely used to induce type 2 diabetes in rats [13].

The choice for a 10% fructose solution was based on a previous study that compared the effects of the administration of 10, 20, 30 and 40% fructose solutions on the development of the disease [13]. Once some animals belonging to the three highest dose groups ended up dying during the trial, to ensure the success of the experiment, the concentration of 10% fructose was chosen. The same study was used as a reference to define the STZ dose. In our study, as the objective was to validate the humane endpoints table, the animals were euthanized after 5 weeks of STZ administration and not after 9 weeks, as reported by Wilson and Islam [13]. Those doses and time points guaranteed diabetes development without compromising severely animals' welfare. In this study, no deaths were recorded, inversely to that previously described in an animal model of diabetes type 1 induced by the single administration of STZ at 50 mg/kg [25]. The study by Wilson and Islam also reported no animal deaths with the same doses of fructose and streptozotocin [13]. After fructose and STZ administration, the beds and cages of induced animals were changed daily; the animals were handled extensively before and during the experiments. Moreover, the animals were observed daily to check their welfare, available water was verified daily, and scores were attributed to each parameter weekly.

Over the weeks of fructose ingestion, the induced animals consumed more water, probably due to the sweet taste, and ate less food. These animals began to drink and eat more water and food after receiving STZ. Previous research in both male Wistar and Sprague-Dawley rats found similar results [13,26]. Indeed, polydipsia and polyphagia are symptoms of diabetes [27]. Polydipsia (excessive thirst) is commonly linked to high blood glucose levels, as the kidneys produce more urine to eliminate the excess sugar. Consequently, animals tend to drink more water to replace fluid loss [28]. This event can lead to increase kidney damage. Indeed, the kidneys' relative weight was increased in the induced group when compared to the control group, suggesting that morphological changes may have occurred in these organs. In fact, hypertrophy of cortical tubules due to the vacuolation of the cytoplasm of degenerative epithelial cells was observed in tissue sections of fructose-fed and STZ-injected rats. Polyphagia causes the body to lose energy, which triggers the brain to increase appetite to compensate [29]. Interestingly, Wilson and Islam [13] reported that their type 2 diabetes-induced model did not display polyphagia. According to Chu and collaborators [30], analyzing the circulating levels of leptin is important for validating polyphagia, as this hormone is responsible for balancing food intake and energy expenditure. Polyuria was also observed in the induced rats. This symptom was observed by looking at the bed, soaked in urine and by its smell. It became necessary to change it every day, compared to the weekly bedding change required for the control rat cages. Our results indicate that animals from induced groups showed renal histological changes associated with polyuria and suggest the presence of early signs of diabetes-related nephropathy. In advanced diabetes, the decline of renal function can be attributed to a prolonged state of nitric oxide deficiency that consequently may exacerbate polyuria [31]. In future studies, the urinary levels of albumin should be assessed to better screen the onset of kidney disease. The dehydration observed in the induced animals throughout the weeks was also linked to polyuria and polydipsia [32].

Some altered parameters in the table of humane endpoints, such as lack of grooming, narrowing of the orbital area, and hunched posture, were indicative of animals' suffering and stress, most likely because of diabetes development. The Grimace scale described in the table was demonstrated as a scale that, analyzing altered facial expressions in rodents, provides a precise and reliable assessment of the animals' pain status [9]. Throughout our trial, the symptoms related to diabetic neuropathic pain identified by the Grimace scale were identified, including altered eyes position, slight discomfort of the animals during handling and curved posture. Lack of grooming was also analyzed by Yanlin Wang-Fischer and Tina Garyantes in a diabetic rat model as an indicator of disease or stress [33]. Diarrhea, as well as visualization and palpation of the abdomen, were included as typical clinical signs of diabetes mellitus in our humane endpoints table. High blood sugar levels can damage the body's tiny blood vessels and neurons, resulting in occurrences like the ones described above [34,35]. In fact, in 7th week of protocol, the animals showed a distended abdomen, probably caused by gastrointestinal constipation and the presence of liquid and pasty diarrhea. Other parameters stated in the table remained unchanged throughout the experiment; however, if the protocol were to be extended for a longer duration, there would likely be an increase in both the total score attributed to the animals and several altered parameters, such as body weight loss in fructose-fed and STZ-injected rats. Considering the humane endpoints evaluated, the most useful parameters to assess humane endpoints in this type 2 diabetes induction model were dehydration, absence of grooming, the posture of the animals, abdominal visualization and palpation, and fecal appearance. These results suggest that the parameters used to evaluate animal welfare were sensitive in this model. More than 20 years ago, David Morton proposed a table to determine humane endpoints in an STZ-induced diabetes model in Wistar rats [4]. In that table, each parameter was evaluated as negative or positive and assigned a score. In addition, Morton stated that this assessment should be performed in critical periods, such as after administration and in the middle of the study. Although we did not include parameters such as temperature and food and water consumption in our table of humane endpoints, they were evaluated separately. Effectively, our purpose was to create a table that was relatively simple to evaluate, with detailed information and with a direct score assignment. Moreover, instead of dividing the parameters at a distance and through handling, we used the main parameters, and through the attribution of a score, a brief description of the changes was carried out so that the table can be easily reproducible and the subjectivity reduced. In addition, our observations were recorded weekly, allowing better control of the animal's health status. A study performed by Wang-Fischer and Garyantes evaluated animal welfare in an animal model of type 1 diabetes. However, the authors only attributed a score to the grooming and the feces appearance. Parameters such as dehydration, sleepiness, unkempt appearance, anemia, skin or eye infection, and morbidity were only subjectively evaluated by a veterinarian, and no score was applied [33]. Furthermore, to the best of our knowledge, no humane endpoint studies have been performed in a rat model of type 2 diabetes with the addition of fructose in drinking water.

We strongly believe that this table will be helpful in future experiments on fructose-fed and STZ-induced type 2 diabetes, and depending on the study's goals, it can be extended to other animal models of diabetes. To better monitor animal welfare, human endpoints can also include the body temperature of animals [36]. Diabetes is associated with damage to blood vessels and nerves, which can lead to rises in body temperature [37]. However, no differences were seen in the induced animals until the end of the study. Maybe some difference in body temperature between groups may be observed in longer protocols.

The fasting blood glucose levels revealed statistically significant variations at the end of the protocol when the analysis was performed in an automated autoanalyzer, confirming the hyperglycemia associated with diabetes development [38]. Fructose-feeding is known to trigger insulin resistance, islet dysfunction, renal hypertrophy and tubulointerstitial disease, and cataracts, among other complications [39]. In this animal model, serum insulin concentrations were found to be lower in this animal model 9 weeks after STZ administration, which was attributed to fructose-induced oxidative stress and β-cell damage [13]. The same authors observed histological signs of partial β-cell damage and noted sensitivity to anti-diabetic drugs, supporting insulin resistance in this model. Using the same animal model but with access to a diet supplemented with bread, we did not notice significant changes in insulin levels, despite the observed hyperglycemia, 5 weeks after STZ injection. Our data suggest that at this stage of the disease, insulin is not able to stimulate glucose uptake in striated muscles and adipose tissues [40]. As the disease progresses, islet cells became severely damaged and unable to secrete insulin [13]. At the end of the experiment, the induced animals (fasted or not) had a lower mean body weight compared to the controls. These findings are in accordance with previous studies in this diabetes induction model [33,41]. Effectively, these results are not related to food and water consumption, as the induced animals had a higher consumption, but they may be associated with increased energy expenditure [42]. The induced groups had lower Lee's index, BMI, and specific rate of body weight gain when compared with control animals. The feed efficiency coefficient is a scale that represents the conversion of the consumed food into animals' weight gain [23]. The lower values of this variable can be explained by the higher food consumption of induced animals compared to controls. This result can be explained by the typical weight loss seen in diabetic rats, as stated before, as well as the polyphagia phenomena.

5. Conclusions

Our results suggest that our table of humane endpoints is appropriate to monitor the animals' welfare in this rat model of type 2 diabetes STZ-induced following fructose ingestion and can be used by other researchers using this model and eventually expanded to other models of diabetes to successfully ensure animals' welfare. In this study, none of the biological parameters appeared to justify the premature sacrifice of any of the animals; however, if the sum of the assigned scores is equal to or higher than four, the animal must be carefully evaluated and considered its sacrifice.

Supplementary Materials: The following supporting information can be downloaded at: https://www.mdpi.com/article/10.3390/ani13081397/s1, Table S1: The final score assigned to each animal throughout the trial.

Author Contributions: Experiments conducted with live animals, sacrifice and sample processing, R.S.-R., A.I.F.-R., J.S., A.V., T.A., L.A. and P.A.O.; writing of the original draft preparation, R.S.-R.; writing of review and editing, A.I.F.-R., M.d.L.P., L.G., R.F., A.M.S.S., S.M.C. and P.A.O.; statistical analysis, R.S.-R. and L.G.; histology analysis, M.d.L.P.; design of the experiments and supervision, R.F., S.M.C. and P.A.O.; funding acquisition, A.M.S.S. All authors have read and agreed to the published version of the manuscript.

Funding: This work was supported by National Funds by FCT—Portuguese Foundation for Science and Technology under the project UIDB/04033/2020. Thanks are also due to the University of Aveiro and Fundação para a Ciência e para a Tecnologia/Ministério da Ciência, Tecnologia e Ensino Superior (FCT/MCTES) for financial support and the associated laboratory LAQV-REQUIMTE (project reference UIDB/50006/2020) through national funds and co-financed by Fundo Europeu de Desenvolvimento Regional (FEDER) within the PT2020 Partnership Agreement. Project NETDIAMOND (POCI-01-0145-FEDER-016385), funded by the European Structural and Investment Funds (FEEI) through the Competitiveness and Internationalization Operational Program—COMPETE 2020 and by national funds through the FCT. R.S.-R thanks FCT/MCTES (Fundação para a Ciência e Tecnologia and Ministério da Ciência, Tecnologia e Ensino Superior) and ESF (European Social Fund) through NORTE 2020 (Programa Operacional Região Norte) for her Ph.D. grant, ref. 2022.14518.BD.

Institutional Review Board Statement: The experimental protocol was approved by the University of Trás-os-Montes and Alto Douro Ethics Committee ("ORBEA—Órgão Responsável pelo Bem-Estar e Ética Animal; approval no 852-e-CITAB-2020 and A_1-e-CITAB-2021). All the experiments performed on the animals were carried out under the European (European Directive 2010/63/EU) and national legislation (Decree-Law no. 113/2013) on the protection of animals used for scientific purposes.

Informed Consent Statement: Not applicable.

Data Availability Statement: The data that support the findings of this study are available on request from the corresponding author.

Conflicts of Interest: The authors declare no conflict of interest.

References

1. King, A.J. The Use of Animal Models in Diabetes Research. *Br. J. Pharmacol.* **2012**, *166*, 877–894. [CrossRef]
2. Russell, W.M.S.; Burch, R.L. *The Principles of Humane Experimental Technique*; Methuen. Co., Ltd.: London, UK, 1959; Volume 1, p. 500. [CrossRef]

3. Olfert, E.; Bhasin, J.; Latt, R.; Macallum, E.; McCutcheon, K.; Rainnie, D.; Schunk, M. *Guidelines on: Choosing an Appropriate Endpoint in Experiments Using Animals for Research, Teaching and Testing*; Canadian Council on Animal Care: Ottawa, ON, Canada, 1998.
4. Morton, D.B. A Systematic Approach for Establishing Humane Endpoints. *ILAR J.* **2000**, *41*, 80–86. [CrossRef] [PubMed]
5. Lloyd, M.H.; Wolfensohn, S.E. Practical Use of Distress Scoring Systems in the Application of Humane Endpoints. *Hum. Endpoints Anim. Exp. Biomed. Res.* **1999**, *1999*, 48–53.
6. Ullman-Cullere, M.H.; Foltz, C.J. Body Condition Scoring: A Rapid and Accurate Method for Assessing Health Status in Mice. *Lab. Anim. Sci.* **1999**, *49*, 319–323. [PubMed]
7. Kalueff, A.V.; Stewart, A.M.; Song, C.; Berridge, K.C.; Graybiel, A.M.; Fentress, J.C. Neurobiology of Rodent Self-Grooming and Its Value for Translational Neuroscience. *Nat. Rev. Neurosci.* **2016**, *17*, 45–59. [CrossRef] [PubMed]
8. Mason, G.; Wilson, D.; Hampton, C.; Würbel, H. Non-Invasively Assessing Disturbance and Stress in Laboratory Rats by Scoring Chromodacryorrhoea. *Altern. Lab. Anim.* **2004**, *32*, 153–159. [CrossRef] [PubMed]
9. Sotocina, S.G.; Sorge, R.E.; Zaloum, A.; Tuttle, A.H.; Martin, L.J.; Wieskopf, J.S.; Mapplebeck, J.C.; Wei, P.; Zhan, S.; Zhang, S.; et al. The Rat Grimace Scale: A Partially Automated Method for Quantifying Pain in the Laboratory Rat via Facial Expressions. *Mol. Pain.* **2011**, *7*, 55. [CrossRef]
10. Faustino-Rocha, A.I.; Ginja, M.; Ferreira, R.; Oliveira, P.A. Studying Humane Endpoints in a Rat Model of Mammary Carcinogenesis. *Iran. J. Basic. Med. Sci.* **2019**, *22*, 643–649. [CrossRef]
11. Warn, P.A.; Brampton, M.W.; Sharp, A.; Morrissey, G.; Steel, N.; Denning, D.W.; Priest, T. Infrared Body Temperature Measurement of Mice as an Early Predictor of Death in Experimental Fungal Infections. *Lab. Anim.* **2003**, *37*, 126–131. [CrossRef]
12. World Health Organization Diabetes. Available online: https://www.who.int/health-topics/diabetes#tab=tab_1 (accessed on 4 April 2023).
13. Wilson, R.D.; Islam, M.S. Fructose-Fed Streptozotocin-Injected Rat: An Alternative Model for Type 2 Diabetes. *Pharmacol. Rep.* **2012**, *64*, 129–139. [CrossRef]
14. Wang, Y.-J.; Xie, X.-S.; Feng, S.-G.; Long, Q.-X.; Ai, N.; Wang, B.-F. Causes of Death in STZ-Induced Rat Models of Diabetes Mellitus. *Sichuan Da Xue Xue Bao Yi Xue Ban. J. Sichuan Univ.* **2014**, *45*, 691–695.
15. Dang, J.-K.; Wu, Y.; Cao, H.; Meng, B.; Huang, C.-C.; Chen, G.; Li, J.; Song, X.-J.; Lian, Q.-Q. Establishment of a Rat Model of Type II Diabetic Neuropathic Pain. *Pain. Med.* **2014**, *15*, 637–646. [CrossRef] [PubMed]
16. Mostafavinia, A.; Amini, A.; Ghorishi, S.K.; Pouriran, R.; Bayat, M. The Effects of Dosage and the Routes of Administrations of Streptozotocin and Alloxan on Induction Rate of Type1 Diabetes Mellitus and Mortality Rate in Rats. *Lab. Anim. Res.* **2016**, *32*, 160–165. [CrossRef]
17. Silva-Reis, R.; Faustino-Rocha, A.I.; Gonçalves, M.; Ribeiro, C.C.; Ferreira, T.; Ribeiro-Silva, C.; Gonçalves, L.; Antunes, L.; Venâncio, C.; Ferreira, R.; et al. Refinement of Animal Model of Colorectal Carcinogenesis through the Definition of Novel Humane Endpoints. *Animals* **2021**, *11*, 985. [CrossRef] [PubMed]
18. Oliveira, M.; Nascimento-gonçalves, E.; Silva, J.; Oliveira, P.A.; Ferreira, R.; Antunes, L.; Arantes-rodrigues, R.; Faustino-rocha, A.N.A.I. Implementation of Human Endpoints in a Urinary Bladder Carcinogenesis Study in Rats. *In Vivo* **2017**, *31*, 1073–1080. [CrossRef]
19. Miedel, E.L.; Hankenson, F.C. Biology and Diseases of Hamsters. In *Laboratory Animal Medicine*; Elsevier: Amsterdam, The Netherlands, 2015; pp. 209–245.
20. Mayer, J.; Donnelly, T.M. *Clinical Veterinary Advisor*; Elsevier: Amsterdam, The Netherlands, 2013; pp. 242–252.
21. Parasuraman, S.; Raveendran, R.; Kesavan, R. Blood Sample Collection in Small Laboratory Animals. *J. Pharmacol. Pharmacother.* **2010**, *1*, 87–93. [CrossRef]
22. Zhu, G.; Wang, Q.; Lu, S.; Niu, Y. Hydrogen Peroxide: A Potential Wound Therapeutic Target. *Med. Princ. Pract.* **2017**, *26*, 301–308. [CrossRef]
23. da Silva Nery, C.; Lins Pinheiro, I.; de Santana Muniz, G.; Antônio Alves de Vasconcelos, D.; Pereira de França, S.; do Nascimento, E. Murinometric Evaluations and Feed Efficiency in Rats from Reduced Litter during Lactation and Submitted or Not to Swimming Exercise. *Rev. Bras. Med. Esporte* **2011**, *17*, 49–55.
24. Novelli, E.L.B.; Diniz, Y.S.; Galhardi, C.M.; Ebaid, G.M.X.; Rodrigues, H.G.; Mani, F.; Fernandes, A.A.H.; Cicogna, A.C.; Novelli Filho, J.L.V.B. Anthropometrical Parameters and Markers of Obesity in Rats. *Lab. Anim.* **2007**, *41*, 111–119. [CrossRef]
25. Cruz, P.L.; Moraes-Silva, I.C.; Ribeiro, A.A.; Machi, J.F.; de Melo, M.D.T.; dos Santos, F.; da Silva, M.B.; Strunz, C.M.C.; Caldini, E.G.; Irigoyen, M.C. Nicotinamide Attenuates Streptozotocin-Induced Diabetes Complications and Increases Survival Rate in Rats: Role of Autonomic Nervous System. *BMC Endocr. Disord.* **2021**, *21*, 133. [CrossRef]
26. Wei, M.; Ong, L.; Smith, M.T.; Ross, F.B.; Hoey, A.J.; Burstow, D.; Fracp, M. The Streptozotocin-Diabetic Rat as a Model of the Chronic Complications of Diabetes. *Asia Pac. Heart J.* **2003**, *12*, 44–50. [CrossRef] [PubMed]
27. Dewi, N.; Syaify, A.; Wahyudi, I.A. Effect of Gestational Diabetes Mellitus on the Expression of Amelogenin in Rat Offspring Tooth Germ. *Dent. J. (Majalah Kedokteran Gigi)* **2013**, *46*, 135. [CrossRef]
28. Garrahy, A.; Moran, C.; Thompson, C.J. Diagnosis and Management of Central Diabetes Insipidus in Adults. *Clin. Endocrinol.* **2019**, *90*, 23–30. [CrossRef] [PubMed]
29. Okon, U.A.; Owo, D.U.; Udokang, N.E.; Udobang, J.A.; Ekpenyong, C.E. Oral Administration of Aqueous Leaf Extract of Ocimum Gratissimum Ameliorates Polyphagia, Polydipsia and Weight Loss in Streptozotocin-Induced Diabetic Rats. *Am. J. Med. Med. Sci.* **2012**, *2*, 45–49. [CrossRef]

30. Xu, J.; Bartolome, C.L.; Low, C.S.; Yi, X.; Chien, C.H.; Wang, P.; Kong, D. Genetic Identification of Leptin Neural Circuits in Energy and Glucose Homeostases. *Nature* **2018**, *556*, 505–509. [CrossRef]
31. Cipriani, P.; Kim, S.L.; Klein, J.D.; Sim, J.H.; von Bergen, T.N.; Blount, M.A. The Role of Nitric Oxide in the Dysregulation of the Urine Concentration Mechanism in Diabetes Mellitus. *Front. Physiol.* **2012**, *3*, 176. [CrossRef] [PubMed]
32. Milles, J.J.; Spruce, B.; Baylis, P.H. A Comparison of Diagnostic Methods to Differentiate Diabetes Insipidus from Primary Polyuria: A Review of 21 Patients. *Acta Endocrinol.* **1983**, *104*, 410–416. [CrossRef]
33. Wang-Fischer, Y.; Garyantes, T. Improving the Reliability and Utility of Streptozotocin-Induced Rat Diabetic Model. *J. Diabetes Res.* **2018**, *2018*, 8054073. [CrossRef]
34. FELDMAN, M. Disorders of Gastrointestinal Motility Associated with Diabetes Mellitus. *Ann. Intern. Med.* **1983**, *98*, 378. [CrossRef]
35. Valdovinos, M.A.; Camilleri, M.; Zimmerman, B.R. Chronic Diarrhea in Diabetes Mellitus: Mechanisms and an Approach to Diagnosis and Treatment. *Mayo Clin. Proc.* **1993**, *68*, 691–702. [CrossRef]
36. Mei, J.; Banneke, S.; Lips, J.; Kuffner, M.T.C.; Hoffmann, C.J.; Dirnagl, U.; Endres, M.; Harms, C.; Emmrich, J.V. Refining Humane Endpoints in Mouse Models of Disease by Systematic Review and Machine Learning-Based Endpoint Definition. *ALTEX* **2019**, *36*, 555–571. [CrossRef] [PubMed]
37. Ramos-Lobo, A.M.; Buonfiglio, D.C.; Cipolla-Neto, J. Streptozotocin-Induced Diabetes Disrupts the Body Temperature Daily Rhythm in Rats. *Diabetol. Metab. Syndr.* **2015**, *7*, 39. [CrossRef] [PubMed]
38. Opherk, C.; Tronche, F.; Kellendonk, C.; Kohlmüller, D.; Schulze, A.; Schmid, W.; Schütz, G. Inactivation of the Glucocorticoid Receptor in Hepatocytes Leads to Fasting Hypoglycemia and Ameliorates Hyperglycemia in Streptozotocin-Induced Diabetes Mellitus. *Mol. Endocrinol.* **2004**, *18*, 1346–1353. [CrossRef] [PubMed]
39. Kretowicz, M.; Johnson, R.J.; Ishimoto, T.; Nakagawa, T.; Manitius, J. The Impact of Fructose on Renal Function and Blood Pressure. *Int. J. Nephrol.* **2011**, *2011*, 315879. [CrossRef]
40. Soto-Greene, M.L.; Salas-Lopez, D.; Sanchez, J.; Like, R.C. Antecedents to Effective Treatment of Hypertension in Hispanic Populations. *Clin. Cornerstone* **2004**, *6*, 30–38. [CrossRef] [PubMed]
41. Gundala, N.K.V.; Naidu, V.G.M.; Das, U.N. Amelioration of Streptozotocin-Induced Type 2 Diabetes Mellitus in Wistar Rats by Arachidonic Acid. *Biochem. Biophys. Res. Commun.* **2018**, *496*, 105–113. [CrossRef]
42. di Sebastiano, K.M.; Yang, L.; Zbuk, K.; Wong, R.K.; Chow, T.; Koff, D.; Moran, G.R.; Mourtzakis, M. Accelerated Muscle and Adipose Tissue Loss May Predict Survival in Pancreatic Cancer Patients: The Relationship with Diabetes and Anaemia. *Br. J. Nutr.* **2013**, *109*, 302–312. [CrossRef]

Disclaimer/Publisher's Note: The statements, opinions and data contained in all publications are solely those of the individual author(s) and contributor(s) and not of MDPI and/or the editor(s). MDPI and/or the editor(s) disclaim responsibility for any injury to people or property resulting from any ideas, methods, instructions or products referred to in the content.

animals

MDPI

Review

Challenges with Assessing and Treating Pain in Research Primates: A Focused Survey and Literature Review

Emilie A. Paterson [1] and Patricia V. Turner [1,2,*]

1 Department of Pathobiology, University of Guelph, Guelph, ON N1G 0C4, Canada
2 Global Animal Welfare and Training, Charles River, Wilmington, MA 01887, USA
* Correspondence: pvturner@uoguelph.ca

Simple Summary: It is crucial that research primates receive adequate pain treatment from ethical, animal welfare, and research-related perspectives. There is limited research on current pain management in research primates. A survey was administered to primate veterinarians (n = 93 respondents) to investigate a veterinary approach to pain recognition and alleviation as well as the pain management challenges that primate veterinarians may face. Survey results were used to guide a subsequent literature review on the topic. This review discusses current evidence and challenges in research primate pain management such as limited pharmacokinetic data and efficacy testing as well as a lack of validated pain assessment tools to recognize and evaluate pain in primates. Both the survey and literature review demonstrate gaps and challenges in primate pain management, and suggest science-based recommendations for improving current management guidance as well as future areas of research.

Abstract: Research primates may undergo surgical procedures making effective pain management essential to ensure good animal welfare and unbiased scientific data. Adequate pain mitigation is dependent on whether veterinarians, technicians, researchers, and caregivers can recognize and assess pain, as well as the availability of efficacious therapeutics. A survey was conducted to evaluate primate veterinary approaches to pain assessment and alleviation, as well as expressed challenges for adequately managing primate pain. The survey (n = 93 respondents) collected information regarding institutional policies and procedures for pain recognition, methods used for pain relief, and perceived levels of confidence in primate pain assessment. Results indicated that 71% (n = 60) of respondents worked at institutions that were without formal experimental pain assessment policies. Pain assessment methods were consistent across respondents with the majority evaluating pain based on changes in general activity levels (100%, n = 86) and food consumption (97%, n = 84). Self-reported confidence in recognizing and managing pain ranged from slightly confident to highly confident, and there was a commonly expressed concern about the lack of objective pain assessment tools and science-based evidence regarding therapeutic recommendations of analgesics for research primates. These opinions correspond with significant gaps in the primate pain management literature, including limited specific pharmacokinetic data and efficacy testing for commonly used analgesics in research primate species as well as limited research on objective and specific measures of pain in research primates. These results demonstrate that there are inconsistencies in institutional policies and procedures surrounding pain management in research primates and a lack of objective pain assessment methods. Demonstrating the gaps and challenges in primate pain management can inform guideline development and suggest areas for future research.

Keywords: pain assessment; 3Rs; veterinary medicine; analgesia; macaque; animal welfare

Citation: Paterson, E.A.; Turner, P.V. Challenges with Assessing and Treating Pain in Research Primates: A Focused Survey and Literature Review. *Animals* **2022**, *12*, 2304. https://doi.org/10.3390ani12172304

Academic Editor: Garikoitz Azkona

Received: 7 August 2022
Accepted: 27 August 2022
Published: 5 September 2022

Copyright: © 2022 by the authors. Licensee MDPI, Basel, Switzerland. This article is an open access article distributed under the terms and conditions of the Creative Commons Attribution (CC BY) license (https://creativecommons.org/licenses/by/4.0/).

1. Introduction

Research primates may undergo surgical and other invasive procedures for experimental purposes as well as periodically for veterinary care, which may result in pain [1,2].

To meet ethical obligations and ensure good animal welfare and quality of scientific data, prompt pain mitigation is necessary [3–5] and is required by research regulatory and oversight bodies [6–8]. Effective pain treatment is dependent on being able to recognize pain and assess its intensity [9–11]. While pain management has been an ongoing subject of research in human and veterinary medicine, pain management in research primates is less well studied and documented.

The International Association for the Study of Pain defines pain as "An unpleasant sensory and emotional experience associated with, or resembling that associated with, actual or potential tissue damage" [12]. Many factors can influence pain perception in animals, such as age [13], sex [14], and the environment [15], as well as past experiences with pain [16], and social ranking [17]. This demonstrates that pain is a subjective experience in animals as in humans, and emphasizes the importance of individualized pain assessment and treatment. Conversely, pain alters many physiologic, psychologic, and behavioural parameters that in turn can be used to identify if an animal is experiencing pain [6,9,18–20]. Some of these changes, such as weight loss, decreased food intake, altered mentation, and others, are not specific for pain and need to be interpreted in context [11], [21–23]. Among the most commonly used pain indicators in research primates are behavioural assessment, in which activity levels, locomotion, posture, and species-typical behaviours are assessed [6,20]. However, unless baseline measures and guidelines are provided, behavioural observations can be highly subjective [24]. It can also be challenging to directly assess pain in primates given that they are prey species and tend to hide signs of pain in front of human observers [25].

Pain management in human and veterinary medicine has continued to be an ongoing topic of research given that issues of inadequate pain mitigation are universal. Primate pain management practices are largely based on empirical evidence and given that primates are a nondomestic and nonverbal species, it can be difficult to recognize, assess, and treat pain effectively [9]. A survey conducted almost two decades ago in the UK examining the recognition and assessment of pain in laboratory animals demonstrated that there is a lack of objective pain assessment methods and that pain management practices are poorly reported [26]. A more recent survey on pain management in research animals also demonstrated that there are gaps in pain management due to the lack of evidence-based knowledge and that existing pain management protocols are generally not species-specific or discipline-specific [27]. These two surveys pertain to all laboratory species and there is limited knowledge of the current pain management practices for research primates and the views of primate veterinarians on this topic. In the process of developing a new Association of Primate Veterinarians (APV) primate pain mitigation position statement [28], there was interest in knowing more about routine practices for assessing pain. To address this issue, a cross-sectional anonymous online survey was developed on pain management in research primates and issued to members of APV and European Primate Veterinarians (EPV). The objective of the survey was to better understand: (1) pain assessment practices and institutional policies regarding pain assessment methods and treatments; and (2) primate veterinarian perceptions about pain management, including challenges and confidence levels in recognizing and treating primates in pain. From the survey, recurrent themes were used to conduct a literature search on primate pain treatment and management, specifically, looking for validated pain assessment tools for research primates as well as studies reporting analgesic pharmacokinetics and efficacy for common New and Old World research primates.

2. Survey about Primate Veterinarians' Perspectives on Pain Management in Research Primates

2.1. Narrative Review Methods

The literature review was performed using the following databases: EBSCOhost (MEDLINE, Academic Search Premier, Ipswich, MA, USA), Elsevier ScienceDirect Journals, GALE ACADEMIC ONEFILE, Google Scholar, JSTOR, ProQuest (Science database),

PubMed, SAGE Premier Journals, Scholars Portal Journals, SpringerLink, Web of Science (all databases), and Wiley InterScience Journals. Themes were identified from survey results and divided into 2 major topics: pain treatment and/or analgesics and pain assessment and/or recognition. Keywords were identified for both topics and entered into the databases in which all date ranges were included and sorted by relevance.

2.2. *Survey Methods*

APV is largely a U.S.-based not-for-profit organization of veterinarians who provide care for and oversee the health of multiple species of primates in research settings. EPV is an EU-based not-for-profit organization of veterinarians that also provides care for research primates. Both organizations deliver educational content to their members and promote the informal exchange of experiences, knowledge, and research data to facilitate ongoing professional development on primate medicine, care, and welfare. The survey was sent to members of both organizations.

A 21-item questionnaire was developed (see Supplementary Materials). The APV Pain Assessment Subcommittee members reviewed the survey and the APV Board of Directors approved the final version. No identifying information was collected of participants and the survey was administered by APV association management personnel, with only aggregate data provided to the researchers. The survey was deemed exempt from REB approval because of this. Participants were informed before answering the survey that their participation was voluntary, all answers would be anonymous, no incentives were used, and they could choose to leave questions unanswered at their discretion. To be eligible to participate in this study, individuals had to be members of APV or EPV. The survey was conducted from 28 January 2019 to 15 February 2019. APV members received the survey by email through the APV listserve. EPV members received the survey through the EPV LinkedIn group.

2.3. *Survey Data Analysis*

Descriptive statistics were conducted using Microsoft Excel (version 2017) on all questions except open-ended questions. The final open-ended questions were only answered by 29% of respondents; non-response rates on questions ranged from 0 to 14, with an average of 87 (94%) responses per question. Percentages given for responses are from the total number of responses received. Participants working in two types of facilities were grouped as follows: if academia and a private research or government facility were reported, they were categorized as the latter. Participants with two positions (example: clinical veterinarian and facility director), were classified using the more senior position as the primary position. For the three questions relating to type of pain control drugs used for primates, responses that included name brands were classified according to the US Food and Drug Administration generic name. A Likert scale was used to assess confidence levels of respondents on a particular topic; a score of 1 indicating not very confident and a score of 5 indicating highly confident.

2.4. *Survey Results*

2.4.1. Demographics of Respondents and Primate Species

A total of 93 members of APV and EPV responded to the survey, representing at least a 20% response rate, given that there is a membership overlap between these two societies. Eighty-four (90%) respondents were from the USA, two (2%) were from Canada, two (2%) were from Germany, and five (4%) were single responses from other countries (Barbados, the Netherlands, China, and Israel). Forty-one (44%) respondents worked in an academic setting and thirty (32%) in a private or contract research institution. Seventy-four (81%) respondents indicated that their main role was as a clinical, research, or attending veterinarian at their facility. Eighty-nine (96%) respondents worked with primates at the time of the survey and the majority with macaques (genus *Macaca*). More in-depth demographic data can be found in Table 1.

Table 1. Demographics of primate pain survey respondents and primate species.

Parameter Assessed	No. (%) of Respondents	
Country		
United States	83 (90)	
Canada	2 (2)	
Germany	2 (2)	
China	1 (1)	$n = 92$ [a]
Netherlands	1 (1)	
Barbados	1 (1)	
Puerto Rico	1 (1)	
Israel	1 (1)	
Institution type		
University/Academic research	41 (44)	
Private or contract research	30 (33)	
Pharmaceutical research	7 (8)	
National Primate Research Center	6 (7)	
Hospital	2 (2)	$n = 92$ [a]
Non-profit/Sanctuary	2 (2)	
Primate breeding facility/Supplier	2 (2)	
Military	1 (1)	
Private consulting	1 (1)	
Primary job function [b]		
Veterinarian (Clinical/Attending/Research)	74 (81)	
Director	8 (9)	
Administrative	3 (3)	
Management	2 (2)	$n = 91$ [a]
Consultant	2 (2)	
IACUC member	1 (1)	
Post-Doctoral fellow	1 (1)	
Primary job includes working with living primates		
Yes	89 (96)	$n = 93$ [a]
No	4 (4)	
Primate species currently worked with [b]		
Macaques	84 (96)	
Baboons	17 (19)	
Squirrel monkeys	16 (18)	
African green monkeys	12 (14)	$n = 88$ [a]
Chimpanzees	7 (8)	
Owl monkeys	6 (7)	
Sooty mangabeys	2 (2)	
Other	25 (29)	

[a] Total number of survey participants = 93; [b] participants could select more than one answer.

2.4.2. Policies or Procedures for Pain Recognition in Research Primates

Respondents were asked if their animal ethics committee had a formal experimental pain assessment policy for research primates, and if so, whether the Standard Operating Procedures (SOP) were generic or specific to primates. Sixty (65%) respondents indicated that their institutional animal ethics committee did not have a formal pain assessment policy for primates (Figure 1). Twenty-five (27%) respondents indicated that their institution had a formal pain assessment policy for primates, fifteen (16%) had SOPs specific to primate pain assessment, and nine (10%) had generic SOPs related to pain assessment for all species housed in their facility. Respondents were also asked what type of pain assessment methods they used for primates. Eighty-five (91%) respondents reported using direct observation (e.g., cage side), thirty-two (34%) reported assessing pain indirectly (e.g., video camera), and four (4%) reported that the primates could not be closely observed at their facility due to their housing situation. Over 90% of respondents used general activity levels, food consumption, disuse or guarding of a body part, posture, and lameness to identify pain (Figure 2).

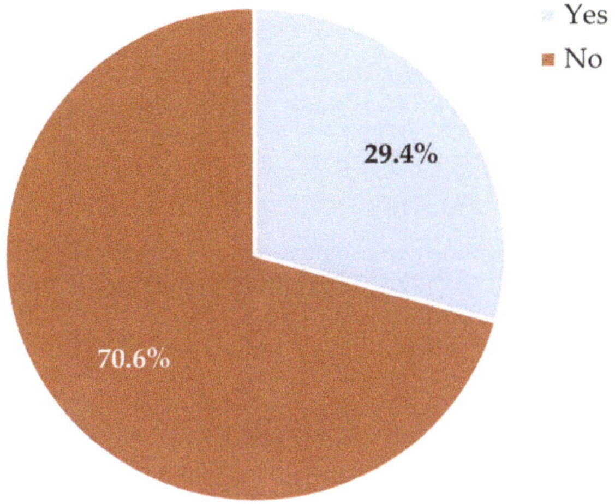

Figure 1. Percentage of animal ethics committees that have (blue) or do not have (red) a formal experimental pain assessment policy for research primates at their institution (n = 85).

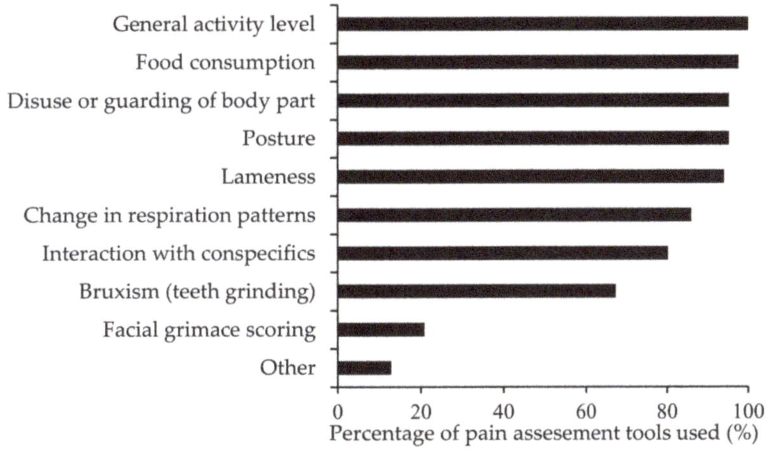

Figure 2. The different pain assessment tools in use (expressed as a percentage) for primates at the respondents' facilities (n = 86).

Respondents were asked who was responsible for assessing pain in primates at their facility and if responses to pain treatment were evaluated. Based on responses received, veterinarians and veterinary technicians were primarily responsible for conducting pain assessments at all facilities. Seventy-eight (84%) respondents indicated that primates are routinely monitored in the post-procedural period to evaluate the effectiveness of analgesia. Respondents were also asked how often unplanned top-ups of analgesics occur. Forty-five (48%) respondents reported that unplanned top-ups of analgesics occur sometimes and thirty-four (37%) reported that this rarely occurs (Table 2).

Table 2. Policies and protocols for research primate pain assessment and treatment.[a]

Parameters Assessed	No. (%) of Respondents	
Individual responsible for making pain assessment in primates [a]		
Veterinarian	83 (98)	$n = 85$ [b]
Veterinarian technicians	77 (91)	
Animal care personnel	54 (64)	
Other research staff	36 (42)	
Principal investigators	30 (35)	
Students	8 (9)	
Other	3 (4)	
Primates monitored after treatment to evaluate effectiveness of analgesia		
Yes	78 (91)	$n = 86$ [b]
Sometimes	8 (9)	
No	0 (0)	
Frequency of unplanned top-ups in analgesic medication to primates		
Often	3 (4)	$n = 85$ [b]
Sometimes	45 (53)	
Rarely	34 (40)	
Never	1 (1)	
N/A (no procedures requiring analgesic)	2 (2)	

[a] Total number of survey participants = 93; [b] participants could select more than one answer.

2.4.3. Methods Used to Alleviate Pain in Research Primates

In terms of methods used to alleviate pain in primates, most respondents indicated that analgesic drugs were generally given. Eighty-six (92%) respondents reported using nonsteroidal anti-inflammatory drugs (NSAID), eighty-four (90%) used opioids, and eighty (90%) reported using topical/local anesthetics (Figure 3). Meloxicam was the most widely used NSAID and was reported to be used by 77 (83%) respondents (Figure 4A). Of these respondents, 13 indicated also using the sustained-release formulation of meloxicam. For opioids, 80 (86%) respondents reported using buprenorphine (Figure 4B), with 36 further indicating that they also used the sustained-release formulation of buprenorphine. In terms of topical/local anesthetic treatment, 70 (75%) respondents reported using bupivacaine and 69 (74%) respondents reported using lidocaine (Figure 4C). Other analgesic agents were also used, but less frequently than those indicated above (Figure 4).

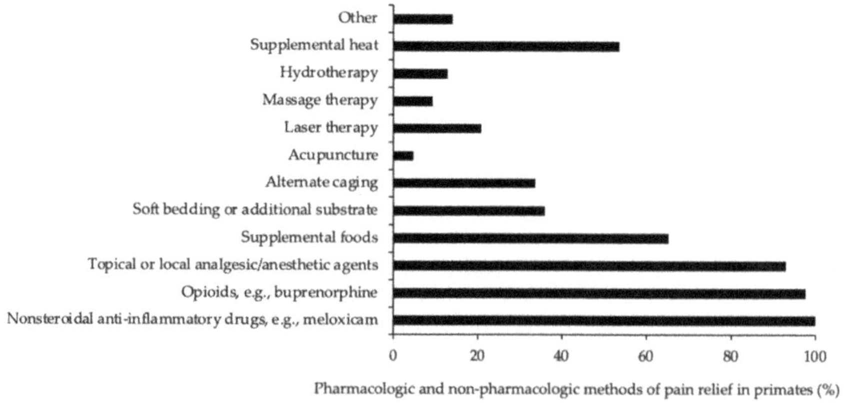

Figure 3. Methods used to treat and manage pain in research primates (expressed as a percentage of responses) ($n = 86$).

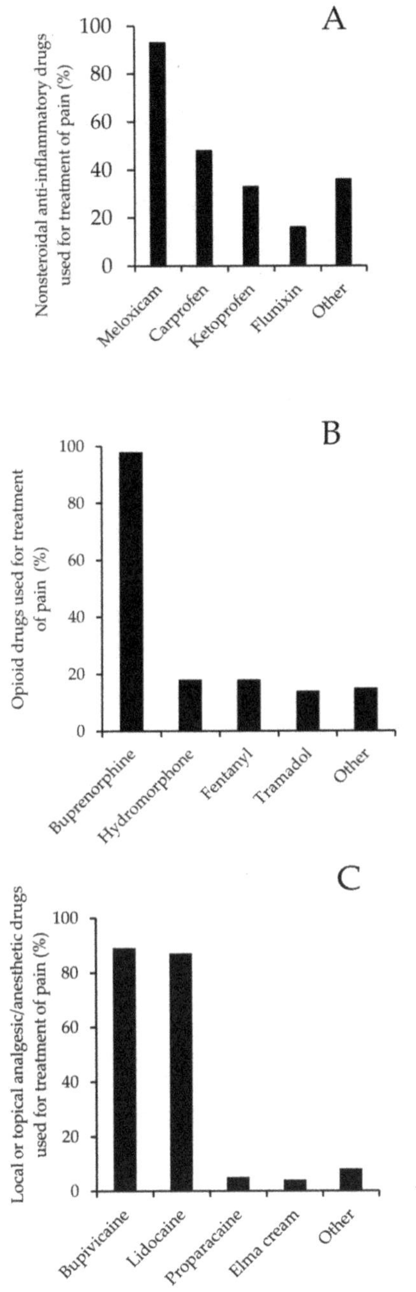

Figure 4. Pharmacologic agents used for the treatment of pain in primates (expressed as percentage of use): (**A**) NSAIDs (n = 83), (**B**) opioids (n = 82), and (**C**) local or topical anesthetic agents (n = 79).

The most common methods of non-pharmacologic care included acupuncture (4%, n = 4), hydrotherapy (12%, n = 11), and massage therapy (9%, n = 8) (see Figure 3).

2.4.4. Quality of Pain Assessments in Research Primates

Respondents were asked to self-report their confidence level in recognizing and managing pain in primates vs. that of their associates. Forty-two (45%) respondents indicated that they were highly confident in recognizing and managing pain whereas forty-four (47%) respondents indicated that they were somewhat confident that research personnel at their facility could recognize pain in research primates (Figure 5).

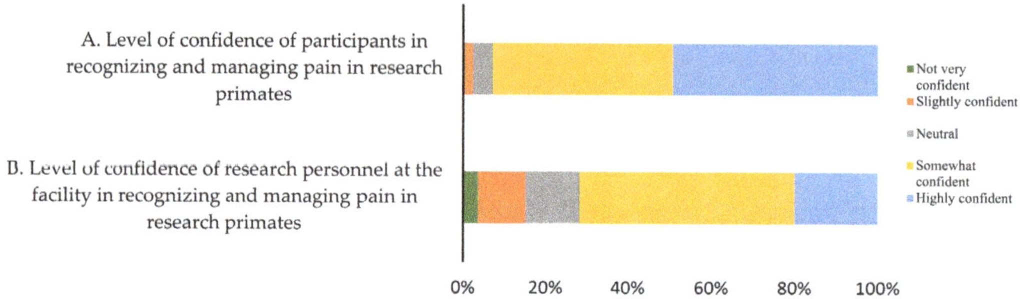

Figure 5. The level of confidence of recognizing and managing pain in research primates using a 5-score Likert scale (n = 85).

Additional pertinent comments provided by participants included, " ... use a multimodal approach, whenever possible", " ... I think we need a shift in pain management which focuses on pre-emptive analgesia and intraoperative analgesic methods". "To me, postoperative analgesic protocols are well established but relied on too much". and "Research staff are sometimes the first to pick up on subtle signs, but trained veterinary technicians are also very good at assessments". "They are key components in the monitoring of all". Finally, one respondent noted, " ... unfortunately we still don't have objective tools to score pain and the effectiveness of the provided analgesics".

2.5. Discussion

This study summarized the results of 93 laboratory animal veterinarians, largely based in the U.S.A. across a range of employment sectors, and all with significant experience working with primates in research settings. The most significant finding of this survey indicates that primate pain management may be less than optimal due to inconsistencies in institutional policies and procedures and a lack of objective pain assessment tools in research primates. Macaques were the most reported species of primates housed from responding facilities, as expected from a 2014 survey on the use of research primates in North America in which rhesus and cynomolgus macaques were the most commonly reported species [29].

These survey results suggest that the majority of primate research facilities do not currently have a formal experimental pain assessment policy and even institutions that have a formal experimental pain assessment policy are not generally species-specific. A general survey on pain management in research animals demonstrated that there are inconsistencies in pain management across institutions and species as a consequence of not having specific guidelines in place [27]. Similarly, a survey conducted in the UK on pain recognition and treatment identified that only 6 of 25 institutions had a written policy for pain management [26]. The lack of formal pain management policies for research animals could be due to a lack of resources (i.e., time and scientific evidence) to effectively retrieve information and communicate pain assessment methods [24,26,30]. It may also be because there is a dearth of information published on the topic. As demonstrated in a number of recent literature surveys, researchers publishing in mainstream scientific journals have not been rigorous about reporting pain assessment and mitigation strategies, making it difficult for those searching for evidence of effective treatments [31–34].

The majority of respondents reported that veterinarians and veterinary technicians are largely responsible for conducting pain assessments at their facility and that these assessments are mostly conducted by direct observation. Rhesus macaques have been reported to suppress signs of illness following direct observation compared to indirect observation (i.e., video camera) [25]. Thus, exclusive use of direct observation methods may result in reduced detection of pain in primates. The predominant pain assessment technique reported was behavioural observation including general activity levels, disuse, or guarding of a body part, lameness, posture, and interactions with conspecifics [34]. As for health indicators of pain, food consumption and respiratory patterns were used most often as a measure of pain. All of the reported indicators are in line with the recommended guidelines for primate welfare assessments [28]. More recently, primate welfare assessment indicators have been identified using a Delphi method; although, the indicators are not specific to pain, rather they are indicators of general wellness and evaluate different categories of welfare including physical, environmental, and input-based measures [35,36]. As these guidelines state, a reference point (i.e., the individual's normal behaviour) should be quantified so that when pain is assessed it is as objective as possible. Most indicators used are not specific to pain and need to be interpreted in context. Formalized score sheets are a good means to quantify pain behaviour, track frequency of evaluation, and can be kept in health records as a reference for the individual animal as well as the procedure.

Procedures that are thought to cause pain in humans need to be treated accordingly in animals unless proven otherwise. Pharmacologic methods were reported as the primary method of treatment. It can be difficult for veterinarians to choose which drug to use as well as the appropriate dosage due to the limited scientific evidence and possible research model pharmacologic restrictions [5,27,30,37]. It is a common practice in laboratory animal medicine to use multimodal pain treatment—that is, combining different drug categories to target different mechanisms of pain development [38,39]. In this survey, buprenorphine, meloxicam, and bupivacaine were the most commonly used opioids, NSAIDs, and topical anesthetics, respectively, for pain management in primates. These results are similar to the findings in a recent review of the analgesics and anesthetics reported in experimental surgical procedures in primates [31]. The only difference is in the NSAID category in which carprofen was reported more than meloxicam. Both meloxicam and carprofen have a similar mechanism of action and both are cyclooxygenase-2 selective [40]. Meloxicam has been demonstrated to be effective for postoperative pain mitigation in primates for orthopedic surgery [41,42] and neurosurgery [42,43] and is also used in combination with opioids [44]. A sustained-release formulation of meloxicam (0.6 mg/kg) is reported to result in therapeutic blood drug levels in cynomolgus macaques for 48–72 h compared to intramuscular or oral administration, which may last up to 24 h or 8–12 h, respectively [44]. Similarly, sustained release formulations of buprenorphine may last up to 96 h vs. 6–8 h for intramuscular formulations in macaques [45]. In the current survey, a small proportion of respondents reported using the sustained-release formulations of NSAIDs and opioids. It is unknown whether this is due to a lack of availability, concerns surrounding adverse side effects, a lack of knowledge about these formulations, or other reasons.

Perceived confidence can have an impact on pain management practices. This is a common phenomenon in human and veterinary medicine and recent surveys in both fields have queried levels of confidence in pain assessment and mitigation. Results from those surveys demonstrate that human nurses and veterinary technicians can have diminished levels of confidence due to limited knowledge on pain assessment and the appropriate analgesics to use, a lack of appropriate tools to assess pain objectively, and a lack of continuing education [27,30,37,46–48]. These factors also may have an impact on the confidence levels reported in this study, in which approximately half the respondents reported being somewhat confident in assessing and managing pain and the other half reported being highly confident. In a recent survey similar to this study, primate veterinarians were asked about their level of confidence and to associate their level of confidence with certain statements [49]. It was found that primate veterinarians who have higher levels of confidence will be more likely to use behaviour and facial expressions as pain indicators and to opt for an increased use of pain medication [49]. Conversely, in the current survey, when asked to report the perceived level of confidence in research personnel, the majority reported that they were only somewhat or less confident. We need to consider that confidence levels do not reflect skill level and that this survey assessed the participants' self-reported confidence and not their objectively assessed ability to identify and evaluate pain in primates.

Study limitations include a small sample size, and thus, the results may not be reflective of the views and opinions of all APV and EPV members. Furthermore, the majority of respondents were from the U.S.A, and may not be reflective of the views and opinions of primate veterinarians in other countries. Finally, due to participant anonymity, answers could not be linked to the participant demographics, and thus it was not possible to assess the relationship between these parameters (for example, years of clinical experience vs. confidence in recognizing and treating pain in macaques).

Subsequent to the administration of this survey, the APV Guidelines on Pain Management were published [50]. It would be interesting to conduct a follow-up survey study to examine the impact of APV guidelines on pain management practices in different institutions. As demonstrated by this survey, there is a lack of objective pain assessment tools in research primates, and thus future research should focus on validating pain assessment tools for these animals.

3. Pain Assessments in Research Primates—A Review

To provide effective pain treatment it is important to recognize and evaluate the intensity of pain. There are various assays used for pain identification and assessment that fluctuate in objectivity, reliability, and practicality. Pain can be difficult to identify and quantify in primates as they are prey species and often hide signs of pain in front of human observers, unless severe [25]. In this section, we will present commonly used pain assessment methods or assays employed to evaluate pain in primates and discuss their pros and cons in research settings (see Table 3).

Table 3. Reported assays and methods used to recognize and evaluate pain in research primates.

Assay Category	Assay or Method	Description	Reference
Reflex-based	Application of noxious stimuli (i.e., chemical, thermal, or mechanic)	Dose-related increase in pain	[51–53]
Physiologic	Cage-side observation Thermometer/infrared thermography Stethoscope Urine, fecal, blood samples Telemetry	Blood pressure Respiratory rate or changes in respiratory rate Body temperature Heart rate Cortisol/adrenocorticotropin	[18,54–58]
Clinical	Cage-side observation Scale Quantify food intake	Body weight/body condition score Appetite	[59,60]
Behavioural	Cage-side observation Scoring grid Daily activity budget Ethograms Behavioural scoring using software (i.e., Observer XT)	General activity levels Posture Changes in species-typical behaviour Social behaviour	[19,20,61–63]
Facial expression	Cage-side observation (no validated grimace scale)	Pain grimace	[20]

3.1. *Reflex-Based Assays*

Reflex-based assays are commonly used to evaluate dose-related increases in pain threshold to assess the efficacy of analgesics, but they are rarely used in clinical practice [64]. The approach usually involves the application of a standard noxious stimulus (i.e., chemical, thermal, or mechanic) followed by quantification of the animal's reflex response. For example, a study using squirrel monkeys performing an operant behaviour (i.e., pulling on a thermal rod that increased in temperature) demonstrated that administration of several opioids resulted in dose-related increases in temperature threshold [52]. Similarly, in rhesus macaques, thermode behavioural testing has been validated pharmacologically as a tool to determine analgesic efficacy with commonly used opioids (i.e., tramadol, morphine, and fentanyl) [51]. This assay is a simple method to conduct and can be attributed a value. However, this type of assay only captures the sensory component, more specifically, hypersensitivity of the nociceptors; thus, this method does not capture the learned and emotional components of pain [53]. This suggests that the clinical significance may be poor, but these assays may be useful for the preliminary determination of the therapeutic efficacy of novel analgesics.

3.2. Physiologic Parameters

Pain causes a cascade of physiologic events that can be measured and quantified. Acute pain and breakthrough pain activate the sympathetic nervous system resulting in increased blood pressure, respiratory rate, body temperature, and heart rate [54]. Pain also activates the hypothalamic–pituitary–adrenal system causing an increase in certain hormones in serum (i.e., cortisol and adrenocorticotropin) [65]. Physiologic parameters are objective; however, there are currently no predefined, validated values that are specific to pain in primates. Moreover, measuring these parameters usually requires the capture and restraint of primates, which can skew the results by increasing arousal and stress [61]. Field research has made use of techniques to attempt to measure these parameters remotely, such as imaging photoplethysmography to measure heart rate from a distance [55]. Other possibilities include measuring cortisol levels in saliva, urine, or feces [18,56–58]. For example, Salimetrics® Oral Swabs were validated for cortisol measurement in marmosets, in which animals are given a swab to chew (saliva can be extracted from the swab) [66]. Physiologic measures are not specific to pain and need to be interpreted using other measures and information.

3.3. Clinical Indicators

Clinical signs are representative of the outcome of animal care. In a research setting, body weight or body condition scores are recorded for study purposes but also to evaluate overall animal health. When body weight or body condition scores drop this usually signals a need for a further veterinary exam. Weight loss or a decrease in body condition score is an indirect measure of pain that may reflect a behavioural change related to pain; however, it can also be linked to other sources such as illness (i.e., chronic disease, cancer, or infection), social conditions, and the environment [67]. Body condition scoring may be a preferred measure in primates since it can be performed cage side. Currently, validated body condition score scales have been validated for macaques and chimpanzees [59]. However, there is no literature on body weight or body condition score changes in relation to acute or chronic pain in research primates. This method should be used as an indirect indicator of potential chronic pain, general health, or as a human endpoint [60].

Similar to body weight, food intake or appetite are clinical indicators related to general health. In a research setting, food consumption or food evaluations are generally values recorded for study purposes and for health monitoring [68]. When an animal experiences acute or chronic pain, it may lead to a reduced appetite [69]. It is important to note that both body weight and appetite will most likely be reduced in the days following a moderate to highly invasive procedure; thus, these values need to be interpreted in context [9]. These types of clinical measures are representative of a long duration of negative states, such as pain, and they should be used in conjunction with other measures of pain, to avoid prolonged welfare compromise. Usually when pain is anticipated or present animals are treated with analgesic agents. The administration of analgesics or anesthetics alone can also result in diminished food intake and body weight demonstrating the importance of context and variables that can influence these two clinical signs [70].

3.4. Behaviour

Cage-side behavioural observations are the most commonly reported wellness assessment in research primates [71]. Many welfare frameworks and uni/multidimensional scales have been created to quantify pain and behaviour. For example, the Extended Welfare Assessment Grid (EWAG) for the assessment of welfare and cumulative suffering in experimental animals evaluates the following components: clinical condition, experimental/clinal events, environment, and behavioural deviations [72]. Having a grid with specific welfare descriptors aids in assessing objective behavioural measures. However, this tool is not specific to pain assessment and there are currently no pain assessment tools for primates. Similarly, primate welfare assessment indicators have been identified and developed into a tool to assess general primate welfare in a research setting [35,36]. Again, these indicators are not specific to pain, but can be influenced by pain or pain treatment, for example, reduced appetite [35]. Another example more specific to pain is the Melbourne Pain Scale used in a clinical setting for cats and dogs. This tool assesses mostly behaviour indicators of pain, including activity levels, vocalizations, response to palpation, posture, mental status, and physiology measures [19]. These tools demonstrate that it is possible to measure behaviour objectively; however, as pain is a subjective and transient experience, knowledge of the animal's normal state is necessary. This can be challenging in a research setting if regular behaviour and temperament assessments are not conducted or recorded. This demonstrates the importance of communicating with technical staff given their close, daily experience with animals.

Various behaviours can be indicative of potential pain in primates. General activity levels can be assessed to detect potential pain. A recent study examining wellness indicators in rhesus macaques in the post-procedure period found a significant reduction in overall activity levels and a decrease in the behavioural repertoire including arboreal behaviours (i.e., climbing, hanging, and standing up straight) [20]. Another behaviour indicative of potential pain found in the latter study was slouched posture, in which the head is positioned below the level of the shoulders. These findings are similar to a study that evaluated the efficacy of different analgesics following abdominal surgery in olive baboons, supported by telemetry data [43]. As with other pain indicators, decreases in general activity and hunched posture may occur due to other states, such as depression [63]. Furthermore, behaviours may be influenced by the location and type of pain. Assessing guarding and disuse of a body part, vocalization in response to touch or palpation, as well as lameness, may be indicative of pain in a specific area. This demonstrates that pain should be interpreted in context and emphasizes the importance of keeping an observation log even when primates are not on study.

Changes in species-typical behaviour relative to baseline can also be indicative of potential pain. Primates are social species, thus evaluating social behaviour with conspecifics or humans can be helpful. In a worksheet developed to assess behaviour as a quality of life assessment in nonhuman primates, researchers incorporate several social behaviours such as affiliate behaviour (i.e., grooming, huddling, embracing, or proximity to others), play behaviour (i.e., wrestling, pulling, tickling, chasing, or play biting), aggressive behaviour (i.e., threatening, chasing, hitting, attacking, fighting, or biting), submissiveness to other (i.e., pant-grunting, lip-smacking, bobbing, avoiding, crying, or grimacing), and interest in a novel situation that includes humans [73]. Using a primate's natural daily activity budget as a benchmark for measuring welfare can be useful. For example, primates spend ~40 to 60% of their daily activity budget foraging for food; thus, a significant alteration in this time proportion can be indicative that the animal is feeling unwell [74]. This study also emphasizes the importance of creating an environment that provides primates with the opportunity to perform species-motivated behaviours so that changes can be used as a measure of welfare and, potentially, be indicative of the presence of pain.

To quantify daily activity budgets, detailed ethograms can be created [75,76]. In the context of pain, ethograms can be helpful to quantify the reduction in normal or species-typical behaviour and the appearance of pain-related behaviours [77]. Software systems such as Observer XT can facilitate behavioural scoring and statistical analysis; however, conducting these assessments is laborious, inter- and intra-observer reliability needs to be assessed, and assessments need to be conducted in real time to be useful. To attempt to address the problem of real-time scalability, new technology is being developed to automate pain behaviour recognition for some species, such as mice [77–79]; however, none currently exist for research primates.

3.5. Facial Expressions

In the past decade, facial grimace scales have been developed for many species, including domestic [80,81], agricultural [82–85], and research [86–88] animals. The facial action units within these scales are similar among species, generally focusing on the eye, mouth, and nose areas. Generally, facial grimace scales are composed of 3–5 facial action units scored on a numerical scale resulting in a score that reflects the level of pain an animal is experiencing, ideally resulting in prompt decisions for pain treatment.

Grimace scales are relatively simple to conduct and a rapid measure of pain that is readily available and requires minimal training; however, this method should be used alongside other pain assessments to offer a holistic approach and make the most accurate decision for pain treatments. Although no validated grimace scale exists for primates, as reported in the survey (see Section 2), this method is used by primate veterinarians. In a recent study conducted on macaques, it was found that eye narrowing and lip tightening were present in the postoperative period compared to baseline [20]. However, due to the methodology and levels of variations between subjects these changes were not significant [20]. This shows that there may be similarities with other species and that future research may yet validate a facial grimace scale for research primates.

Facial grimacing is a novel measure of pain in animals and past research has focused on scoring images retrospectively. Currently, research is emphasizing the use and validation of the grimace scale for real-time use. The mice and rat grimace scales have been validated for cage-side use, and this has helped to demonstrate that recommended analgesic doses for certain procedures in mice and rats were insufficient, leading to new recommendations [61,89–91].

As with the other pain assessment methods, there are some considerations when using facial expressions to evaluate pain. Human observers can influence an animal's facial expressions; thus, when possible, indirect observations are preferred in primates [25]. In a recent review, the main barriers to the widespread clinical application of grimace scales are discussed [92]. This includes not being able to compare the suspected painful animal to their baseline (i.e., the methods used to create grimace scales do not lead to practical implementation); statistical significance in parameter changes may not translate to clinical significance, and thus it is important to set an intervention threshold; and the variance between observer and their experience with a given species, emphasizing the importance of having a robust training [92]. Grimace scales have great potential for clinical use, but need to be used alongside clinical assessment and behaviour.

4. Pain Treatments in Research Primates—A Review

The standard means to treat pain in research primates is through pharmacologic methods. Most therapeutic pain treatments fall into one of the following classes: opioids, nonsteroidal anti-inflammatory drugs, and local anesthetics. In this section, we present the challenges of treating pain in research primates, the primary analgesics used with pharmacokinetic values and efficacy evidence (where available), and the routes of administration and potential side effects.

4.1. Research Primates and Analgesics

Treating pain in research animals can be challenging due to the balance between scientific outcome (i.e., analgesic interaction with test articles or models and the confounding effect of unalleviated pain) and animal welfare (i.e., ethical duty to minimize pain and distress) [5]. There is a lack of evidence-based knowledge for pain management in primates specifically for pharmacokinetics and analgesic efficacy as well as a lack of reporting detailed analgesic protocols [31,32].

Consequently, current recommendations are often extrapolated from other species, even though pharmacokinetics can have significant interspecies and intraspecies variability. To effectively treat pain, it is necessary to know the appropriate analgesic, dosage, frequency, and period of action. Furthermore, pain is a subjective experience and varies in intensity and duration depending on the individual's sex [93], previous experience with pain [16], psychological state [94], social status [17], and genetics [95]. Considering this variability, pain treatments should be tailored to the individual, which can be difficult in a research setting when working with large groups of animals.

4.2. Opioids

Opioids are generally used to treat moderate to severe anticipated pain in research primates. As reported in Section 2, the most commonly used opioids are buprenorphine, hydromorphone, fentanyl, and tramadol. These opioids have various mechanisms of action and potencies. For example, buprenorphine is a partial mμ agonist and can reach a plateau, limiting analgesic effects, whereas fentanyl is an mμ agonist and serum concentrations rise as dosages increase [96]. These properties should be considered when developing a pain management protocol.

The most studied analgesic therapeutic class in research primates, in terms of pharmacokinetics, is opioids (see Table 4). When comparing the efficacy and pharmacokinetics reported in other laboratory species, such as rodents, empirical evidence in primates is very limited [97]. Another gap in the primate literature is efficacy testing. Most rodent studies conduct efficacy testing with reflex or chemical-based assays [97], whereas there are few validated reflex-based assays for research primates.

Studies evaluating efficacy testing in primates have usually assessed behaviour and physical indicators of pain in the post-operative period, but rarely report pharmacokinetic values. For example, the efficacy of buprenorphine (0.01 mg/kg given every 12 h) was assessed in olive baboons undergoing a surgical procedure with and without combination with carprofen [43]. Some individuals had elevated heart rates and reduced activity levels in the post-operative period when compared to the multimodal approach potentially indicative of pain [43]. Although this study did not report pharmacokinetic values, it can be interpreted with other studies that used the same dosage of buprenorphine in primates. For example, a study in cynomolgus macaques demonstrated that buprenorphine (0.01 mg/kg) had a half-life of 2.6 ± 0.7 h and a Cmax of 8.1 ng/mL with a recommended dose interval of 6–8 h [45]. This study also examined the pharmacokinetics of buprenorphine at 0.03 mg/kg with a very high standard deviation (i.e., Cmax 40.7 ± 48.7 ng/mL) indicating high individual variability with the same dosage [45]. The half-life at this dosage was 5.3 h—suggesting that waiting 8–12 h before redosing may leave the animal without sufficient analgesic coverage. This emphasizes that more research that examines both efficacy and pharmacokinetics is needed in research primates.

Similarly, there is evidence of high variability for efficacy and pharmacokinetics between species. A recent study examining two formulations of transdermal fentanyl patches in cynomolgus macaques based on reported doses in dogs demonstrated adverse effects at 2.6 mg/kg, which had been demonstrated to be effective and safe in dogs [98]. This highlights some of the challenges in creating adequate pain management protocols in primates.

Table 4. Pharmacokinetics of common analgesics reported in primates.

Species	Class	Agent	Dosage	Route	Duration of Action	Cmax	Half-Life	AUC	Efficacy	Reference
Rhesus macaque	Opioid	Bupr	0.03 mg/kg	IM IV bolus	12 h 24 h	11.8 ng/mL C0: 33.0 ng/mL	- -	0–24: 1519 min*ng/mL 2188 min*ng/mL	No	[79]
Rhesus macaque	Opioid	Bupr (HCBS)	0.24 mg/kg 0.72 mg/kg	SC	48 h 72 h	19.1 ng/mL 65.2 ng/mL	α5.64 h β19.6 h α3.49 h β20.6	0–120: 236.4 ng*h/mL 641.3 ng*h/mL	No	[99]
Cynomolgus macaque	Opioid	Bupr	10 μg/h 20 μg/h	TDP TDP	5 d 6 d	3.3 ng/mL 8.1 ng/mL	64.2 h 42.4 h	0–168: 300.8 ng*h/mL 678.5 ng*h/mL	No	[100]
Cynomolgus and rhesus macaque	Opioid	Bupr	0.01 mg/kg 0.03 mg/kg	IM	6–8 h 12 h	8.1 ng/mL 40.7 ng/mL	2.6 h 5.3 h	0–120: 9.1 ng*h/mL 39.0 ng*h/mL	No	[45]
Cynomolgus and rhesus macaque	Opioid	Bupr-SR	0.2 mg/kg	SC	5 d	15.3 ng/mL	42.6 h	0–120: 177.0 ng*h/mL	No	[45]
Olive baboons	Opioid	Bupr	0.01 mg/kg	IM	12 h	-	-	-	Behaviour Heart rate Cortisol	[43]
Rhesus macaque	Opioid	Liposomal Hydr	2 mg/kg	SC	-	55.3 ng/mL	30.4 h	0–144: 424.7 ng*h/mL	No	[101]
Rhesus macaques	Opioid	Hydr	0.1 mg/kg	SC IV	-	26.4 ng/mL C0: 35.6 ng/mL	0.7 h 0.6 h	0–144: 32.5 ng*h/mL 36.3 ng*h/mL	No	[101]
Rhesus macaques	Opioid	Hydr	0.075 mg/kg	IM IV	2 h	12.0 ng/mL 77.6 ng/mL	81.5 min 17.7 min	- -	No	[72]
Rhesus macaques	Opioid	Fentanyl	1.3 mg/kg 2.6 mg/kg	TFS	7 d 10 d	1.95 μg/mL 4.2 μg/ml	90.9 h 97.4 h	0–504: 221.0 h/μg/mL 433.0 h/μg/mL	No	[102]
Rhesus macaques	Opioid	Fentanyl	0.005 mg/kg 0.01 mg/kg	SC	-	-	-	-	Thermode stimulation	[51]
Cynomolgus macaques	Opioid	Fentanyl	25 μg/h	TDP	4 d	2.4 ng/mL	45.2 h	0–96 h: 8.5 ng*h/mL	No	[98]

Table 4. *Cont.*

Species	Class	Agent	Dosage	Route	Duration of Action	Cmax	Half-Life	AUC	Efficacy	Reference
Cynomolgus macaques	Opioid	Fentanyl	1.95 mg/kg	TDS	4 d	177.1 ng/mL	32.8 h	0–96 h: 646.8 ng*h/mL -	No	[98]
Cynomolgus macaques	Opioid	Fentanyl	25μg/h	TDP	4 d	2.2 ng/mL	16.6 h	0–168: 110.3 ng*h/mL	No	[100]
Rhesus macaque	Opioid	Tram	1.5 mg/kg 3.0 mg/kg	IV PO	- -	C0: 540 ng/mL 15.2 ng/mL	111 min 133 min	- -	No	[72]
Rhesus macaque	Opioid	Tram	2.5 mg/kg 5 mg/kg	SC	-	-	-	-	Thermode stimulation	[51]
Cynomolgus macaque	NSAID	Mel	0.2 mg/kg 0.1 mg/kg	IM P0	24 h 8–12 h	2134.2 ng/mL 440.7 ng/mL	13.6 h 14.1 h	- -	No	[44]
Cynomolgus macaque	NSAID	Mel-SR	0.6 mg/kg	SC	48–72 h	3183.2 ng/mL	13.1 h	0–120: 80,407.4 ng*h/mL	No	[44]
Olive baboons	NSAID	Car	2.2 mg/kg	IM	12 h	-	-	-	Behaviour Heart rate Cortisol	[43]
Olive baboons	NSAID + Opioid	Car + Bupr	0.01 mg/kg + 2.2 mg/kg	IM	12 h	-	-	-	Behaviour Cortisol Heart rate	[43]

HCBS: highly concentrated buprenorphine solution; TDP: transdermal patches; TDS: transdermal solution; d: days; h: hour; Tram: tramadol; Mel: meloxicam; Car: carprofen; Bupr: buprenorphine; Hydr: hydromorphone: SR: sustained release.

4.3. Nonsteroidal Anti-Inflammatory Drugs (NSAID)

NSAIDs are frequently used to treat mild to moderate pain in primates or are added as part of a multimodal regimen. Typically, NSAIDs act on cyclooxygenase-1 (COX-1) and COX-2 receptors to control the inflammatory response and provide analgesia [95,103]. As reported in section A, the most used NSAIDs in research primates are meloxicam, carprofen, ketoprofen, and flunixin. Through this review, it was determined that of these commonly used NSAIDs, only meloxicam has pharmacokinetic values specific to research primates and only carprofen has been studied for efficacy in primates (see Table 4). No pharmacokinetic studies are available for the other common analgesics reported in Section 2 (carprofen, ketoprofen, and flunixin); thus, data from other species are provided in Table 5. As mentioned, most information about dosages and dosing intervals for research primates contain values extrapolated from other species, such as rodents, cats, and dogs [60]. Oral and injectable preparations of both meloxicam and carprofen are available and recommended to be given once a day [104,105]. However, research in rodents shows that previously recommended doses of NSAIDs, such as carprofen were insufficient for common surgical procedures, demonstrated by elevated facial grimace scores [106,107]. This highlights the importance of pain recognition and assessment outside of reflex-based assays and the need to evaluate pain when using recommended doses of analgesic agents. Furthermore, as mentioned above, there is interspecies variability; thus, analgesics need to be used with care in primates with effects monitored frequently.

Table 5. Pharmacokinetics of analgesics used in other common research species.

Species	Class	Agent	Dosage	Route	Dosing Interval	Cmax	Half-Life	AUC	Efficacy	Reference
Dog	NSAID	Carprofen	25 mg	PO SC	12 h	16.5 µg/mL 8.08 µg/mL	4.95 h 7.07 h	0–12: 71.7 µg*h/mL 64.9 µg*h/mL	No	[108]
Goat	NSAID	Ketoprofen	3 mg/kg	IV	12 h	13.6 µg/mL	3.10	0–24: 7.71 µg*h/mL	No	[109]
Dog	NSAID	Ketoprofen	1.0 mg/kg	PO	-	2.02 µg/mL	1.65 h	0–12.5: 6.06 µg*h/mL	No	[110]
Rat	NSAID	(S)-Ketoprofen	3.2 mg/kg	PO	-	2.73 µg/mL	26.9 h	0–24: 34.5 µg*h/mL	Pain-induced function impairment	[111]
Sow	NSAID	Flunixin	3.3 mg/kg	TD	-	14.61 ng/mL	9.76 h	214.78 ng*h/mL	No	[112]
Mice	NSAIDS	Flunixin	2.0 mg/kg	SC	-	4553.4 ng/mL	0.95 h	0–6: 4742 ng*h/mL	No	[113]

4.4. Multimodal Analgesia

Combining different classes of analgesic agents to target different pain pathways is often beneficial [39,114]. Although there are limited data on the pharmacokinetics of multimodal regimens in primates, it is known that when an opioid, an NSAID, and a regional block with a local anesthetic are combined, this allows for the reduction in the dosages/frequency of the individual drugs and consequently their side effects (multimodal analgesia regimes in rodents, reviewed in [96]). This review did not evaluate local anesthetics in depth; their use was queried in the survey in Section 2, and the most used local anesthetics were reported to be bupivacaine, lidocaine, proparacaine, and EMLA® (a prilocaine/lidocaine mixture) cream. Local anesthetics, such as bupivacaine and lidocaine typically have a short duration of action of 60 min and 30 min, respectively, and are used around the surgical site to reduce peripherical nociceptor activity [104,115]. Thoughtful perioperative planning, for example, by administering NSAIDs prior to surgery, as well as opioids and local anesthetics during surgery to treat pain before its onset, has demonstrated a faster recovery, minimizing the potential for breakthrough pain [116]. More information is needed to optimize perioperative analgesia protocols in research primates.

4.5. Route of Analgesic Administration

There are many factors to consider when choosing the route of administration of therapeutics in primates, such as the stress associated with handling or immobilization, the frequency of administration to achieve therapeutic levels, the level of absorption/bioavailability, and the desired effect. Below we discuss the common routes used in primates as well as the advantages and disadvantages from a practical and physiological standpoint.

Oral administration of analgesics is common and some primates will voluntarily take the medication cage side if it is palatable. This is especially true when positive reinforcement training is employed with primates [117]. Another option for voluntary consumption is through chewable commercially available tablets. Common opioids used in research primates such as tramadol and hydromorphone as well as NSAIDs such as meloxicam, carprofen, ketoprofen, and flunixin are available in commercially produced oral preparations [96]. Oral administration may cause more efficacy variability compared to other routes of administration based on fed or fasted conditions of animals [118]. Furthermore, some therapeutic agents may irritate the gastrointestinal mucosa when given orally [59].

Subcutaneous (SC) injections are given between the layer of skin and muscle and can be administered over a large portion of the body. Typically, the rate of absorption is slower, which may be desirable for prolonged action [105]. Subcutaneous injections can cause depot accumulation; thus, injection sites need to be changed when multiple injections are given [105]. In primates, common opioids such as buprenorphine, hydromorphone, and fentanyl and NSAIDs such as meloxicam have been reported to be used SC [45,51,99,101,119].

Intramuscular (IM) injection is the most common route of analgesic administration in primates because it is an easy technique that requires minimal handling and restraint. IM injections are given in deep muscle tissue and the high vascularisation permits rapid absorption [105]. The volume per injection site should be limited based on the primate's weight to minimize the potential for injury and necrosis. For example, a primate weighing approximately 3 kg or 13 kg should receive no more than 0.5 mL or 1.0 mL per site, respectively [120]. The standard IM injection sites for primates include the caudal thigh, the deltoids, and the longissimus (paralumbar) muscles to avoid major blood vessels and nerves [120]. Opioids such as buprenorphine and hydromorphone as well as NSAIDS, including carprofen and meloxicam, have been reported to be used via IM injection in research primates [43–45,121].

Intravenous (IV) injections are usually given in a superficial vein with a needle or via continuous infusion with a catheter. The rate of infusion/administration is controlled for a given time and smaller doses are generally required since agents are administered directly into the bloodstream [105]. Without sedation and/or training, primates may be unwilling to cooperate for a long duration. In primates, common opioids such as buprenorphine, hydromorphone, and tramadol have been administered through IV injection [51,72,101].

Finally, transdermal drug delivery via patch or other protected depot can be beneficial for long-term pain treatments through slow epidermal absorption and requires minimal handling after the first application [122]. Primates usually require a jacket (which can require further pre-study habituation) to avoid self or partner ingestion. Inadvertent ingestion of a fentanyl patch with fatal consequences has been reported in primates [123]. More recently, research on the use of transdermal fentanyl solution and patches has demonstrated prolonged serum concentrations at therapeutic levels over 3–5 days [97,100,102].

4.6. Adverse Effects

The goal of therapeutic pain treatment in animals is to create a balanced state and minimize the pain experienced without producing substantial adverse effects. Multimodal approaches are recommended when possible, as these techniques generally result in reduced dosages and dosing frequency when compared to individual analgesics. Each class of therapeutic analgesics has side effects based on their different action mechanisms, chemical structures, formulation pH, etc. It is important to know these side effects when treating pain to recognize them and modify treatment accordingly to ensure that a toxic level is not reached. From a research perspective, knowledge of adverse effects can be used to distinguish analgesic effects from test article effects or study outcomes. There is limited research into analgesic adverse effects in research primates; thus, these are mostly identified based on the literature for other species, such as dogs.

Opioids are potent drugs with a narrow window for therapeutic safety; consequently, they must be used with caution and doses must be calculated for the individual animal [104]. Opioids have been reported to cause respiratory depression, bradycardia, and when administered at high doses or via IV injection can cause hypertension [124]. Opioids can also affect the gastrointestinal tract by reducing mobility and emptying as well as inducing nausea and vomiting [125,126]. Opioids may induce behaviour changes, specifically, sedation. For example, in a study comparing the behavioural and physiologic effects of morphine versus fentanyl in dogs, significantly higher sedation scores were seen when fentanyl was the chosen analgesic [127]. The primary adverse effects of NSAIDs occur in the gastrointestinal tract, inducing ulceration, perforations, diarrhea, vomiting, and reduced appetite [128]. In extreme cases in dogs, some COX-2-specific NSAIDs have also been reported to cause hepatic failure and lethargy [129]. These side effects must be assessed at the individual level as different animals will react differently.

Different routes of administration can also create adverse effects. Skin puncture can be mildly painful, and it is important to regularly check injection sites as some animals may have an adverse reaction. For example, a small proportion of cynomolgus macaques injected SC with meloxicam SR showed adverse injection site reactions including redness, sloughing of superficial tissue, and abscess formation, whereas other animals in the same study did not [44]. If injections are frequent, consider recording injection sites to ensure that specific sites are not over-used to avoid tissue damage. If injections cause a severe reaction consider an alternate route of administration.

Distinguishing pain behaviour from sedation behaviour in research primates is important as signs can be similar. Furthermore, in some circumstances primates are immobilized with anesthetics such as ketamine to be manipulated; thus, when administered in conjunction with analgesics, side effects may be difficult to distinguish. The main side effect that anesthetics and analgesics have in common is reduced appetite. A study in rhesus macaques and African green monkeys evaluated the association between ketamine injections (10 mg/kg) and appetite, 24, 48, 72, and 120 h post-injection. The researchers demonstrated a significant decrease in food intake at all timepoints with 24 h post-dose being the most significant (mean % intake reduction: African green monkeys: 57%; rhesus males: 48%; rhesus females: 40%, respectively) [70]. Decreased food intake has also been reported following the use of analgesics in healthy animals, likely due to sedation [39].

5. Recommendations and Considerations for Refinement of Pain Management Guidance for Research Primates

5.1. Institutional Policy to Implement Pain Management Guidance

Based on our survey results (Section 2), there is evidence of the need to implement guidance within and between research institutions on primate pain assessment, pain treatment, and general pain management procedures. The Canadian Council on Animal Care (CCAC) has released new guidelines encouraging research facilities to create and implement welfare assessments for laboratory species including the need to incorporate indicators related to pain assessment and management [130]. There is a critical need for pharmacokinetic and efficacy testing (based on objective pain assessment methods) to inform treatment protocols for research primates as most evidence stems from anecdotal evidence.

5.2. Analgesic Administration Based on Empircal Evidence

There are many possible routes of administration for therapeutics, and it is important to consider the required handling/restraint for each method. Incorporating slow-release formulations may reduce the need for handling and restraint. These formulations also help to minimize the risk of breakthrough pain [131]. Given the variability of absorption between different dose routes, there should be frequent monitoring in the immediate hours following presumed pain with the administration of analgesics to ensure adequate pain relief.

5.3. *Appropriate Use of Pain Assessment Tools*

Research institutions should have a standardized pain assessment protocol that integrates two or more methods identified in Table 5. These protocols should have an objective scoring system that can be replicated by multiple users and that demonstrate consistent results over time. Integrating an analgesic threshold associated with the outcome of pain assessment should be established. The gold standard pain assessment method for research primates is indirect observation of behaviour. Combining this with other methods, such as physiologic and clinical markers, ensures a more reliable assessment of pain and thus better management and mitigation.

6. Conclusions

Results from a survey administered to primate veterinarians demonstrated inconsistencies in research primate pain management as well as a general lack of objective pain assessment tools. Information in this review may be used by research institutions to evaluate primate care as well as for creating primate-specific internal guidance. These inconsistencies correspond with gaps in the research primate pain literature, which includes limited pharmacokinetic studies and efficacy testing for commonly used analgesics as well as limited objective measures of pain. These findings should encourage researchers and veterinarians to study and report more detailed methods of pain management practices to further improve research primate welfare and the quality of scientific data.

Supplementary Materials: The following supporting information can be downloaded at: https://www.mdpi.com/article/10.3390ani12172304/s1, Supplementary Material: The use of analgesics in research primates questionnaire.

Author Contributions: Conceptualization and methodology, P.V.T., original draft preparation, E.A.P., writing—review and editing, E.A.P. and P.V.T. All authors have read and agreed to the published version of the manuscript.

Funding: This research received no external funding.

Institutional Review Board Statement: Institutional ethical review was not required for secondary use of survey data collected anonymously by a third party.

Informed Consent Statement: Not applicable.

Data Availability Statement: Dataset available from authors upon request.

Acknowledgments: We thank the APV Primate Pain Working Group for review of the draft survey and John Farrar for assistance with survey dissemination. We would also like to thank Raina Hubley for her support in the literature search.

Conflicts of Interest: The authors declare no conflict of interest.

References

1. Walker, G.; Wilcock, A.; Manderson, C.; Weller, R.; Crosby, V. The Acceptability of Different Routes of Administration of Analgesia for Breakthrough Pain. *Palliat. Med.* **2003**, *17*, 219–221. [CrossRef] [PubMed]
2. Carlsson, H.E.; Schapiro, S.J.; Farah, I.; Hau, J. Use of Primates in Research: A Global Overview. *Am. J. Primatol.* **2004**, *63*, 225–237. [CrossRef] [PubMed]
3. Friedman, H.; Ator, N.; Haigwood, N.; Newsome, W.; Allen, J.S.; Golos, T.G.; Kordower, J.H.; Shade, R.E.; Goldberg, M.E.; Bailey, M.R.; et al. The Critical Role of Nonhuman Primates in Medical Research—White Paper. *Pathog. Immun.* **2017**, *2*, 352. [CrossRef] [PubMed]
4. Coleman, K. Caring for Nonhuman Primates in Biomedical Research Facilities: Scientific, Moral and Emotional Considerations. *Am. J. Primatol.* **2011**, *73*, 220–225. [CrossRef]
5. Lloyd, M.H.; Foden, B.W.; Wolfensohn, S.E. Refinement: Promoting the Three Rs in Practice. *Lab. Anim.* **2008**, *42*, 284–293. [CrossRef]

6. Peterson, N.C.; Nunamaker, E.A.; Turner, P.V. To Treat or Not to Treat: The Effects of Pain on Experimental Parameters. *Comp. Med.* **2017**, *67*, 469–482.
7. CCAC. *CCAC Guidelines: Nonhuman Primates*; CCAC: Ottawa, ON, Canada, 2019.
8. European Commission. *Directive 2010/63/EU of the European Parliament and of the Council of 22 September 2010 on the Protection of Animals Used for Scientific Purposes*; Official Journal of the European Union: Maastricht, The Netherlands, 2010; pp. 33–79.
9. United States Department of Agriculture. *Animal Welfare Act and Animal Welfare Regulations*; United States Department of Agriculture: Washington, DC, USA, 2019.
10. NRC. *Recognition and Alleviation of Pain in Laboratory Animals, 44*; National Academies Press (US): Washington, DC, USA, 2009. [CrossRef]
11. Stasiak, K.L.; Maul, D.; French, E.; Hellyer, P.W.; Vandewoude, S. Species-Specific Assessment of Pain in Laboratory Animals. *Contemp. Top Lab Anim. Sci.* **2003**, *42*, 13–20.
12. Carstens, E.; Moberg, G.P. Recognizing Pain and Distress in Laboratory Animals. *ILAR J.* **2000**, *41*, 62–71. [CrossRef]
13. IASP's Proposed New Definition of Pain Released for Comment—IASP. Available online: https://www.iasp-pain.org/PublicationsNews/NewsDetail.aspx?ItemNumber=9218 (accessed on 12 January 2020).
14. Yezierski, R.P. The Effects of Age on Pain Sensitivity: Preclinical Studies. *Pain Med.* **2012**, *23*, S27. [CrossRef]
15. Da Silva, J.T.; Tricou, C.; Zhang, Y.; Seminowicz, D.A.; Ro, J.Y. Brain Networks and Endogenous Pain Inhibition Are Modulated by Age and Sex in Healthy Rats. *Pain* **2020**, *161*, 1371–1380. [CrossRef]
16. Parent-Vachon, M.; Vachon, P. Environmental Enrichment Alleviates Chronic Pain in Rats Following a Spared Nerve Injury to Induce Neuropathic Pain. A Preliminary Study. *Vet. Med. Res. Rep.* **2018**, *9*, 69–72. [CrossRef] [PubMed]
17. Guesgen, M.J.; Beausoleil, N.J.; Stewart, M. Effects of Early Human Handling on the Pain Sensitivity of Young Lambs. *Vet Anaesthe Analg.* **2013**, *40*, 55–62. [CrossRef] [PubMed]
18. Langford, D.J.; Tuttle, A.H.; Brown, K.; Deschenes, S.; Fischer, D.B.; Mutso, A.; Root, K.C.; Sotocinal, S.G.; Stern, M.A.; Mogil, J.S.; et al. Social Approach to Pain in Laboratory Mice. *Soc. Neurosci.* **2010**, *5*, 163–170. [CrossRef]
19. Chen, H.; Yao, H.; Yang, W.; Fan, P.; Xiang, Z. Assessing the Utility of Urinary and Fecal Cortisol as an Indicator of Stress in Golden Snub-Nosed Monkeys (Rhinopithecus Roxellana). *PeerJ* **2017**, *8*, e3648. [CrossRef] [PubMed]
20. Saberi Afshar, F.; Shekarian, M.; Baniadam, A.; Avizeh, R.; Najafzadeh, H.; Pourmehdi, M. Comparison of Different Tools for Pain Assessment Following Ovariohysterectomy in Bitches. *Iran. J. Vet. Med.* **2017**, *11*, 255–265. [CrossRef]
21. Descovich, K.A.; Richmond, S.E.; Leach, M.C.; Buchanan-Smith, H.M.; Flecknell, P.; Farningham, D.A.H.; Witham, C.; Gates, M.C.; Vick, S.J. Opportunities for Refinement in Neuroscience: Indicators of Wellness and Post-Operative Pain in Laboratory Macaques. *ALTEX-Altern. Anim. Exp.* **2019**, *36*, 535–554. [CrossRef]
22. Novak, M.A.; Hamel, A.F.; Kelly, B.J.; Dettmer, A.M.; Meyer, J.S. Stress, the HPA Axis, and Nonhuman Primate Well-Being: A Review. *Appl. Anim. Behav. Sci.* **2013**, *143*, 135–149. [CrossRef]
23. Coleman, K.; Pierre, P.J. Assessing Anxiety in Nonhuman Primates. *ILAR J.* **2014**, *55*, 333–346. [CrossRef]
24. Tiefenbacher, S.; Novak, M.A.; Marinus, L.M.; Chase, W.K.; Miller, J.A.; Meyer, J.S. Altered Hypothalamic-Pituitary-Adrenocortical Function in Rhesus Monkeys (Macaca Mulatta) with Self-Injurious Behavior. *Psychoneuroendocrinology* **2004**, *29*, 501–515. [CrossRef]
25. Tardif, S.D.; Coleman, K.; Hobbs, T.R.; Lutz, C. IACUC Review of Nonhuman Primate Research. *ILAR J.* **2013**, *54*, 234–245. [CrossRef]
26. Gaither, A.M.; Baker, K.C.; Gilbert, M.H.; Blanchard, J.L.; Liu, D.X.; Luchins, K.R.; Bohm, R.P. Videotaped Behavior as a Predictor of Clinical Outcome in Rhesus Macaques (Macaca Mulatta). *Comp. Med.* **2014**, *64*, 193–199. [PubMed]
27. Hawkins, P. Recognizing and Assessing Pain, Suffering and Distress in Laboratory Animals: A Survey of Current Practice in the UK with Recommendations. *Lab. Anim.* **2002**, *36*, 378–395. [CrossRef] [PubMed]
28. Fenwick, N.; Duffus, S.E.G.; Griffin, G. Pain Management for Animals Used in Science: Views of Scientists and Veterinarians in Canada. *Animals* **2014**, *4*, 494–514. [CrossRef] [PubMed]
29. Association of Primate Veterinarians Guidelines for Nonhuman Primate Restraint. *J. Am. Assoc. Lab. Anim. Sci.* **2019**, *58*, 282–284.
30. Lankau, E.W.; Turner, P.V.; Mullan, R.J.; Galland, G.G. Materials and Methods Use of Nonhuman Primates in Research in North America. *J. Am. Assoc. Lab. Anim. Sci.* **2014**, *53*, 278–282.
31. Karas, A.Z. Barriers to Assessment and Treatment of Pain in Laboratory Animals. *Lab. Anim.* **2006**, *35*, 38–45. [CrossRef]
32. Bertrand, H.G.M.J.; Sandersen, C.; Flecknell, P.A. Reported Analgesic and Anaesthetic Administration to Non-Human Primates Undergoing Experimental Surgical Procedure: 2010–2015. *J. Med. Primatol.* **2018**, *47*, 217–225. [CrossRef]
33. Coulter, C.A.; Flecknell, P.A.; Richardson, C.A. Reported Analgesic Administration to Rabbits, Pigs, Sheep, Dogs and Non-Human Primates Undergoing Experimental Surgical Procedures. *Lab. Anim.* **2009**, *43*, 232–238. [CrossRef]
34. Coulter, C.A.; Flecknell, P.A.; Leach, M.C.; Richardson, C.A. Reported Analgesic Administration to Rabbits Undergoing Experimental Surgical Procedures. *BMC Vet. Res.* **2011**, *7*, 12. [CrossRef]
35. Stokes, E.L.; Flecknell, P.A.; Richardson, C.A. Reported Analgesic and Anaesthetic Administration to Rodents Undergoing Experimental Surgical Procedures. *Lab. Anim.* **2009**, *43*, 149–154. [CrossRef]
36. Prescott, M.J.; Leach, M.C.; Truelove, M.A.; Vitale, A.; Superiore, I.; Sanità, D. Harmonisation of Welfare Indicators for Macaques and Marmosets Used or Bred for Research. *F1000Research* **2022**, *11*, 272. [CrossRef]

37. Truelove, M.A.; Martin, J.E.; Langford, F.M.; Leach, M.C. The Identification of Effective Welfare Indicators for Laboratory-Housed Macaques Using a Delphi Consultation Process. *Sci. Rep.* **2020**, *10*, 20402. [CrossRef] [PubMed]
38. Hugonnard, M.; Leblond, A.; Keroack, S.; Cadoré, J.L.; Troncy, E. Attitudes and Concerns of French Veterinarians towards Pain and Analgesia in Dogs and Cats. *Vet. Anaesth. Analg.* **2004**, *31*, 154–163. [CrossRef] [PubMed]
39. Di Vincenti, L. Analgesic Use in Nonhuman Primates Undergoing Neurosurgical Procedures. *J. Am. Assoc. Lab. Anim. Sci.* **2013**, *52*, 10–16.
40. Chou, R.; Gordon, D.B.; De Leon-Casasola, O.A.; Rosenberg, J.M.; Bickler, S.; Brennan, T.; Carter, T.; Cassidy, C.L.; Chittenden, E.H.; Degenhardt, E.; et al. Management of Postoperative Pain: A Clinical Practice Guideline From the American Pain Society, the American Society of Regional Anesthesia and Pain Medicine, and the American Society of Anesthesiologists' Committee on Regional Anesthesia, Executive Committee, and Administrative Council. *J. Pain* **2016**, *17*, 131–157. [CrossRef]
41. Brideau, C.; Van Staden, C.; Chan, C.C. In Vitro Effects of Cyclooxygenase Inhibitors in Whole Blood of Horses, Dogs, and Cats. *Am. J. Vet. Res.* **2001**, *62*, 1755–1760. [CrossRef]
42. Lee, J.I.; Kim, Y.S.; Kim, M.J.; Hong, S.H. Idiopathic New Bone Formation in the Femoral Shafts of a Cynomolgus Monkey (Macaca Fascicularis). *J. Am. Assoc. Lab. Anim. Sci.* **2008**, *47*, 68–71.
43. Saleem, K.S.; Tanaka, K. Divergent Projections from the Anterior Inferotemporal Area TE to the Perirhinal and Entorhinal Cortices in the Macaque Monkey. *J. Neurosci.* **1996**, *16*, 4757–4775. [CrossRef]
44. Allison, S.O.; Halliday, L.C.; French, J.A.; Novikov, D.D.; Fortman, J.D. Assessment of Buprenorphine, Carprofen, and Their Combination for Postoperative Analgesia in Olive Baboons (Papio Anubis). *J. Am. Assoc. Lab. Anim. Sci.* **2007**, *46*, 24–31.
45. Bauer, C.; Frost, P.; Kirschner, S. Pharmacokinetics of 3 Formulations of Meloxicam in Cynomolgus Macaques (Macaca Fascicularis). *J. Am. Assoc. Lab. Anim. Sci.* **2014**, *53*, 502–511.
46. Nunamaker, E.A.; Halliday, L.C.; Moody, D.E.; Fang, W.B.; Lindeblad, M.; Fortman, J.D. Pharmacokinetics of 2 Formulations of Buprenorphine in Macaques (Macaca Mulatta and Macaca Fascicularis). *J. Am. Assoc. Lab. Anim. Sci.* **2013**, *52*, 48–56. [PubMed]
47. Phillips, J.L.; Heneka, N.; Hickman, L.; Lam, L. Self-Perceived Pain Assessment Knowledge and Confidence (Self-PAC) Scale for Cancer and Palliative Care Nurses: A Preliminary Validation Study. *Pain Manag. Nurs.* **2018**, *19*, 619–626. [CrossRef] [PubMed]
48. Amponsah, A.K.; Oduro, E.; Bam, V.; Kyei-Dompim, J.; Ahoto, C.K.; Axelin, A. Nursing Students and Nurses' Knowledge and Attitudes Regarding Children's Pain: A Comparative Cross-Sectional Study. *PLoS ONE* **2019**, *14*, e0223730. [CrossRef] [PubMed]
49. Coleman, D.L.; Slingsby, L.S. Attitudes of Veterinary Nurses to the Assessment of Pain and the Use of Pain Scales. *Vet. Rec.* **2007**, *160*, 541–544. [CrossRef] [PubMed]
50. Miyabe-Nishiwaki, T.; Gris, V.N.; Muta, K.; Nishimura, R.; Mills, D.S. Primate Veterinarians' Knowledge and Attitudes Regarding Pain in Macaques. *J. Med. Primatol.* **2021**, *50*, 259–269. [CrossRef]
51. Association of Primate Veterinarians' Guidelines for Assessment of Acute Pain in Nonhuman Primates. *J. Am. Assoc. Lab. Sci.* **2019**, *58*, 748–749.
52. Vardigan, J.D.; Houghton, A.K.; Lange, H.S.; Adarayan, E.D.; Pall, P.S.; Ballard, J.E.; Henze, D.A.; Uslaner, J.M. Pharmacological Validation of a Novel Nonhuman Primate Measure of Thermal Responsivity with Utility for Predicting Analgesic Effects. *J. Pain Res.* **2018**, *11*, 735. [CrossRef]
53. Kangas, B.D.; Bergman, J. Operant Nociception in Nonhuman Primates. *Pain* **2014**, *155*, 1821–1828. [CrossRef]
54. Graham, D.M.; Hampshire, V. Methods for Measuring Pain in Laboratory Animals. *Lab. Anim.* **2016**, *45*, 99–101. [CrossRef]
55. Crockett, A.; Panickar, A. Role of the Sympathetic Nervous System in Pain. *Anaesth. Intensive Care Med.* **2011**, *12*, 50–54. [CrossRef]
56. Unakafov, A.M.; Möller, S.; Kagan, I.; Gail, A.; Treue, S.; Wolf, F. Using Imaging Photoplethysmography for Heart Rate Estimation in Non-Human Primates. *PLoS ONE* **2018**, *13*, e0202581. [CrossRef]
57. Palme, R. Monitoring Stress Hormone Metabolites as a Useful, Non-Invasive Tool for Welfare Assessment in Farm Animals. *Anim. Welf.* **2012**, *21*, 331–337. [CrossRef]
58. Pfefferle, D.; Plümer, S.; Burchardt, L.; Treue, S.; Gail, A. Assessment of Stress Responses in Rhesus Macaques (Macaca Mulatta) to Daily Routine Procedures in System Neuroscience Based on Salivary Cortisol Concentrations. *PLoS ONE* **2018**, *13*, e0190190. [CrossRef] [PubMed]
59. Behie, A.M.; Pavelka, M.S.M.; Chapman, C.A. Sources of variation in fecal cortisol levels in howler monkeys in Belize. *Am. J. Primatol.* **2010**, *72*, 600–606. [CrossRef] [PubMed]
60. Clingerman, K.J.; Summers, L. Validation of a Body Condition Scoring System in Rhesus Macaques (Macaca Mulatta): Inter- and Intrarater Variability. *J. Am. Assoc. Lab. Anim. Sci.* **2012**, *51*, 31–36.
61. Chapman, K.; Sewell, F.; Allais, L.; Delongeas, J.L.; Donald, E.; Festag, M.; Kervyn, S.; Ockert, D.; Nogues, V.; Palmer, H.; et al. A Global Pharmaceutical Company Initiative: An Evidence-Based Approach to Define the Upper Limit of Body Weight Loss in Short Term Toxicity Studies. *Regul. Toxicol. Pharmacol.* **2013**, *67*, 27–38. [CrossRef]
62. Lambeth, S.P.; Hau, J.; Perlman, J.E.; Martino, M.; Schapiro, S.J. Positive Reinforcement Training Affects Hematologic and Serum Chemistry Values in Captive Chimpanzees (Pan Troglodytes). *Am. J. Primatol.* **2006**, *68*, 245–256. [CrossRef] [PubMed]
63. Honess, P.; Wolfensohn, S. The Extended Welfare Assessment Grid: A Matrix for the Assessment of Welfare and Cumulative Suffering in Experimental Animals. *ATLA* **2010**, *38*, 205–212. [CrossRef]
64. Hennessy, M.B.; Chun, K.; Capitanio, J.P. Depressive-like Behavior, Its Sensitization, Social Buffering, and Altered Cytokine Responses in Rhesus Macaques Moved from Outdoor Social Groups to Indoor Housing. *Soc. Neurosci.* **2017**, *12*, 65–75. [CrossRef]
65. Barrot, M. Tests and Models of Nociception and Pain in Rodents. *Neuroscience* **2012**, *211*, 39–50. [CrossRef]

66. Cambridge, A.J.; Tobias, K.M.; Newberry, R.C.; Sarkar, D.K. Subjective and Objective Measurements of Postoperative Pain in Cats. *J. Am. Vet. Med. Assoc.* **2000**, *217*, 685–690. [CrossRef] [PubMed]
67. Ash, H.; Smith, T.E.; Knight, S.; Buchanan-Smith, H.M. Measuring Physiological Stress in the Common Marmoset (Callithrix Jacchus): Validation of a Salivary Cortisol Collection and Assay Technique. *Physiol. Behav.* **2018**, *185*, 14–22. [CrossRef] [PubMed]
68. Turner, P.V.; Pang, D.S.; Lofgren, J.L. A Review of Pain Assessment Methods in Laboratory Rodents. *Comp. Med.* **2019**, *69*, 451–467. [CrossRef] [PubMed]
69. Weary, D.M.; Niel, L.; Flower, F.C.; Fraser, D. Identifying and Preventing Pain in Animals. *Appl. Anim. Behav. Sci.* **2006**, *100*, 64–76. [CrossRef]
70. Liles, J.H.; Flecknell, P.A. The Effects of Surgical Stimulus on the Rat and the Influence of Analgesic Treatment. *Br. Vet. J.* **1993**, *149*, 515–525. [CrossRef]
71. Springer, D.A.; Baker, K.C. Effect of Ketamine Anesthesia on Daily Food Intake InMacaca Mulatta AndCercopithecus Aethiops. *Am. J. Primatol.* **2007**, *69*, 1080–1092. [CrossRef]
72. Krugner-Higby, L.; Kukanich, B.; Schmidt, B.; Heath, T.D.; Brown, C. Pharmacokinetics and Behavioral Effects of Liposomal Hydromorphone Suitable for Perioperative Use in Rhesus Macaques. *Psychopharmacology* **2011**, *216*, 511–523. [CrossRef]
73. Lutz, C.K.; Coleman, K.; Hopper, L.M.; Novak, M.A.; Perlman, J.E.; Pomerantz, O. Nonhuman Primate Abnormal Behavior: Etiology, Assessment, and Treatment. *Am. J. Primatol.* **2022**, *84*, e23380. [CrossRef]
74. Lambeth, S.P.; Schapiro, S.J.; Bernacky, B.J.; Wilkerson, G.K. Establishing "quality of Life" Parameters Using Behavioural Guidelines for Humane Euthanasia of Captive Non-Human Primates. *Anim. Welf.* **2013**, *22*, 429–435. [CrossRef]
75. Howell, C.P.; Cheyne, S.M. Complexities of Using Wild versus Captive Activity Budget Comparisons for Assessing Captive Primate Welfare. *J. Appl. Anim. Welf. Sci.* **2019**, *22*, 78–96. [CrossRef]
76. Xu, F.; Xie, L.; Li, X.; Li, Q.; Wang, T.; Ji, Y.; Kong, F.; Zhan, Q.; Cheng, K.; Fang, L.; et al. Construction and Validation of a Systematic Ethogram of Macaca Fascicularis in a Free Enclosure. *PLoS ONE* **2012**, *7*, e37486. [CrossRef] [PubMed]
77. Hage, S.R.; Ott, T.; Eiselt, A.-K.; Jacob, S.N.; Nieder, A. Ethograms Indicate Stable Well-Being during Prolonged Training Phases in Rhesus Monkeys Used in Neurophysiological Research. *Lab. Anim.* **2014**, *48*, 82–87. [CrossRef] [PubMed]
78. Ellen, Y.; Flecknell, P.; Leach, M. Evaluation of Using Behavioural Changes to Assess Post-Operative Pain in the Guinea Pig (Cavia Porcellus). *PLoS ONE* **2016**, *11*, e0161941. [CrossRef] [PubMed]
79. Roughan, J.V.; Wright-Williams, S.L.; Flecknell, P.A. Automated Analysis of Postoperative Behaviour: Assessment of Homecages-can as a Novel Method to Rapidly Identify Pain and Analgesic Effects in Mice. *Lab. Anim.* **2009**, *43*, 17–26. [CrossRef]
80. Andresen, N.; Wöllhaf, M.; Hohlbaum, K.; Lewejohann, L.; Hellwich, O.; Thöne-Reineke, C.; Belik, V. Towards a Fully Automated Surveillance of Well-Being Status in Laboratory Mice Using Deep Learning: Starting with Facial Expression Analysis. *PLoS ONE* **2020**, *15*, e0228059. [CrossRef]
81. Evangelista, M.C.; Watanabe, R.; Leung, V.S.Y.; Monteiro, B.P.; O'Toole, E.; Pang, D.S.J.; Steagall, P.V. Facial Expressions of Pain in Cats: The Development and Validation of a Feline Grimace Scale. *Sci. Rep.* **2019**, *9*, 1–11. [CrossRef]
82. Dalla Costa, E.; Minero, M.; Lebelt, D.; Stucke, D.; Canali, E.; Leach, M.C. Development of the Horse Grimace Scale (HGS) as a Pain Assessment Tool in Horses Undergoing Routine Castration. *PLoS ONE* **2014**, *9*, e92281. [CrossRef]
83. Guesgen, M.J.; Beausoleil, N.J.; Leach, M.; Minot, E.O.; Stewart, M.; Stafford, K.J. Coding and Quantification of a Facial Expression for Pain in Lambs. *Beh. Process.* **2016**, *132*, 49–56. [CrossRef]
84. Mclennan, K.M.; Rebelo, C.J.B.; Corke, M.J.; Holmes, M.A.; Leach, M.C.; Constantino-Casas, F. Development of a Facial Expression Scale Using Footrot and Mastitis as Models of Pain in Sheep. *Appl. Anim. Behav. Sci.* **2016**, *176*, 19–26. [CrossRef]
85. Viscardi, A.V.; Hunniford, M.; Lawlis, P.; Leach, M.; Turner, P.V. Development of a Piglet Grimace Scale to Evaluate Piglet Pain Using Facial Expressions Following Castration and Tail Docking: A Pilot Study. *Front. Vet. Sci.* **2017**, *4*, 51. [CrossRef]
86. Di Giminiani, P.; Brierley, V.L.M.H.; Scollo, A.; Gottardo, F.; Malcolm, E.M.; Edwards, S.A.; Leach, M.C. The Assessment of Facial Expressions in Piglets Undergoing Tail Docking and Castration: Toward the Development of the Piglet Grimace Scale. *Front. Vet. Sci.* **2016**, *3*, 100. [CrossRef] [PubMed]
87. Langford, D.J.; Bailey, A.L.; Chanda, M.L.; Clarke, S.E.; Drummond, T.E.; Echols, S.; Glick, S.; Ingrao, J.; Klassen-Ross, T.; Lacroix-Fralish, M.L.; et al. Coding of Facial Expressions of Pain in the Laboratory Mouse. *Nat. Methods* **2010**, *7*, 447–449. [CrossRef] [PubMed]
88. Sotocinal, S.G.; Sorge, R.E.; Zaloum, A.; Tuttle, A.H.; Martin, L.J.; Wieskopf, J.S.; Mapplebeck, J.C.C.S.; Wei, P.; Zhan, S.; Zhang, S.; et al. The Rat Grimace Scale: A Partially Automated Method for Quantifying Pain in the Laboratory Rat via Facial Expressions. *Mol. Pain* **2011**, *7*, 55. [CrossRef] [PubMed]
89. Keating, S.C.J.; Thomas, A.A.; Flecknell, P.A.; Leach, M.C. Evaluation of EMLA Cream for Preventing Pain during Tattooing of Rabbits: Changes in Physiological, Behavioural and Facial Expression Responses. *PLoS ONE* **2012**, *7*, e44437. [CrossRef] [PubMed]
90. Leung, V.; Zhang, E.; Pang, D.S.J. Real-Time Application of the Rat Grimace Scale as a Welfare Refinement in Laboratory Rats. *Sci. Rep.* **2016**, *6*, 1–12. [CrossRef]
91. Miller, A.L.; Leach, M.C. The Mouse Grimace Scale: A Clinically Useful Tool? *PLoS ONE* **2015**, *10*, e0136000. [CrossRef]
92. Philips, B.H.; Weisshaar, C.L.; Winkelstein, B.A. Use of the Rat Grimace Scale to Evaluate Neuropathic Pain in a Model of Cervical Radiculopathy. *Comp. Med.* **2017**, *67*, 34–42.

93. Mota-Rojas, D.; Olmos-Hernández, A.; Verduzco-Mendoza, A.; Hernández, E.; Martínez-Burnes, J.; Whittaker, A.L. The Utility of Grimace Scales for Practical Pain Assessment in Laboratory Animals. *Animals* **2020**, *10*, 1838. [CrossRef] [PubMed]
94. Nielsen, C.S.; Stubhaug, A.; Price, D.D.; Vassend, O.; Czajkowski, N.; Harris, J.R. Individual Differences in Pain Sensitivity: Genetic and Environmental Contributions. *Pain* **2008**, *136*, 21–29. [CrossRef]
95. Starr, C.J.; Houle, T.T.; Coghill, R.C. Psychological and Sensory Predictors of Experimental Thermal Pain: A Multifactorial Model. *J. Pain* **2010**, *11*, 1394–1402. [CrossRef]
96. Kim, H.; Neubert, J.K.; San Miguel, A.; Xu, K.; Krishnaraju, R.K.; Iadarola, M.J.; Goldman, D.; Dionne, R.A. Genetic Influence on Variability in Human Acute Experimental Pain Sensitivity Associated with Gender, Ethnicity and Psychological Temperament. *Pain* **2004**, *109*, 488–496. [CrossRef] [PubMed]
97. Flecknell, P. Analgesia and Post-Operative Care. In *Laboratory Animal Anaesthesia*; Elsevier: Amsterdam, The Netherlands, 2016; pp. 141–192. [CrossRef]
98. Foley, P.L.; Kendall, L.V.; Turner, P.V. Clinical Management of Pain in Rodents. *Comp. Med.* **2019**, *69*, 468–489. [CrossRef] [PubMed]
99. Kelly, K.R.; Pypendop, B.H.; Christe, K.L. Pharmacokinetics of Buprenorphine Following Intravenous and Intramuscular Administration in Male Rhesus Macaques (Macaca Mulatta). *J. Vet. Pharmacol. Ther.* **2014**, *37*, 480–485. [CrossRef] [PubMed]
100. MacKiewicz, A.L.; Salyards, G.W.; Knych, H.K.; Hill, A.E.; Christe, K.L. Pharmacokinetics of a Long-Lasting, Highly Concentrated Buprenorphine Solution after Subcutaneous Administration in Rhesus Macaques (Macaca Mulatta). *J. Am. Assoc. Lab. Anim. Sci.* **2019**, *58*, 501–509. [CrossRef] [PubMed]
101. Smith, A.A.; Halliday, L.C.; Lindeblad, M.O.; Fortman, J.D. Evaluation of Analgesic Patches in Cynomolgus Macaques (Macaca Fascicularis). *J. Am. Assoc. Lab. Anim. Sci.* **2019**, *58*, 356–361. [CrossRef]
102. Kelly, K.R.; Pypendop, B.H.; Christe, K.L. Pharmacokinetics of Hydromorphone after Intravenous and Intramuscular Administration in Male Rhesus Macaques (Macaca Mulatta). *J. Am. Assoc. Lab. Anim. Sci.* **2014**, *53*, 512–516.
103. Salyards, G.W.; Lemoy, M.J.; Knych, H.K.; Hill, A.E.; Christe, K.L. Pharmacokinetics of a Novel, Transdermal Fentanyl Solution in Rhesus Macaques (Macaca Mulatta). *J. Am. Assoc. Lab. Anim. Sci.* **2017**, *56*, 443–451.
104. Praveen Rao, P.N.; Knaus, E.E. Evolution of Nonsteroidal Anti-Inflammatory Drugs (NSAIDs): Cyclooxygenase (COX) Inhibition and Beyond. *J. Pharm. Pharm. Sci.* **2008**, *11*, 81s–110s. [CrossRef]
105. Murphy, K.L.; Baxter, M.G.; Flecknell, P.A. Anesthesia and Analgesia in Nonhuman Primates. In *Nonhuman Primates in Biomedical Research*; Elsevier Inc: Amsterdam, The Netherlands, 2012; Volume 1, pp. 403–435. [CrossRef]
106. Jain, K.K. An Overview of Drug Delivery Systems. In *Methods in Molecular Biology*; Humana Press Inc.: Totowa, NY, USA, 2020; Volume 2059, pp. 1–54. [CrossRef]
107. Cho, C.; Michalidis, V.; Lecker, I.; Collymore, C.; Hanwell, D.; Loka, M.; Danesh, M.; Pham, C.; Urban, P.; Bonin, R.P.; et al. Evaluating Analgesic Efficacy and Administration Route Following Craniotomy in Mice Using the Grimace Scale. *Sci. Rep.* **2019**, *9*, 359. [CrossRef]
108. Matsumiya, L.C.; Sorge, R.E.; Sotocinal, S.G.; Tabaka, J.M.; Wieskopf, J.S.; Zaloum, A.; King, O.D.; Mogil, J.S. Using the Mouse Grimace Scale to Reevaluate the Efficacy of Postoperative Analgesics in Laboratory Mice. *J. Am. Assoc. Lab. Anim. Sci.* **2012**, *51*, 42–49.
109. Clark, T.P.; Chieffo, C.; Huhn, J.C.; Nimz, E.L.; Wang, C.; Boy, M.G. The steady-state pharmacokinetics and bioequivalence of carprofen administered orally and subcutaneously in dogs. *J. Vet. Pharmacol. Ther.* **2003**, *26*, 187–192. [CrossRef] [PubMed]
110. Ur-Rehman, Z.; Ashraf, M.; Rasheed, M.A. Pharmacokinetic Evaluation and Dosage Optimization of Ketoprofen in Healthy Beetal Goats. *Pakistan Vet J.* **2019**, *39*, 511–514. [CrossRef]
111. Montoya, L.; Ambros, L.; Kreil, V.; Bonafine, R.; Albarellos, G.; Hallu, R.; Soraci, A. A pharmacokinetic comparison of meloxicam and ketoprofen following oral administration to healthy dogs. *Vet. Res. Commun.* **2004**, *28*, 415–428. [CrossRef]
112. Medina-López, N. Vara-Gama, O.; Soria-Arteche, L.; Moreno-Rocha, and F.; López-Muñoz. Pharmacokinetics and Pharmacodynamics of (S)-Ketoprofen Co-Administered with Caffeine: A Preclinical Study in Arthritic Rats. *Pharmaceutics.* **2018**, *10*, 20. [CrossRef]
113. Cramer, M.C.; Pairis-Garcia, M.D.; Bowman, A.S.; Moeller, S.J.; Zhang, Y.; Sidhu, P.K.; Magnin, G.; Coetzee, J.F. Pharmacokinetics of transdermal flunixin in sows. *J. Vet. Pharmacol. Ther.* **2019**, *42*, 492–495. [CrossRef]
114. Ogino, T.; Arai, T. Pharmacokinetic Interactions of Flunixin Meglumine and Enrofloxacin in ICR Mice. *Am. J. Vet Res.* **2007**, *66*, 1209–1213. [CrossRef]
115. Lamont, L.A. Multimodal Pain Management in Veterinary Medicine: The Physiologic Basis of Pharmacologic Therapies. *Vet. Clin. N. Am. Small Anim. Pract.* **2008**, *38*, 1173–1186. [CrossRef] [PubMed]
116. Frankhuijzen, A.L. Pharmacology of Local Anaesthetics. In *Local Anaesthesia in Dentistry*; Springer International Publishing: Berlin/Heidelberg, Germany, 2017; pp. 37–50. [CrossRef]
117. Dermot, K.J.; Ahmad, M.; Brull, S.J. Preemptive Analgesia I: Physiological Pathways and Pharmacological Modalities. *Can. J. Anesth.* **2001**, *48*, 1000–1010. [CrossRef]
118. Perlman, J.E.; Bloomsmith, M.A.; Whittaker, M.A.; McMillan, J.L.; Minier, D.E.; McCowan, B. Implementing Positive Reinforcement Animal Training Programs at Primate Laboratories. *Appl. Anim. Behav. Sci.* **2012**, *138*, 114–126. [CrossRef]
119. Moller, P.L.; Sindet-Pedersen, S.; Petersen, C.T.; Juhl, G.I.; Dillenschneider, A.; Skoglund, L.A. Onset of Acetaminophen Analgesia: Comparison of Oral and Intravenous Routes after Third Molar Surgery. *Br. J. Anaesth.* **2005**, *94*, 642–648. [CrossRef]

120. Bekker, A.; Kloepping, C.; Collingwood, S. Meloxicam in the Management of Post-Operative Pain: Narrative Review. *J. Anaesthesiol. Clin. Pharmacol.* **2018**, *34*, 450–457. [CrossRef]
121. Carlson, A.M.; Kelly, R.; Fetterer, D.P.; Rico, P.J.; Bailey, E.J. Pharmacokinetics of 2 Formulations of Transdermal Fentanyl in Cynomolgus Macaques (Macaca Fascicularis). *J. Am. Assoc. Lab. Anim. Sci.* **2016**, *55*, 436–442. [PubMed]
122. Wolf, R.F.; White, G.L. Clinical Techniques Used for Nonhuman Primates. In *Nonhuman Primates in Biomedical Research*; Elsevier Inc.: Amsterdam, The Netherlands, 2012; pp. 323–337. [CrossRef]
123. Jain, K.K. (Ed.) Methods in Molecular Biology. In *Drug Delivery Systems*; Springer: New York, NY, USA, 2020; Volume 2059. [CrossRef]
124. Deschamps, J.-Y.; Gaulier, J.-M.; Podevin, G.; Cherel, Y.; Ferry, N.; Roux, F.A. Fatal Overdose after Ingestion of a Transdermal Fentanyl Patch in Two Non-Human Primates. *Vet. Anaesth. Analg.* **2012**, *39*, 653–656. [CrossRef]
125. Pugsley, M.K. The Diverse Molecular Mechanisms Responsible for the Actions of Opioids on the Cardiovascular System. *Pharmacol. Ther.* **2002**, *240*, 51–75. [CrossRef]
126. Smith, H.S.; Laufer, A. Opioid Induced Nausea and Vomiting. *Eur. J. Pharmacol.* **2014**, *1*, 67–78. [CrossRef]
127. Hay Kraus, B.L. Effect of Dosing Interval on Efficacy of Maropitant for Prevention of Hydromorphoneinduced Vomiting and Signs of Nausea in Dogs. *J. Am. Vet. Med. Assoc.* **2014**, *245*, 1015–1020. [CrossRef]
128. Pekcan, Z.; Koc, B. The Post-Operative Analgesic Effects of Epidurally Administered Morphine and Transdermal Fentanyl Patch after Ovariohysterectomy in Dogs. *Vet. Anaesth. Analg.* **2010**, *37*, 557–565. [CrossRef] [PubMed]
129. Mansa, S.; Palmér, E.; Grøndahl, C.; Lønaas, L.; Nyman, G. Long-Term Treatment with Carprofen of 805 Dogs with Osteoarthritis. *Vet. Rec.* **2007**, *160*, 427–430. [CrossRef]
130. MacPhail, C.M.; Lappin, M.R.; Meyer, D.J.; Smith, S.G.; Webster, C.R.; Armstrong, P.J. Hepatocellular Toxicosis Associated with Administration of Carprofen in 21 Dogs. *J. Am. Vet. Med. Assoc.* **1998**, *212*, 1895–1901. [PubMed]
131. Council on Animal Care. *CCAC Guidelines: Animal Welfare Assessment*; CCAC: Ottawa, ON, Canada, 2021.

animals

MDPI

Article

Severity Classification of Laboratory Animal Procedures in Two Belgian Academic Institutions

Stéphanie De Vleeschauwer [1,*], Kathleen Lambaerts [1], Sophie Hernot [2] and Karlijn Debusschere [3,4]

[1] Laboratory Animal Center, KU Leuven, 3000 Leuven, Belgium
[2] Laboratory for In Vivo Cellular and Molecular Imaging (ICMI-MIMA/BEFY), Vrije Universiteit Brussel, 1090 Brussels, Belgium; sophie.hernot@vub.be
[3] Core Facility ANIM, Vrije Universiteit Brussel, 1090 Brussels, Belgium; karlijn.debusschere@ugent.be
[4] Core ARTH, Animal Facility, Ghent University, 9000 Ghent, Belgium
* Correspondence: stephanie.devleeschauwer@kuleuven.be

Simple Summary: According to European regulations, the severity of the suffering of animals during animal experiments should be assessed. Regulatory documents and guidelines provide recommendations on how to approach this severity assessment; however, they are often not specific enough, resulting in inconsistencies between different institutes performing the same procedures. To overcome this, two Belgian academic institutions with a focus on biomedical research, collaborated to develop and align the severity classification for all procedures performed. This was performed based on the available literature and guidelines, as well as the professional judgment of the designated veterinarians, animal welfare bodies and animal ethics committees. Throughout the manuscript, we motivate which criteria were used to classify procedures or groups of procedures within a specific category. Our collaborative classification includes many procedures and disease models in a variety of animal species for which a severity classification was not reported so far, or the terms that assign them to a different severity were too vague. We believe this extensive list of procedures and the approach described in this paper could be of great value to other research institutions.

Abstract: According to the EU Directive 2010/63, all animal procedures must be classified as non-recovery, mild, moderate or severe. Several examples are included in the Directive to help in severity classification. Since the implementation of the Directive, different publications and guidelines have been disseminated on the topic. However, due to the large variety of disease models and animal procedures carried out in many different animal species, guidance on the severity classification of specific procedures or models is often lacking or not specific enough. The latter is especially the case in disease models where the level of pain, suffering, distress and lasting harm depends on the duration of the study (for progressive disease models) or the dosage given (for infectious or chemically induced disease models). This, in turn, may lead to inconsistencies in severity classification between countries, within countries and even within institutions. To overcome this, two Belgian academic institutions with a focus on biomedical research collaborated to develop a severity classification for all the procedures performed. This work started with listing all in-house procedures and assigning them to 16 (sub)categories. First, we determined which parameters, such as clinical signs, dosage or duration, were crucial for severity classification within a specific (sub)category. Next, a severity classification was assigned to the different procedures, which was based on professional judgment by the designated veterinarians, members of the animal welfare body (AWB) and institutional animal ethics committee (AEC), integrating the available literature and guidelines. During the classification process, the use of vague terminology, such as 'minor impact', was avoided as much as possible. Instead, well-defined cut-offs between severity levels were used. Furthermore, we sought to define common denominators to group procedures and to be able to classify new procedures more easily. Although the primary aim is to address prospective severity, this can also be used to assess actual severity. In summary, we developed a severity classification for all procedures performed in two academic, biomedical institutions. These include many procedures and disease models in a variety of animal species for which a severity classification was not reported so far, or the terms that assign them to a different severity were too vague.

Citation: De Vleeschauwer, S.; Lambaerts, K.; Hernot, S.; Debusschere, K. Severity Classification of Laboratory Animal Procedures in Two Belgian Academic Institutions. *Animals* **2023**, *13*, 2581. https://doi.org/10.3390/ani13162581

Academic Editor: Garikoitz Azkona

Received: 18 July 2023
Revised: 7 August 2023
Accepted: 8 August 2023
Published: 10 August 2023

Copyright: © 2023 by the authors. Licensee MDPI, Basel, Switzerland. This article is an open access article distributed under the terms and conditions of the Creative Commons Attribution (CC BY) license (https://creativecommons.org/licenses/by/4.0/).

Keywords: severity classification; animal procedures; EU Directive; animal ethics committee; animal welfare body

1. Introduction

According to EU Directive 2010/63, a 'procedure' means any use, invasive or non-invasive, of an animal for experimental or other scientific purposes, with known or unknown outcomes, or educational purposes, which may cause the animal a level of pain, suffering, distress or lasting harm equivalent to, or higher than, what could be caused by the introduction of a needle in accordance with good veterinary practice. It includes any course of action intended, or liable, to result in the birth or hatching of an animal or the creation and maintenance of a genetically modified animal line in any such condition but excludes the killing of animals solely for the use of their organs or tissue. The Directive also states that all procedures on animals should be classified as 'non-recovery', 'mild', 'moderate' or 'severe'. Severity classification of animal procedures does not only create opportunities to refine procedures, but it is also crucial in the harm–benefit analysis performed during the project's evaluation and authorization [1,2]. The severity of procedures should be classified prospectively and retrospectively. The prospective severity classification is performed upfront during project writing and will thus help in the harm–benefit analysis. The retrospective, or actual severity, is estimated for each individual animal and is based on that animal's experience during the course of the procedure. The latter enters the annual statistics that promote transparency and dictate public opinion.

However, the estimation of the level of pain, suffering, distress and lasting harm, and thus classifying the severity of a procedure, is not easy. Annex VIII of the EU Directive gives some examples of the severity classification of procedures. Although some classifications are very straight forward, e.g., categorizing 'non-invasive imaging of animals with appropriate sedation or anesthesia' as mild, other examples remain vague with room for interpretation. For example, 'breeding of genetically altered (GA) animals, which is expected to result in a phenotype with mild effects' does not provide clear guidance on what these mild effects are. Since the implementation of the Directive by the different EU member states, several articles have been published on the topic [3–5]. In addition, different working groups (Working Group of Berlin Animal Welfare Officers [6], EU severity assessment framework [7]) and countries (Switzerland [8] and UK [9]) have written guidelines on the severity classification of animal procedures. However, due to the large variety of disease models and animal procedures carried out in multiple animal species, guidance on the severity classification of specific procedures or models is often lacking or not specific enough.

Furthermore, the reported severity classifications per country within the EU depict large differences in the percentage of mild, moderate and severe procedures (Table 1).

Table 1. Severity classifications reported for the EU; Germany, France and the UK (the three countries using the most animals in Europe) and Belgium in 2020.

	Mild (%)	Moderate (%)	Severe (%)	Non-Recovery (%)	Total Number of Animals Used	Reference
EU	48.7	37.3	9.9	4.1	8,054,930	[10]
Germany	66.9	23.7	3.8	5.6	1,850,443	
France	30.1	49.8	13.7	6.4	1,643,787	
UK	51	24	4	7	1,681,383	
Belgium	55.23	31.851	10.76	2.16	437,275	

These differences may be due to different research areas but also due to different severity classifications by the competent authorities. Within the EU, Belgium is one of the few countries where the competent authority has delegated project evaluation and authorization, and thus severity classification, to the institutional level instead of the regional or national level. As a result, the differences in severity classification for the same procedure probably occur even between different institutions within the same country.

Here, we want to share the prospective severity classification we have developed and aligned for two Belgian academic institutions. Severity classification was based on existing guidelines and the professional judgment of all parties involved. This was performed for all procedures performed in (laboratory) animals in two Belgian academic institutions with a focus on biomedical research.

2. Materials and Methods

2.1. Establishing a Framework for Severity Classification

In 2016, procedures and disease models performed at KU Leuven were listed by reviewing all ethical projects approved in the period 2013–2015. Although this catalog was not made within the scope of this project, it formed the basis for our severity classification. In 2019, the Norwegian Consensus-Platform for the Replacement, Reduction and Refinement of animal experiments (Norecopa) compiled available severity classifications from the EU Directive Annex VIII [7], the FELASA/ECLAM/ESLAV report [3], the UK Home Office [9], the Swiss FSVO (Federal Food Safety and Veterinary Office) [8] and the Working Group of Berlin Animal Welfare Officers [6]. At that time, KU Leuven decided to develop an in-house severity classification system for all procedures and disease models within KU Leuven. This work started by reviewing the literature and the previously cited legislation and guidelines that were consulted via the Norecopa database [11]. By doing this, it became clear that not all procedures from KU Leuven were listed and that some procedures were classified differently by the different guidelines. For example, gavage is classified as mild in the EU Directive while the Swiss FSVO considers this as moderate; a single subcutaneous (SC) injection is considered mild according to UK guidelines, while the Swiss FSVO considers this below threshold. Therefore, we decided to develop a more detailed severity classification at KU Leuven. During the process, another academic institution (VUB) was approached as we wanted to align the severity classification system with another Belgian academic institution. Together, we defined the following key points for our severity classification system:

- It must be clear with defined cut-offs, avoiding (where possible) vague terms;
- It should not only be a list of procedures within a severity classification but must also define common denominators so that it can easily be adapted when new models are adopted in an institution;
- It must represent the severity of a procedure that is performed once by a well-trained person using the most refined technique. It must be applicable to all species and procedures.

2.2. Approach to an Institutional Severity Classification System

The development of the severity classification system is based on existing guidelines and professional judgment and is performed in a stepwise approach. First, all procedures from the catalog of procedures were divided into categories. Next, for each (sub)category, the aforementioned references and guidelines were extensively reviewed. Based on this, we decided which format of the available criteria and guidelines could be useful to approach the severity classification (see results). As mentioned above, guidelines were inconsistent for some procedures or missing for other procedures. When differences occurred between existing guidelines, the experts decided on the appropriate severity classification. When procedures were not reported in existing guidelines, the experts first defined criteria and common denominators to classify the severity of these (see results). If necessary (i.e., when experience with or knowledge of certain procedures or models was insufficient), research experts were consulted. This was mainly performed for the behavioral tests, pain tests and infectious disease models. Next, per (sub)category, an initial draft was drawn containing the severity classification proposal, references from other guidelines and, if necessary, references from the literature describing, e.g., clinical signs in certain disease models. The content of this initial draft was discussed until consensus was reached. If needed, additional references and (research) experts were consulted. Notably, consensus was usually reached easily. Once the initial draft was finalized, a final draft was sent to the whole institutional AEC (and AWB for VUB) for approval. AECs only evaluated procedures performed in their own institution. Once approved, the severity classification was adopted in the institutions.

2.3. *Experts Involved*

The following experts were involved in drafting and evaluating the severity classification:

- The designated veterinarians from both institutions. They both have been working as designated veterinarians in academic institutions for more than 5 years. As members of different institutional AECs, they are closely involved in severity classification of procedures and research projects;
- Subgroup of the KU Leuven AEC: to facilitate the process, a subgroup of the EC was appointed for this work. This subgroup consisted of the designated veterinarian, the animal welfare officer of the aquatic species and one of the external members of the AEC, a veterinarian who has been working as a designated veterinarian and consultant in laboratory animals for many years;
- AWB: at VUB there is one central AWB in contrast to KU Leuven where each department has its own AWB. This means that the involvement of the AWB of the KU Leuven, in contrast to the VUB, is limited to some members of the AWB that were represented in other categories of experts;
- AECs of both institutions. At VUB, the AEC consists of 32 members (members + substitutes), at KU Leuven of 27 members (members + substitutes);
- Research experts: Researchers that were contacted to obtain more information on disease models were considered experts when they worked with a specific animal model for many years and had multiple peer-reviewed papers describing and utilizing a specific animal model.

The involvement of the different experts in the different steps of the process is indicated in Table 2.

Table 2. Involvement of the different experts in developing the severity classification system.

		Catalogue of Procedures	Extensive Review of Existing Literature and Guidelines	Decide on Approach to Severity Classification	Drawing First Draft	Reviewing First Draft Classification per Category	Drawing Final Draft per Category	Reviewing Final Draft
KU Leuven	Designated veterinarian	x	x	x	x	x	x	
	Subgroup of AEC			x		x		
	AEC							x
	Research experts				x			
VUB	Designated veterinarian	x			x	x	x	
	AWB	x				x		x
	AEC							x
	Research experts					x		

3. Results

In the period of 2013–2015, a total of 804 projects involving animals were approved by the institutional AEC of KU Leuven. From these projects, 673 'technical' procedures in

14 species were identified. These species were as follows: mouse, rat, rabbit, pig, sheep, rhesus macaque, zebrafish, killifish, chicken, xenopus, gerbil, guinea pig, hamster and calf. The technical procedures were divided into the following (sub)categories: administration, sampling, surgery and surgical induction of disease, behavioral testing, pain tests, imaging and function measurements. To cover the disease models, the following categories were added: chemical disease, wherein a disease is induced by the administration of a chemical, neoplasm, infectious disease and 'disease models—others', for those models not qualifying the aforementioned subcategories. GA lines with a harmful phenotype, abnormal housing and nutrition and clinical signs formed separate categories. These 16 subcategories were assigned to the following four larger categories entitled: procedures, measurements and tests, disease models and animals (Figure 1). As in vivo pharmacokinetics/pharmacodynamics and toxicity tests are not often performed in both institutions, no separate category was made for these tests in contrast to most available guidelines. Where necessary, the animal species are specified. If species is not specified, the severity level is considered to be the same across species, although not all procedures are carried out in all species.

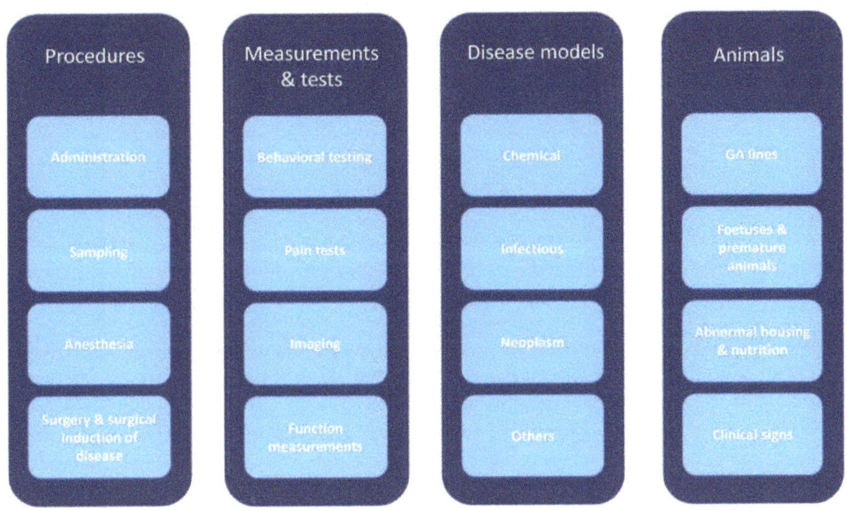

Figure 1. Diagram of main categories (dark blue) and subcategories (light blue)—animal procedures are subdivided into 16 subcategories. Based on commonalities, these subcategories were assigned to four main categories of procedures.

As mentioned before, the factors that were thought to be crucial for the severity classification of a specific (sub-)category were determined, such as clinical signs, dosage or duration. Based on these factors, defined cut-offs were established, and procedures were assigned to different severity levels. The factors used and/or the final severity classification are not always according to the existing guidelines. This is especially true in the disease models, as they often depend on different parameters and can be progressive. Therefore, clinical signs were used as a means to classify the severity of the different disease models.

In the following chapters, we will describe in more detail which factors were used to assign a severity classification to a procedure. As some subcategories are classified similarly, the classification is discussed in the same section of the text. Tables S1–S14 in the Supplementary Material provide an overview of all procedures with their severity classification. We have underlined the procedures/models for which severity has not been described elsewhere.

3.1. Severity Classification of Administration

To determine the severity of compound administration, we focused on the technical procedures required to administer a compound and not on the effects of the compound given. For these compound-specific effects, we refer to the chapters and Supplementary Tables S5–S8 on disease models (chemical, infectious, neoplasm and others).

Following the EU Directive and UK guidelines, we classified all routes of administration in all species as mild. This includes conventional routes, like intravenous (IV) and SC injections, but also less conventional routes, such as intragastric administration in mouse pups. Some routes, such as intranasal administration, are considered mild as long as they are performed under anesthesia. Hydrodynamic tail vein injection (HDTVI) and injection in the footpad were classified as moderate as HDTVI can lead to transient, but severe, cardiovascular effects [12,13], and injections in the footpad are painful due to swelling in weight bearing structures. Interestingly, our classification diverged most from the Swiss guidelines, in which some administration methods, such as a single SC or IV injection, are considered below threshold and gavage is considered moderate.

Table S1 describes the different severity classifications of administration.

3.2. Severity Classification of Sampling

All technical procedures involving taking fluid or tissue samples, including sampling for genotyping, were grouped into the category 'sampling.' Fluid sampling includes all fluids that can be sampled in different species, such as blood, urine and cerebrospinal fluid. For blood sampling, the withdrawn volume, whether it is replaced or not, and the technique used, were taken into account to assign the severity level. However, we did not determine a severity classification for serial blood sampling, as the frequency, time interval between samples and volume taken can all influence the final severity and, as such, need to be evaluated on a case-by-case basis. According to the Commission Implementing Decision (EU) 2020/569 of 16 April 2020 [14] and the EU Framework for the genetically altered animals [15], tissue sampling for genotyping is not considered a procedure if the sample obtained is a by-product from identification. However, if it is not a by-product from identification, it is considered a procedure and has thus been classified accordingly. Furthermore, 'oocyte collection in xenopus by gentle squeezing' and 'non-invasive mucus sampling for genotyping zebrafish' were included, both technical procedures for which a severity classification was not reported so far. Overall, most techniques are classified according to EU Directive and legislation.

Table S2 describes the different severity classifications of fluid and tissue sampling.

3.3. Severity Classification of Anesthesia, Surgery and Surgical Induction of Disease

Following EU Directive and UK guidelines, anesthesia as such is prospectively classified as mild.

To classify the surgical procedures, including those used to induce a specific disease model, a differentiation between minor and major surgery was made. Although the definition of minor and major surgery is still under debate [16,17], we defined minor surgery as surgery not opening body cavities and major surgery as surgery opening body cavities, such as the abdomen or thorax. Most types of minor surgery are classified as mild, and most types of major surgery, with appropriate analgesia, as moderate. However, the consequences of the (minor or major) surgical procedures should also be taken into account. Therefore, we also considered the following: impairment of locomotion, loss of function, rejection of organ transplanted and failure of the device implanted. As a result, we have stratified the severity of cardiac assist device implantation, stroke and myocardial infarction into two different severity categories, i.e., moderate and severe, rather than assigning them to a single severity classification as the examples provided by the EU Directive and the Swiss guidelines. In the case of cardiac device implantation, the presence or absence of a functional heart determines if the procedure is moderate or severe, respectively. In the case of myocardial infarction and stroke, the size, and, thus, effects on the animal determine whether this procedure is classified as moderate or severe.

Table S3 describes the different severity classifications of surgery and surgical induction of disease.

3.4. Severity Classification of Clinical Signs

Disease models are challenging to classify, as often diseases are progressive, i.e., worsening over time. Depending on the study objective, researchers may be interested in the early or late stages of a disease. To classify the severity of disease models, we therefore, decided to estimate the severity of disease models mainly based on the severity of clinical signs. Hence, much effort was put into the severity classification of clinical signs. Not only do these enable us to classify disease models we currently have at our institutions, but they will also facilitate severity assessment in case new models are developed. Furthermore, they also aid in the assessment of actual severity.

Clinical signs and their severity classification were developed for mammals, zebrafish larvae (up to 12 days post fertilization) and sexually mature zebrafish. We have defined clear cut-offs between severity levels, taking into consideration different parameters, such as duration, effect on behavior, etc. For the body weight loss of mammals, we took into consideration the developmental stage and timeframe wherein the animal loses weight. This is in line with UK guidelines. The severity classification of the other clinical signs of mammals is largely consistent with the available guidelines, although we included duration for some clinical signs. As at our institute, we also have disease models involving zebrafish larvae (independently feeding and thus covered by Directive 2010/63/EU); we established parameters for clinical follow-up of zebrafish larvae up to 12 days post fertilization. The following clinical signs are described and assigned to mild, moderate or severe: overall morphology, necrosis, swim bladder, posture, cardiac function and touch response. To our knowledge, clinical signs and severity classification for zebrafish larvae have not yet been reported. At our institute, there are currently no disease models involving adult zebrafish. Nonetheless, clinical signs and severity for adult zebrafish were included in the table. Therefore, we mainly followed Sabrautzki et al. [18], who defined and assigned scores to different clinical signs in adult zebrafish.

Table S4 describes the different severity classifications of clinical signs.

3.5. Severity Classification of Disease Models

As mentioned above, we have defined disease models as follows: chemical, infectious, neoplasm or others. As we do not have a separate category for toxicity tests, we included these in the chemically induced models. Although very different in etiology, they all result in the animals developing general or organ-specific clinical signs. Although the clinical signs seen in a disease model are usually the same, the severity of these signs may depend on several factors.

In the case of chemically induced disease models, the severity of the clinical signs may depend on the dosage and the type of chemical given. As this may differ depending on the study objective, we chose to prospectively classify this type of model according to the severity of the clinical signs expected using a specific dosage of a compound rather than assigning a fixed severity level. For example, intestinal inflammation in dextrane sulfate sodium (DSS) colitis in mice depends on many factors, such as DSS molecular weight and dosage, mouse strain, etc. [19,20]. Depending on the study objective, mice may experience mild to severe clinical signs. The severity classification should therefore be based on the severity of these clinical signs. The same is true for, e.g., diabetes. For well-established chemically induced models that always give similar clinical signs, the severity level is fixed. For instance, models of LPS-induced acute respiratory distress are always classified as severe. To classify the severity of toxicity tests at our institution, mainly conducted on zebrafish larvae, we took the following factors into consideration: clinical signs and, as described by Hawkins et al. [5], predictability and death as a possible outcome.

In infectious diseases, the clinical signs depend on the pathogen strain and dosage, the route of administration and animal species and the strain, making severity classification even more challenging. For the standardized models (using a specific pathogen and animal strain, dosage and route of administration), the clinical signs are known, and a specific prospective severity level is assigned. For example, Zika, dengue and Japanese encephalitis in AG129 mice always lead to severe clinical signs and are thus classified accordingly. Some infectious disease models are not so standardized, for instance, models of new, emerging pathogen strains. Prospective severity classification for these models is difficult, and we have therefore chosen not to include these.

For severity classification of (mouse) cancer models, we follow the same reasoning, i.e., assign severity level based on the clinical signs expected. As cancer is a progressive disease, these clinical signs will depend on study duration, objective and humane endpoints applied. In contrast to other disease models, we here refer to a specific set of clinical signs that can be expected in most cancer models.

The disease models categorized in 'others', are those that do not fit into one of the above categories as they are induced by laser, hypoxia or hyperoxia, mechanically or by irradiation. Again, in this subcategory, severity is classified for different models that were not described before.

Table S5 describes the different severity classifications of chemically induced disease models.

Table S6 describes the different severity classifications of infectious disease models.

Table S7 describes the different severity classifications of neoplasm.

Table S8 describes the different severity classifications of other disease models.

3.6. Severity Classification of Abnormal Housing and Nutrition

Some procedures require changes in housing and/or nutrition. For alterations in both housing and nutrition, we mainly followed the Swiss FSVO guidelines, taking into consideration the duration of the abnormal housing and several other factors, such as social isolation. For food deprivation, the severity is based on body weight loss. For water restriction, a differentiation is made between feeding dry food and food containing fluids with different cut-offs in time. In the latter case, food restriction is only classified as severe if it is accompanied by dehydration. In general, instead of using non-specific terms, such as 'a short period of time', we rather specify the duration of the abnormal housing/nutrition, linking this to the appropriate severity. In our classification, no specific examples per species are given to make the use of these tables as broad as possible. As mentioned before, this does not necessarily mean that all these abnormal housing and nutrition conditions are used in all species.

Table S9 describes the different severity classifications of abnormal housing and nutrition.

3.7. Severity Classification of Behavioral Testing, Function Measurements and Imaging

There is a plethora of behavioral tests and function measurements. The difference between both terminologies is not always clear nor well described. We have therefore defined behavioral tests as all tests measuring fear, cognition, memory, etc. Function measurements, on the other hand, were defined as all tests measuring a body function, including motor function.

Starting from an overview of both behavioral tests and function measurements performed at our institutions, we searched for common denominators to classify them into different severity categories. The criteria taken into account are the following: handling and change in the environment during the test, changes in housing and nutrition (see above), invasiveness and the duration of restraint in case the procedure is being performed in an awake animal.

All imaging techniques, including those requiring injection of tracers or contrast, are classified as mild when performed under anesthesia. In the case no anesthesia is used and restraint of the animal is thus required, the severity is determined by the duration of restraint. A cut-off of one hour was determined to shift from mild to moderate severity.

Table S10 describes the different severity classifications of behavioral testing.

Table S11 describes the different severity classifications of function measurements.

3.8. Severity Classification of Pain Tests

For the pain tests, we took the duration and intensity of pain caused into account. By doing this, we mainly followed the Swiss FSVO guidelines and added just a few changes. Based on the retrospective severity assessment, we decided to classify nerve crush and ligation as moderate and not severe. Furthermore, writhing is always considered to cause severe pain [21]. For the footpad injections, we considered the effects caused by the injected compound. This results in a severe classification for the footpad injection of Complete Freund's Adjuvant, capsaicin and acrolein and a moderate severity for footpad injection of saline and pregnolone sulfate.

Table S12 describes the different severity classifications of pain tests.

3.9. Severity Classification of GA Lines with Harmful Phenotype

As stated in the EU Directive, GA animals with a harmful phenotype must be classified as mild, moderate or severe. The determination of the severity of GA animals remains, however, very difficult. Defining what is considered harmful and what is not is the first challenge. Alterations may result in phenotypes that are macroscopically visible but do not necessarily affect the welfare or wellbeing of the animals. Therefore, we choose to classify all genetic alterations causing macroscopic changes not affecting welfare, as well as all phenotypic changes that can only be detected using specific testing (e.g., by behavioral tests or blood analysis), as not harmful. Another difficulty in assessing the severity of GA lines is that, in contrast to other categories, the assessment should be performed on the line and not the individual animal. Consequently, in the severity assessment, we consider the life-long effects of the GA, not taking into account the humane endpoints applied. Naturally, this does not mean that humane endpoints should not be applied. As a result, we classify some lines, especially those with progressive diseases such as cancer, as severe instead of stratifying the severity based on humane endpoints applied as Zintsch et al. [4] or the Swiss guidelines [8]. Here again, the classification is mainly based on clinical signs seen during the assessment of the line. Of note, our institutions only have GA rodent and zebrafish lines.

Table S13 provides an overview of the severity classification of GA lines with harmful phenotypes.

3.10. Severity Classification of Fetuses and Premature Animals

Article 2 of the EU Directive states that the Directive should apply to live, non-human vertebrate animals, including the following: independently feeding larval forms and the fetal forms of mammals from the last third of their normal development. Consequently, the fetuses of mammals in the first and second semester of gestation, birds before hatching and non-independently feeding larval forms from aquatic species are not considered experimental animals. Therefore, their use is not regulated unless procedures carried out could result in pain, suffering, distress or lasting harm if the fetuses are allowed to live beyond the first two-thirds of their development. According to Commission Implementing Decision (EU) 2020/569 of 16 April 2020 [14], fetal and embryonic forms of mammalian species shall be excluded from the provision of annual statistical data. Only animals that are born, including by cesarean section, and live are to be counted. To our knowledge, this means that for procedures on mammalian fetuses and early developmental stages in other species, severity should only be assessed for those animals that are born and in which pain, suffering or distress occurs or is likely. We report here the severity of two such models for which the severity was not reported before, i.e., in utero creation of spina bifida in lambs and growth retardation in rabbits.

Table S14 provides an overview of the severity classification of procedures carried out in early developmental stages.

4. Discussion

This manuscript describes the severity classification of laboratory animal procedures performed in two Belgian academic institutions with a focus on biomedical research. We explain how the process of severity classification is performed, and, more importantly, we describe the severity classifications for procedures and disease models through the tables presented in the Supplementary Information. Our goal was to be as specific as possible and provide clear cut-offs between categories. In addition, we included many disease models for which severity classification was not yet described. Those novelties are made clear by underlining them in the Supplementary Tables S1–S14.

Although performed with great care, this study has some shortcomings. Evidence-based severity classification is upcoming and has been performed for different procedures and models, such as epilepsy [22–24], repeated anesthesia [25,26], depression [27] and models of gastrointestinal diseases [28]. Recently, a mathematical model to estimate the severity of animal procedures has been described by Morton [29]. These methods use objective parameters, scores and tests to assess the severity of animal procedures. The severity classification described in this manuscript is based on available guidelines, in-house expertise as well as retrospective analyses, and although we are aligned with other guidelines, it is inevitably partly subjective. Further evidence-based severity classification of the different animal procedures is necessary, especially since the severity of procedures is an essential part of the harm–benefit analysis performed during project evaluation. It should thus give a correct reflection of the expected suffering of an animal during a certain procedure.

As we have covered all species used at our institutions and mostly used a one-fits-all approach (i.e., not distinguishing between species), species-specific behavior and sensory elements might not have been highlighted sufficiently. For example, we do not distinguish between the different (social) species when classifying abnormal housing and nutrition. However, the well-being of an individual animal might be affected differently depending on the species and sometimes even the sex. Indeed, some studies show no difference in the behavior between single-housed versus group-housed male mice [30], while others consider the single housing of male rats as severe [31]. These differences might be even more pronounced in non-rodent species. Future scientific research is needed to give us more insight into how different animals perceive certain signals or procedures. This could aid in the classification of different procedures, and additionally, these insights might help in refining certain procedures.

Throughout the tables, we aimed to give clear guidelines, with specific time periods and/or measurable clinical signs, to aid researchers, AWB and AEC members in the severity classification of animal procedures. However, the severity classification only applies to the procedure being performed once. Therefore, re-evaluation is necessary when different procedures are being combined or repeated, and a cumulative severity score needs to be given. We chose to not incorporate cumulative scores, as there are too many variables within an experimental set-up that can influence severity, e.g., frequency of and time interval between procedures. And, although repeating procedures or performing different procedures does not automatically increase severity, it might become more severe. This must, therefore, carefully be considered in order to correctly assess the severity.

Our severity classification of disease models is mostly based on clinical signs. This implies a good knowledge of the anatomy, physiology and normal behavior of the species involved. Especially the recognition of more subtle clinical signs, such as paresis and pain, might be challenging, especially as researchers in biomedical sciences may come from varying backgrounds. They thus need proper education and guidance. This applies to all species and might even be more challenging in aquatic species, such as xenopus and zebrafish, especially in their early developmental stages. We therefore included clinical signs of adult zebrafish and reported for the first-time signs to assess the early developmental stages of zebrafish (up to 12 days post fertilization). Signs to assess zebrafish between 12 days and sexual maturity and xenopus still need to be developed.

Although the aim of the EU Directive is to provide guidance, uniformity and clarity for animals involved in research, some articles of the Directive, such as the classification of procedures performed on fetuses and premature animals, remain controversial and difficult to interpret. According to the Directive, from a certain age pre-birth, a mammal fetus or larval form is considered an experimental animal as there is evidence it might experience pain. However, according to Commission Implementing Decision (EU) 2020/569 of 16 April 2020, the fetal and embryonic forms of mammalian species shall be excluded from the provision of annual statistical data. This seems in contrast with the definition of an experimental animal for which annual statistics should be provided by each EU member state. The severity classification of these early forms is therefore performed to the best of our knowledge. Similarly, some procedures in animals are not standardized in the community, making the severity classification difficult. For instance, the insulin tolerance test in mice requires a period of fasting. However, different fasting times have been reported [32]. Our severity classification takes both fasting and the influence of glycemic levels into consideration. Both these factors are affected by fasting time and, as such, might/should be classified differently. Not only for the severity estimation but also for the reproducibility of in vivo experiments, an important concern in biomedical research [33–36], further standardization of procedures is mandatory.

As new disease models, tests and GA models are developed on a daily basis, it would be interesting if more emphasis is put on how to perform a good retrospective analysis and assess actual severity. For new models or drugs, it is sometimes hard to classify prospective severity based on the available knowledge. Clear communication on the retrospective severity and, more importantly, the methods used to assess the severity level would be very informative for the research community.

As discussed above, severity classification remains a difficult task for the researchers and competent authorities during project application and evaluation, respectively. Although difficult, both prospective and actual or retrospective severity classification and reporting are important. Prospective severity is crucial in performing correct harm–benefit analysis during the project evaluation and in refining procedures. As actual severity is reported, it promotes transparency and dictates public opinion. In Belgium, project evaluation has been delegated from the competent authorities to the institutional AEC. As shown in Table 1, severity levels reported within the different countries significantly differ, and although not reported, this difference may also exist between institutions. The aim of this work was twofold. On the one hand, we wanted to align severity between two different institutions. On the other hand, we wanted to report the severity of all procedures and models used at our institutions, many of which were not covered in the available guidelines.

5. Conclusions

The severity classification of animal procedures remains a challenging task as it is often mostly subjective. The large variety of procedures in many different species makes it even more challenging. This is reflected in the reported number of procedures within a specific severity classification by the different EU member states, with large variations between the member states. Further evidence-based severity classification of animal procedures is needed. In the meantime, sharing experience in the field will help in further aligning the severity classification of animal procedures. This manuscript summarizes the severity classification of two Belgian academic institutions wherein animals are mainly used in biomedical research. We have provided a severity classification for all procedures and models used at our institutions and discussed the process of that classification. Many of those procedures were not covered in the available guidelines, and we hereby thus report severity classification for many procedures for the first time.

Supplementary Materials: The following supporting information can be downloaded at: https://www.mdpi.com/article/10.3390/ani13162581/s1, Table S1: Severity classification of administration; Table S2: Severity classification of sampling; Table S3: Severity classification of surgery and surgical induction of disease; Table S4: Severity classification of clinical signs; Table S5: Severity classification of chemical disease models; Table S6: Severity classification of infectious diseases; Table S7: Severity classification of neoplasm; Table S8: Severity classification of other disease models; Table S9: Severity classification of abnormal housing and nutrition; Table S10: Severity classification of behavioral tests; Table S11: Severity classification of function measurements; Table S12: Severity classification of pain tests; Table S13: Severity classification of genetically altered (GA) lines; Table S14: Severity classification of fetuses and premature animals.

Author Contributions: Conceptualization: S.D.V., K.L. and K.D.; methodology: S.D.V.; writing—original draft preparation, S.D.V. and K.D.; writing—review and editing: K.L. and S.H. All authors have read and agreed to the published version of the manuscript.

Funding: This research received no external funding.

Institutional Review Board Statement: Not applicable.

Informed Consent Statement: Not applicable.

Data Availability Statement: Not applicable.

Acknowledgments: We would like to thank all EC and AWB members who critically evaluated severity classification of all procedures and all researchers who helped in classifying specific procedures.

Conflicts of Interest: The authors declare no conflict of interest.

References

1. Bronstad, A.; Newcomer, C.E.; Decelle, T.; Everitt, J.I.; Guillen, J.; Laber, K. Current concepts of Harm-Benefit Analysis of Animal Experiments—Report from the AALAS-FELASA Working Group on Harm-Benefit Analysis—Part 1. *Lab. Anim.* **2016**, *50*, 1–20. [CrossRef] [PubMed]
2. Laber, K.; Newcomer, C.E.; Decelle, T.; Everitt, J.I.; Guillen, J.; Bronstad, A. Recommendations for Addressing Harm-Benefit Analysis and Implementation in Ethical Evaluation—Report from the AALAS-FELASA Working Group on Harm-Benefit Analysis—Part 2. *Lab. Anim.* **2016**, *50*, 21–42. [CrossRef]
3. Smith, D.; Anderson, D.; Degryse, A.D.; Bol, C.; Criado, A.; Ferrara, A.; Franco, N.H.; Gyertyan, I.; Orellana, J.M.; Ostergaard, G.; et al. Classification and reporting of severity experienced by animals used in scientific procedures: FELASA/ECLAM/ESLAV Working Group report. *Lab. Anim.* **2018**, *52*, 5–57. [CrossRef] [PubMed]
4. Zintzsch, A.; Noe, E.; Reissmann, M.; Ullmann, K.; Kramer, S.; Jerchow, B.; Kluge, R.; Gosele, C.; Nickles, H.; Puppe, A.; et al. Guidelines on severity assessment and classification of genetically altered mouse and rat lines. *Lab. Anim.* **2017**, *51*, 573–582. [CrossRef] [PubMed]
5. Hawkins, P.; Dennison, N.; Goodman, G.; Hetherington, S.; Llywelyn-Jones, S.; Ryder, K.; Smith, A.J. Guidance on the severity classification of scientific procedures involving fish: Report of a Working Group appointed by the Norwegian Consensus-Platform for the Replacement, Reduction and Refinement of animal experiments (Norecopa). *Lab. Anim.* **2011**, *45*, 219–224. [CrossRef] [PubMed]
6. Working Group of Berlin Animal Welfare Officers. Empfehlung aus dem Arbeitskreis Berliner Tierschutzbeauftragte zu Score Sheets, Belastungskategorien und Abbruchkriterien. Available online: https://www.ak-tierschutzbeauftragte.berlin/galerie (accessed on 12 June 2023).
7. European Commission; Directorate-General for Environment. *Caring for Animals Aiming for Better Science: Directive 2010/63/EU on Protection of Animals used for Scientific Purposes: Severity Assessment Framework*; European Commission: Brussels, Belgium, 2019.
8. Federal Food Safety and Veterinary Office. Technical Information Animal Experimentation, Severity Degrees 1.04. Available online: https://norecopa.no/3r-guide/swiss-guidance-for-severity-assessment (accessed on 12 June 2023).
9. UK Home Office. Recording and Reporting the Actual Severity of Regulated Procedures. Available online: https://www.gov.uk/guidance/animal-research-technical-advice (accessed on 12 June 2023).
10. European Commission. Statistics and Non-Technical Project Summaries. Available online: https://ec.europa.eu/environment/chemicals/lab_animals/alures_en.htm (accessed on 13 June 2023).
11. Norwegian Consensus-Platform for the Advancement of the 3Rs, Compilation of Severity Classifications Across Europe. Available online: https://norecopa.no/more-resources/severity-classification/compilation-of-severity-classifications-across-europe/ (accessed on 12 June 2023).

12. Sawyer, G.J.; Dong, X.; Whitehorne, M.; Grehan, A.; Seddon, M.; Shah, A.M.; Zhang, X.; Fabre, J.W. Cardiovascular function following acute volume overload for hydrodynamic gene delivery to the liver. *Gene Ther.* **2007**, *14*, 1208–1217. [CrossRef] [PubMed]
13. Suda, T.; Gao, X.; Stolz, D.B.; Liu, D. Structural impact of hydrodynamic injection on mouse liver. *Gene Ther.* **2007**, *14*, 129–137. [CrossRef]
14. European Commission. Commission Implementing Decision (EU) 2020/569 of 16 April 2020 Establishing a Common Format and Information Content for the Submission of the Information to be Reported by Member States Pursuant to Directive 2010/63/EU of the European Parliament and of the Council on the Protection of Animals Used for Scientific Purposes and Repealing Commission Implementing Decision 2012/707/EU. Available online: https://op.europa.eu/en/publication-detail/-/publication/1fa78fbf-85f8-11ea-bf12-01aa75ed71a1/language-en#:~:text=Home-,Commission%20Implementing%20Decision%20(EU)%202020%2F569%20of%2016%20April,of%20animals%20used%20for%20scientific (accessed on 12 June 2023).
15. European Commission; Directorate-General for Environment. *Framework for the Genetically Altered Animals under Directive 2010/63/EU on the Protection of Animals Used for Scientific Purposes*; Publications Office of the European Union: Luxembourg, 2022.
16. Newsome, K.; McKenny, M.; Elkbuli, A. Major and minor surgery: Terms used for hundreds of years that have yet to be defined. *Ann. Med. Surg.* **2021**, *66*, 102409. [CrossRef]
17. Martin, D.; Mantziari, S.; Demartines, N.; Hubner, M.; Group, E.S.A.S. Defining Major Surgery: A Delphi Consensus Among European Surgical Association (ESA) Members. *World J. Surg.* **2020**, *44*, 2211–2219. [CrossRef]
18. Sabrautzki, S.; Miller, M.; Kague, E.; Brielmeier, M. Welfare Assessment of Adult Laboratory Zebrafish: A Practical Guide. *Zebrafish* **2021**, *18*, 282–292. [CrossRef]
19. Perse, M.; Cerar, A. Dextran sodium sulphate colitis mouse model: Traps and tricks. *J. Biomed. Biotechnol.* **2012**, *2012*, 718617. [CrossRef] [PubMed]
20. Vowinkel, T.; Kalogeris, T.J.; Mori, M.; Krieglstein, C.F.; Granger, D.N. Impact of dextran sulfate sodium load on the severity of inflammation in experimental colitis. *Dig. Dis. Sci.* **2004**, *49*, 556–564. [CrossRef] [PubMed]
21. Gawade, S.P. Acetic acid induced painful endogenous infliction in writhing test on mice. *J. Pharmacol. Pharmacother.* **2012**, *3*, 348. [CrossRef] [PubMed]
22. Boldt, L.; Koska, I.; Maarten van Dijk, R.; Talbot, S.R.; Miljanovic, N.; Palme, R.; Bleich, A.; Potschka, H. Toward evidence-based severity assessment in mouse models with repeated seizures: I. Electrical kindling. *Epilepsy Behav.* **2021**, *115*, 107689. [CrossRef]
23. Buchecker, V.; Koska, I.; Pace, C.; Talbot, S.R.; Palme, R.; Bleich, A.; Potschka, H. Toward Evidence-Based Severity Assessment in Mouse Models with Repeated Seizures: (II.) Impact of Surgery and Intrahippocampal Kainate. *Eur. Surg. Res.* **2023**, *64*, 89–107. [CrossRef]
24. van Dijk, R.M.; Koska, I.; Bleich, A.; Tolba, R.; Seiffert, I.; Moller, C.; Di Liberto, V.; Talbot, S.R.; Potschka, H. Design of composite measure schemes for comparative severity assessment in animal-based neuroscience research: A case study focussed on rat epilepsy models. *PLoS ONE* **2020**, *15*, e0230141. [CrossRef] [PubMed]
25. Hohlbaum, K.; Bert, B.; Dietze, S.; Palme, R.; Fink, H.; Thone-Reineke, C. Severity classification of repeated isoflurane anesthesia in C57BL/6JRj mice-Assessing the degree of distress. *PLoS ONE* **2017**, *12*, e0179588. [CrossRef]
26. Hohlbaum, K.; Bert, B.; Dietze, S.; Palme, R.; Fink, H.; Thone-Reineke, C. Impact of repeated anesthesia with ketamine and xylazine on the well-being of C57BL/6JRj mice. *PLoS ONE* **2018**, *13*, e0203559. [CrossRef]
27. Mallien, A.S.; Hager, C.; Palme, R.; Talbot, S.R.; Vogt, M.A.; Pfeiffer, N.; Brandwein, C.; Struve, B.; Inta, D.; Chourbaji, S.; et al. Systematic analysis of severity in a widely used cognitive depression model for mice. *Lab. Anim.* **2020**, *54*, 40–49. [CrossRef]
28. Talbot, S.R.; Kumstel, S.; Schulz, B.; Tang, G.; Abdelrahman, A.; Seume, N.; Wendt, E.H.U.; Eichberg, J.; Hager, C.; Bleich, A.; et al. Robustness of a multivariate composite score when evaluating distress of animal models for gastrointestinal diseases. *Sci. Rep.* **2023**, *13*, 2605. [CrossRef]
29. Morton, D.B. A Model Framework for the Estimation of Animal 'Suffering': Its Use in Predicting and Retrospectively Assessing the Impact of Experiments on Animals. *Animals* **2023**, *13*, 800. [CrossRef]
30. Buckinx, A.; Van Schuerbeek, A.; Bossuyt, J.; Allaoui, W.; Van Den Herrewegen, Y.; Smolders, I.; De Bundel, D. Exploring Refinement Strategies for Single Housing of Male C57BL/6JRj Mice: Effect of Cage Divider on Stress-Related Behavior and Hypothalamic-Pituitary-Adrenal-Axis Activity. *Front. Behav. Neurosci.* **2021**, *15*, 743959. [CrossRef]
31. Manouze, H.; Ghestem, A.; Poillerat, V.; Bennis, M.; Ba-M'hamed, S.; Benoliel, J.J.; Becker, C.; Bernard, C. Effects of Single Cage Housing on Stress, Cognitive, and Seizure Parameters in the Rat and Mouse Pilocarpine Models of Epilepsy. *eNeuro* **2019**, *6*. [CrossRef] [PubMed]
32. Carper, D.; Coue, M.; Laurens, C.; Langin, D.; Moro, C. Reappraisal of the optimal fasting time for insulin tolerance tests in mice. *Mol. Metab.* **2020**, *42*, 101058. [CrossRef] [PubMed]
33. Hay, M.; Thomas, D.W.; Craighead, J.L.; Economides, C.; Rosenthal, J. Clinical development success rates for investigational drugs. *Nat. Biotechnol.* **2014**, *32*, 40–51. [CrossRef] [PubMed]
34. Henderson, V.C.; Demko, N.; Hakala, A.; MacKinnon, N.; Federico, C.A.; Fergusson, D.; Kimmelman, J. A meta-analysis of threats to valid clinical inference in preclinical research of sunitinib. *Elife* **2015**, *4*, e08351. [CrossRef] [PubMed]

35. Pritt, S.L.; Hammer, R.E. The Interplay of Ethics, Animal Welfare, and IACUC Oversight on the Reproducibility of Animal Studies. *Comp. Med.* **2017**, *67*, 101–105.
36. Cheleuitte-Nieves, C.; Lipman, N.S. Improving Replicability, Reproducibility, And Reliability In Preclinical Research: A Shared Responsibility. *ILAR J.* **2019**, *60*, 113–119. [CrossRef]

Disclaimer/Publisher's Note: The statements, opinions and data contained in all publications are solely those of the individual author(s) and contributor(s) and not of MDPI and/or the editor(s). MDPI and/or the editor(s) disclaim responsibility for any injury to people or property resulting from any ideas, methods, instructions or products referred to in the content.

animals

Concept Paper

A Model Framework for the Estimation of Animal 'Suffering': Its Use in Predicting and Retrospectively Assessing the Impact of Experiments on Animals

David B. Morton

School Bioscience, University of Birmingham, Birmingham B15 2TT, UK; dbmgm2@gmail.com

Simple Summary: Assessing the impact of scientific procedures on animals is not always easy, especially when considering the variety of species, ages and experimental conditions to which research animals are exposed. It is important to do so, as legislation demands that humane endpoints or intervention points be implemented by scientists when appropriate during an experiment. Herein, I describe a scheme that is objective and avoids subjective assessments. It can be applied to many types of experiments and for most species of animals used in research. It has the additional advantage that the type of adverse effect does not have to be specified, e.g., pain, distress, suffering or lasting harm, as it measures the impact of the experiment on the animals, which is more likely to reflect their emotional state. The scheme can also be used to assess the effectiveness of any alleviative therapy.

Abstract: This paper presents and illustrates, with a working example, a hypothesis for the assessment of ongoing severity before and during an experiment that will enable humane endpoints and intervention points to be applied accurately and reproducibly, as well as helping to implement any national legal severity limits in subacute and chronic animal experiments, e.g., as specified by the competent authority. The underlying assumption of the model framework is that the degree of deviation from normality of specified measurable biological criteria will reflect the level of pain, suffering, distress and lasting harm incurred by or during an experiment. The choice of criteria will normally reflect the impact on an animal and have to be chosen by scientists and those caring for the animals. They will usually include measurements of good health such as temperature, body weight, body condition and behaviour, which vary according to the species, husbandry and experimental protocols and, in some species, unusual parameters such as time of the year (e.g., migrating birds). In animal research legislation, endpoints or severity limits may be specified so that individual animals do not suffer unnecessarily or endure severe pain and distress that is long-lasting (Directive 2010/63/EU, Art.15.2). In addition, the overall severity is estimated and classified as part of the harm: benefit licence assessment. I present a mathematical model to analyse the measurement data to determine the degree of harm (or severity) incurred. The results can be used to initiate alleviative treatment if required or if permitted during the course of an experiment. In addition, any animal determined to have breached the severity classification of a procedure can be humanely killed, treated or withdrawn from the experiment. The system incorporates the flexibility to be used in most animal research work by being tailored to the research, the procedures carried out and the species under investigation. The criteria used to score severity can also be used as additional scientific outcome criteria and for an analysis of the scientific integrity of the project.

Keywords: assessment; adverse effects; mental pain; distress; severity limits

Citation: Morton, D.B. A Model Framework for the Estimation of Animal 'Suffering': Its Use in Predicting and Retrospectively Assessing the Impact of Experiments on Animals. *Animals* **2023**, *13*, 800. https://doi.org/10.3390/ani13050800

Academic Editor: Garikoitz Azkona

Received: 5 January 2023
Revised: 15 February 2023
Accepted: 19 February 2023
Published: 22 February 2023

Copyright: © 2023 by the author. Licensee MDPI, Basel, Switzerland. This article is an open access article distributed under the terms and conditions of the Creative Commons Attribution (CC BY) license (https://creativecommons.org/licenses/by/4.0/).

1. Introduction

Morton and Griffiths [1] pointed out that the very first step in any assessment for avoidance or alleviation of pain and distress in animals is to recognise when they are suffering or experiencing adverse states in some way. (I am using the term suffering in a broad way to encompass all adverse physical, physiological and mental states in animals, which may be exacerbated further through some form of cognitive reflection and memory ('mentation' as termed by De Grazia [2]). Other ways of looking at suffering include the notion of an impaired wellbeing and measuring the impact of the procedures on an animal. While pain and mental distress may be the commonest adverse states for most experimental animals, there are other ways in which animals experience adverse states that can lead to poor physical, physiological and mental welfare and, consequentially, a poor quality of life. In animal research, the impact of an experiment on an animal's homeostasis (as measured by deviation from 'normality' that is not transient) will most likely reflect the harmful consequences of experimental procedures, as well as the extent to which the animal is suffering [1]. As we humans cannot verbally communicate with animals, this approach is as good as (and probably better than) most other scientific measures. It should be appreciated that all scientific measures of welfare, whether physiological, biochemical, hormonal, behavioural or genomic, need to be interpreted for assessment of the degree of harm that an animal is experiencing, which can be further categorised as a mental or physical state (and the possibility and need to provide some form of alleviation).

An assessment may be carried out for the purposes of avoidance of suffering (through refinement of the experimental design), to determine the effectiveness of alleviative therapies (e.g., analgesics for pain or anxiolytics for fear) or, in the case of animal research, for assessment of the level of severity (adverse effects) for ethical and legal reasons. Severity comprises two components: the intensity of the adverse state such as pain or distress and, secondly, its duration. Another factor to consider when calculating the overall animal welfare impact of an experiment is any cumulative suffering, for example, inadequate husbandry, in addition to the experimental procedures, the number of times an animal is subjected to a repeated technique (e.g., injections) or procedure or any additional procedures in more complex protocols. Short-term intense stressors may also have long-term impacts, e.g., on the immune system. It is possible to determine the overall animal 'cost' in terms of harms in a series of experiments by including the number of animals affected.

In competent humans, the assessment of suffering usually relies on verbal reporting of some sort, whereas in babies and other non-verbal or non-communicating humans and non-human animals, other methods have to be used. In animals, an estimate of the intensity of suffering can be made by measuring the impact of an experimental protocol on an animal, i.e., the degree to which it has deviated from normality compared with a control naïve group kept in a similar or the same way. The duration of any suffering is the second most important component of severity and is far easier to measure objectively, although it is often not specified. The combination of intensity and duration is used to estimate 'severity'. The advantage of measuring impact by deviation from normal in this way is that it does not require the type of adverse state experienced to be explicit, that is, it does not need to be defined as pain, physical dystress (a state of stressor overstimulation of the endocrine system as first described by Moberg 1985) [3] or mental distress, such as fear, anxiety, boredom, frustration, grief, etc. If there is no measurable impact, then it is unlikely that the animal will be suffering to any serious extent. That said, short-term suffering is very difficult to assess, as many of the usual markers such as weight loss or appetite will not be measurable, and other, usually behavioural, measures need to be used. Consequently, the approach outlined in this paper is not applicable to acute/hyperacute suffering and is more applicable when the duration extends over a day or more. Scorable clinical signs have to be selected in accordance with the cause and the type of suffering involved, which are partly determined by the experimental procedures being carried out and the species. These clinical signs (sometimes mistakenly referred to as symptoms, which are feelings reported by patients to doctors and not observable) in animals must be observable and assessed as to whether they are within the range of normality or whether they are 'abnormal'. Recognition can be grouped into five categories: relating to appearance, posture, natural behaviour, provoked behaviour and clinical measurements [1].

Considerable advances have been made in the details of these sorts of abnormalities and their observational robustness in many areas of research. For example, in one year in one laboratory animal journal, four papers were published in the fields of rat models of osteoarthritis [4–7]. Three or four decades ago, no mention would have been made of the adverse effects on an animal in an experiment, let alone any measurement of them [8–11]. This serves to emphasise that with the increasing awareness that pain and distress in animals is important, the measurements used are become increasingly more accurate, reliable and generally accepted. Moreover, the Federation of European Laboratory Animal Science Associations (FELASA) provides comprehensive guidelines on the classification and reporting of severity experienced by animals used in scientific procedures [12].

What to Measure?

In a research setting, it has to be decided what measurements to take to assess any suffering. It is always better to use several criteria to ensure a better estimate, as just one or two may be misleading. It may be possible to carry out measurements of the body's reaction to adverse effects such as effects on the sympathetic nervous system (catecholamine levels and their corresponding end-organ responses, e.g., heart rate, blood pressure and acute-phase proteins) or the pituitary–adrenal cortex axis (particularly adrenal corticosteroids). However, these tests can be expensive and may themselves require invasive blood sampling methods, whereas the behaviour of an animal can be simply observed without cost, including how an animal responds to a particular stimulus (provoked behaviour). Furthermore, some of the affected parameters may be compounded by an inherent variation and vary according to the time of day, the time at which an animal is observed or sampled and the specific experimental procedures being carried out. The time and method of sampling, as well as behavioural observation, have to be practical. For example, in terms of time, one would expect a greater change in pain and distress parameters around the time of surgery rather than 2–4 weeks later. For experimental arthritis, adverse feelings of pain such as dull aches may persist for weeks and not just hours or days as for a surgical procedure; therefore, measuring body weight may be more appropriate. For an experiment with several techniques or procedures being carried out over time, several time points within the whole experimental period may have to be scored to develop an accurate picture of the overall 'suffering' from the animal's viewpoint. In conclusion, the time of observation and the key criteria to be measured have to be carefully selected.

In EU Directive 2010/63/EU [13], harm or the adverse effects are defined as "any pain, suffering, distress or lasting harm equivalent to or higher than that caused by the introduction of a needle in good veterinary practice" experienced by an animal as a result of the scientific procedures performed ([13] Art. 3.1). There is an obligation for an assessment of the harm likely to occur as a result of the experimental protocols to be conducted at the outset of a research project and for this to be recorded on the project licence as the severity classification for a procedure or series of procedures, i.e., for the whole project. Severity may also have to be recorded in hindsight as the harm that actually occurred, referred to in the Directive as a "Retrospective assessment" ([13] Art. 39); in specific cases, this is a legal requirement. If an animal is to be reused, then the severity of any previous use has to be determined ([13] Art. 16.1. (a)). Finally, almost all international legislative measures to control animal experimentation worldwide require the successful application of the Three Rs. This paper deals with the aspect of refinement and provides a model whereby animal suffering can be recognised, reduced, reviewed and even terminated on objective grounds when necessary.

This model (first drawn up in the UK prior to the revision of the Directive [13] in 2010 and presented at various seminars) concerns measuring severity retrospectively and comparing it with the predicted severity in order to determine if the severity classification is being or has been exceeded. It is based on the author's practical experience as a laboratory animal veterinarian and close observation of experimental animals in a variety of research settings. In addition, FELASA has produced some valuable and complementary guidelines

(12) that expand further on this theme and provides several more comprehensive and complicated examples. In this paper, I have tried to provide a more practical and accessible approach that is simpler and easier to apply through the selection of the key criteria that fit the purpose of assessing the level of harm and that can be applied to all vertebrates and even invertebrates.

2. Methodology and Results

The first stage is to identify the criteria that can be used to assess any harm and then to select from those the ones that commonly occur and are easy to score. The second stage is to select the time at which to score those criteria. There is no limit to the number of criteria scored or the number of time points chosen, but in order to reduce workload, only the key and most relevant criteria should be used. These key criteria should also best reflect an animal's adverse state and not simply repeat other closely correlated signs. In addition, the measured signs should reliably indicate severity, be robust, subject to little inter- or intraobserver variation, and be economical and easy to measure. Furthermore, the criteria measured should be independent of each other, for example, a failure to eat and body weight are usually closely interdependent.

The aim of this semiquantitative assessment is to be able to compare predicted and actual suffering and to express these estimates mathematically. In the following example, the deviation from normality or intensity is determined for each independent criterion for each animal in the group at a set time. This figure is then averaged to obtain a score for that criterion. The average score per animal is made up of the averages for each criterion. It is then possible to compare, for example, different experimental groups at the same time point, as well as at different time points in an experiment. The easiest way to understand this is to provide a simplified working example.

Example: A Hypothetical Comparison of Two Dose Groups in a Toxicity Test in Rats

In a study of drug safety and effectiveness, several doses were given, and the rats were observed for adverse effects. Pilot studies and previous data with respect to closely related compounds had shown the chemical's likely adverse effects, and the relevant criteria were chosen according to the criteria described above. In this example, only two dose groups and two criteria were compared at the same time point. The project licence for the work had predicted the severity classification, and severity limits had been set for individual animals in a dose-level group. For the Low-dose group, the severity limit was mild, and for the High-dose group, the severity limit was severe. The accuracy of these severity limit predictions can be determined by assessing the actual severity outcomes in real time as illustrated below.

2.1. Choice of Criteria

Affected animals were predicted to show signs, to varying degrees, of lethargy, fever, reduced appetite, a failure to groom or preen, lachrymal secretions from the eyes and nose, diarrhoea and occasional vomiting. The control group (in this case, drug vehicle only) kept under similar conditions was not expected to show any adverse effects and would be classified as the 'normal comparator' for the dosed groups. Among these clinical signs, body temperature (Criterion 1) and body weight (Criterion 2) were the easiest and most accurate to measure.

2.2. Scoring the Criteria

The next step was to classify the criteria to reflect the 'intensity' of the adverse state, i.e., the impact on the animal based on the extent to which the two criteria deviated from normality (see Tables 1 and 2). Three major bands were chosen to reflect the three recognised legal severity categories of mild, moderate and severe. However, in practice, there are two other recognisable and necessary categories: no change (normal variation) and excessive change that is beyond the upper severity limit (i.e., more than severe and long lasting, which has to be described and be quantitatively and qualitatively defined and would include death). These bands were also scored, resulting in a total of five bands. For body temperature, the five bands corresponded to no significant change (0–0.2C, i.e., normal), an increase of 0.2 to 1C (mild), and increase of 1 to 2C (moderate), an increase of 2 to 3C (severe) and an increase of more than 3C above normal (greater than severe).

For body weight, the five bands are no significant change (0–5%, i.e., normal); a mild change of 5–10%, a moderate change of 10–20%, a severe change of 20–30% and a loss of more than 30% (greater than severe). Each of the five recognisable bands for each criterion is assigned a score of 0, 1, 2, 3 or 4, and in this way, the scores are converted into a mathematical model for analysis. Each animal is scored at the same time point for each criterion at the same time of day in an attempt to keep everything constant except for the measurement. This figure is then averaged to obtain a group average for that criterion at that specific time.

Part A in Table 1 (body temperature, low dose) shows that two animals scored within the range considered normal, and three animals had a rise of between 0.2 and 1C. This resulted in a group score of 3 and a group average of 0.6 (i.e., total group score (3) divided by 5 animals) for that criterion.

Table 1. (A) The proposed scoring matrix for body temperature (low dose). (B) The proposed scoring matrix for body weight (low dose).

			A		
Change	**<0.2**	**>0.2 to 1C**	**>1 to 2C**	**>2 to 3C**	**>3C**
Control	**Normal range**	**Mild**	**Moderate**	**Severe**	**>Severe or death**
SCORE:	**0**	**1**	**2**	**3**	**4**
1	X				
2		X			
3	X				
4		X			
5		X			
Total score	(2 × 0)	(3 × 1) = 3	0	0	0
			B		
Change	**0 to 5%**	**>5 to 10%**	**>10 to 20%**	**>20 to 30%**	**> 30%**
Control	**Normal range**	**Mild**	**Moderate**	**Severe**	**> Severe or death**
SCORE:	**0**	**1**	**2**	**3**	**4**
1		X			
2		X			
3		X			
4			X		
5		X			
Total score	0	(4 × 1) = 4	2	0	0

Group total = 0 + 3 + 0 + 0 + 0 = 3; group average = 3/5 = 0.6.

For body weight (body weight (low dose), Part B in Table 1) four animals lost between 5 and 10% body weight, and one animal lost between 10 and 20% body weight, resulting in a group score of 6 (0 + 4 + 2 + 0 + 0) and a group average of 1.2 (i.e., total group score (6) divided by 5 animals). The same process is now applied to the high-dose group for the two criteria (see Table 2).

Table 2. (A) The proposed scoring matrix for body temperature (high dose). (B) The proposed scoring matrix for body weight (high dose).

			A		
Change	**<0.2**	**>0.2 to 1C**	**>1 to 2C**	**>2 to 3C**	**>3C**
Control	**Normal range**	**Mild**	**Moderate**	**Severe**	**>Severe or death**
SCORE:	**0**	**1**	**2**	**3**	**4**
1				X	
2				X	
3					X
4			X		
5				X	
Total score	0	0	1 × 2 = 2	3 × 3 = 9	1 × 4 = 4
			B		
Change	**0 to 5%**	**>5 to 10%**	**>10 to 20%**	**>20 to 30%**	**>30%**
Control	**Normal range**	**Mild**	**Moderate**	**Severe**	**>Severe or death**
SCORE:	**0**	**1**	**2**	**3**	**4**
1				X	
2					X
3					X
4			X		
5				X	
Total score	0	0	1 × 2 = 2	2 × 3 = 6	2 × 4 = 8

Group total = 0 + 0 + 2 + 9 + 4 = 15; group average = 15/5 = 3.0.

The scores for the calculation of actual overall impact assessment can now be performed by summing the two criteria to derive an average impact intensity score for each dose level.

2.3. Low-Dose Group

$$\textit{Overall score} := \frac{\textit{Number of animals} \times (\textit{group average criterion } 1 + \textit{group average criterion } 2)}{\textit{Number criteria used}}$$

$$= \frac{5 \times (0.6 + 1.2)}{2} = \frac{5 \times 1.8}{2} = 4.5$$

Therefore, the average impact score for each criterion for each animal in the low-dose group is 4.5/5 = **0.9**.

2.4. High-Dose Group

$$\textit{Overall score} := \frac{\textit{Number of animals} \times (\textit{group average criterion } 1 + \textit{group average criterion } 2)}{\textit{Number criteria used}}$$

$$= \frac{5 \times (3 + 3.2)}{2} = \frac{5 \times 6.2}{2} = 5 \times 3.1 = 15.5$$

Therefore, the average impact score for each criterion of each animal in the high-dose group is 15.5/5 = **3.1**.

3. Interpretation of the Results

The data obtained from this example can be used to determine whether the severity limit has been exceeded during the study, as well as whether the overall severity classification for the project has also been exceeded.

3.1. Severity Limit

The severity limit is imposed to protect the individual animal; any animal that exceeds the severity limit on the average of the specific criteria should be reported to the project licence holder to take some sort of action. For example, the animal can be withdrawn from the study or, in some circumstances, attempts can be made to alleviate the suffering. An alternative is to contact the competent authority to change the limit and for the study to continue as planned.

3.2. Severity Classification of the Project

A comparison can now be made with the prediction of mild severity in the low-dose group and severe in the high-dose group. In theory, the average score for each animal in a mild band should be 1; however, the actual average was 0.9. Therefore, in the final analysis for that time point, the suffering was less than predicted. Using the same approach, the average score for each animal in the high-dose group should have been 3 (severe) but it was actually 3.1. Therefore, in the high-dose case, the severity classification was exceeded. As the severity limit applies to an individual animal (see [13]. Dir.Annex VIII), some of these animals should have been killed or withdrawn from the study, although their scores should still be recorded.

4. More Complex Experiments: Continued Use and Reuse (Directive 2010/63/EU, Art 16.1 (a))

If a project is more complex with several phases comprising a continued use, each phase can be scored separately and assigned a severity limit, e.g., surgery to implant a monitor followed by a treatment and removal of the monitor. The cumulative severity would then be the severity of each phase of the project, i.e., the severity classification for the whole project. This may be useful scientifically, as a grossly abnormal animal during one phase may not yield reliable data at a later stage because the results could have been confounded by physiological, homeostatic or behavioural responses during an earlier phase.

In terms of reuse, no animal may be reused if it has experienced suffering of mild or moderate severity during a previous use; this scheme will be useful for such assessment.

5. Discussion

5.1. Legal Aspects

Current EU [13] and UK [14] legislation requires project licence holders to conduct an ongoing assessment of the levels of pain, suffering, distress and lasting harm (severity) experienced by animals in an experiment. It also requires that any such suffering be reduced to the minimum necessary to achieve the scientific objective. The scheme described in this paper will help to do this for subacute and chronic experiments (i.e., more than 2 days) but not always for acute experiments, in which the intensity of suffering may exceed the severity limit but not result in any significant physiological or physical changes that can easily be observed and scored. These hyperacute changes, indicating pain and distress, require a different approach such as observing the behaviour of the animals (e.g., vocalisation, escape behaviours, focal attention to a particular body site, etc). It would also be possible to measure blood hormone hormones (e.g., a rise in corticosteroid and/or catecholamine levels) and end-organ responses (e.g., increases in heart rate, blood pressure and respiration); however, these are complicated by the measurement procedures involved, e.g., handling and blood sampling, which in and of themselves, can cause extra suffering [15].

5.2. Choice of Signs to Score

The framework outlined above allows the licence holder to use whatever signs are most appropriate for the specific experimental procedures being carried out. They can be clinical signs, a change in normal behaviour, exhibition of abnormal behaviour or an experimental variable that itself may indicate an adverse effect. The key points are that the signs being scored should be robust in the sense that they reliably indicate a scorable change; any change is easy and convenient to observe; scoring does not cause any further suffering to the animals concerned, i.e., measurement should be non-invasive; and is economical to implement. It is best if the signs also reflect some biological significance; for example, a ruffled/harsh/starey coat is a more general sign of poor wellbeing than measuring tumour size or joint pain, which are specific to the experiment. General signs may be of help in diagnosing an unspecified adverse state, whereas specific signs can be used to make a better scientific severity assessment and may also be used to help with the scientific objectives.

Some signs can relate more to mental distress than pain, and these two states, i.e., pain and distress, often go hand in hand; all painful states are likely to cause mental distress, but not all signs of mental distress indicate pain.

5.3. Number of Scorable Signs

How many scorable signs are needed to make an assessment? In one sense, the fewer signs to score, the easier; therefore, it is useful to try to restrict the number to those that are most important or are most likely to reflect animal suffering. The signs to be scored can be evaluated using a statistical elimination/reduction analysis and the minimum number of signs that closely correlate with the final severity outcome. This may be a circular argument and open to bias, but the overall direction of changes and their magnitude make it less likely to be wildly inaccurate. Using the approach described above, any number of signs can be scored cumulatively and scored in pairs or trios, e.g., A + B, or A + B + C, etc., to provide the best fit; that is, A should normally always accompany B, B should not normally be measured in combination with C, etc.

The final crucial assessment is to compare and rank the experimental group(s) with the control group. The control group may be a scientific control (e.g., vehicle only or sham-operated); the most valid type of control would probably be to compare the experimental group with a group of naïve animals kept under the same conditions. Scoring the 'extra' severity or suffering is a true indication of the impact of the experimental procedures on an animal. It is worth noting that even the scientific control may sometimes experience some pain and distress, e.g., injection of vehicle only or sham surgery. It is therefore important to always measure and score the control group as well as the experimental group. A comparison with naïve animals kept under the same housing and husbandry conditions provides the most objective indicator of all extra adverse effects experienced by the animals in the experiment, although it will not include any mental distress that may be caused by the housing and husbandry alone.

5.4. Categories of Adverse States

Adverse states inevitably include both a physical and a mental component; for example, all painful states will also have a mental component of suffering. Distress, on the other hand, is more difficult to determine and quantify and can be broken down into two types. The classical type described by Moberg [3,16], where there is a marked perturbance in the pituitary–hypothalamic axis that affects end-organ responses such as the adrenal cortex, thyroid and reproductive organs can lead to a marked loss of homeostasis and, ultimately, a failure to thrive. This type of stress response is termed *physiological dystress*. The other sort of distress, which was originally described by Selye, involves behavioural changes and is more mental than physical. It can be caused by negative emotional states of, e.g., frustration, boredom, fear or anxiety, and is usually transient, although not always; this type of distress is termed *mental distress*. The choice of controls mentioned in the preceding paragraph dealing with naïve animals will inevitably include a minimum of mental distress as a result of the housing and husbandry, e.g., frustration and boredom. Extra suffering as a result of the procedures is the difference between the control group and the experimental group(s), e.g., pain and fear. This further illustrates the difference between any physiological dystress and any mental distress.

5.5. Change in Score Direction

Some changes in scorable signs may be positive or negative and can go in either direction depending on the experiment and the species being scored. For example, in experimental studies on infection, an increase in body temperature is normally seen in most species, but for those species with a high metabolic rate, e.g., small rodents such as mice, infection may lead to a decrease in temperature; exactly the same principle applies to scoring the severity deviation from normality [1,17,18]. In terminally ill animals, the body temperature will gradually fall if they are left to die. When using body weight as a measurement, the age and maturity of an animal has to be taken into account. That is why the control should be chosen carefully. A failure to eat will lead to body weight loss depending on the level of intake, but in growing animals, the comparison should be age-matched, as a slowing of growth may be just as important. Using body weight to measure tumour growth may result in appetite being decreased or increased, whereas body weight may increase or decrease or not change. It has to be coupled with other signs such as tumour size. Body weight may decrease as a result of lowered intake or increased metabolic rate, or an increase in body weight may be observed as a result of fluid accumulation, e.g., ascites with liver tumours or decreased activity. All confounding factors need to be taken into account and interpreted biologically when selecting appropriate clinical signs of suffering.

5.6. Scoring of Death

Death should always be scored as a criterion in its own right, as it represents major change (usually unexpected) and should considered a serious adverse effect regardless of deviation from normality. If death is expected, stringent efforts must be made to predict it and to apply a humane endpoint, as death often indicates prior hyperacute suffering of some sort [19]. However, death can also occur very rapidly between inspections. The legislation permits death as an endpoint in exceptional circumstances [20].

5.7. What Is Normal?

Sometimes it can be challenging to decide what is the normal range for a particular sign, as there is so much biological variation. Each animal is scored as an individual and not as a group; therefore, investigators look for a change in the individual rather than an absolute number or standardised qualitative estimate. Depending on the species, change relative to normal can vary in rodents from 0–10% for body weight depending on the time of day, whereas for other species, the range is much smaller. This is the reason why scoring should take into account the circadian rhythm of animals and ideally be carried out at the same time each day. An animal's reaction to its housing and husbandry vary as much as the husbandry, the housing and the personnel. That confounded by individual biological variation means that 'normal' will inherently have some variation, which is why +/−5% leeway is given for that category.

In the examples I have chosen for this study, I have given a score for each level of change that correlates with severity classification in the legislation. The classification of normal can be difficult in some species due to the natural variation in some species, especially in non-mammals. Normal can vary with ambient temperature (poikilotherms such as retiles amphibia, fish, including and the mole rat) and season of the year (e.g., body weight in migratory birds); however, for laboratory mammals kept under research conditions and husbandry, 'normal' parameters are relatively stable for a given age and sex. The normal baseline forms the basis for assessing the impact of an experiment on an animal. The degrees of change above or below the normal threshold are set to reflect the severity classification from 'mild' to greater than 'severe', which includes death; although changes greater than severe are not permitted, they will inadvertently occur between scoring inspections unless an adequate safety margin is implemented (see the section above on scoring death).

5.8. Assessment of the Severity Limit and Severity Band

The average score for an individual animal in a group at a particular time is the sum of all the scores for a specific sign divided by the number of animals. This score is then added to the scores for all the other signs, which provides an estimate of the actual severity being experienced by that animal at that time. For the total score in an experiment, all the scores for all the animals in that group are added together, and that sum can then be used to obtain the average score for each animal in that experimental group. This figure can them be used to determine any breach of the project severity classification, which can be an overestimate or an underestimate. However, for the application of the severity limit and the implementation of a humane endpoint, it is the estimated suffering of an individual animal that is important. *The project severity classification*, on the other hand, is based on the overall average for the worst-case scenario of the project. The estimation of the level of suffering (severity) can then be used to corroborate the estimated projected severity of the project.

5.9. Inspection Intervals

It is a legal requirement that animals be inspected for health every day ([12] Art 13.1.(c)). Animals used in an experiment should be checked more regularly depending on to how likely they are to undergo a change in their severity classification; the frequency of inspection should be increased for higher severity levels to three to four times or more daily or if there is a likelihood of exceeding the upper severe category during the night and outside normal working hours (e.g., weekends) (see unpredicted endpoints in Ashall and Millar, 2014) [21].

5.10. Uses of Scoring Results Other than Severity Estimates

Ongoing contemporaneous scoring of the signs can help to achieve the goals of ascertaining when an animal has exceeded an allocated individual severity classification or limit, in addition to helping to define a humane endpoint, scientific endpoint or intervention point depending on which is being sought. It will also help in any retrospective analysis of any suffering experienced by animals, as well as the effectiveness of any alleviating strategies. Scoring the adverse effects of scientific procedures during an experiment will help in the validation of the scientific data being measured, as confounding covariables may negate or modify the interpretation of the scientific results. It must always be remembered that an animal's physiological responses to the experimental procedures being carried out are likely to affect the scientific data being harvested and may confound the interpretation of the data. Furthermore, these physiological responses may even be of greater significance than any alleviative treatments; therefore, attempts should always be made to treat pain and distress.

Training: For some clinical signs used for scoring, e.g., behaviours, appearance, stance and posture and provoked responses, it is often vitally important that there is consistency between those scoring the signs; this may require some serious training. Such training should be extended to all appropriate staff, including those staff working at weekends, holidays and out-of-hours during the evening and the night.

Finally, it should not be forgotten that such an approach may be useful to assess positive mental states, as well as negative ones, although the choice of specific measures will obviously vary, e.g., use of any enrichment or attractions and time spent playing in young animals.

6. Conclusions

This paper describes the principles of a simplified system of analysis of the impact of an experiment on an animal that is objective, robust and reproducible. It will permit a contemporaneous assessment of the adverse effects and a scheme for the implementation of the severity classification limit. It will also provide a mechanism for the retrospective assessment and accuracy assessment of the predicted severity classification.

Funding: This research received no external funding.

Institutional Review Board Statement: Not applicable as theoretical and no animals were used.

Informed Consent Statement: Not applicable as theoretical and no humans were used.

Data Availability Statement: Not applicable as a theoretical concept.

Conflicts of Interest: The author declares no conflict of interest.

References

1. Morton, D.B.; Griffiths, P.H.M. Guidelines on the recognition of pain, distress and discomfort in experimental animals and an hypothesis for assessment. *Vet. Rec.* **1985**, *116*, 431–436. [CrossRef] [PubMed]
2. DeGrazia, D. *Taking Animals Seriously: Mental Life and Moral Status*; Cambridge University Press: Cambridge, UK, 1996.
3. Moberg, G.P. *Animal Stress*; American Physiological Society: Bethesda, MD, USA, 1985; ISBN 0-683-06101-1.
4. Jacobs, B.Y.; Allen, K.D. Factors affecting the reliability of behavioral assessments for rodent osteoarthritis models. *Lab. Anim.* **2020**, *54*, 317–329. [CrossRef] [PubMed]
5. van Aken, M.A.; Groothuis, P.G.; Panagiotou, M.; van Duin, M.; Nap, A.W.; van Rijn, T.C.; Kozicz, T.; Braat, D.D.; Peeters, A.B. An objective and automated method for evaluating abdominal hyperalgesia in a rat model of endometriosis. *Lab. Anim.* **2020**, *54*, 365–372. [CrossRef]
6. Rix, A.; Drude, N.; Mrugalla, A.; Mottaghy, F.M.; Tolba, R.H.; Kiessling, F. Performance of severity parameters to detect chemotherapy-induced pain and distress in mice. *Lab. Anim.* **2020**, *54*, 452–460. [CrossRef]
7. Zieglowski, L.; Kümmecke, A.; Ernst, L.; Schulz, M.; Talbot, S.R.; Palme, R.; Czaplik, M.; Tolba, R.H. Severity assessment using three common behavioural or locomotor tests after laparotomy in rats: A pilot study. *Lab. Anim.* **2020**, *54*, 525–535. [CrossRef]
8. Townsend, P.; Morton, D.B. Practical assessment of adverse effects and its use in determining humane endpoints. In Proceedings of the 1993 5th FELASA Symposium: Welfare & Science, Brighton, UK, 8–11 June 1993; pp. 19–23.
9. Townsend, P.; University of Birmingham, Birmingham, UK. Personal communication, 1990.
10. Flecknell, P. Practical Postoperative Pain Relief in Small Animals. *Vet. Nurs. J.* **1992**, *7*, 55–60. [CrossRef]
11. Flecknell, P. Advances in the Assessment and Alleviation of Pain in Laboratory and Domestic Animals. *J. Vet. Anaesth.* **1994**, *21*, 98–105. [CrossRef]
12. Smith, D.; Anderson, D.; Degryse, A.-D.; Bol, C.; Criado, A.; Ferrara, A.; Franco, N.H.; Gyertyan, I.; Orellana, J.M.; Ostergaard, G.; et al. Classification and reporting of severity experienced by animals used in scientific procedures: FELASA/ECLAM/ESLAV Working Group report. *Lab. Anim.* **2018**, *52*, 5–57. [CrossRef] [PubMed]
13. Directive 2010/63/EU of the European Parliament and of the Council of 22 September 2010 on the Protection of Animals Used for Scientific Purposes. *Off. J. Eur. Union* **2010**, *276*, 33–79.
14. *ASPA HMSO 2000 HO Guidance on the Operation of the Animals (Scientific Procedures) Act*; The Stationary Office: London, UK, 1986; ISBN 0 10 556706 X X. Available online: https://www.gov.uk/government/publications/consolidated-version-of-aspa-1986 (accessed on 18 February 2023).
15. Blenkuš, U.; Gerós, A.F.; Carpinteiro, C.; Aguiar, P.D.C.; Olsson, I.A.S.; Franco, N.H. Non-invasive assessment of mild stress-induced hyperthermia by infrared thermography in laboratory mice. *Animals* **2022**, *12*, 177. [CrossRef] [PubMed]
16. Moberg, G.P.; Mench, J.A. *The Biology of Animal Stress. Basic Principles and Implications for Animal Welfare*; CABI: Wallingford, UK, 2000; p. 377. ISBN 0 85199 359 1.
17. Morton, D.B. A Systematic Approach for Establishing Humane Endpoints. *ILAR J.* **2000**, *41*, 80–86. [CrossRef] [PubMed]
18. OECD. Environmental Health and Safety Publications Series on Testing and Assessment No. 19 Guidance Document on the Recognition, Assessment, and Use of Clinical Signs as Humane Endpoints for Experimental Animals Used in Safety Evaluation Environment Directorate. 2000. Available online: https://www.oecd.org/officialdocuments/publicdisplaydocumentpdf/?cote=env/jm/mono(2018)6&doclanguage=en (accessed on 18 February 2023).
19. Mellor, D.J.; Morton, D.B. Humane Endpoints in research and testing. In *Animals Alternatives, Welfare and Ethics*; Elsevier: Amsterdam, The Netherlands, 1997.
20. Beauchamp, T.B.; Morton, D.B. The Upper Limits of Pain and Suffering in Animal Research: A Moral Assessment of the European Union's Legislative Framework. *Camb. Q. Healthc. Ethics* **2015**, *24*, 431–447. [CrossRef] [PubMed]
21. Ashall, V.; Millar, K. Endpoint Matrix: A Conceptual Tool to Promote Consideration of the Multiple Dimensions of Humane Endpoints. *Altex* **2014**, *31*, 209–213. [CrossRef] [PubMed]

Disclaimer/Publisher's Note: The statements, opinions and data contained in all publications are solely those of the individual author(s) and contributor(s) and not of MDPI and/or the editor(s). MDPI and/or the editor(s) disclaim responsibility for any injury to people or property resulting from any ideas, methods, instructions or products referred to in the content.

animals

MDPI

Article

Rehoming and Other Refinements and Replacement in Procedures Using Golden Hamsters in SARS-CoV-2 Vaccine Research

Malan Štrbenc [1,*], Urška Kuhar [2], Duško Lainšček [3], Sara Orehek [3], Brigita Slavec [4], Uroš Krapež [4], Tadej Malovrh [2] and Gregor Majdič [1]

[1] Institute for Preclinical Sciences, Veterinary Faculty, University of Ljubljana, 1000 Ljubljana, Slovenia; gregor.majdic@vf.uni-lj.si
[2] Institute for Microbiology and Parasitology, Veterinary Faculty, University of Ljubljana, 1000 Ljubljana, Slovenia; urska.kuhar@vf.uni-lj.si (U.K.); tadej.malovrh@vf.uni-lj.si (T.M.)
[3] Department of Synthetic Biology and Immunology, National Institute of Chemistry, 1000 Ljubljana, Slovenia; dusko.lainscek@ki.si (D.L.); sara.orehek@ki.si (S.O.)
[4] Institute of Poultry, Birds, Small Mammals and Reptiles, Veterinary Faculty, University of Ljubljana, 1000 Ljubljana, Slovenia; brigita.slavec@vf.uni-lj.si (B.S.); uros.krapez@vf.uni-lj.si (U.K.)
* Correspondence: malan.strbenc@vf.uni-lj.si

Simple Summary: In 2020, Slovenia joined the global effort to develop effective vaccines and drugs to treat COVID-19. Two vaccine candidates developed in previous studies were selected and tested in the golden hamster model using four different vaccination protocols. We followed the required 3Rs principle when performing the procedures on the animals: we *refined* animal housing, handling, and measurements, including the introduction of pilot animal infection tests, and we *reduced* the total number of animals used primarily through the *replacement* procedure. Replacement was conducted by using a virus neutralisation test on cell cultures prior to infecting and killing the animals. We determined that the antibodies produced by the tested vaccines did not have sufficient neutralising properties, and the project was terminated. Approximately half of the golden hamsters that were no longer needed in the procedures were rehomed and we received very encouraging feedback from adopters.

Abstract: Effective vaccines are needed to fight the COVID-19 pandemic. Forty golden hamsters were inoculated with two promising vaccine candidates and eighteen animals were used in pilot trials with viral challenge. ELISA assays were performed to determine endpoint serum titres for specific antibodies and virus neutralisation tests were used to evaluate the efficacy of antibodies. All tests with serum from vaccinated hamsters were negative even after booster vaccinations and changes in vaccination protocol. We concluded that antibodies did not have sufficient neutralising properties. Refinements were observed at all steps, and the in vitro method (virus neutralisation test) presented a replacement measure and ultimately lead to a reduction in the total number of animals used in the project. The institutional animal welfare officer and institutional designated veterinarian approved the reuse or rehoming of the surplus animals. Simple socialization procedures were performed and ultimately 19 animals were rehomed, and feedback was collected. Recently, FELASA published recommendations for rehoming of animals used for scientific and educational purposes, with species-specific guidelines, including mice, rats, and rabbits. Based on our positive experience and feedback from adopters, we concluded that the rehoming of rodents, including hamsters, is not only possible, but highly recommended.

Keywords: vaccine research; golden hamster; animal models; SARS-CoV-2

Citation: Štrbenc, M.; Kuhar, U.; Lainšček, D.; Orehek, S.; Slavec, B.; Krapež, U.; Malovrh, T.; Majdič, G. Rehoming and Other Refinements and Replacement in Procedures Using Golden Hamsters in SARS-CoV-2 Vaccine Research. *Animals* **2023**, *13*, 2616. https://doi.org/10.3390/ani13162616

Academic Editors: Melanie L. Graham and Garikoitz Azkona

Received: 4 July 2023
Revised: 10 August 2023
Accepted: 11 August 2023
Published: 14 August 2023

Copyright: © 2023 by the authors. Licensee MDPI, Basel, Switzerland. This article is an open access article distributed under the terms and conditions of the Creative Commons Attribution (CC BY) license (https://creativecommons.org/licenses/by/4.0/).

1. Introduction

In 2020, researchers in Slovenia joined a global race to find effective therapies and vaccines against the COVID-19 pandemic. In this context, small animal models are essential, as preclinical studies in animals are crucial for basic and applied research in most infection studies. Several SARS-CoV-2 animal models based on ACE2 receptor homology have been studied, including nonhuman primates, transgenic mice, ferrets, cats, and hamsters [1]. Among wild-type rodents, only hamsters—*Cricetinae*—show this homology, and project proposals using hamsters have nearly tripled in Germany, for example, compared with pre-pandemic years [2]. The golden hamster, also known as the Syrian hamster (*Mesocriccetus auratus*), is the most common in publications, although it has been shown that the Chinese hamster (*Cricetulus griseus*) actually develops more severe signs of disease [3], infection in the Roborovski dwarf hamster (*Phodopus roborovskii*) can sometimes be fatal [4], and both species seem to mimic human pathology better. However, the golden hamster is well studied and readily available from laboratory animal breeders, having been recognized as a good model for many emerging infectious diseases [5], served as a model for SARS-CoV infection as early as 2005, and is essential for some specific studies [6]. They are also inexpensive, easy to handle and keep in captivity.

Vaccine efficacy (and safety) testing has been characterized by extensive use of laboratory animals. Working groups have been formed and recommendations made to develop the required 3R—Replacement, Reduction, and Refinement—methods that include a non-clinical endpoint, ultimately reducing the number of animals and decreasing severity levels of procedures on animals (refinement) [7]. Determination of antibody efficacy involves multiple testing platforms, including ELISA, lateral flow immunoassay, microsphere immunoassay, and pseudovirus systems. These assays measure the binding of antibodies to the SARS-CoV-2 spike, RBD, pre-fusion, and N proteins. However, because not all binding antibodies can block viral infection, these platforms do not measure antibody inhibition of SARS-CoV-2 infection. One possible in vitro replacement is a virus neutralisation test (VNT)—a highly sensitive and specific serological assay for detecting the presence and amount of functional antibodies that prevent viral infectivity [8]. However, for initial efficacy tests, it is not yet possible to replace the challenge procedure. Challenge of animals is used as a closed system to examine the impact of the virus and the interaction of vaccines or drugs in a living model. At least improvements—including re-evaluation of humane endpoints and overall animal welfare—must be considered. One of these aspects is the final fate of the animals. According to Articles 17 and 19 of Directive 2010/63/EU, the most appropriate decision should be taken as regards the future of the animal on the basis of animal welfare and potential risks to the environment. It is up to Member States to state under which conditions a reuse, setting free, or rehoming of the animals used in scientific procedures or bred for scientific purposes may be allowed. In recent years, there has been a particular focus on the rehoming of animals, especially appealing species such as dogs, cats, horses, and camelids, but reports of the successful rehoming of poultry, rabbits, and rodents have also been presented on institutional websites and congresses [9]. The FELASA Workgroup recently issued related guidelines [10], and designated surveys shed some light on practice in Europe and the UK [11,12].

In the present study, we tested two promising vaccine candidates based on previous research [13] on an appropriate animal model—the golden hamster—testing different vaccination protocols on 40 animals. We describe measures for refinement of all procedures on animals and show examples of replacement of the in vivo method. VNT was introduced to decide which of the vaccination protocols were promising, and with negative results (no neutralizing antibodies detected), the in vitro test actually replaced the viral challenge to test antibody efficiency. We also introduced a rehoming procedure at the end of the study.

2. Materials and Methods

2.1. Animals

Golden hamsters (*Mesocricetus auratus*) of strain HsdHan AURA were supplied by Envigo (Udine, Italy) from a UK-based breeding stock (40 animals: 20 males, 20 females) and 18 animals (8 males, 10 females) were supplied by Janvier Laboratories (Le Genest-Saint-Isle, France) of strain RjHan: AURA. Both outbred lines originate from Zentralinstitut fur Versuchstiere, Hannover, Germany. The animals were 7–8 weeks old at purchase and housed in same-sex pairs in conventional polypropylene cages with filter covers (Techniplast Eurostandard type III). Between week 8 and 10, all females and some males started to fight despite increased enrichment. Between the 10th and 12th week, we separated all animals to prevent fighting and they remained in single housing until the end of the vaccination project. A total of 40 were used for vaccine testing: 32 with the first vaccine candidate and 8 animals with another vaccine candidate. The animals were kept in dedicated rodent rooms at the Faculty of Veterinary Medicine, University of Ljubljana, which provided a controlled environment with a relative humidity of 45–60%, a temperature between 21 and 23 °C, and a 12:12 light—dark cycle. Feed and water were given ad libitum; pelleted irradiated feed contained 21% crude protein (Sniff diets S8189-S098). Irradiated wood fibre bedding was used (Lignocel, Rosenberg, Germany); for enrichment, sterilized paper strip nesting material (Sizzelenest, Datesand, Manchester, UK), sterilized aspen wood gnawing blocks (Datesand, Manschester, UK), and cardboard or polypropylene tubes for hiding were offered.

For viral challenge tests, a pilot trial was planned and performed on 18 animals in 3 separate trials in a high-containment laboratory (Biosafety level 3) using an animal bio-containment system (Techniplast IsoRat900 N with Teklad Isoplast bedding)—designated A-BSL3 (animal biosafety level 3). For the second and third trials, animals were 8 weeks old and kept in pairs or threes as no territorial aggression occurred yet. In total, 58 animals were used for the project.

All animal procedures were performed in accordance with the EU Directive (2010/63/EU), approved by the Administration of the Republic of Slovenia for Food Safety, Veterinary and Plant Protection (Ministry of Agriculture, Forestry and Foods) and its Ethical Committee (Decisions U34401-18/2020/8 and U34401-11/2021/9) and reports were prepared according to PREPARE and ARRIVE guidelines.

2.2. Vaccination

Two different routes of administration and two concentrations of the best vaccine candidate identified from previous in vitro and mouse immune response tests [13] were selected, namely, the plasmid DNA RBD-bann. A total of 32 animals of both sexes, 4 months old, were immunized and divided into 4 groups (4 males and 4 females each), which were administered either 20 or 50 μg of the vaccine plasmid DNA intranasally (i.n.) or intramuscularly (i.m.). For intramuscular administration, 50 μL was injected into the lateral thigh (biceps muscle); for intranasal administration, 50 μL was slowly pipetted into both nostrils (2–3 drops per nostril of plasmid DNA in 0.9% saline). All vaccinations were repeated after 2 weeks and specific antibody titres in serum were determined by ELISA and neutralisation properties by VNT assay. Twelve animals with the highest titres received a second booster dose of the same product i.m. after 4 weeks. The other 20 animals that were vaccinated with a lower dose and/or the i.n. route had lower antibody titres and were vaccinated 4 weeks after the second dose with two additional booster doses 3 weeks apart: 10 animals with an increased dose of naked plasmid DNA (40 μg per animal i.m.) and 10 animals with the second-best candidate from previous research: RBD-bann recombinant protein (100 μg/animal in 50 μL i.m.) coupled with squalene adjuvant (2:1 ratio) AddaVaxTM (Invivogen; vac-adx-10, San Diego, CA, USA) in order to boost the antibody production.

For the final vaccination protocol, 8 naive animals (7 months old) were used and vaccinated with recombinant RBD protein and Complete Freund's adjuvant (Calbiochem,

Merck) in a 1:1 ratio, 50 μL subcutaneously (s.c.) in two skin folds—just caudal to the elbow joint (retroaxillary) and cranial to the stifle (regio plicae lateralis).

Blood samples were collected from the animals immediately before each vaccination and 2 and 4 weeks after the last booster vaccination, always under general anaesthesia (procedure described in Refinements down the page). Blood was transferred to a mini-tube containing a gel clot activator (Microvette® 500 Serum Gel, Sarstedt, Germany), allowed to coagulate for 8 h or overnight in the refrigerator, and centrifuged at 9000× *g* for 10 min. The separated serum was frozen at −22 °C until analysis.

2.3. *Refinements in Animal Work*

After transport, the animals were allowed to acclimate to the new environment for at least 1 week. If they were transferred to a BSL3 facility, they remained undisturbed for another 3–4 days. Hamsters were provided with environmental enrichments and handled by hand cupping or tunnels (especially when changing cages). Half of the food pellets were always offered on the floor of the cage to allow for hamster-specific storage behaviour. Blood samples were collected under general anaesthesia at the junction of the cranial vena cava, external jugular vein, and subclavian vein. Isoflurane was used as 4.5% in the induction chamber, maintained by 2.5–3.5% via the face mask: the animal was placed in dorsal recumbency, the puncture site was located on the left or right craniolateral to the manubrium sterni, and the needle was inserted in a caudal direction toward the contralateral hip joint [14]. The puncture site was disinfected and moistened with Spitaderm (Ecolab, Monheim am Rhein, Germany). An amount of 200–400 μL of blood per animal was collected, gentle pressure with a cotton swab for 30 s was applied on the puncture site, the gas supply was switched to an oxygen mixture, and respiration and cardiac activity were observed for a few minutes. After recovery from anaesthesia, animals were examined periodically (30 min, 2 h, 6 h, and daily thereafter) for obvious hematoma, discomfort, or pain. Initial administrations of the vaccine (intranasal—i.n. and intramuscular—i.m.) were performed under the same anaesthesia: for intramuscular administration, in the lateral position and with a face mask; for intranasal administration, the mask was removed, the hamster was grasped around the chest with the head tilted upwards, and 3–5 small drops were pipetted onto the nose, which were then inhaled. If the animal began to awaken, it was placed back under the mask (isoflurane 3%) and the remaining amount of vaccine was pipetted after the animal had lost reflexes. For booster s.c. application, only manual restraint was used, as no intermediate blood sampling was performed, thus no anaesthesia was needed and no local reaction was seen after the first dose. For both i.m. injections and blood sampling, 0.5 mL insulin syringes with fixed 30 G needles were used (BD Micro Fine Plus). For subcutaneous injection of protein suspensions and Freund adjuvant, a 1 mL insulin syringe with a 27 G needle was used (BD Microlance 3).

Randomization was performed during acclimation of the animals—one animal caretaker randomly assigned animals to cages in pairs; when pairs were separated, cages and animals were renumbered by the second caretaker. Ear notches for individual ID were made under general anaesthesia at the first blood draw. The allocation of the first vaccination protocol was conducted by the researcher, who did not know the allocation of each animal ID, except for the sex of the animals. The remaining vaccinations were allocated based on preliminary results. The researchers who performed titration, ELISA, PCR, VNT tests, or histology were blinded during the analysis.

2.4. *ELISA Test*

ELISA was performed to determine endpoint titres for designated specific antibodies as described before [13]. Sera from hamsters after the first and second vaccinations were compared, and sera from the same hamsters before vaccination were used as negative control. Details are given in the supplementary materials.

2.5. Pseudovirus Neutralisation System

To test the general neutralisation properties of the antibodies detected in hamster sera, a pseudovirus (PV) neutralisation assay was carried out as described before [13]. Details are given in the supplementary materials under Figure S2.

2.6. Detection of Neutralising Properties of SARS-CoV-2 Antibodies

Serum samples were tested for the presence of neutralising SARS-CoV-2 antibodies using a virus neutralisation test (VNT). The SARS-CoV-2 strain Slovenia/SI-4265/20 was provided by the European Virus Archive and the Vero E6 cell line (African green monkey kidney cells) was provided by ATCC: VERO C1008; CRL-1586 [15]. The positive reference serum (EURM-018 human serum, JRC, EC) was used as a positive control. More details are given in Supplementary Figure S3.

2.7. Pilot Trial for Viral Challenge Test

In the first pilot trial, 6 naïve hamsters (2 males, 4 females) were intranasally inoculated with 25 µL of the SARS-CoV-2 virus strain Slovenia/SI-4265/20 (European Virus Archive) with a titre of 5×10^4 TCID 50/mL diluted in Eagle's Minimum Essential Medium (EMEM), observed daily for clinical signs, and killed with CO_2 and exsanguination on days 2 (2 female), 4 (1 male, 1 female), and 7 (1 male, 1 female) after inoculation to determine the best time point and tissue samples. Nasal conchae, trachea, lung tissue, duodenum, and whole brain were collected. FLOQSwabs (Copan, Italia) from the caudal nasal cavity, trachea, lung's cutting surface, and duodenum were also taken for RT-qPCR. Based on the initial results, a second trial with a group of 6 animals (3 males and 3 females) was conducted. The animals were inoculated intranasally (i.n.) with 50 µL (25 µL per nostril) of the virus with a titre of 5×10^4 $TCID_{50}$/mL, monitored daily for clinical signs, and killed with exsanguination (cardiocentesis) under inhalation anaesthesia on the fourth day after inoculation. Because clinical and pathologic signs were low, we increased the viral load and tested on an additional 6 animals (3 female, 3 male)—inoculated i.n. with 50 µL (25 µL per nostril) of the virus with a titre of 1×10^6 $TCID_{50}$/mL—designated as the third pilot trial. Animals were monitored and weighed every 24 h and killed by exsanguination under inhalation anaesthesia on day 4. Blood and lung tissue samples were collected—the entire left lung lobe was placed in 4% buffered paraformaldehyde for histological processing, and the entire right lung was weighed and homogenized in sterile Eagle's Minimum Essential Medium (EMEM). The clinical score sheet was used for signs of lethargy, ruffled coat, hunched posture, laboured breathing, nasal discharge, cyanosis, and facial grimace: 1 for mild and 2 for severe manifestation (daily score above 6 was considered as humane endpoint; none of the animals reached this). The same trained person observed the animals and filled in the evaluation forms. She was not blind to the study, but she was blind to animal ID each time to start with as the cage position might have changed at the first morning inspection by the animal caretaker. Animals in A-BSL3 were identified by cage cards and abdominal marking (blue pen) to separate between 2 or 3 animals in the cage only upon close inspection.

2.8. Determination of Virus Titres

Virus titres were determined by titration of sample homogenate suspensions on Vero E6 cells and measured as the 50% infectious tissue culture dose ($TCID_{50}$/mL) [16–18]. Some details are available in the supplementary materials.

2.9. RNA Extractions and Quantitative RT-PCR

The swabs of organs were individually vortexed in 2 mL phosphate-buffered saline for 2 min prior to genomic nucleic acid extraction. Total RNA and DNA were extracted from 140 µL of sample supernatant by the QIAamp Viral RNA Mini Kit (Qiagen, Hilden, Germany) according to the manufacturer's instructions. Viral RNA was detected with the real-time assay RT-qPCR targeting the E gene of SARS-CoV-2 using the primers and probe described by Corman et al. [19]. RT-qPCR was performed by QuantStudio 5 (Thermo Fisher Scientific, Waltham, MA, USA) using 2 µL of the extracted total RNA.

2.10. Immunohistochemistry

The left lung was fixed in 4% buffered paraformaldehyde for 3–4 days, embedded in paraffin, and cut into 5 µm horizontal sections and stained either with HE or the location of the virus with SARS-CoV-1/2 Spike Protein (2B3E5) mouse mAb (#52342, Cell Signaling Technology, Danvers, MA, USA) was labelled. More details on this procedure can be found in the supplementary materials (Figure S5).

2.11. Reuse and Rehome

The prospect of reusing or rehoming the animals was not included in the project proposal, as all animals were expected to enter the A-BSL3 facility for viral challenge and be killed there. Based on the results during the vaccination process, it was decided to omit the viral challenge procedure on vaccinated animals because antibodies produced did not have (detectable) neutralising properties and infecting the animals would not bring any new data needed for the purpose of the project. The facility animal welfare officer was consulted to approve the reuse or rehoming of the 40 surplus animals—the animals that were vaccinated but not exposed to the virus. The institutional designated veterinarian certified that most of the animals were in good health and posed no threat to the environment or human health, so they could be reused or rehomed, even though they were about halfway through their life expectancy at that point: 16 months old and life expectancy is between 2 and 3 years.

A total of 16 animals in suboptimal condition were sacrificed for tissue samples for basic research and controls. In the meantime, simple socialization measures were implemented, primarily through giving fresh food treats, frequent handling, and alternative environmental enrichment. Another 5 animals that did not adapt well to handling were deemed unsuitable for rehoming as pets. Adopters were initially institutional and later external. Prospective adopters contacted the person responsible for the hamsters, an interview was conducted to confirm the adopter's knowledge of hamster care, a simple adoption contract was signed, and voluntary follow-up meetings were held over the following months. In the adoption contract, there was also basic information on procedures the animals were subjected to and approvals of the designated veterinarian and welfare officer. DNA vaccines to date do not persist or even biodistribute throughout the body of the vaccine recipient when delivered parenterally into muscle, subcutaneous tissue, or dermal layers. The local response to plasmid DNA inoculation is that cells take up the plasmid and then express the immunogen(s)-encoded mechanisms; the nucleic acid is degraded by normal molecular mechanisms. As a consequence, the plasmid DNA clears from the injection site over time [20]. A minimum of 2 months passed between the last vaccination and adoption and adopters expressed no concerns on vaccination methods used.

Out of 40 animals, 19 animals were designated fit for rehoming and all found new homes. Eighteen months after the last adoption, a simple anonymous online survey was conducted via a local popular webpage, www.enka.si. The link to the questionnaire was sent to 17 adopters, and 11 responded. The (translated) questionnaire is added in the supplementary materials.

3. Results

Sixty hamsters were planned to test the plasmid DNA RBD-bann vaccine (in low and high doses and in two modes of administration, i.m. and i.n.) as well as the protein-based RBD-bann. The supplier was only able to supply forty hamsters, which we divided into four groups for the plasmid DNA vaccine and one group for the protein-based vaccine, with each group consisting of four male and four female hamsters. Prior to vaccination, the hamsters tested negative for SARS-CoV-2 (oropharyngeal swab to rule out possible exposure to the virus in the environment) and a blood sample was taken. This was used as a control—all antibody titres were compared with a control serum from each individual.

Protein RBD bann was tested on 8 naïve hamsters after the initial ELISA and VNT results and also on 10 animals in the form of booster doses after two doses of the DNA vaccine. The diagram in Figure 1 explains the workflow.

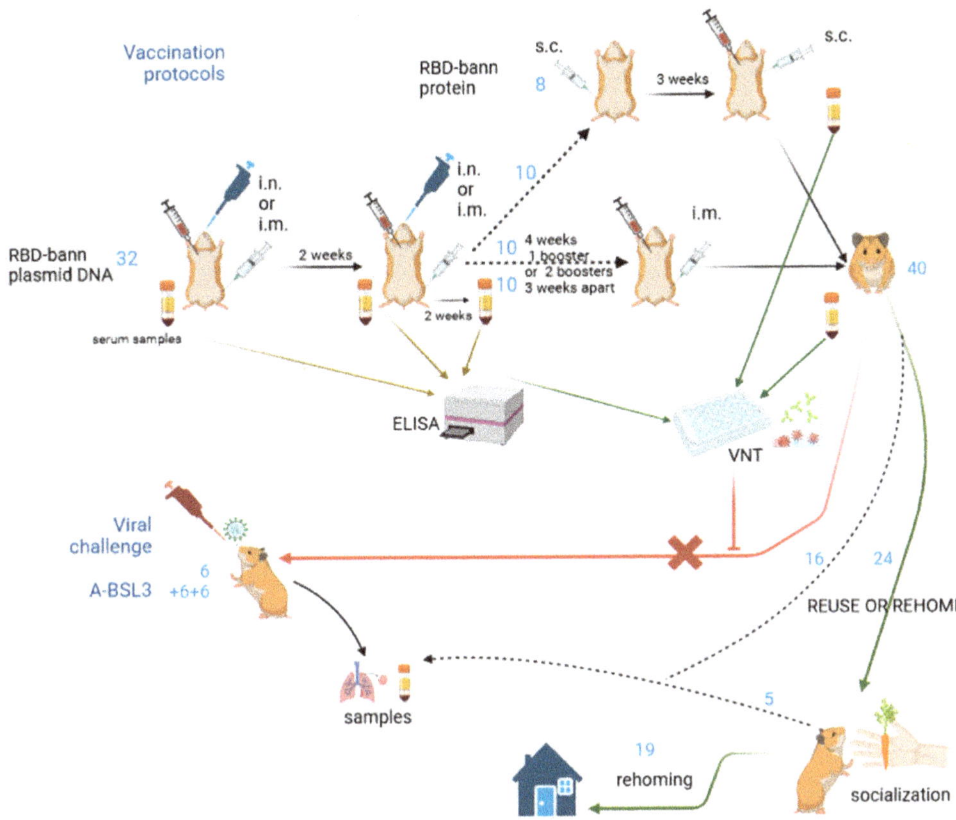

Figure 1. Diagram of workflow for vaccination, sampling, and pilot viral challenges. The blue numbers indicate the number of animals used in each run, initial vaccinations were half i.m. (syringe symbol) and half i.n. (pipette). ELISA testing of serum was crucial for further steps in the vaccination protocols: 10 animals with highest IgG titres received a second booster of 50 μg plasmid DNA, 10 animals received 2 boosters (3 weeks apart) with a higher dose of plasmid DNA than that initially (40 μg instead of 20 μg) and 10 animals received 2 boosters of protein RBD-bann 3 weeks apart, all i.m. An extra 8 animals were vaccinated with protein RBD and complete Freund adjuvant (top line). The virus neutralisation test (VNT) served as the eventual replacement method for viral challenge. There were no neutralization properties, the CPE effect was seen in all dilutions, and the continuation of procedures to viral challenge was stopped (red line). Health checks were performed on 40 surviving hamsters and a decision was made as to whether they should be reused or rehomed. A total of 24 hamsters underwent a simple socialization process and 19 eventually found new homes. Viral challenge (i.n. inoculation, red pipette) was tested in pilot trials on non-vaccinated animals. Diagram drawn in BioRender.

With 18 animals from the new supplier, we conducted pilot tests with the virus in an animal biosafety level 3 (A-BSL3) laboratory. None of the vaccinated animals entered A-BSL3 and were not infected with the virus.

3.1. Serum Antibodies

As expected, only the booster dose resulted in high levels of IgG titres in the hamster serum presented in the graph in supplementary material Figure S1. At a lower dose of the tested vaccine (20 μg per animal), some animals showed an insufficient immune response, which was more evident with the plasmid DNA product.

3.2. Pseudovirus Neutralisation Test

Neutralisation of the pseudovirus showed little effect, and insignificant lower or higher luciferase signal levels did not correspond to group assignment (Supplementary Figure S2).

3.3. Virus Neutralisation Test

No neutralising antibodies were detected by VNT in any of the hamster serum samples tested; all dilutions resulted in cytopathogenic effect (CPE—Supplementary Figure S3). The positive control serum VNT titre was 1:32.

3.4. Refinement of Viral Challenge Procedure—Pilot Control Groups

To determine the best time for euthanasia, tissue, and method of sampling, pilot experiments were conducted with a minimal number of animals. Non-vaccinated animals were infected with 25 μL 5×10^4 $TCID_{50}$/mL SARS-CoV-2 strain Slovenia/SI-4265/20, day 4 after inoculation was selected as the best time for the clinical endpoint, and swabs from the caudal nasal cavity were more representative than swabs from the oropharynx but were time-consuming to obtain in an A-BSL3 facility. In the duodenum and trachea, viral RNA was either not detected or was very low (Ct values above 33—see Supplementary Table S1). In the second trial, we doubled the amount of viral particles by applying 50 μL (25 μL per nostril) of 5×10^4 $TCID_{50}$/mL to 6 unvaccinated hamsters and all were euthanized on day 4. RT-qPCR of viral RNA and viral titres on Vero E6 cells roughly correlated, but levels were low and clinical signs were minor; in addition, contrary to expectations, females gained weight (Table 1). By increasing the viral load 100× (50 μL of undiluted stock—5×10^6 $TCID_{50}$/mL) in the third trial, we observed more obvious clinical signs: lethargy and ruffled coat, and 50% of animals had a mild nasal discharge and showed laboured breathing. The amount of virus in the lung tissue detected by viral titre from lung homogenate increased, viral RNA RT-qPCR Ct values dropped, and the ΔCt between the second and third trial were significantly different at $p = 0.002$ as presented in the supplementary materials (Figure S4). However, the increased viral load could not be confirmed with higher immunohistochemistry scoreof spike protein antigen (Supplementary Figure S5 and scoring in Table 1).

Table 1. Second and third trials for pilot viral challenge control group with intranasal inoculation of 50 μL of 5×10^4 or 5×10^6 $TCID_{50}$/mL SARS-CoV-2 strain Slovenia/SI-4265/20 to achieve evident clinical scoring and tissue viral load. Differences between male and female animals were noted especially in body mass gain/loss.

Trial	Sex	Lung Virus Titre $TCID_{50}$/mL	Lung Histology *	Body Mass Change	Clinical Score **
Viral challenge 50 μL 5×10^4 $TCID_{50}$/ml	females (n = 3)	detected 1/3 (8.62×10^2)	range − to +	average +4%	1
	males (n = 3)	detected 1/3 (1.86×10^4)	range −/+ to +	average −2%	1
Viral challenge 50 μL 5×10^6 $TCID_{50}$/ml	females (n = 3)	detected 3/3 (1.56×10^4–6.34×10^4)	range − to ++	average +5%	2–3
	males (n = 3)	detected 3/3 (1.49×10^4–1.07×10^5)	all +	average −2.5%	1–3

* Semi-quantification of immunohistochemistry, colorimetric detection of spike protein in lung bronchi and interstitial tissue on 5 semi-serial sections of 1 lung lobe. Maximum score is +++. ** Clinical scores 1 point for mild and 2 for prominent: lethargy or slow arousal response, ruffled coat, abnormal posture, dyspnoea, ocular or nasal discharge, diarrhoea. Score above 6/day was considered humane endpoint.

The use of 6 extra animals (third trial) or pilot trials as a whole (18 animals) might contradict the reduction principle of 3R but was considered as refinement as the virulence of specific SARS-CoV-2 strains for golden hamsters was not known. None of the vaccinated animals were subjected to the viral challenge procedure.

3.5. Rehoming

Of the 40 animals vaccinated, 1 developed otitis media, was treated at the Faculty's Clinic for birds, small mammals, and reptiles, and was eventually adopted by the attending veterinarian with the approval of the institutional animal welfare officer. The designated veterinarian performed the clinical examination of the remaining animals. Three were in worse condition, likely due to their advanced age (over 1 year old), and were euthanized. Another 12 animals were selected for reuse for scientific or educational purposes, humanely killed, and tissue samples collected. Socialization measures were taken for the remaining 24 animals. Five of the animals did not respond well to frequent handling, usually hiding in the tunnel when staff were present, so they were not candidates for rehoming as pets. A total of 19 hamsters were eventually adopted; the first six adopters were internal, the next seven semi-internal (veterinary students and trainees in the LAS course of October 2021), and the rest external. The word was spread through friends and relatives. Internal adopters provided frequent feedback, and a year and a half later, a general opinion was obtained through an anonymous online survey (translation available in the supplementary materials). Of 17 invitations sent (2 contacts were lost), 11 responded; the main results are shown in Figure 2. All feedback received was positive and adopters would recommend continuing the practice. A slight dissatisfaction of four respondents (they would not recommend the next adoption and were not completely satisfied with the adoption procedure) was related to the result that their four hamsters had a shorter life than they expected and had some health problems, e.g., the appearance of tumours in one. For confidentiality reasons, no further analysis was carried out. The translation of the questionary with consent request can be found in Supplementary Material.

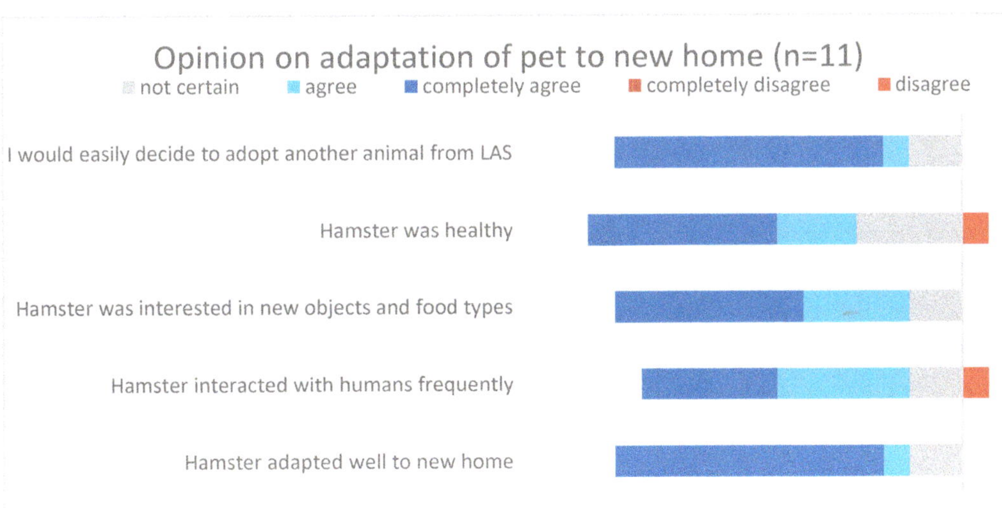

Figure 2. *Cont.*

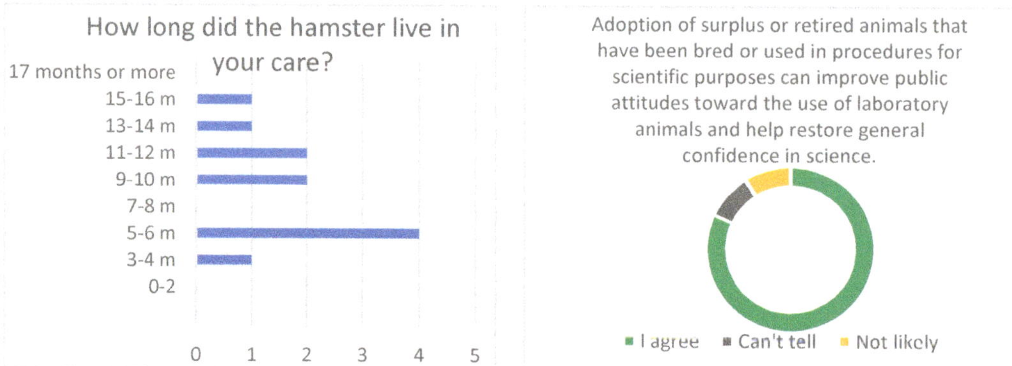

Figure 2. The main findings from the survey among adopters 1.5 years after the rehoming procedure. None of the responders choose the option of "completely disagree" in the first panel. No. of responses = 11 (65% success rate).

4. Discussion

Effective vaccines are needed to combat the emerging viral epidemics. In this study, we evaluated different deliveries of promising vaccine candidates against SARS-CoV-2 in the Syrian hamster model. Unfortunately, the goal of our application-oriented research project was not achieved—the vaccine candidates elicited an immunological response, but we could not confirm the neutralising properties of the antibodies produced in the golden hamster. In 2021, almost every country was trying to develop a vaccine for SARS-CoV-2, the vast majority focusing on spike protein as the most immunogenic structural protein of the virus. The success of a vaccine to fight a pandemic is dependent on durability and stability of neutralization antibody titres [21]. Mutations in antibody epitopes on the spike protein can result in increased viral resistance to neutralizing antibodies and have been associated with reduced vaccine effectiveness with recent variants like omicron. Although variant-specific vaccines and novel monoclonal antibodies are being developed for optimal activity, booster immunizations with the old products are also investigated as they can significantly increase serum neutralizing activity [22]. Humoral immune response—the antibodies—is the best-defined correlate of protection against the SARS-CoV-2 infection; however, nowadays with emerging variants, more focus is put also on cellular components: specific B cells are important for long-term protection from the disease and T cells specific for SARS-CoV-2 can protect from severe disease [23]. At the end of 2020, however, the most important goal was to slow down the spread of the disease, and only efficient neutralisation of the virus was the aim of the project funded by the Slovenian Research and Innovation Agency and the Ministry of Defence. We have no explanation for our unexpected result—no neutralising properties of the antibodies—and no further studies on the causes and other types of immune response were conducted with additional animals, as this would create an unauthorised deviation from the official decision of the Slovenian Ministry of Agriculture, Forestry and Food and its Ethics Committee. However, we did emphasize the importance of the 3R principle in practice. The alternative procedure reduced the total number of animals in the study: the sero-neutralisation test or VNT turned out as a complete replacement for the efficacy test with viral challenge of the animals as no new data would be obtained for the scope of the project. Because VNT continued to show a negative result despite booster vaccinations and changes in the vaccination protocol, and was backed up with a negative PV test, we decided to discontinue the trials. None of the vaccinated animals underwent viral challenge (moderate procedure according to classification of the severity of procedures on animals—Directive 2010/63/EU) and they were not killed for this purpose, meaning the reduction in severity was achieved. Also, the total expected number of hamsters in the project (120) was reduced because none of the vaccination protocols were repeated (i.e., larger groups for better statistical evaluation and control group with placebo injections), no alternative modes of administration were tried (oral patches and enhanced intranasal application), and no product safety testing was required.

In the meantime, while the vaccination protocol was ongoing, we subjected a small group of animals to a pilot viral challenge test in order to establish an appropriate protocol for the planned tests of vaccine efficacy. Pilot studies are recommended to design a study with a needed sample size estimation when the general results of infection procedures are unknown [24] and when the laboratory/institution is confronted with new methods, equipment, animal species, etc. [25]. In our study, a pilot trial was planned to test the workflow in a newly equipped A-BSL3 facility and to determine endpoint measurements; 12 animals in two consecutive trials were initially used for this purpose. We followed the protocol descriptions from the publications on the use of Syrian hamsters at that time and infected the animals with similar viral loads between 1×10^4 and 1×10^5 $TCID_{50}/mL$ [26,27]. Because the observed clinical signs and virus isolates from lung tissue were very low, we tested a higher dose in six additional animals with better results (higher clinical score and lung viral load). This dose would be used in further procedures to achieve robust infection in control animals, because we want to avoid difficulties in distinguishing between mild infection and possibly only mild improvement. Pilot trials seem to contradict the reduction principle if all goes well. But if animals with a promising vaccination response (and control groups) are infected with suboptimal viral load, the efficiency parameters will likely be insignificant, requiring the whole protocol to be repeated, thus using even more animals. Higher doses have been used by some other investigators [28], and even the oropharyngeal route could more effective [29]. Alternatively, a more infectious viral strain or a completely different model animal could be considered. The results of our pilot tests confirmed the need to consider this as a refinement method when planning research projects including animal procedures.

The mild clinical signs observed, including very little decrease in body mass (or even increase in females), must not only be due to the (initially) low inoculum dose [30] but are also likely strain-dependent. Increase in body weight has been noticed when animals were infected with a lower dose, but the lung pathology was comparable to higher doses, where animals lost weight [31]. The local specific strain Slovenia/SI-4265/20 has not been tested elsewhere and we have not used any other strain for infecting the animals for comparison. Although this is one of the early isolates in 2020, its virulence for golden hamsters could be low, as has been shown for some later variants [32,33]. Body mass is a robust measurable parameter; on the other hand, observation of respiratory patterns and general well-being is highly subjective, and standardized clinical scoring for golden hamsters or similar models is difficult to develop [30]. For the best refinement solution, automated home cage monitoring systems would be a perfect addition, especially because hamsters are nocturnal and the prominent changes in their behaviour can easily go unnoticed.

The immunohistochemistry score in our pilot control group was weak and only slightly improved after increasing the inoculation dose. Most of the positive signals were located in the bronchi and only sparse interstitial locations in two individual animals. It should be noted that pathologic changes are not evenly distributed and, therefore, a wide transverse section of the entire lung lobe, ideally the left non-lobulated one, is needed [34]. At least five semi-serial sections are required for reliable quantification with immunohistochemistry, which is slightly more time-consuming than histopathology scoring, where a grid scoring was proposed on one slide [31]. For optimal pathologic evaluation, euthanasia and sampling should be rather conducted on day 5 or 6 post-inoculation, as the virus enters/replicates interstitially after day 4 [35] and exact pathological scoring was not used at this stage in our experiment. For appropriate refinement, decision making for the clinical endpoint should be discussed—for optimal histopathology and clinical score results, hamsters should be left in (moderate) distress for longer (4–7 days). For therapies and vaccines targeting viral replication, sample analysis (virology) at the peak of viral replication between 2 and 3 dpi may be more useful.

Researchers always strive to minimize the number of animals used as long as a relevant statistical analysis can be performed. When working with dangerous infectious agents, downsizing animal numbers may even be enforced due to the limited capacity of A-BSL3 facilities. The risk of over-reduction exists and thus challenges the reliability and reproducibility of the studies—the factors that are often referred to "Rs beyond the 3R". For our study, we managed to organise a capacity to house up to 60 hamsters simultaneously in a renovated A-BSL3 but were unexpectedly faced with limited availability of the species. Instead of the planned 60 hamsters for different vaccination protocols and three control groups with placebo injections, we could initially acquire only 40 animals. A decision was made to optimise the experiment design: test all the planned protocols, but for the controls/blanks, use the sera of the same hamsters before the vaccination. Such a reduction principle—comparing responses at later time points to a baseline level—is commonly used in toxicology [36]. We did plan to retest only one most promising protocol, including more animals depending on variability and adjusting the dose or adjuvant if needed as well simultaneously using the placebo group, e.g., DNA vector without the insert. As none of the vaccination regimens even came close to the wanted result, the repeats and further steps (i.e., safety) were not conducted, and this ultimately reduced the number of animals.

Regarding the reliability of the studies, we need to highlight the sex differences, as we have observed more pronounced clinical symptoms in males and this has subsequently been well documented [37,38]. Studies examining only one sex (either males only or females only with no clear explanation of selection) may therefore overestimate or underestimate vaccine efficacy, dangers of new variants, or transmission through domestic animals [27,29,39]. Studying both sexes has been a guiding principle in biomedical research for so long that the decision to use only one sex can hardly be justified [40,41], even if it seems to contradict the reduction principle. Furthermore, for real vaccine or drug efficacy, the number of animals used should likely increase by including different ages of animals, as disease progression is different in aged animals [42]. We started with vaccinations when animals were 4 months old, and eight in the last protocol tried were 7 months old. Most publications have used very young hamsters, probably to speed up research when faced with pandemics. We can argue this is a drawback; using prepubescent animals for this kind of translational study is not exactly beneficial and rather animals in their prime adult age should be considered. Therefore, also the wanted (starting) age requires more careful planning in the inclusion of model animals in the future.

We found no reports of negative results that would explain the lack of neutralising properties of hamster antibodies, and overall, not that many papers describing vaccine testing in golden hamsters were found. It is likely that most laboratories started using genetically modified mice (hACE2 mice and further variants) when they became more readily available, because after the initial boom, the number of studies using hamsters declined slightly [2]. However, we can surmise that more researchers encountered obstacles in using this animal model but did not report it. The VNT test should not yield false-negative results because it has not been shown to be species-dependent [43]. Withholding negative results from publication—known as publication bias—can seriously distort the literature and consume scarce resources through lengthy and perhaps even futile research. The research community is ethically obligated to make the best use of the results of animal studies, which is not the case when negative results are not published [44]. To overcome this publication bias, initiatives to pre-register animal studies have been proposed [45]. To date, there are only two reliable animal registries: Preclinicaltrials.eu and the Animal Study Registry (animalstudyregistry.org). Unfortunately, neither of them show any results with the keyword "hamster" (last accessed 15 June 2023). Also, while vaccine efficacy in the first weeks after vaccination can be readily studied in animal models, monitoring the duration of vaccine-induced immunity remains a challenge due to differences in metabolism and life expectancy between rodents and humans [35], and the more data published, the better.

Refinements such as environmental enrichments, friendly handling, and general anaesthesia for all inoculations and blood sampling have been introduced for the general welfare of the animals, even though minimal pain procedures were unlikely to affect the final immunological outcome. We hypothesize that minimizing the pain and other negative associations with humans contributed to rapid and smooth socialization in most animals, as confirmed by adopter feedback. It is important to note that this species exhibits burrowing behaviour [46], and they all used the "Sizzlenest" paper strips to build nests but left cotton fibre "Nestlet" mostly intact. Also, most hamsters used the bent (at right angle) plastic tunnels to seek refuge when startled but rarely slept in them. Some hamsters preferred to nibble on aspen blocks and others the cardboard tunnels; therefore, the choice should be given. An important aspect of hamster welfare is to check that they can reach food through the grates as their snouts are shorter and wider than those of rats or mice. It is best to offer them the food inside the cage [47]. If the food is scattered in the bedding, this encourages the hamsters' natural behaviour of gathering and storing their food in a specific location. They regularly urinate in one corner of the cage only, so the food is not spoiled by urine, and this also enables partial cage cleaning. In the wild, the golden hamster is a solitary animal, but juveniles are supplied by breeders conditioned to group housing, and housing with littermates may actually be recommended, at least for males [48]. In our experience, it was possible to keep hamsters in same-sex pairs or groups of three for short-term experiments, such as virus challenge trials in A-BSL3, where acclimatization lasted a week, the experiment lasted a maximum of another week, and the cage floor surface exceed 1800 cm^2. With vaccination protocols, animals were kept longer in the conventional facility, and soon after reaching sexual maturity, all females began to fight, inflicting bite wounds. To create comparable conditions, we separated both sexes and switched to single housing in these procedures.

The final refinement represented the decision to rehome some of the animals. The rehoming process was initially met with some scepticism by the research team but was then supported by the animal welfare officer, approved by the designated veterinarian, and accepted with great satisfaction in the end. Most adopters were overjoyed to give the animal a second chance in life and would gladly adopt a similar animal next time. We do not know the opinion of the adopters who did not respond to the survey (35%)—their experience may be different but would not be the deciding factor in this case. To date, there are no specific institutional (or national) guidelines or protocols regarding rehoming animals used or bred for research purposes. In Slovenia up until now, only farm and wild animals that have undergone non-invasive or minimal harm procedures have been considered suitable for "return to herd" or "set free" in the case of wildlife. Once the ice was broken, future project proposals may consider rehoming laboratory animals at this and other institutions as public interest—including for rodents—appears to be greater than scientists would generally hope for. Success stories, including rehoming rodents, come from other European countries. Some have chosen to work with animal welfare NGOs (animal protection and animal rights organizations) [9]. This is probably the best solution as long as the NGO staff can establish a good relationship of trust and communication and work hand in hand with the scientists and regulators to ensure that they are sending an appropriate message to the public.

As laboratory animal science faces significant challenges, openness and transparency with the public is now encouraged. Citing MacArthur Clark [49], "A balance is demanded between the needs of the science and the needs of the animals since this balance supports and sustains the public's confidence in ethical review processes and the regulatory oversight of the use of animals in science. Tinkering with this balance, either excessively in favour of science or excessively in favour of animal welfare, will lead to loss of public confidence and erosion of the conditional permission society grants to enable animal research to continue." We believe that a process of rehoming surplus or retired laboratory animals with success stories published in the dedicated media can be a step forward in confirming this balance.

5. Conclusions

Refinement is a ubiquitous tool in almost every step of research projects involving procedures on animals. Pilot trials are highly recommended when new techniques, equipment, infectious agents, etc., are used because it confirms the observation of the reduction principle. In our pilot viral challenge test, we found that we needed to use a higher viral load of a particular virus strain to elicit a sufficient clinical response as originally planned—and this would prevent the need to repeat the entire experiment with vaccinated animals. The virus neutralisation test is an example of an in vitro prerequisite that may reduce the severity or even total number of animals in the project. Because it could not demonstrate the neutralising properties of the hamster antibodies on vaccine candidates tested in our project, it served as a replacement for the in vivo viral challenge. The surviving golden hamsters were successfully rehomed in our study. Positive attitudes among both scientists and adopters of retired or surplus laboratory animals can improve public attitudes toward the use of animals for scientific purposes and help restore general confidence in science.

Supplementary Materials: The following supporting information can be downloaded at: https://www.mdpi.com/article/10.3390/ani13162616/s1, Figure S1: results of ELISA test for specific IgG; Figure S2: results of pseudovirus neutralisation test; Figure S3: demonstration of CPE in VNT test; Table S1 and Figure S4: results of RT-qPCR; Figure S5: photomicrographs of histology and immunohistology of infected animals from pilot viral challenge tests; all with additional method description. A translation of the questionnaire sent to adopters is also added.

Author Contributions: Conceptualization, G.M. and T.M.; methodology, T.M., U.K. (Uroš Krapež), and D.L.; investigation and analysis, D.L., S.O., M.Š., B.S., U.K. (Uroš Krapež), U.K. (Urška Kuhar), and T.M.; writing—original draft preparation, M.Š.; writing—review and editing, M.Š., D.L., U.K. (Uroš Krapež), B.S., U.K. (Urška Kuhar), and G.M.; supervision and funding acquisition, G.M. and T.M. All authors have read and agreed to the published version of the manuscript.

Funding: This research was funded by the Slovenian Research and Innovation Agency core programs (Javna agencija za znanstvenoraziskovalno in inovacijsko dejavnost Republike Slovenije, ARIS) P4-0053, P4-0092, and P4-0176 and grants V4-2038 and J4-4563.

Institutional Review Board Statement: The animal study protocol was approved by the Institutional animal welfare officer and by the Administration of the Republic of Slovenia for Food Safety, Veterinary and Plant Protection (Ministry of Agriculture, Forestry and Foods) and its Ethical Committee (Decisions U34401-18/2020/8 and U34401-11/2021/9).

Informed Consent Statement: Informed consent was obtained from all individuals who participated in the survey and gave feedback on adopted animals.

Data Availability Statement: The data presented in this study are available on request from the corresponding author.

Acknowledgments: Authors thank Nina Šterman, Sanja Bogićević, and Patricija Tandara for animal care, Nina Kočar and Joško Račnik from the Clinic for small animals, birds and reptiles for valuable advice on anaesthesia and blood draws, as well Alenka Dovč and Jelka Zabavnik Piano for their advice and help with animal rehoming.

Conflicts of Interest: The authors declare no conflict of interest.

References

1. Shou, S.; Liu, M.; Yang, Y.; Kang, N.; Song, Y.; Tan, D.; Liu, N.; Wang, F.; Liu, J.; Xie, Y. Animal Models for COVID-19: Hamsters, Mouse, Ferret, Mink, Tree Shrew, and Non-Human Primates. *Front. Microbiol.* **2021**, *12*, 626553.
2. Schwedhelm, P.; Kusnick, J.; Heinl, C.; Schönfelder, G.; Bert, B. How Many Animals Are Used for SARS-CoV-2 Research? *EMBO Rep.* **2021**, *22*, e53751. [CrossRef]
3. Bertzbach, L.D.; Vladimirova, D.; Dietert, K.; Abdelgawad, A.; Gruber, A.D.; Osterrieder, N.; Trimpert, J. SARS-CoV-2 Infection of Chinese Hamsters (*Cricetulus Griseus*) Reproduces COVID-19 Pneumonia in a Well-Established Small Animal Model. *Transbound. Emerg. Dis.* **2021**, *68*, 1075–1079. [CrossRef]

4. Trimpert, J.; Vladimirova, D.; Dietert, K.; Abdelgawad, A.; Kunec, D.; Dökel, S.; Voss, A.; Gruber, A.D.; Bertzbach, L.D.; Osterrieder, N. The Roborovski Dwarf Hamster Is A Highly Susceptible Model for a Rapid and Fatal Course of SARS-CoV-2 Infection. *Cell Rep.* **2020**, *33*, 108488. [CrossRef]
5. Warner, B.M.; Safronetz, D.; Kobinger, G.P. Syrian Hamsters as a Small Animal Model for Emerging Infectious Diseases: Advances in Immunologic Methods. In *Advances in Experimental Medicine and Biology*; Springer: New York, NY, USA, 2017; Volume 972, pp. 87–101.
6. Valentine, H.; Daugherity, E.K.; Singh, B.; Maurer, K.J. The Experimental Use of Syrian Hamsters. In *The Laboratory Rabbit, Guinea Pig, Hamster, and Other Rodents*; Elsevier: Amsterdam, The Netherlands, 2012; pp. 875–906, ISBN 9780123809209.
7. Hendriksen, C.F.M. Humane Endpoints in Vaccine Potency Testing. *Procedia Vaccinol.* **2011**, *5*, 221–226. [CrossRef]
8. Focosi, D.; Maggi, F.; Mazzetti, P.; Pistello, M. Viral Infection Neutralization Tests: A Focus on Severe Acute Respiratory Syndrome-Coronavirus-2 with Implications for Convalescent Plasma Therapy. *Rev. Med. Virol.* **2021**, *31*, e2170.
9. 15th FELASA Congress 2022—Abstract Book. SAGE-Book-of-Abstracts. *Lab. Anim.* **2022**, *56*, 20–203. [CrossRef]
10. Ecuer, E.; Boxall, J.; Louwerse, A.L.; Mikkelsen, L.F.; Moons, C.P.H.; Roth, M.; Spiri, A.M. FELASA Recommendations for the Rehoming of Animals Used for Scientific and Educational Purposes. *Lab. Anim.* **2023**. [CrossRef]
11. Moons, C.P.; Spiri, A.M.; Boxall, J.; Louwerse, A.; Mikkelsen, L.F.; Roth, M.; Ecuer, E. Survey among FELASA Members about Rehoming of Animals Used for Scientific and Educational Purposes. *Lab. Anim.* **2023**, 002367722311537. [CrossRef]
12. Skidmore, T.; Roe, E. A Semi-Structured Questionnaire Survey of Laboratory Animal Rehoming Practice across 41 UK Animal Research Facilities. *PLoS ONE* **2020**, *15*, e0234922. [CrossRef]
13. Lainšček, D.; Fink, T.; Forstnerič, V.; Hafner-Bratkovič, I.; Orehek, S.; Strmšek, Ž.; Manček-Keber, M.; Pečan, P.; Esih, H.; Malenšek, Š.; et al. A Nanoscaffolded Spike-Rbd Vaccine Provides Protection against Sars-Cov-2 with Minimal Anti-Scaffold Response. *Vaccines* **2021**, *9*, 431. [CrossRef]
14. Picazo, M.G.; Benito, P.J.; García-Olmo, D.C. Efficiency and Safety of a Technique for Drawing Blood from the Hamster Cranial Vena Cava. *Lab. Anim.* **2009**, *38*, 211–216. [CrossRef]
15. Ogando, N.S.; Dalebout, T.J.; Zevenhoven-Dobbe, J.C.; Limpens, R.W.A.L.; van der Meer, Y.; Caly, L.; Druce, J.; de Vries, J.J.C.; Kikkert, M.; Barcena, M.; et al. SARS-Coronavirus-2 Replication in Vero E6 Cells: Replication Kinetics, Rapid Adaptation and Cytopathology. *J. Gen. Virol.* **2020**, *101*, 925–940. [CrossRef] [PubMed]
16. Souf, S. Recent Advances in Diagnostic Testing for Viral Infections. *Biosci. Horiz.* **2016**, *9*, hzw010. [CrossRef]
17. Pellett, P.E.; Mitra, S.; Holland, T.C. Basics of Virology. In *Handbook of Clinical Neurology*; Elsevier: Amsterdam, The Netherlands, 2014.
18. Lei, C.; Yang, J.; Hu, J.; Sun, X. On the Calculation of TCID50 for Quantitation of Virus Infectivity. *Virol. Sin.* **2021**, *36*, 141–144. [CrossRef] [PubMed]
19. Corman, V.M.; Landt, O.; Kaiser, M.; Molenkamp, R.; Meijer, A.; Chu, D.K.W.; Bleicker, T.; Brünink, S.; Schneider, J.; Schmidt, M.L.; et al. Detection of 2019 Novel Coronavirus (2019-NCoV) by Real-Time RT-PCR. *Eurosurveillance* **2020**, *25*, 2000045. [CrossRef]
20. WHO Expert Committee on Biological Standardization. Guidelines on the Quality, Safety and Efficacy of Plasmid DNA Vaccines. In *Seventy-First Report*; WHO Technical Report Series, No. 1028; WHO: Geneva, Switzerland, 2021.
21. Gupta, S.L.; Jaiswal, R.K. Neutralizing Antibody: A Savior in the Covid-19 Disease. *Mol. Biol. Rep.* **2022**, *49*, 2465–2474. [CrossRef]
22. Gruell, H.; Vanshylla, K.; Tober-Lau, P.; Hillus, D.; Schommers, P.; Lehmann, C.; Kurth, F.; Sander, L.E.; Klein, F. MRNA Booster Immunization Elicits Potent Neutralizing Serum Activity against the SARS-CoV-2 Omicron Variant. *Nat. Med.* **2022**, *28*, 477–480. [CrossRef]
23. Pušnik, J.; Monzon-Posadas, W.O.; Zorn, J.; Peters, K.; Baum, M.; Proksch, H.; Schlüter, C.B.; Alter, G.; Menting, T.; Streeck, H. SARS-CoV-2 Humoral and Cellular Immunity Following Different Combinations of Vaccination and Breakthrough Infection. *Nat. Commun.* **2023**, *14*, 572. [CrossRef]
24. Laws, T.R.; Maishman, T.C. Considerations in the Design of Animal Infection Pilot Studies. *Front. Cell. Infect. Microbiol.* **2022**, *12*, 948464. [CrossRef]
25. Festing, M.F.W.; Altman, D.G. Guidelines for the Design and Statistical Analysis of Experiments Using Laboratory Animals. *ILAR J.* **2002**, *43*, 244–258. [CrossRef]
26. Lee, A.C.Y.; Zhang, A.J.; Chan, J.F.W.; Li, C.; Fan, Z.; Liu, F.; Chen, Y.; Liang, R.; Sridhar, S.; Cai, J.P.; et al. Oral SARS-CoV-2 Inoculation Establishes Subclinical Respiratory Infection with Virus Shedding in Golden Syrian Hamsters. *Cell Rep. Med.* **2020**, *1*, 100121. [CrossRef] [PubMed]
27. Abdelnabi, R.; Boudewijns, R.; Foo, C.S.; Seldeslachts, L.; Sanchez-Felipe, L.; Zhang, X.; Delang, L.; Maes, P.; Kaptein, S.J.F.; Weynand, B.; et al. Comparing Infectivity and Virulence of Emerging SARS-CoV-2 Variants in Syrian Hamsters. *EBioMedicine* **2021**, *68*, 103403. [CrossRef] [PubMed]
28. Toomer, G.; Burns, W.; Garcia, L.; Henry, G.; Biancofiori, A.; George, A.; Duffy, C.; Chu, J.; Sides, M.; Muñoz, M.; et al. Characterization of Three Variants of SARS-CoV-2 In Vivo Shows Host-Dependent Pathogenicity in Hamsters, While Not in K18-HACE2 Mice. *Viruses* **2022**, *14*, 2584. [CrossRef] [PubMed]
29. Blaurock, C.; Breithaupt, A.; Weber, S.; Wylezich, C.; Keller, M.; Mohl, P.; Görlich, D.; Groschup, M.H.; Sadeghi, B.; Höper, D.; et al. Compellingly High SARS-CoV-2 Susceptibility of Golden Syrian Hamsters Suggests Multiple Zoonotic Infections of Pet Hamsters during the COVID-19 Pandemic. *Sci. Rep.* **2022**, *12*, 15069. [CrossRef]

30. Braxton, A.M.; Creisher, P.S.; Ruiz-Bedoya, C.A.; Mulka, K.R.; Dhakal, S.; Ordonez, A.A.; Beck, S.E.; Jain, S.K.; Villano, J.S. Hamsters as a Model of Severe Acute Respiratory Syndrome Coronavirus-2. *Comp. Med.* **2021**, *71*, 398–410. [CrossRef]
31. Lien, C.E.; Lin, Y.J.; Chen, C.; Lian, W.C.; Kuo, T.Y.; Campbell, J.D.; Traquina, P.; Lin, M.Y.; Liu, L.T.C.; Chuang, Y.S.; et al. CpG-Adjuvanted Stable Prefusion SARS-CoV-2 Spike Protein Protected Hamsters from SARS-CoV-2 Challenge. *Sci. Rep.* **2021**, *11*, 8761. [CrossRef]
32. Abdelnabi, R.; Foo, C.S.; Zhang, X.; Lemmens, V.; Maes, P.; Slechten, B.; Raymenants, J.; André, E.; Weynand, B.; Dallmeier, K.; et al. The Omicron (B.1.1.529) SARS-CoV-2 Variant of Concern Does Not Readily Infect Syrian Hamsters. *Antivir. Res.* **2022**, *198*, 105253. [CrossRef]
33. Cochin, M.; Luciani, L.; Touret, F.; Driouich, J.S.; Petit, P.R.; Moureau, G.; Baronti, C.; Laprie, C.; Thirion, L.; Maes, P.; et al. The SARS-CoV-2 Alpha Variant Exhibits Comparable Fitness to the D614G Strain in a Syrian Hamster Model. *Commun. Biol.* **2022**, *5*, 225. [CrossRef]
34. Gruber, A.D.; Osterrieder, N.; Bertzbach, L.D.; Vladimirova, D.; Greuel, S.; Ihlow, J.; Horst, D.; Trimpert, J.; Dietert, K. Standardization of Reporting Criteria for Lung Pathology in SARS-CoV-2 Infected Hamsters: What Matters? *Am. J. Respir. Cell Mol. Biol.* **2020**, *63*, 856–859. [CrossRef]
35. Gruber, A.D.; Firsching, T.C.; Trimpert, J.; Dietert, K. Hamster Models of COVID-19 Pneumonia Reviewed: How Human Can They Be? *Vet. Pathol.* **2022**, *59*, 528–545. [PubMed]
36. Bate, S.; Karp, N.A. A Common Control Group—Optimising the Experiment Design to Maximise Sensitivity. *PLoS ONE* **2014**, *9*, e114872. [CrossRef]
37. Yuan, L.; Zhu, H.; Zhou, M.; Ma, J.; Chen, R.; Chen, Y.; Chen, L.; Wu, K.; Cai, M.; Hong, J.; et al. Gender Associates with Both Susceptibility to Infection and Pathogenesis of SARS-CoV-2 in Syrian Hamster. *Signal Transduct. Target. Ther.* **2021**, *6*, 136. [CrossRef]
38. Castellan, M.; Zamperin, G.; Franzoni, G.; Foiani, G.; Zorzan, M.; Drzewnioková, P.; Mancin, M.; Brian, I.; Bortolami, A.; Pagliari, M.; et al. Host Response of Syrian Hamster to SARS-CoV-2 Infection Including Differences with Humans and between Sexes. *Viruses* **2023**, *15*, 428. [CrossRef]
39. Fischer, R.J.; van Doremalen, N.; Adney, D.R.; Yinda, C.K.; Port, J.R.; Holbrook, M.G.; Schulz, J.E.; Williamson, B.N.; Thomas, T.; Barbian, K.; et al. ChAdOx1 NCoV-19 (AZD1222) Protects Syrian Hamsters against SARS-CoV-2 B.1.351 and B.1.1.7. *Nat. Commun.* **2021**, *12*, 5868. [CrossRef]
40. Clayton, J.A. Studying Both Sexes: A Guiding Principle for Biomedicine; Studying Both Sexes: A Guiding Principle for Biomedicine. *FASEB J.* **2015**, *30*, 519–524. [CrossRef] [PubMed]
41. Clayton, J.A.; Collins, F.S. Policy: NIH to Balance Sex in Cell and Animal Studies. *Nature* **2014**, *509*, 282–283. [CrossRef]
42. Osterrieder, N.; Bertzbach, L.D.; Dietert, K.; Abdelgawad, A.; Vladimirova, D.; Kunec, D.; Hoffmann, D.; Beer, M.; Gruber, A.D.; Trimpert, J. Age-Dependent Progression of SARS-CoV-2 Infection in Syrian Hamsters. *Viruses* **2020**, *12*, 779. [CrossRef] [PubMed]
43. Tan, C.W.; Chia, W.N.; Qin, X.; Liu, P.; Chen, M.I.C.; Tiu, C.; Hu, Z.; Chen, V.C.W.; Young, B.E.; Sia, W.R.; et al. A SARS-CoV-2 Surrogate Virus Neutralization Test Based on Antibody-Mediated Blockage of ACE2–Spike Protein–Protein Interaction. *Nat. Biotechnol.* **2020**, *38*, 1073–1078. [CrossRef]
44. Joober, R.; Schmitz, N.; Annable, L.; Boksa, P. Publication Bias: What Are the Challenges and Can They Be Overcome? *J. Psychiatry Neurosci.* **2012**, *37*, 149–152. [CrossRef] [PubMed]
45. Bert, B.; Heinl, C.; Chmielewska, J.; Schwarz, F.; Grune, B.; Hensel, A.; Greiner, M.; Schönfelder, G. Refining Animal Research: The Animal Study Registry. *PLoS Biol.* **2019**, *17*, e3000463. [CrossRef] [PubMed]
46. Sørensen, D.B.; Krohn, T.; Hansen, H.N.; Ottesen, J.L.; Hansen, A.K. An Ethological Approach to Housing Requirements of Golden Hamsters, Mongolian Gerbils and Fat Sand Rats in the Laboratory—A Review. *Appl. Anim. Behav. Sci.* **2005**, *94*, 181–195. [CrossRef]
47. National Research Council. *Nutrient Requirements of Laboratory Animals*, 4th Revised ed.; The National Academies Press: Washington, DC, USA, 1995.
48. Arnold, C.E.; Estep, D.Q. Effects of Housing on Social Preference and Behaviour in Male Golden Hamsters (*Mesocricetus Auratus*). *Appl. Anim. Behav. Sci.* **1990**, *27*, 253–261. [CrossRef]
49. Clark, J.M. The 3Rs in Research: A Contemporary Approach to Replacement, Reduction and Refinement. *Br. J. Nutr.* **2017**, *120*, S1–S7. [CrossRef] [PubMed]

Disclaimer/Publisher's Note: The statements, opinions and data contained in all publications are solely those of the individual author(s) and contributor(s) and not of MDPI and/or the editor(s). MDPI and/or the editor(s) disclaim responsibility for any injury to people or property resulting from any ideas, methods, instructions or products referred to in the content.

MDPI
St. Alban-Anlage 66
4052 Basel
Switzerland
www.mdpi.com

Animals Editorial Office
E-mail: animals@mdpi.com
www.mdpi.com/journal/animals

Disclaimer/Publisher's Note: The statements, opinions and data contained in all publications are solely those of the individual author(s) and contributor(s) and not of MDPI and/or the editor(s). MDPI and/or the editor(s) disclaim responsibility for any injury to people or property resulting from any ideas, methods, instructions or products referred to in the content.

www.ingramcontent.com/pod-product-compliance
Lightning Source LLC
Chambersburg PA
CBHW041939260926
51850CB00018B/299
* 9 7 8 3 0 3 6 5 9 8 3 7 6 *